Analysis of Framed Structures

JAMES M. GERE
Professor of Structural Engineering
Stanford University

and

WILLIAM WEAVER, Jr.
Associate Professor of Structural Engineering
Stanford University

Analysis of Framed Structures

VAN NOSTRAND REINHOLD COMPANY
NEW YORK CINCINNATI TORONTO LONDON MELBOURNE

VAN NOSTRAND REINHOLD COMPANY Regional Offices:
New York Cincinnati Chicago Millbrae Dallas

VAN NOSTRAND REINHOLD COMPANY International Offices:
London Toronto Melbourne

Published by VAN NOSTRAND REINHOLD COMPANY
450 West 33rd Street, New York, N. Y. 10001

Published simultaneously in Canada by
VAN NOSTRAND REINHOLD LTD.

10 9 8 7 6 5 4

PREFACE

The advent of the digital computer has made it necessary to reorganize the theory of structures into matrix form. The emphasis is now upon the flexibility and stiffness methods of analysis, which are considered to be the most fundamental and all-inclusive of the available theories. These two complementary methods are especially suitable for matrix formulation and machine computation.

This book was written as a text for college students on the subject of the analysis of framed structures by matrix methods. The material has been used in courses that the authors have developed over a period of several years. The subject has been taught regularly to both undergraduate and graduate students, and they appear to be equally enthusiastic about it. Since the material is relatively new to structural engineers, the authors also have kept in mind the needs of those in practice whose college programs of a few years ago did not include this subject. It is believed that the presentation is sufficiently complete and detailed that a good knowledge of the methods can be obtained by personal study from the book.

The preparation needed before undertaking to study this subject is that which is normally obtained during the first portion of an undergraduate engineering program; specifically, the reader should be familiar with statics and mechanics of materials, as well as algebra and introductory calculus. A prior course in elementary structural analysis naturally would be of benefit, although it is not a necessary prerequisite for the subject matter of the book. Elementary matrix algebra is used throughout the book, and therefore the reader must have an introduction to this subject as a prerequisite. Since the topics from matrix algebra that are needed are of an elementary nature, the reader can acquire the necessary knowledge by self-study during a period of two or three weeks. The authors customarily have devoted about the first two weeks of their course to this subject, and have found that sufficient knowledge can be obtained by the students in that amount of time. A separate mathematics course in matrix algebra is not necessary, although some students will wish to take such a course in preparation for more advanced work. To assist those who need only an introduction to matrix algebra, without benefit of a formal course, the authors have written a supplementary book on the subject.*

* *Matrix Algebra for Engineers,* New York: Van Nostrand Reinhold Company, 1965.

There are a number of reasons why matrix analysis of structures is important to the structural analyst. One of the most important is that it makes possible a unified and inclusive approach to the subject—an approach that is valid for structures of all types. A second reason is that it provides an efficient means of describing the various steps in the analysis, so that these steps can be more easily programmed for a digital computer. The use of matrices is natural when performing calculations with a computer, since they permit large groups of numbers to be manipulated in a simple and effective manner. The reader will find that the methods of analysis developed in this book are highly organized, and that the same basic procedures can be followed in the analysis of different types of structures, such as beams, trusses, and frames.

To some extent it is tacitly assumed in this book that the final goal to be reached is the preparation of computer programs capable of analyzing any framed structure. While the details of coding programs in a particular computer language are not discussed in this book, the development nevertheless is carried to the point where that would be the next step. In other words, the reader will gain sufficient information to enable him to prepare a program for his particular computer. For this purpose, Chapter 5 contains a series of flow charts for the analysis of the six basic types of framed structures, and these will assist the reader in preparing his own programs. One of the authors' purposes, however, has been to emphasize the theory and organization of the analysis, and a companion book is devoted primarily to the more advanced and detailed aspects of programming.*

For the most part, this book attempts to develop the subject of matrix analysis from the point of view of the structural engineer. Thus only those concepts that are directly useful in solving structures are presented. (The reader with further interests will find much material available for study in the references at the end of the book.) The first chapter and the appendices cover certain fundamental concepts of structural analysis, and this material is prerequisite to the main part of the book. Those who have previously studied theory of structures will find that this material is largely review; however, those who are encountering structural theory for the first time will find it necessary to become thoroughly familiar with these basic matters.

The flexibility and stiffness methods are discussed and compared in Chapter 2, and numerous illustrative problems are given for both methods. In Chapters 3 and 4 each of the methods is expanded, and more detail (leading toward computer programming) is provided. Computer programs for the analysis of framed structures by the stiffness method are presented in Chapter 5. In order to maintain the emphasis in the

* W. Weaver, Jr., *Computer Programs for Structural Analysis,* New York: Van Nostrand Reinhold Company, 1967.

earlier chapters on the fundamental matters, many special topics are postponed until Chapter 6. Then, in this chapter, such topics as non-prismatic members, temperature effects, and elastic connections are considered. All of these subjects can be viewed as modifications to the basic procedures described in the earlier chapters.

Problems for solution are given at the ends of some of the chapters, and generally these problems are placed in order of increasing difficulty. At the end of the book will be found references for further study, answers to all of the problems, and the Appendices, which contain tables of useful information.

The authors are grateful to Dr. Winfred O. Carter and Mr. Eduardo Calcaño for their assistance in writing computer programs and solving problems for the book. Thanks are also due to the students who offered many helpful suggestions for improving the presentation, and to Miss Elizabeth Ritchie and Mrs. David Morison for typing the manuscript.

June 1965

JAMES M. GERE
WILLIAM WEAVER, JR.

CONTENTS

Chapter 1

ELEMENTARY CONCEPTS OF STRUCTURAL ANALYSIS

1.1 Introduction. The flexibility and stiffness methods of structural analysis are the principal subject matter of this book. They are the most fundamental methods available to the structural analyst and are applicable to structures of all types. However, only framed structures (as described in the next article) will be discussed in this book.

Before beginning the discussion of the flexibility and stiffness methods, several preliminary subjects will be considered. These subjects include a description of the types of structures under investigation and discussions of various concepts, such as actions, displacements, static indeterminacy, kinematic indeterminacy, flexibilities, and stiffnesses.

1.2 Types of Framed Structures. All of the structures that are analyzed in later chapters are called *framed structures* and can be divided into six categories: beams, plane trusses, space trusses, plane frames, grids, and space frames. These types of structures are illustrated in Fig. 1-1 and described later in detail. These categories are selected because each represents a class of structures having special characteristics. Furthermore, while the basic principles of the flexibility and stiffness methods are the same for all types of structures, the analyses for these six categories are sufficiently different in the details to warrant separate discussions of them.

Every framed structure consists of members that are long in comparison to their cross-sectional dimensions. The *joints* of a framed structure are points of intersection of the members, as well as points of support and free ends of members. Examples of joints are points A, B, C, and D in Figs. 1-1a and 1-1d. Supports may be *fixed,* as at support A in the beam of Fig. 1-1a, or *pinned,* as shown for support A in the plane frame of Fig. 1-1d, or there may be *roller supports,* illustrated by supports B and C in Fig. 1-1a. In special instances the connections between members or between members and supports may be elastic (or semirigid). However, the discussion of this possibility will be postponed until later (see Arts. 6.7 and 6.10). *Loads* on a framed structure may be concentrated forces, distributed loads, or couples.

Consider now the distinguishing features of each type of structure shown in Fig. 1-1. A *beam* (Fig. 1-1a) consists of a straight member having one or more points of support, such as points A, B, and C. Forces applied to a beam are assumed to act in a plane which contains an axis of symmetry of the cross-section of the beam (an axis of symmetry is also a principal axis of the cross-section). Moreover, all external couples acting on the beam have their moment vectors normal to this plane, and the beam deflects in the same plane (the *plane of bending*) and does not twist. (The case of a beam which does not fulfill these criteria is discussed in Art. 3.11.) Internal stress resultants may exist at any cross-section of the beam and, in the general case, may include an axial force, a shearing force, and a bending couple.

A *plane truss* (Fig. 1-1b) is idealized as a system of members lying in a plane and interconnected at hinged joints. All applied forces are assumed to act in the plane of the structure, and all external couples have their moment vectors normal to the plane, just as in the case of a beam. The loads may consist of concentrated forces applied at the joints, as well as loads that act on the members themselves. For purposes of analysis, the latter loads may be replaced by statically equivalent loads acting at the joints. Then the analysis of a truss subjected only to joint loads will result in axial forces of tension or compression in the members. In addition to these axial forces, there will be bending moments and shear forces in those members having loads that act directly upon them. The determination of all such stress resultants constitutes the complete analysis of the forces in the members of a truss.

A *space truss* (see Fig. 1-1c) is similar to a plane truss except that the members may have any directions in space. The forces acting on a space truss may be in arbitrary directions, but any couple acting on a member must have its moment vector perpendicular to the axis of the member. The reason for this requirement is that a truss member is incapable of supporting a twisting moment.

A *plane frame* (Fig. 1-1d) is composed of members lying in a single plane and having axes of symmetry in that plane (as in the case of a beam). The joints between members (such as joints B and C) are rigid connections. The forces acting on a frame and the translations of the frame are in the plane of the structure; all couples acting on the frame have their moment vectors normal to the plane. The internal stress resultants acting at any cross-section of a plane frame member may consist in general of a bending couple, a shearing force, and an axial force.

A *grid* is a plane structure composed of continuous members that either intersect or cross one another (see Fig. 1-1e). In the latter case the connections between members are often considered to be hinged, whereas in the former case the connections are assumed to be rigid. While in a plane frame the applied forces all lie in the plane of the structure, in the case of a grid all forces are normal to the plane of the structure and all couples have their vectors in the plane of the grid. This orientation of loading may result in torsion as well as bending in some of the members. Each member is assumed to have two axes of symmetry in the cross-section, so that bending and torsion occur independently of one another (see Art. 3.11 for a discussion of unsymmetrical members).

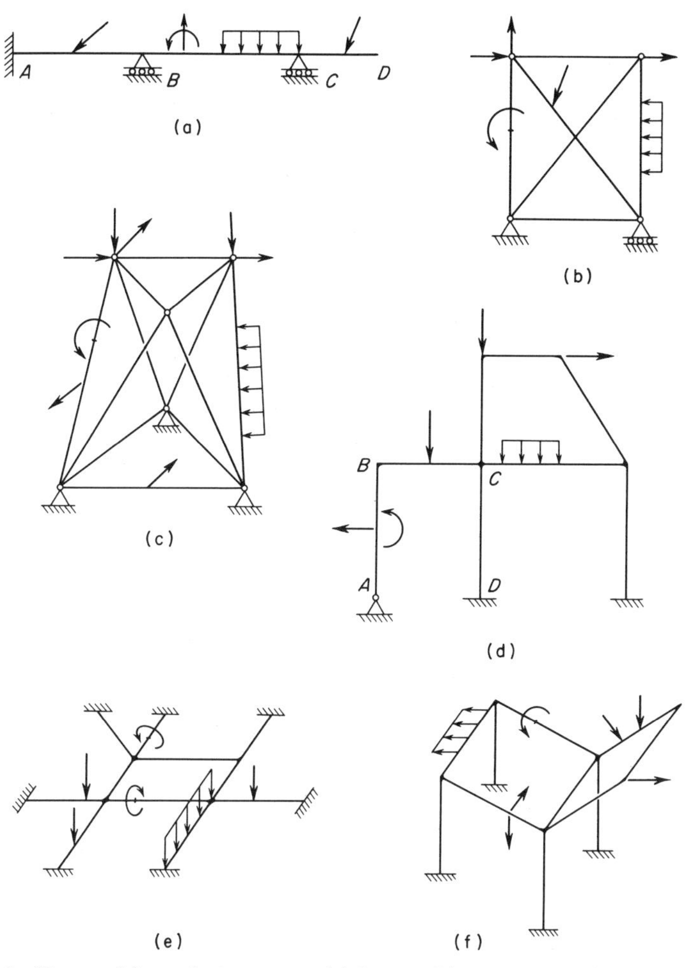

FIG. 1-1. Types of framed structures: (a) beam, (b) plane truss, (c) space truss, (d) plane frame, (e) grid, and (f) space frame.

The final type of structure is a *space frame* (Fig. 1-1f). Space frames are the most general type of framed structure, inasmuch as there are no restrictions on the locations of joints, directions of members, or directions of loads. The individual members of a space frame may carry internal axial forces, torsional couples, bending couples in both principal directions of the cross-section, and shearing forces in both principal directions. The members are assumed to have two axes of symmetry in the cross-section, as explained above for a grid.

It will be assumed throughout most of the subsequent discussions that the structures being considered have prismatic members, that is, each member has a straight axis and uniform cross-section throughout its length. Nonprismatic members are treated later by a modification of the basic approach (see Art. 6.8).

1.3 Deformations and Displacements. When a structure is acted upon by forces, the members of the structure will undergo *deformations* (or small changes in shape) and, as a consequence, points within the structure will be displaced to new positions. In general, all points of the structure except immovable points of support will undergo such displacements. The calculation of these displacements is an essential part of structural analysis, as will be seen later in the discussions of the flexibility and stiffness methods. However, before considering the displacements, it is first necessary to have an understanding of the deformations that produce the displacements.

To begin the discussion, consider a segment of arbitrary length cut from a member of a framed structure, as shown in Fig. 1-2a. For simplicity the bar is assumed to have a circular cross-section. At any cross-section, such as at the right-hand end of the segment, there will be stress resultants that in the general case consist of three forces and three couples. The forces are the axial force N_x and the shearing forces V_y and V_z; the couples are the twisting couple T, and the bending couples M_y and M_z. Note that moment vectors are shown in the figure with doubled-headed arrows, in order to distinguish them from force vectors. The deformations of the bar can be analyzed by taking separately each stress resultant and determining its effect on an element of the bar. The element is obtained by isolating a portion of the bar between two cross-sections a small distance dx apart (see Fig. 1-2a).

The effect of the axial force N_x on the element is shown in Fig. 1-2b. Assuming that the force acts through the centroid of the cross-sectional area, it is found that the element is uniformly extended, the significant strains in the element being normal strains in the x direction. In the case of a shear force (Fig. 1-2c) one cross-section of the bar is displaced laterally with respect to the other. There may also be some distortion of the cross-sections, but this has a negligible effect on the determination of displacements and can be ignored. A bending couple (Fig. 1-2d) causes relative rotation of the two cross-sections so that they are no longer parallel to one another. The resulting strains in the element are in the longitudinal direction of the bar, and they consist of contraction on the compression side and extension on the tension side. Finally, the twisting couple T causes a relative rotation of the two cross-sections about the x axis (see Fig. 1-2e) and, for example, point A is displaced to A'. In the case of a circular bar, twisting produces only shearing strains and the cross-sections remain plane. For other cross-sectional shapes, distortion of the cross-sections will occur.

The deformations shown in Figs. 1-2b, 1-2c, 1-2d, and 1-2e are called axial, shearing, flexural, and torsional deformations, respectively. Their

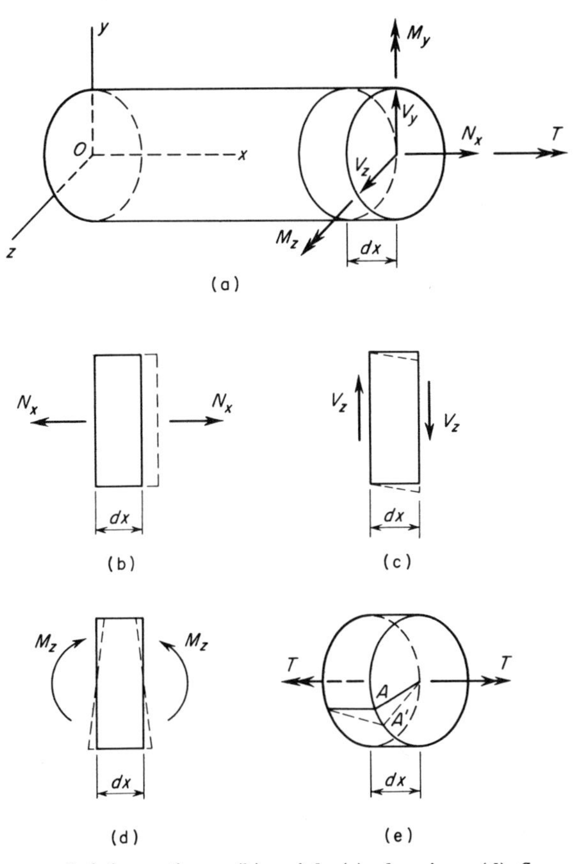

FIG. 1-2. Types of deformations: (b) axial, (c) shearing, (d) flexural, and (e) torsional.

evaluation is dependent upon the cross-sectional shape of the bar and the mechanical properties of the material. This book is concerned only with materials that are linearly elastic, that is, materials that follow Hooke's law. For these materials the various formulas for the deformations, as well as those for the stresses and strains in the element, are given for reference purposes in Appendix A, Art. A.1.

The *displacements* in a structure are caused by the cumulative effects of the deformations of all the elements. There are a number of ways of calculating these displacements in framed structures, depending upon the type of deformation being considered as well as the type of structure. For example, deflections of beams considering only flexural deformations can be found by direct integration of the differential equation for bending of a beam. Another method, which can be used for all types of framed structures including beams, trusses, grids, and frames, is the unit-load method. In both of these methods, as well as others in common use, it is assumed that the displacements of the structure are small. For purposes of review and convenient reference in the work to follow, a brief discussion of the calculation of displacements by the unit-load method is given in Art. A.2 of Appendix A. In addition, a table of beam displacements for several common conditions of loading is given in Art. A.3. The reader should be familiar with this material in order to solve the examples and problems that appear later.

In any particular structure under investigation, not all types of deformations will be of significance in calculating the displacements. For example, in beams the flexural deformations normally are the only ones of importance, and it is usual to ignore the axial deformations. Of course, there are exceptional situations in which beams are required to carry large axial forces, and under such circumstances the axial deformation must be included in the analysis. It is also possible for axial forces to produce a beam-column action which has a nonlinear effect on the displacements (see Art. 6.12).

For truss structures of the types shown in Figs. 1-1b and 1-1c, the analyses are made in two parts. If the joints of the truss are idealized as hinges and if all loads act only at the joints, then the analysis involves only axial deformations of the members. The second part of the analysis is for the effects of the loads that act on the members between the joints, and this part is essentially the analysis of simply supported beams. If the joints of a truss-like structure actually are rigid, then bending occurs in the members even though all loads may act at the joints. In such a case, flexural deformations could be important, and in this event the structure may be analyzed as a plane or space frame.

In plane frames (see Fig. 1-1d) the significant deformations are flexural and axial. If the members are slender and are not triangulated in the fashion of a truss, the flexural deformations are much more important than the axial ones. However, the axial contributions should be included in the analysis of a plane frame if there is any doubt about their relative importance.

In grid structures (Fig. 1-1e) the flexural deformations are always important, but the cross-sectional properties of the members and the method of fabricating joints will determine whether or not torsional deformations must be considered. If the members are thin-walled open sections, such as I-beams, they are likely to be very flexible in torsion and, hence, large twisting moments will not develop in the members. Also, if the members of a grid are not rigidly connected at crossing points, there will be no interaction between the flexural and torsional moments. In either of these cases, only flexural deformations need be taken into account in the analysis. On the other hand, if the members of a grid are torsionally stiff and rigidly interconnected at crossing points, the analysis must include both torsional and flexural deformations. Normally, there are no axial forces present in a grid because the forces are normal to the plane of the grid. This situation is analogous to that in a beam having all its loads perpendicular to the axis of the beam, in which case there are no axial forces in the beam. Thus, axial deformations are not included in a grid analysis.

Space frames (Fig. 1-1f) represent the most general type of framed structure, both with respect to geometry and with respect to loads. Therefore, it follows that axial, flexural, and torsional deformations all may enter into the analysis of a space frame, depending upon the particular structure and loads.

Shearing deformations are usually very small in framed structures and hence are seldom considered in the analysis. However, their effects may always be included if necessary in the analysis of a beam, plane frame, grid, or space frame (see Arts. 3.4 and 6.11).

There are other effects, such as temperature changes and prestrains, that may be of importance in analyzing a structure. These subjects are discussed in later chapters in conjunction with the flexibility and stiffness methods of analysis.

1.4 Actions and Displacements. The terms "action" and "displacement" are used to describe certain fundamental concepts in structural analysis. An *action* (sometimes called a generalized force) is most commonly a single force or a couple. However, an action may also be a combination of forces and couples, a distributed loading, or a combination of these actions. In such combined cases, however, it is necessary that all the forces, couples, and distributed loads be related to one another in some definite manner so that the entire combination can be denoted by a single symbol. For example, if the loading on the simply supported beam AB shown in Fig. 1-3 consists of two equal forces P, it is possible to consider the combination of the two loads as a single action and to denote it by one symbol, such as F. It is also possible to think of the combination of the two loads plus the two reactions R_A and R_B at the supports as a single action, since all four forces have a unique relationship to one another. In a more general situation, it is possible for a very complicated loading system on a structure to be treated as a single action if all components of the load are related to one another in a definite manner.

In addition to actions that are external to a structure, it is necessary to deal also with internal actions. These actions are the resultants of internal stress distributions, and they include bending moments, shearing forces, axial forces, and twisting couples. Depending upon the particular analysis being made, such actions may appear as one force, one couple, two forces, or two couples. For example, in making static equilibrium analyses of structures these actions normally appear as single forces and couples, as illustrated in Fig. 1-4a. The cantilever beam shown in the figure is subjected at end B to loads in the form of actions P_1 and M_1. At the fixed end A the reactive force and reactive couple are denoted R_A and M_A, respectively. In order to distinguish these reactions from the loads on the structure, they are drawn with a slanted line (or slash) across the arrow. This convention for identifying reactions will be followed throughout the text (see also Fig. 1-3 for an illustration of the use of the convention).* In calculating the axial force N, bending moment M, and shearing force V at any section of the beam in Fig. 1-4a, such as at the midpoint, it is necessary to consider the static equilibrium of a portion of the beam. One possibility is to construct a free-body diagram of the right-hand half of the beam, as shown in Fig. 1-4b. In so doing, it is evident that each of the internal actions appears in the diagram as a single force or couple.

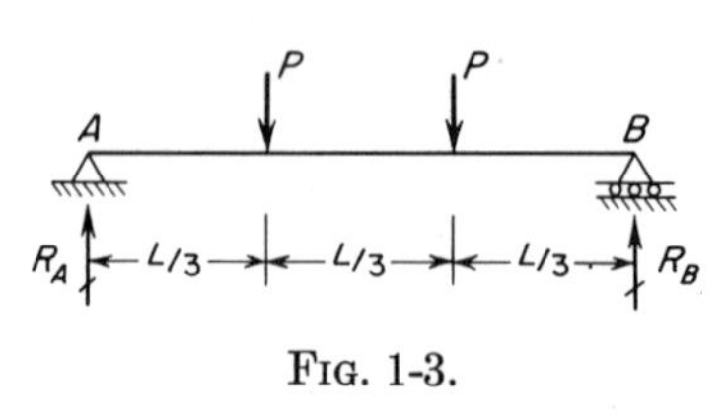

Fig. 1-3.

There are situations, however, in which the internal actions appear as two forces or couples. This case occurs most commonly in structural analysis when a "release" is made at some point in a structure, as shown in Fig. 1-5 for a continuous beam. If the bending moment is released (or eliminated) at joint B of the beam, the result is the same as if a hinge were placed in the beam at that joint (see Fig. 1-5b). Then, in order to take account of the bending moment M_B in the beam, it must be considered as consisting of two equal and opposite couples M_B that act on the left- and right-hand portions of the beam with the hinge, as shown in Fig. 1-5c. In this illustration the moment M_B is assumed positive in the directions shown in the figure, signifying that the couple acting on the left-hand beam is counterclockwise and the couple acting on the right-hand beam is clockwise. Thus, for the purpose of analyzing the beam in Fig. 1-5c, the bending moment at point B may be treated as a single action consisting of two couples. Similar situations are encountered with axial forces, shearing forces, and twisting couples, as

* This convention is used in the well-known textbook by C. H. Norris and J. B. Wilbur, *Elementary Structural Analysis*, 2nd ed., McGraw-Hill Book Co., Inc., New York, 1960, p. 63.

illustrated later in the discussion of the flexibility method of analysis.

A second basic concept is that of a *displacement*, which is most commonly a translation or a rotation at some point in a structure. A translation refers to the distance moved by a point in the structure, and a rotation means the angle of rotation of the tangent to the elastic curve at a point. For example, in the cantilever beam of Fig. 1-4c, the translation Δ of the end of the beam and the rotation Θ at the end are both considered as displacements. Moreover, as in the case of an action, a displacement may also be regarded in a generalized sense as a combination of translations and rotations. As an example, consider the ro-

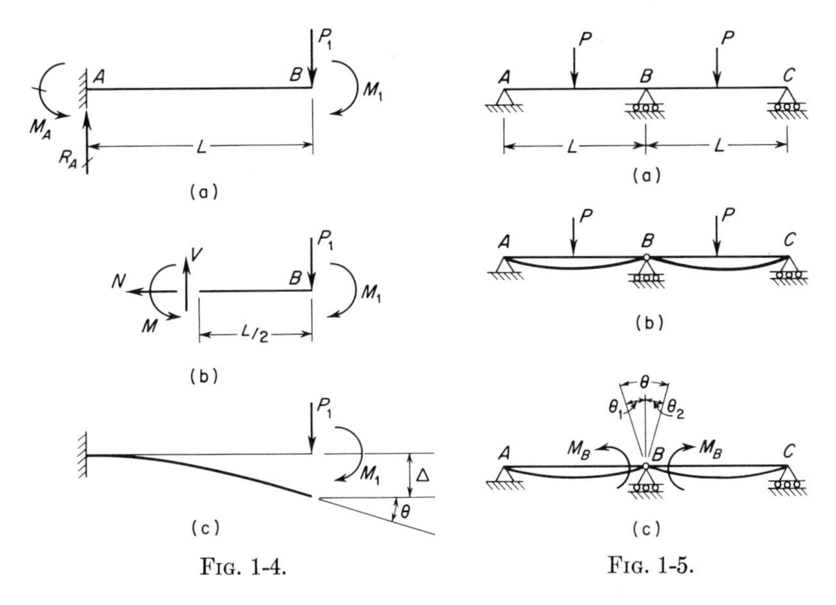

FIG. 1-4.

FIG. 1-5.

tations at the hinge at point B in the two-span beam in Fig. 1-5c. The rotation of the right-hand end of member AB is denoted Θ_1, while the rotation of the left-hand end of member BC is denoted Θ_2. Each of these rotations is considered as a displacement. Furthermore, the sum of the two rotations, denoted as Θ, is also a displacement. The angle Θ can be considered as the relative rotation at point B between the ends of members AB and BC.

Another illustration of displacements is shown in Fig. 1-6, in which a plane frame is subjected to several loads. The horizontal translations Δ_A, Δ_B, and Δ_C of joints A, B, and C, respectively, are displacements, as also are the rotations Θ_A, Θ_B, and Θ_C of these joints. Joint displacements of these types play an important role in the analysis of framed structures.

It is frequently necessary in structural analysis to deal with actions and displacements that *correspond* to one another. Actions and displacements are said to be corresponding when they are of an analogous type and are located at the same point on a structure. Thus, the displacement corresponding to a concentrated force is a translation of the structure at the point where the force acts, although the displacement is not necessarily caused by the force. Furthermore, the corresponding displacement must

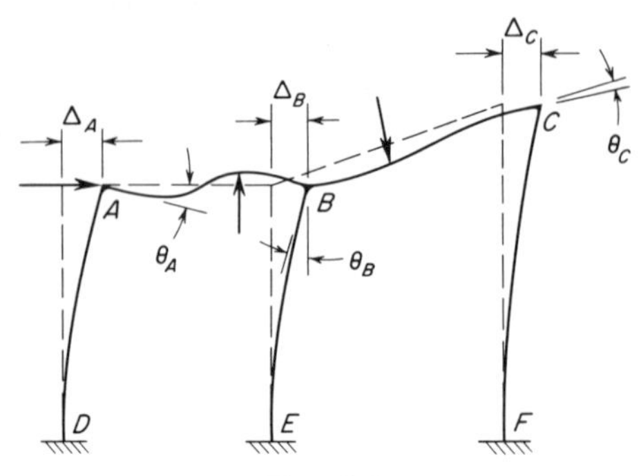

FIG. 1-6.

be taken along the line of action of the force and must have the same positive direction as the force. In the case of a couple, the corresponding displacement is a rotation at the point where the couple is applied and is taken positive in the same sense as the couple. As an illustration, consider again the cantilever beam shown in Fig. 1-4a. The action P_1 is a concentrated force acting downward at the end of the beam, and the downward translation Δ at the end of the beam (see Fig. 1-4c) is the displacement that corresponds to this action. Similarly, the couple M_1 and the rotation Θ are a corresponding action and displacement. It should be noted, however, that the displacement Δ corresponding to the load P_1 is not caused solely by the force P_1, nor is the displacement Θ corresponding to M_1 caused by M_1 alone. Instead, in this example, both Δ and Θ are displacements due to P_1 and M_1 acting simultaneously on the beam. In general, if a particular action is given, the concept of a corresponding displacement refers only to the definition of the displacement, without regard to the actual cause of that displacement. Similarly, if a displacement is given, the concept of a corresponding action will describe a particular kind of action on the structure, but the displacement need not be caused by that action.

As another example of corresponding actions and displacements, refer to the actions shown in Fig. 1-5c. The beam in the figure has a hinge at the middle support and is acted upon by the two couples M_B, which are considered as a single action. The displacement corresponding to the action M_B consists in general of the sum of the counterclockwise rotation Θ_1 of the left-hand beam and the clockwise rotation Θ_2 of the right-hand beam. Therefore, the angle Θ (equal to the sum of Θ_1 and Θ_2) is the displacement corresponding to the action M_B. This displacement is the relative rotation between the two beams at the hinge and has the same positive sense as M_B. If the angle Θ is caused only by the couples M_B, then it is described as the displacement corresponding to M_B and caused by M_B. This displacement can be found with the aid of the table of beam displacements given in Appendix A (see Table A-3, Case 5), and is equal to

$$\Theta = \Theta_1 + \Theta_2 = \frac{M_B L}{3EI} + \frac{M_B L}{3EI} = \frac{2M_B L}{3EI}$$

in which L is the length of each span and EI is the flexural rigidity of the beam.

There are other situations, however, in which it is necessary to deal with a displacement that corresponds to a particular action but is caused by some other action. As an example, consider the beam in Fig. 1-5b, which is the same as the beam in Fig. 1-5c except that it is acted upon by two forces P instead of the couples M_B. The displacement in this beam corresponding to M_B consists of the relative rotation at joint B between the two beams, positive in the same sense as M_B, but due to the loads P only. Again using the table of beam displacements (Table A-3, Case 2), and also assuming that the forces P act at the midpoints of the members, it is found that the displacement Θ corresponding to M_B and caused by the loads P is

$$\Theta = \Theta_1 + \Theta_2 = \frac{PL^2}{16EI} + \frac{PL^2}{16EI} = \frac{PL^2}{8EI}$$

The concept of correspondence between actions and displacements will become more familiar to the reader as additional examples are encountered in subsequent work. Also, it should be noted that the concept can be extended to include distributed actions, as well as combinations of actions of all types. However, these more general ideas have no particular usefulness in the work to follow.

In order to simplify the notation for actions and displacements, it is desirable in many cases to use the symbol A for actions, including both concentrated forces and couples, and the symbol D for displacements, including both translations and rotations. Subscripts can be used to distinguish between the various actions and displacements that may be of interest in a particular analysis. The use of this type of notation is shown in Fig. 1-7, which portrays a cantilever beam subjected to actions A_1, A_2, and A_3. The displacement corresponding to A_1 and due to all loads acting simultaneously is denoted by D_1 in Fig. 1-7a; similarly, the displacements corresponding to A_2 and A_3 are denoted by D_2 and D_3.

Now consider the cantilever beam subjected to action A_1 only (see Fig. 1-7b). The displacement corresponding to A_1 in this beam is denoted by D_{11}. The significance of the two subscripts is as follows. The first subscript indicates that the displacement corresponds to action A_1, and the second indicates that the cause of the displacement is action A_1. In a similar manner, the displacement corresponding to A_2 in this beam is denoted by D_{21}, where the first subscript means that the displacement corresponds to A_2 and the second means that it is caused by A_1. Also shown in Fig. 1-7b is the displacement D_{31} corresponding to the couple A_3.

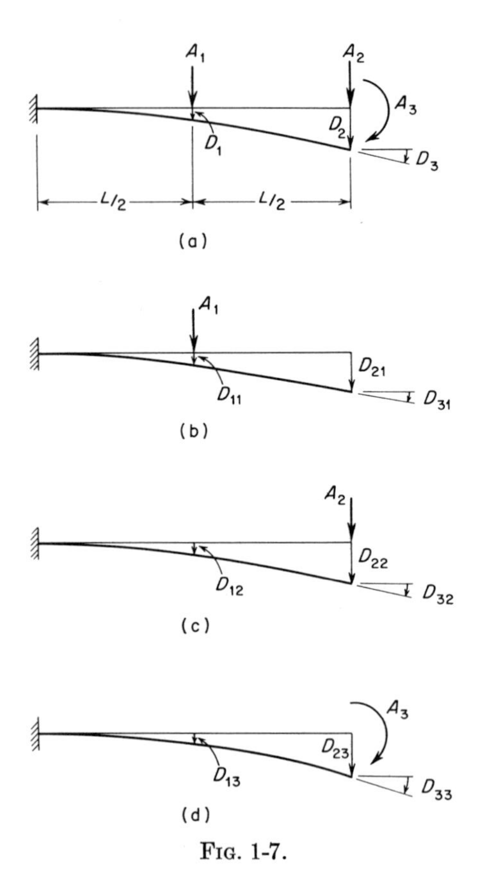

FIG. 1-7.

The displacements caused by action A_2 acting alone are shown in Fig. 1-7c, and those caused by A_3 alone are shown in Fig. 1-7d. In each case the subscripts for the displacement symbols follow the general rule that the first subscript identifies the displacement and the second gives the cause of the displacement. In general, the cause may be a single force, a couple, or an entire loading system. Unless specifically stated otherwise, this convention for subscripts will always be used in later discussions.

For the beams pictured in Fig. 1-7 it is not difficult to determine the various displacements (see Appendix A for methods of calculating displacements of beams). Assuming that the beam has flexural rigidity EI and length L, it is found that the displacements for the beam in Fig. 1-7b are

$$D_{11} = \frac{A_1 L^3}{24EI} \qquad D_{21} = \frac{5A_1 L^3}{48EI} \qquad D_{31} = \frac{A_1 L^2}{8EI}$$

In a similar manner the remaining six displacements in Figs. 1-7c and d ($D_{12}, D_{22}, \ldots, D_{33}$) can be found. Then the displacements in the beam under all loads acting simultaneously (see Fig. 1-7a) are determined by summation:

$$D_1 = D_{11} + D_{12} + D_{13}$$

$$D_2 = D_{21} + D_{22} + D_{23}$$

$$D_3 = D_{31} + D_{32} + D_{33}$$

These summations are expressions of the principle of superposition, which is discussed more fully in Art. 1.9.

1.5 Equilibrium. One of the objectives of any structural analysis is to determine various actions pertaining to the structure, such as reactions at the supports and internal stress resultants (bending moments, shearing forces, etc.). A correct solution for any of these quantities must satisfy all conditions of static equilibrium, not only for the entire structure, but also for any part of the structure taken as a free body.

Consider now any free body subjected to several actions. The resultant of all the actions may be a force, a couple, or both. If the free body is in static equilibrium, the resultant vanishes; that is, the resultant force vector and the resultant moment vector are both zero. A vector in three-dimensional space may always be resolved into three components in mutually orthogonal directions, such as the x, y, and z directions. If the resultant force vector equals zero, then its components also must be equal to zero, and therefore the following equations of static equilibrium are obtained:

$$\sum F_x = 0 \qquad \sum F_y = 0 \qquad \sum F_z = 0 \tag{1-1a}$$

In these equations the expressions $\sum F_x$, $\sum F_y$, and $\sum F_z$ are the algebraic sums of the x, y, and z components, respectively, of all the force vectors acting on the free body. Similarly, if the resultant moment vector equals zero, the moment equations of static equilibrium are

$$\sum M_x = 0 \qquad \sum M_y = 0 \qquad \sum M_z = 0 \tag{1-1b}$$

in which $\sum M_x$, $\sum M_y$, and $\sum M_z$ are the algebraic sums of the moments about the x, y, and z axes, respectively, of all the couples and forces acting on the free body. The six relations in Eqs. (1-1) represent the static equilibrium equations for actions in three dimensions. They may be applied to any free body such as an entire structure, a portion of a structure, a single member, or a joint of a structure.

When all forces acting on a free body are in one plane and all couples have their vectors normal to that plane, only three of the six equilibrium equations will be useful. Assuming that the forces are in the x-y plane, it is apparent that the equations $\sum F_z = 0$, $\sum M_x = 0$, and $\sum M_y = 0$ will be satisfied automatically. The remaining equations are

$$\sum F_x = 0 \qquad \sum F_y = 0 \qquad \sum M_z = 0 \tag{1-2}$$

and these equations become the static equilibrium conditions for actions in the x-y plane.

In the stiffness method of analysis, the basic equations to be solved are those which express the equilibrium conditions at the joints of the structure, as described later in Chapter 2.

1.6 Compatibility. In addition to the static equilibrium conditions, it is necessary in any structural analysis that all conditions of compatibility be satisfied. These conditions refer to continuity of the displacements throughout the structure, and are sometimes referred to as conditions of geometry. As an example, compatibility conditions must be satisfied at all points of support, where it is necessary that the displacements of the structure be consistent with the support conditions. For instance, at a fixed support there can be no rotation of the axis of the member.

Compatibility conditions must also be satisfied at all points throughout the interior of a structure. Usually, it is compatibility conditions at the joints of the structure that are of interest. For example, at a rigid connection between two members the displacements (translations and rotations) of both members must be the same.

In the flexibility method of analysis the basic equations to be solved are equations that express the compatibility of the displacements, as will be described in Chapter 2.

1.7 Static and Kinematic Indeterminacy. There are two types of indeterminacy that must be considered in structural analysis, depending upon whether actions or displacements are of interest. When actions are the unknowns in the analysis, as in the flexibility method, then *static indeterminacy* must be considered. In this case, indeterminacy refers to an excess of unknown actions as compared to the number of equations of static equilibrium that are available. The equations of equilibrium, when applied to the entire structure and to its various parts, may be used for the calculation of reactions and internal stress resultants. If these equations are sufficient for finding all actions, both external and internal, then the structure is statically determinate. If there are more unknown actions than equations, the structure is statically indeterminate. The simply supported beam shown in Fig. 1-3 and the cantilever beam of Fig. 1-4 are examples of statically determinate structures, since in both cases all reactions and stress resultants can be found from equilibrium equations alone. On the other hand, the continuous beam of Fig. 1-5a is statically indeterminate.

The unknown actions in excess of those that can be found by static equilibrium are known as *static redundants,* and the number of such redundants represents the *degree* of static indeterminacy of the structure. Thus, the two-span beam of Fig. 1-5a is statically indeterminate to the first degree, since there is one redundant action. For instance, it can be seen that it is impossible to calculate all of the reactions for the beam by static equilibrium alone. However, after the value of one reaction is obtained (by one means or another), the remaining reactions and all internal stress resultants can be found by statics alone.

Other examples of statically indeterminate structures are shown in Fig. 1-8. The propped cantilever beam in Fig. 1-8a is statically indeterminate to the first degree, since there are four reactive actions (H_A, M_A, R_A, and R_B) whereas only three equations of equilibrium are available for the calculation of reactions (see Eqs. 1-2).

The fixed-end beam of Fig. 1-8b is statically indeterminate to the third degree because there are six reactions to be found in the general case. In the special case when all the concentrated forces on a fixed-end beam act in a direction which is perpendicular to the axis of the beam, there will be no axial forces at the ends of the beam. In such a case the beam can be analyzed as if it were statically indeterminate to the second degree.

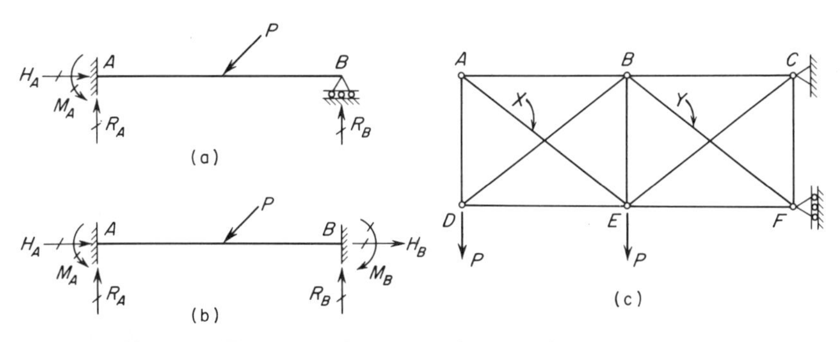

FIG. 1-8. Examples of statically indeterminate structures.

The plane truss in Fig. 1-8c is statically indeterminate to the second degree. This conclusion can be reached by cutting two bars, such as X and Y, thereby releasing the forces in those bars. The truss with the cut bars is then statically determinate, since all reactions and bar forces can be found by a direct application of equations of equilibrium.

Each bar that is cut represents one action, namely, the force in the bar, that is removed from the truss. The number of actions that must be released in order to reduce the statically indeterminate structure to a determinate structure will be equal to the degree of indeterminacy. This method of ascertaining the degree of statical indeterminacy is quite general and can be used with many types of structures.

As another example of this method for determining the degree of indeterminateness of a structure, consider the plane frame shown in Fig. 1-6. The object is to make cuts, or releases, in the frame until the structure has become statically determinate. If bars AB and BC are cut, the structure that remains consists of three cantilever portions (the supports of the cantilevers are at D, E, and F), each of which is statically determinate. Each bar that is cut represents the removal (or release) of three actions (axial force, shearing force, and bending moment) from the original structure. Because a total of six actions were released, the degree of indeterminacy of the frame is six.

A distinction may also be made between external and internal indeterminateness. External indeterminateness refers to the calculation of the reactions for the structure. Normally, there are six equilibrium equations available for the determination of reactions in a space structure, and three for a plane structure. Therefore, a space structure with more than six reactive actions, and a plane structure with more than three reactions, will usually be externally indeterminate. Examples of external indeterminateness can be seen in Fig. 1-8. The propped cantilever beam is externally indeterminate to the first degree, the fixed-end beam is externally indeterminate to the third degree, and the plane truss is statically determinate externally.

Internal indeterminateness refers to the calculation of stress resultants within the structure assuming that all reactions have been found previously. For example, the truss in Fig. 1-8c is internally indeterminate to the second degree, although it is externally determinate, as noted above.

The total degree of indeterminateness of a structure is the sum of the external and internal degrees of indeterminateness. Thus, the truss in Fig. 1-8c is indeterminate to the second degree when considered in its entirety. The beam in Fig. 1-8a is externally indeterminate to the first degree and internally determinate, inasmuch as all stress resultants in the beam can be readily found after all the reactions are known. The plane frame in Fig. 1-6 has nine reactive actions, and therefore is externally indeterminate to the sixth degree. Internally the frame is determinate since all stress resultants can be found if the reactions are known. Therefore, the frame has a total indeterminateness of six, as previously observed for this structure.

Occasionally there are special conditions of construction that affect the degree of indeterminacy of a structure. The three-hinged arched truss shown in Fig. 1-9 has a central hinge at joint B which makes it possible to calculate all four reactions by statics. For the truss shown, the bar forces in all members can be found after the reactions are known, and therefore the structure is statically determinate overall.

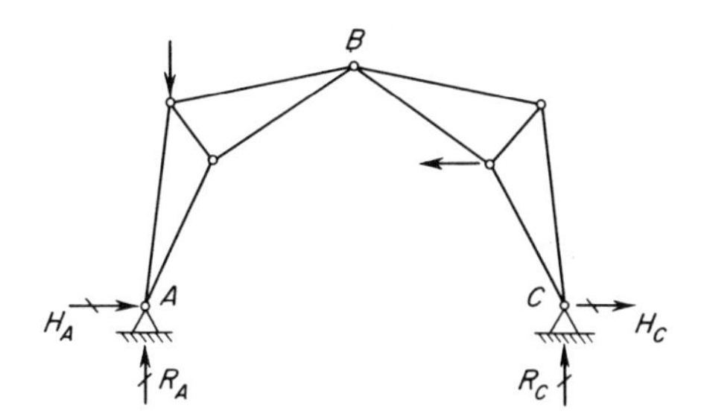

FIG. 1-9. Three-hinged arched truss.

Several additional examples of statically indeterminate structures are given at the end of this article. These examples illustrate how the degree of indeterminacy can be obtained for many structures by intuitive reasoning. Other examples will be encountered in Chapter 2 in connection with the flexibility method of analysis. However, for large structures it is desirable to have more formalized methods of establishing static indeterminacy; such methods are discussed elsewhere and are not repeated here.*

In the stiffness method of analysis the displacements of the joints of the structure become the unknown quantities. Therefore, the second type of indeterminacy, known as *kinematic indeterminacy*, becomes important. In order to understand this type of indeterminacy, it should be recalled that joints in framed structures are defined to be located at all points where two or more members intersect, at points of support, and at free ends. When the structure is subjected to loads, each joint will undergo displacements in the form of translations and rotations, depending upon the configuration of the structure. In some cases the joint displacements will be known because of restraint conditions that are imposed upon the structure. At a fixed support, for instance, there can be no displacements of any kind. However, there will be other joint displacements that are not known in advance, and which can be obtained only by making a complete analysis of the structure. These unknown joint displacements are the kinematically indeterminate quantities, and are sometimes called kinematic redundants. Their number represents the degree of kinematic indeterminacy of the structure, or the number of *degrees of freedom* for joint displacement.

To illustrate the concepts of kinematic indeterminacy, it is useful to

* For a complete discussion of static indeterminacy, see *ibid.*, pp. 64–69, 74–78, 138–147, 153–157, and 271–273.

consider again the examples of Fig. 1-8. Beginning with the beam in Fig. 1-8a, it is seen that end A is fixed and cannot undergo any displacement. On the other hand, joint B has two degrees of freedom for joint displacement since it may translate in the horizontal direction and may rotate. Thus, the beam is kinematically indeterminate to the second degree, and there are two unknown joint displacements to be calculated in a complete analysis of this beam. In many practical analyses it would be permissible to neglect axial deformations of the beam; in such a case, joint B would have only one degree of freedom (rotation), and the structure would be analyzed as if it were kinematically indeterminate to the first degree.

The second example of Fig. 1-8 is a fixed-end beam. Such a beam has no unknown joint displacements, and therefore is kinematically determinate. By comparison, the same beam was statically indeterminate to the third degree.

The third example in Fig. 1-8 is the plane truss which was previously shown to be statically indeterminate to the second degree. Joint A of this truss may undergo two independent components of displacement (such as translations in two perpendicular directions) and hence has two degrees of freedom. Rotation of a joint of a truss has no physical significance because, under the assumption of hinged joints, rotation of a joint produces no effects in the members of the truss. Thus, the degree of kinematic indeterminacy of a truss is always found as if the truss were subjected to loads at the joints only; this is the same philosophy as in the case of static indeterminacy, wherein only axial forces in the members are considered as unknowns. The joints B, D, and E of the truss in Fig. 1-8 also have two degrees of freedom each, while the restrained joints C and F have zero and one degree of freedom, respectively. Thus, the truss has a total of nine degrees of freedom for joint translation and is kinematically indeterminate to the ninth degree.

The rigid frame shown in Fig. 1-6 offers another example of a kinematically indeterminate structure. Since the supports at D, E, and F of this frame are fixed, there can be no displacements at these joints. However, joints A, B, and C each possess three degrees of freedom, since each joint may undergo horizontal and vertical translations and a rotation. Thus, the total number of degrees of kinematic indeterminateness for this frame is nine. If the effects of axial deformations are omitted from the analysis, the degree of kinematic indeterminacy is reduced. There would be no possibility for vertical displacement of any of the joints because the columns would not change length. Furthermore, the horizontal translations of joints A and B would be equal, and the horizontal translation of C would have a known relationship to that of joint B. In other words, if axial deformations are neglected the only independent joint displacements are the rotations of joints A, B, and C

and one horizontal displacement (such as that of joint B). Therefore, the structure could be considered to be kinematically indeterminate to the fourth degree.

Several examples involving both static and kinematic indeterminacy will now be given.

Example 1. The space truss shown in Fig. 1-10 has pin supports at A, B, and C. The degrees of static and kinematic indeterminacy for the truss are to be obtained.

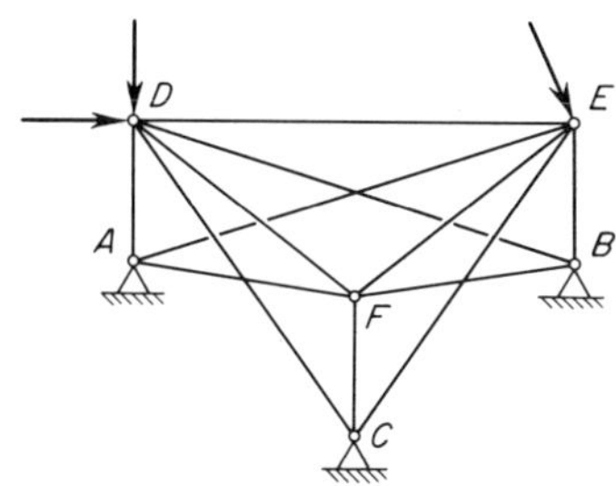

FIG. 1-10. Example 1.

In determining the degree of static indeterminacy, it can be noted that there are three equations of equilibrium available at every joint of the truss for the purpose of calculating bar forces or reactions. Thus, a total of 18 equations of statics is available. The number of unknown actions is 21, since there are 12 bar forces and 9 reactions (three at each support) to be found. The truss is, therefore, statically indeterminate to the third degree. More specifically, the truss is externally indeterminate to the third degree, because there are nine reactions but only six equations for the equilibrium of the truss as a whole. The truss is internally determinate since all bar forces can be found by statics after the reactions are determined.

Each of the joints D, E, and F has three degrees of freedom for joint displacement, because each joint can translate in three mutually orthogonal directions. Therefore, the truss is kinematically indeterminate to the ninth degree.

Example 2. The degrees of static and kinematic indeterminacy are to be found for the space frame shown in Fig. 1-11a.

There are various ways in which the frame can be cut in order to reduce it to a statically determinate structure. One possibility is to cut all four of the bars EF, FG, GH, and EH, thereby giving the released structure shown in Fig. 1-11b. Since each release represents the removal of six actions (axial force, two shearing forces, twisting couple, and two bending moments) the original frame is statically indeterminate to the 24th degree.

The number of possible joint displacements at E, F, G, and H is six at each

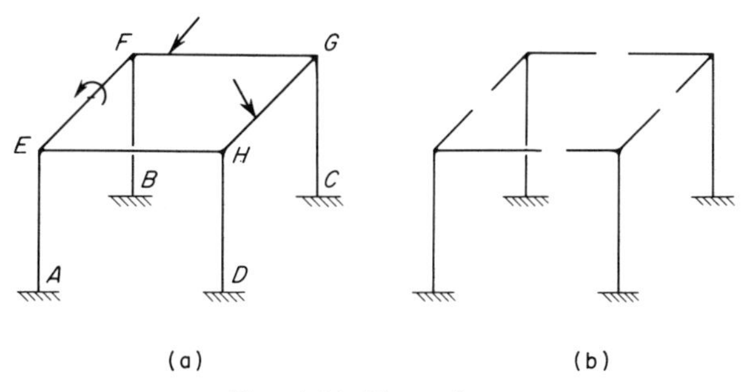

FIG. 1-11. Example 2.

joint (three translations and three rotations); therefore, the frame is kinematically indeterminate to the 24th degree.

Now consider the effect of omitting axial deformations from the analysis. The degree of static indeterminacy is not affected, because the same number of actions will still exist in the structure. On the other hand, there will be fewer degrees of freedom for joint displacement. The columns will not change in length, thereby eliminating four joint translations (one each at E, F, G, and H). In addition, the four horizontal members will not change in length, thereby eliminating four more translations. Thus, it is finally concluded that the degree of kinematic indeterminacy is 16 when axial deformations are excluded from consideration.

Consider next the effect of replacing the fixed supports at A, B, C, and D by immovable pinned supports. The effect of the pinned supports is to reduce the number of reactions at each support from six to three. Therefore, the degree of static indeterminacy becomes 12 less than with fixed supports, or a total of 12 degrees. At the same time, three additional degrees of freedom for rotation have been added at each support, so that the degree of kinematic indeterminacy has been increased by 12 when compared to the frame with fixed supports. It can be seen that removing restraints at the supports of a structure serves to decrease the degree of statical indeterminacy, while increasing the degree of kinematic indeterminacy.

Example 3. The grid shown in Fig. 1-12a lies in a horizontal plane and is supported at A, D, E, and H by simple supports. The joints at B, C, F, and G are rigid connections. What are the degrees of static and kinematic indeterminacy?

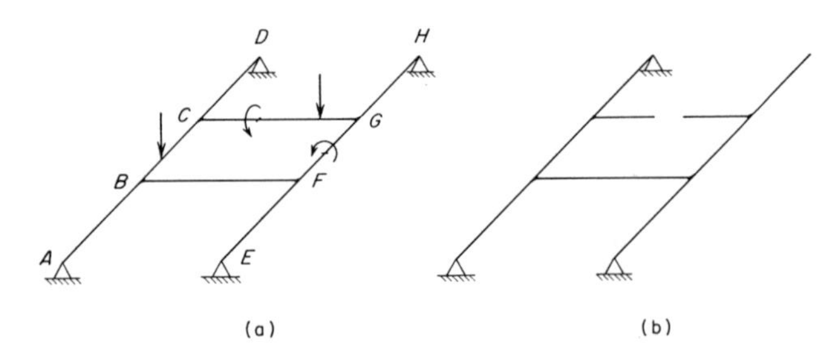

FIG. 1-12. Example 3.

Because there are no axial forces in the members of a grid, only vertical reactions are developed at the supports of this structure. Therefore, the grid is externally indeterminate to the first degree, because only three equilibrium equations are available for the structure in its entirety, but there are four reactions. After removing one reaction, the grid can be made statically determinate by cutting one member, such as CG (see Fig. 1-12b). The release in member CG removes three actions (shearing force in the vertical direction, twisting couple, and bending moment). Thus, the grid can be seen to be internally indeterminate to the third degree, and statically indeterminate overall to the fourth degree.

In general there are three degrees of freedom for displacement at each joint of a grid (one translation and two rotations). Such is the case at joints B, C, F, and G of the grid shown in Fig. 1-12a. However, at joints A, D, E, and H only two joint displacements are possible, inasmuch as joint translation is prevented.

Therefore, the grid shown in the figure is kinematically indeterminate to the 20th degree.

1.8 Mobile Structures. In the preceding discussion of external static indeterminacy, the number of reactive actions for a structure was compared with the number of equations of static equilibrium for the entire structure taken as a free body. If the number of reactions exceeds the number of equations, the structure is externally statically indeterminate; if they are equal, the structure is externally determinate. However, it was tacitly assumed in the discussion that the geometrical arrangement of the reactions was such as to prevent the structure from moving when loads act on it. For instance, the beam shown in Fig. 1-13a has three reactions, which is the same as the number of static equilibrium equations for forces in a plane. It is apparent, however, that the beam will move to the left when the load P is applied. A structure of this type is said to be *mobile.* Other examples of mobile structures are the frame of Fig. 1-13b and the truss of Fig. 1-13c. In the structure of Fig. 1-13b the three reactive forces are concurrent, since their lines of action intersect at point O. Therefore, the frame is mobile, since it cannot support a load such as the force P which does not act through point O. In the truss of Fig. 1-13c there are two bars which are collinear at joint A, and there is no other bar meeting at that joint. Again, the structure is mobile since it is incapable of supporting the load P in its initial configuration.

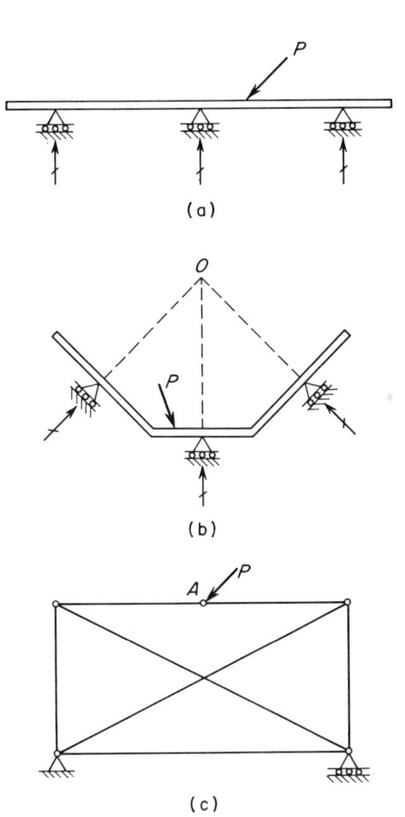

FIG. 1-13. Mobile structures.

From the examples of mobile structures given in Fig. 1-13, it is apparent that both the supports and the members of any structure must be adequate in number and in geometrical arrangement to insure that the structure is not movable. Only structures meeting these conditions will be considered for analysis in subsequent chapters.

1.9 Principle of Superposition. The principle of superposition is one of the most important concepts in structural analysis. It may be used whenever linear relationships exist between actions and displacements (the conditions under which this assumption is valid are described later in this article). In using the principle of superposition it is assumed that certain actions and displacements are imposed upon a structure. These actions and displacements cause other actions and displacements to be developed in the structure. Thus, the former actions and displacements have the nature of causes, while the latter are effects. In general terms the principle states that the effects produced by several causes can be obtained by combining the effects due to the individual causes.

In order to illustrate the use of the principle of superposition when actions are the cause, consider the beam in Fig. 1-14a. This beam is subjected to loads A_1 and A_2, which produce various actions and displacements throughout the structure. For instance, reactions R_A, R_B, and M_B are developed at the supports, and a displacement D is produced at the midpoint of the beam. The effects of the actions A_1 and A_2 acting separately are shown in Figs. 1-14b and 1-14c. In each case there is a displacement at the midpoint of the beam and reactions at the ends. A single prime is used to denote quantities associated with the action A_1, and a double prime is used for quantities associated with A_2.

According to the principle of superposition, the actions and displacements caused by A_1 and A_2 acting separately (Figs. 1-14b and 1-14c) can be combined in order to obtain the actions and displacements caused by A_1 and A_2 acting simultaneously (Fig. 1-14a). Thus, the following *equations of superposition* can be written for the beam in Fig. 1-14:

$$\begin{aligned} R_A &= R'_A + R''_A \qquad & R_B &= R'_B + R''_B \\ M_B &= M'_B + M''_B \qquad & D &= D' + D'' \end{aligned} \tag{1-3}$$

Of course, similar equations of superposition can be written for other actions and displacements in the beam, such as stress resultants at any cross-section of the beam and displacements (translations and rotations) at any point along the axis of the beam. This manner of using superposition was illustrated previously in Art. 1.4.

A second example of the principle of superposition, in which displacements are the cause, is given in Fig. 1-15. The figure portrays again the beam AB with one end simply supported and the other fixed (see Fig. 1-15a). When end B of the beam is displaced downward through a distance Δ and, at the same time, is caused to be rotated through an angle θ (see Fig. 1-15b), various actions and displacements in the beam will be developed. For example, the reactions at each end and the displacement at the center are shown in Fig. 1-15b. The next two sketches (Figs. 1-15c and 1-15d) show the beam with the displacements Δ and θ occurring separately. The reactions at the ends and the displacement at the center are again denoted by the use of primes; a single prime is used to denote quantities caused by the displacement Δ, and double primes are used for quantities caused by the rotation Θ. When the principle of superposition is applied to the reactions and the displacement at the midpoint, the superposition equations again take the form of Eqs. (1-3). This example illustrates how actions and displacements caused by displacements can be superimposed. The same principle applies to any other actions and displacements in the beam.

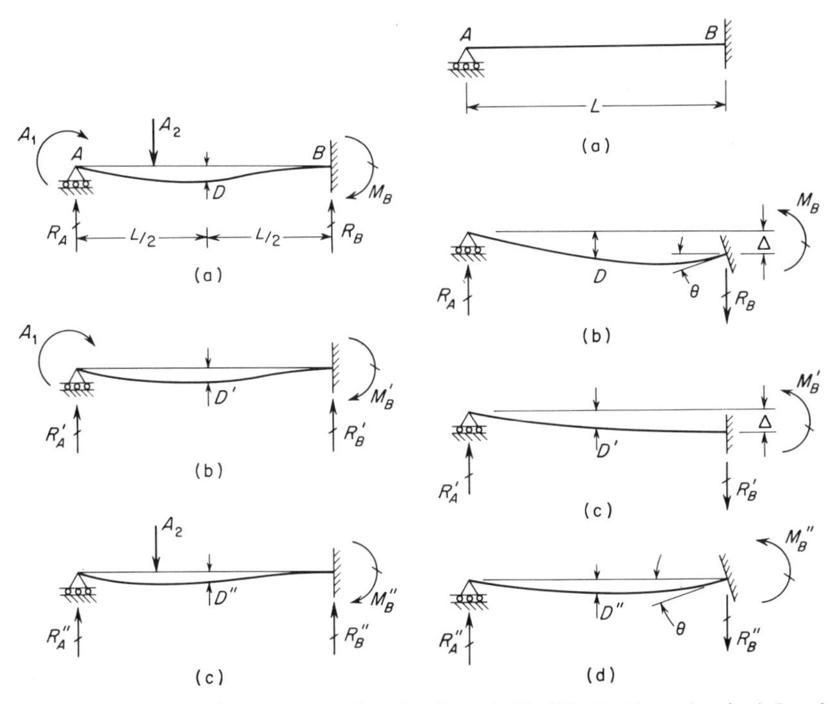

FIG. 1-14. Illustration of principle of superposition.

FIG. 1-15. Illustration of principle of superposition.

The principle of superposition may be used also if the causes consist of both actions and displacements. For example, the beam in Fig. 1-15b could be subjected to various loads as well as to the displacements Δ and Θ. Then the actions and displacements in the beam can be obtained by combining those due to each load and displacement separately.

As mentioned earlier, the principle of superposition will be valid whenever linear relations exist between actions and displacements of the structure. This occurs whenever the following three requirements are satisfied: (1) the material of the structure follows Hooke's law; (2) the displacements of the structure are small; and (3) there is no interaction between axial and flexural effects in the members. The first of these requirements means that the material is perfectly elastic and has a linear relationship between stress and strain. The second requirement means that all calculations involving the over-all dimensions of the structure can be based upon the original dimensions of the structure; it is also a basic assumption for calculating displacements by the methods described in Appendix A. The third requirement implies that the effect of axial forces on the bending of the members is neglected. This requirement refers to the fact that axial forces in a member, in combination with even small deflections of the member, will have an effect on the bending moments. The effect is nonlinear and can be omitted from the analysis when the axial forces (either tension or compression) are not large. (A method of incorporating such effects into the analysis is described in Art. 6.12.)

When all three of the requirements listed above are satisfied, the structure is said to be *linearly elastic*, and the principle of superposition can be used. Since this principle is fundamental to the flexibility and stiffness methods of analysis, it will always be assumed in subsequent discussions that the structures being analyzed meet the stated requirements.

In the preceding discussion of the principle of superposition it was assumed that both actions and displacements were of importance in the analysis, as is generally the case. An exception, however, is the analysis of a statically determinate structure for actions only. Since an analysis of this kind requires the use of equations of equilibrium but does not require the calculation of any displacements, it can be seen that the requirement of linear elasticity is superfluous. An example is the determination of the reactions for a simply supported beam under several loads. The reactions are linear functions of the loads regardless of the characteristics of the material of the beam. It is still necessary, however, that the deflections of the beam be small, since otherwise the position of the loads and reactions would be altered.

1.10 Action and Displacement Equations. The relationships that exist between actions and displacements play an important role in structural analysis and are used extensively in both the flexibility and stiffness methods. A convenient way to express the relationship between the actions on a structure and the displacements of the structure is by means of action and displacement equations. A simple illustration of such equations is obtained by considering the linear, elastic spring shown in Fig. 1-16. The action A will compress the spring, thereby producing a displacement D of the end of the spring. The relationship between A and D can be expressed by a *displacement equation* as follows:

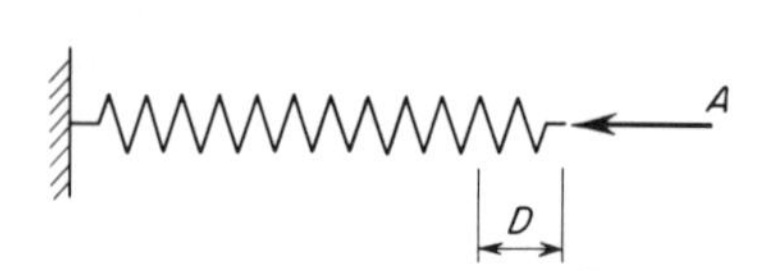

Fig. 1-16. Linear, elastic spring.

$$D = FA \tag{1-4}$$

In this equation F is the *flexibility* of the spring, and is defined as the displacement produced by a unit value of the action A.

The relationship between the action A and the displacement D for the spring in Fig. 1-16 can also be expressed by an *action equation* which gives A in terms of D:

$$A = SD \tag{1-5}$$

In this equation S is the *stiffness* of the spring, which is defined as the action required to produce a unit displacement. It can be seen from Eqs. (1-4) and (1-5) that the flexibility and stiffness of the spring are inverse to one another, as follows:

$$F = \frac{1}{S} = S^{-1} \qquad S = \frac{1}{F} = F^{-1} \tag{1-6}$$

The flexibility of the spring has units of length divided by force, while the stiffness has units of force divided by length.

The same general relationships (Eqs. 1-4 to 1-6) that apply to the spring will hold also for any linearly elastic structure which is subjected to a single action. An example is given in Fig. 1-17a, which shows a simply supported beam acted upon by a concentrated force A at the midpoint. The displacement D shown in the figure is the vertical, downward deflection of the beam at the point where A acts upon the beam Hence, in this example the displacement D not only corresponds to A, but also is caused by A. The action and displacement equations given previously (Eqs. 1-5 and 1-4, respectively) are valid for the beam of Fig. 1-17a, provided that the flexibility F and the stiffness S are determined appropriately. In this case the flexibility F is seen to be the displacement produced by a unit value of the load, as shown in Fig. 1-17b (see Case 2 of Table A-3 in Appendix A):

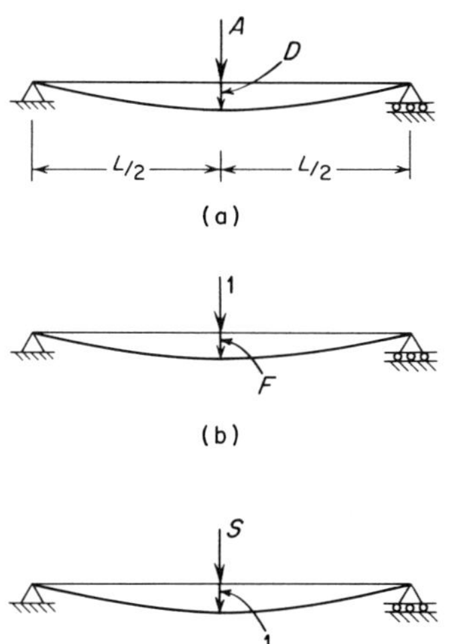

FIG. 1-17. Flexibility and stiffness of a beam subjected to a single load.

$$F = \frac{L^3}{48EI}$$

in which L is the length of the beam and EI is the flexural rigidity. The stiffness S, equal to the inverse of the flexibility, is the action required to produce a unit value of the displacement (see Fig. 1-17c):

$$S = \frac{48EI}{L^3}$$

Note again that S has units of force divided by length. Also, it should be emphasized that the inverse relationship expressed by Eq. (1-6) is valid only when the structure is subjected to a single load.

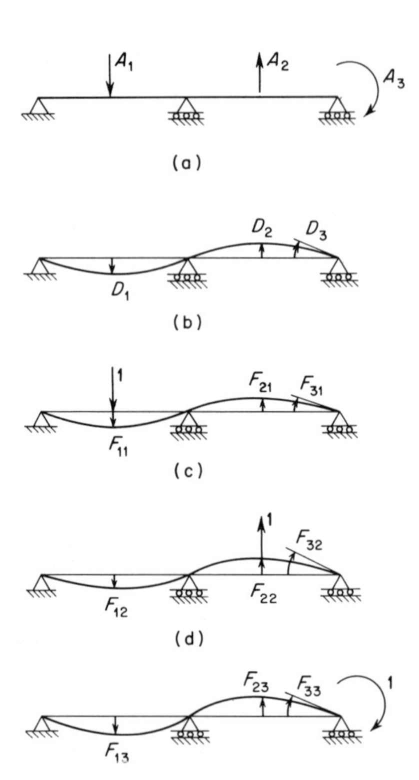

FIG. 1-18. Illustration of flexibility coefficients.

Now consider a more general example in which a structure is acted upon by three loads (Fig. 1-18a). The loads on the beam are denoted by A_1, A_2, and A_3, and are taken positive in the directions shown in the figure. The deflected shape produced by the loads acting on the beam is shown in Fig. 1-18b. In this figure the displacements in the beam corresponding to A_1, A_2, and A_3, and caused by all three loads acting simultaneously, are denoted by D_1, D_2, and D_3, respectively, and are assumed positive in the same directions (or senses) as the corresponding actions.

By using the principle of superposition, each of the displacements in Fig. 1-18b can be expressed as the sum of the displacements due to the loads A_1, A_2, and A_3 acting separately. For example, the displacement D_1 is given by the expression

$$D_1 = D_{11} + D_{12} + D_{13}$$

in which D_{11} is the displacement corresponding to A_1 and caused by A_1, D_{12} is the displacement corresponding to A_1 and caused by A_2, and D_{13} is the displacement corresponding to A_1 and caused by A_3. In a similar manner two additional equations can be written for D_2 and D_3. Each of the displacements appearing on the right-hand sides of such equations is a linear function of one of the loads; that is, each displacement is directly proportional to one of the loads. For example, D_{12} is a displacement caused by A_2 alone, and is equal to A_2 times a certain coefficient. Denoting such coefficients by the symbol F, it is possible to write equations for the displacements D_1, D_2, and D_3 explicitly in terms of the loads, as follows:

$$\begin{aligned} D_1 &= F_{11}A_1 + F_{12}A_2 + F_{13}A_3 \\ D_2 &= F_{21}A_1 + F_{22}A_2 + F_{23}A_3 \\ D_3 &= F_{31}A_1 + F_{32}A_2 + F_{33}A_3 \end{aligned} \tag{1-7}$$

In the first of these equations the expression $F_{11}A_1$ represents the displacement D_{11}, the expression $F_{12}A_2$ represents the displacement D_{12}, and so forth. Each term on the right-hand sides of the above equations is a displacement that is written in the form of a coefficient times the action that produces the displacement. The coefficients are called *flexibility influence coefficients*, or more simply, flexibility coefficients.

Each flexibility coefficient F represents a displacement caused by a unit value of a load. Thus, the coefficient F_{11} represents the displacement corresponding to action A_1 and caused by a unit value of A_1; the coefficient F_{12} is the displacement corresponding to A_1 and caused by a unit value of A_2; and so forth. The physical significance of the flexibility coefficients is shown in Figs. 1-18c, 1-18d, and 1-18e. The displacements of the beam caused by a unit value of the action A_1 are shown in Fig. 1-18c. All of the flexibility coefficients in this figure have a second subscript equal to one, thereby denoting the cause of the displacements. The first subscript in each case identifies the displacement by denoting the action that corresponds to it. Similar comments apply also to the displacements pictured in Figs. 1-18d and 1-18e. The calculation of the flexibility coefficients F may be a simple or a difficult task, depending upon the particular structure being investigated. An example involving a very simple structure is given at the end of Art. 1.11. The more general use of flexibility coefficients in structural analysis, as well as methods for calculating them, will be shown in Chapters 2 and 3.

Instead of expressing the displacements in terms of the actions, as was done in Eqs. (1-7), it is also possible to write action equations expressing the actions in terms of the displacements. Such equations can be obtained, for instance, by solving simultaneously the displacement equations. Thus, if Eqs. (1-7) are solved for the actions in terms of the displacements, the resulting action equations have the form:

$$\begin{aligned} A_1 &= S_{11}D_1 + S_{12}D_2 + S_{13}D_3 \\ A_2 &= S_{21}D_1 + S_{22}D_2 + S_{23}D_3 \\ A_3 &= S_{31}D_1 + S_{32}D_2 + S_{33}D_3 \end{aligned} \tag{1-8}$$

in which each S is a *stiffness influence coefficient* or, more simply, a stiffness coefficient. As stated previously, a stiffness represents an action due to a unit displacement. Thus, the stiffness coefficient S_{11} represents the action corresponding to A_1 when a unit displacement is introduced at D_1 while the other displacements, namely, D_2 and D_3, are kept equal to zero. Similarly, the stiffness coefficient S_{12} is the action corresponding to A_1 caused by a unit displacement at D_2 when D_1 and D_3 are equal to zero. By continuing in this manner, all of the stiffness coefficients can be defined as actions produced by unit displacements.

The physical interpretation of the stiffness coefficients is shown in Fig. 1-19. The first two parts of the figure (Figs. 1-19a and 1-19b) are repeated from Fig. 1-18 in order to show the actions and displacements in the original beam. In Fig. 1-19c the beam is shown with a unit dis-

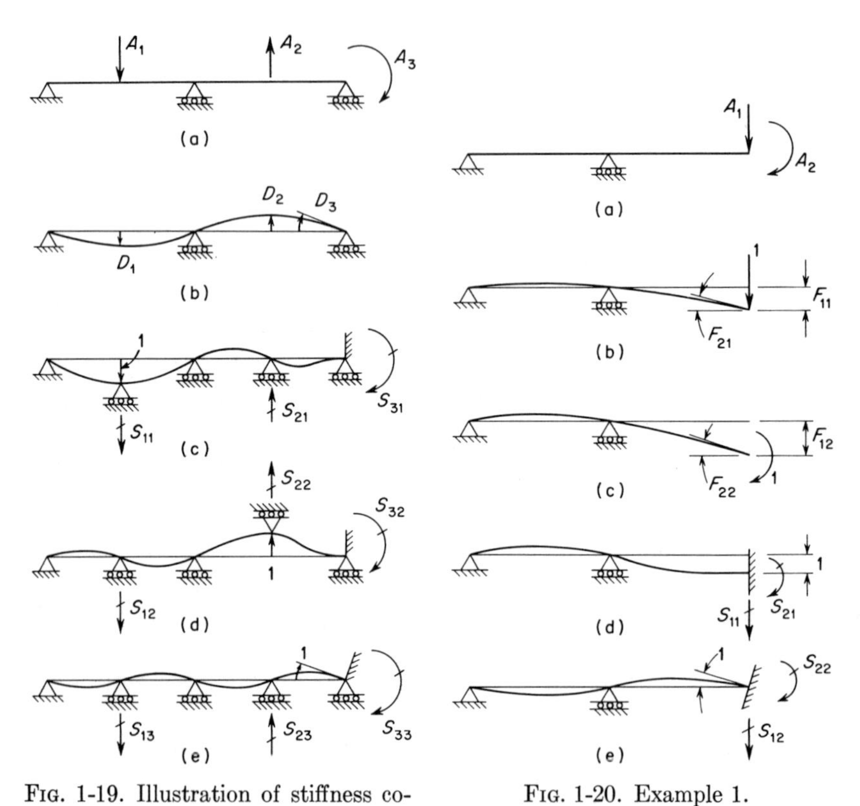

FIG. 1-19. Illustration of stiffness coefficients.

FIG. 1-20. Example 1.

placement corresponding to A_1 induced in the structure while the displacements corresponding to A_2 and A_3 are made equal to zero. To impose these displacements on the beam requires that appropriate artificial restraints be provided. These are shown in the figure by the simple supports corresponding to A_1 and A_2 and the rotational restraint corresponding to A_3. The restraining actions developed by these artificial supports are the stiffness coefficients. For example, it can be seen from the figure that S_{11} is the action corresponding to A_1 and caused by a unit displacement corresponding to A_1 while the displacements corresponding to A_2 and A_3 are retained at zero. The stiffness coefficient S_{21} is the action corresponding to A_2 caused by a unit displacement corresponding to A_1 while the displacements corresponding to A_2 and A_3 are made equal to zero, and so on for the other stiffnesses. Note that each stiffness coefficient is a reaction for the restrained structure, and therefore a slanted line is used across the vector in order to distinguish it from a load vector. Each stiffness coefficient is shown acting in its assumed positive direction, which is automatically the same direction as the corresponding action. If the actual direction of one of the stiffnesses is opposite to that assumed, then the coefficient will have a negative value when it is calculated. The stiffness coefficients caused by unit displacements corresponding to A_2 and A_3 are shown in Figs. 1-19d and 1-19e.

The calculation of the stiffness coefficients for the continuous beam of Fig. 1-19 would be a lengthy task. However, in analyzing a structure by the stiffness method (as is done in Chapters 2 and 4), this difficulty is avoided by limiting the calculation of stiffness coefficients to very special structures that are obtained by completely restraining all the joints of the actual structure. The primary purpose of the preceding discussion and the following two examples is to aid the reader in visualizing the physical significance of stiffness and flexibility coefficients, without regard to matters of practical calculation.

Example 1. The beam shown in Fig. 1-20 is subjected to loads A_1 and A_2 at the free end. The physical significance of the flexibility and stiffness coefficients corresponding to these actions is to be portrayed by means of sketches.

Unit loads corresponding to the actions A_1 and A_2 are shown in Figs. 1-20b and 1-20c, respectively. The displacements produced by these unit loads, and which correspond to the actions A_1 and A_2, are the flexibility coefficients. These coefficients (F_{11}, F_{21}, F_{12}, and F_{22}) are identified in the figures.

The stiffness coefficients (see Figs. 1-20d and 1-20e) are found by imposing unit displacements corresponding to A_1 and A_2, respectively, while at the same time maintaining the other corresponding displacement equal to zero. To accomplish this result requires the introduction of suitable restraints against translation and rotation at the free end of the beam. The restraint actions corresponding to A_1 and A_2 are the stiffness coefficients, and are labeled in the figures.

Example 2. The plane truss illustrated in Fig. 1-21a is subjected to two loads A_1 and A_2. The sketches in Figs. 1-21b, 1-21c, 1-21d, and 1-21e show the physical significance of the flexibility and stiffness coefficients corresponding to A_1 and A_2. Note that the stiffnesses shown in Figs. 1-21d and 1-21e are the restraint actions required when the loaded joint of the truss is displaced a unit distance in the directions of A_1 and A_2, respectively. A restraint of this type is provided by a pinned support.

1.11 Flexibility and Stiffness Matrices. In the preceding article the meaning of action and displacement equations was discussed with reference to particular examples. It is easy to generalize from that discussion and thereby to obtain the equations for a structure subjected to any number of corresponding actions and displacements. Thus, if the

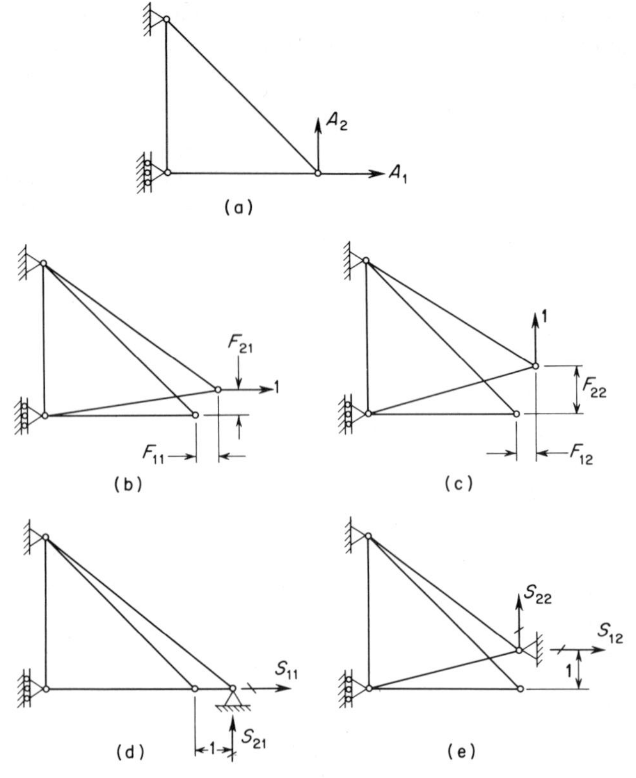

FIG. 1-21. Example 2.

number of actions acting on the structure is n, the equations that give the n corresponding displacements are (compare with Eqs. 1-7):

$$
\begin{aligned}
D_1 &= F_{11}A_1 + F_{12}A_2 + \cdots + F_{1n}A_n \\
D_2 &= F_{21}A_1 + F_{22}A_2 + \cdots + F_{2n}A_n \\
\cdots & \quad \cdots \quad\quad \cdots \quad\quad \cdots \quad\quad \cdots \\
D_n &= F_{n1}A_1 + F_{n2}A_2 + \cdots + F_{nn}A_n
\end{aligned}
\tag{1-9}
$$

In these equations each displacement D corresponds to one of the actions A, and is caused by all of the actions acting simultaneously on the structure. For example, D_1 is the displacement corresponding to A_1 and caused by all of the actions $A_1, A_2, \ldots, A_n$. Each flexibility coefficient F represents a displacement caused by a unit value of one of the actions, while the other actions are zero. For example, F_{21} is the displacement corresponding to action A_2 and caused by a unit value of the action A_1. In general, the flexibility coefficient F_{ij} is the i-th displacement (that is, the displacement corresponding to the i-th action) due to a unit value of the j-th action. The coefficient is taken as positive when it is in the positive direction (or sense) of the i-th action.

In matrix form the displacement equations (Eqs. 1-9) become

$$\begin{bmatrix} D_1 \\ D_2 \\ \cdots \\ D_n \end{bmatrix} = \begin{bmatrix} F_{11} & F_{12} & \cdots & F_{1n} \\ F_{21} & F_{22} & \cdots & F_{2n} \\ \cdots & \cdots & \cdots & \cdots \\ F_{n1} & F_{n2} & \cdots & F_{nn} \end{bmatrix} \begin{bmatrix} A_1 \\ A_2 \\ \cdots \\ A_n \end{bmatrix}$$

or

$$\mathbf{D} = \mathbf{FA} \tag{1-10}$$

in which **D** is a displacement matrix of order $n \times 1$; **F** is a square *flexibility matrix* of order $n \times n$; and **A** is an action or load matrix of order $n \times 1$.* The flexibility coefficients F_{ij} that appear on the principal diagonal of **F** are called *direct flexibility coefficients* and represent displacements caused by unit values of the corresponding actions. The remaining flexibility coefficients are called *cross flexibility coefficients*, and each represents a displacement caused by a unit value of an action that does not correspond to the displacement. It is apparent that $i = j$ for the direct flexibilities and $i \neq j$ for the cross flexibilities.

The action equations for the structure with n actions A acting upon it can be obtained by solving simultaneously Eqs. (1-9) for the actions in terms of the displacements. This operation gives the following action equations:

$$\begin{aligned} A_1 &= S_{11}D_1 + S_{12}D_2 + \cdots + S_{1n}D_n \\ A_2 &= S_{21}D_1 + S_{22}D_2 + \cdots + S_{2n}D_n \\ \cdots & \quad \cdots \quad\quad \cdots \quad\quad \cdots \quad\quad \cdots \\ A_n &= S_{n1}D_1 + S_{n2}D_2 + \cdots + S_{nn}D_n \end{aligned} \tag{1-11}$$

These action equations in matrix form are

$$\begin{bmatrix} A_1 \\ A_2 \\ \cdots \\ A_n \end{bmatrix} = \begin{bmatrix} S_{11} & S_{12} & \cdots & S_{1n} \\ S_{21} & S_{22} & \cdots & S_{2n} \\ \cdots & \cdots & \cdots & \cdots \\ S_{n1} & S_{n2} & \cdots & S_{nn} \end{bmatrix} \begin{bmatrix} D_1 \\ D_2 \\ \cdots \\ D_n \end{bmatrix}$$

or

$$\mathbf{A} = \mathbf{SD} \tag{1-12}$$

As described previously, the matrices **A** and **D** represent the action and displacement matrices of order $n \times 1$. The matrix **S** is a square *stiffness matrix* of order $n \times n$. Each stiffness coefficient S_{ij} can be defined as the i-th action due to a unit value of the j-th displacement, assuming that the remaining displacements are equal to zero. If $i = j$ the coefficient is

* Identifiers for matrices are printed in bold-face type to distinguish them from scalars.

a direct stiffness coefficient; if $i \neq j$, it is a cross stiffness coefficient.

Since Eqs. (1-11) were obtained from Eqs. (1-9) and the actions A and displacements D appearing in these equations are corresponding, it follows that the flexibility and stiffness matrices are related in a special manner. This relationship can be seen by solving Eq. (1-10) for $\mathbf{A}$, giving the expression

$$\mathbf{A} = \mathbf{F}^{-1}\mathbf{D} \tag{1-13}$$

in which $\mathbf{F}^{-1}$ denotes the inverse of the flexibility matrix $\mathbf{F}$. The vectors* $\mathbf{A}$ and $\mathbf{D}$ in this equation are the same as those in Eq. (1-12) and, therefore, it is apparent that

$$\mathbf{S} = \mathbf{F}^{-1} \qquad \text{and} \qquad \mathbf{F} = \mathbf{S}^{-1} \tag{1-14}$$

This relationship shows that the stiffness matrix is the inverse of the flexibility matrix and vice versa, provided the same set of actions and corresponding displacements is being considered in both the action and displacement equations.

A somewhat different situation that occurs in structural analysis is the following. A set of displacement equations relating actions $\mathbf{A}_1$ and corresponding displacements $\mathbf{D}_1$ is obtained for a particular structure, as follows:

$$\mathbf{D}_1 = \mathbf{F}_1\mathbf{A}_1$$

In this equation $\mathbf{F}_1$ is the flexibility matrix relating the displacements $\mathbf{D}_1$ to the actions $\mathbf{A}_1$. Independently, a set of action equations relating another set of actions $\mathbf{A}_2$ to the corresponding displacements $\mathbf{D}_2$ may be written for the same structure; thus:

$$\mathbf{A}_2 = \mathbf{S}_2\mathbf{D}_2$$

It is, of course, not true that the flexibility and stiffness matrices $\mathbf{F}_1$ and $\mathbf{S}_2$ are the inverse of one another. However, it is always possible to obtain the inverse of $\mathbf{F}_1$, and this inverse can correctly be called a stiffness matrix. It is the stiffness matrix $\mathbf{S}_1$ that is associated with the action equation relating $\mathbf{A}_1$ and $\mathbf{D}_1$; specifically, the following equation:

$$\mathbf{A}_1 = \mathbf{F}_1^{-1}\mathbf{D}_1 = \mathbf{S}_1\mathbf{D}_1$$

Similarly, the inverse of the matrix $\mathbf{S}_2$ is a flexibility matrix relating $\mathbf{D}_2$ and $\mathbf{A}_2$, as follows:

$$\mathbf{D}_2 = \mathbf{S}_2^{-1}\mathbf{A}_2 = \mathbf{F}_2\mathbf{A}_2$$

This discussion shows that a flexibility or stiffness matrix is not an array that is determined only by the nature of the structure; it is also related directly to the set of actions and displacements that is under consideration. This idea will be encountered again in the next chapter,

* The term *vector* is used frequently for a matrix of one column or one row.

where the significance of the inverse of a flexibility matrix and the inverse of a stiffness matrix will be discussed more fully in connection with the flexibility and stiffness methods of analysis.

Example. The cantilever beam shown in Fig. 1-22a is subjected to actions A_1 and A_2 at the free end. The corresponding displacements are denoted by D_1 and D_2 in the figure.

The flexibility coefficients are identified in Figs. 1-22b and 1-22c and can be evaluated without difficulty (see Table A-3, Appendix A):

$$F_{11} = \frac{L^3}{3EI} \qquad F_{12} = F_{21} = \frac{L^2}{2EI}$$

$$F_{22} = \frac{L}{EI}$$

Therefore, the displacement equations are

$$D_1 = \frac{L^3}{3EI} A_1 + \frac{L^2}{2EI} A_2$$

$$D_2 = \frac{L^2}{2EI} A_1 + \frac{L}{EI} A_2$$

and it is seen that the flexibility matrix is

$$\mathbf{F} = \begin{bmatrix} \frac{L^3}{3EI} & \frac{L^2}{2EI} \\ \frac{L^2}{2EI} & \frac{L}{EI} \end{bmatrix}$$

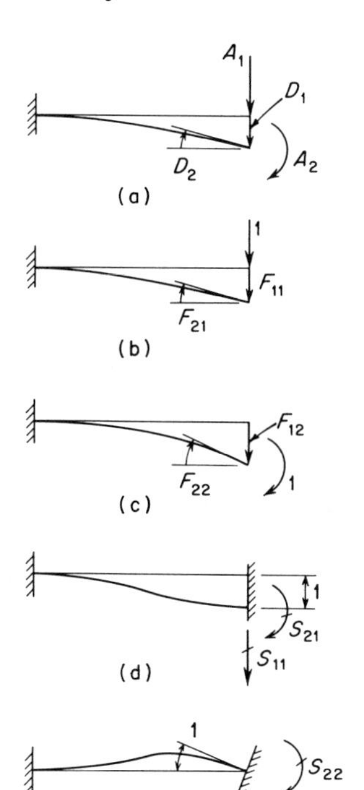

FIG. 1-22. Example of flexibility and stiffness coefficients.

The stiffness coefficients are the restraint actions shown in Figs. 1-22d and 1-22e. In this particular example, the coefficients are reactions for a fixed-end beam, and their expressions can be obtained from a table of fixed-end actions. Such a table is given in Appendix B (see Table B-4), from which the following expressions are obtained:

$$S_{11} = \frac{12EI}{L^3} \qquad S_{12} = S_{21} = -\frac{6EI}{L^2} \qquad S_{22} = \frac{4EI}{L}$$

Thus, the action equations are

$$A_1 = \frac{12EI}{L^3} D_1 - \frac{6EI}{L^2} D_2$$

$$A_2 = -\frac{6EI}{L^2} D_1 + \frac{4EI}{L} D_2$$

and the stiffness matrix is

$$\mathbf{S} = \begin{bmatrix} \frac{12EI}{L^3} & -\frac{6EI}{L^2} \\ -\frac{6EI}{L^2} & \frac{4EI}{L} \end{bmatrix}$$

When the flexibility and stiffness matrices are multiplied together, the identity matrix **I** is obtained:

$$\mathbf{FS} = \mathbf{SF} = \begin{bmatrix} 1 & 0 \\ 0 & 1 \end{bmatrix} = \mathbf{I}$$

Thus, the matrices **F** and **S** are the inverse of one another, as expressed previously by Eq. (1-14).

1.12 Reciprocal Relations. The general form of the action and displacement equations relating a set of actions **A** and corresponding displacements **D** was given in the preceding article. It is convenient in the following discussion to visualize the actions **A** as loads acting upon the structure and to assume that all loads are applied to the structure proportionally. *Proportional loading* means that initially all loads are zero and that the loads are increased gradually in such proportions that all loads reach their final values simultaneously. Thus, at any intermediate stage of the loading process, every load is the same fraction of its final value.

If the loads on a structure are applied proportionally, each load will increase gradually from zero to its final value and, therefore, will do work equal to its average value (one-half of its final value) multiplied by the displacement through which it moves. This displacement will be the displacement corresponding to the load itself, and caused by all of the loads acting simultaneously. Thus, the work W of all forces is

$$\begin{aligned} W &= \tfrac{1}{2}A_1D_1 + \tfrac{1}{2}A_2D_2 + \cdots + \tfrac{1}{2}A_nD_n \\ &= \tfrac{1}{2}(A_1D_1 + A_2D_2 + \cdots + A_nD_n) \end{aligned} \tag{1-15}$$

in which $A_1, A_2, \ldots, A_n$ are the actions on the structure and $D_1, D_2, \ldots, D_n$ are the corresponding displacements. Since the action and displacement matrices **A** and **D** (see Eq. 1-10) are column matrices (or column vectors), an expression for work can be obtained from either of the matrix products shown:

$$\mathbf{W} = \tfrac{1}{2}\mathbf{A}'\mathbf{D} = \tfrac{1}{2}\mathbf{D}'\mathbf{A} \tag{1-16}$$

in which $\mathbf{A}'$ and $\mathbf{D}'$ are the transposes of **A** and **D**, respectively. In the relations given in Eq. (1-16) each of the products $\mathbf{A}'\mathbf{D}$ and $\mathbf{D}'\mathbf{A}$ represents a row matrix times a column matrix, giving a result that is a matrix **W** of one element. The value of the element itself is W, the work defined by

Eq. (1-15). Thus, in the remainder of the discussion the matrix $\mathbf{W}$ and the work W are considered to be interchangeable, and Eq. (1-16) can be considered as an expression that is equivalent to Eq. (1-15).

Since the work of the loads is $\frac{1}{2}\mathbf{A}'\mathbf{D}$ (see Eq. 1-16) and since $\mathbf{D} = \mathbf{FA}$, another expression for the work is

$$W = \tfrac{1}{2}\mathbf{A}'\mathbf{F}\mathbf{A} \tag{a}$$

By taking the transpose of both sides of Eq. (1-10), the expression for the transpose of $\mathbf{D}$ becomes $\mathbf{D}' = \mathbf{A}'\mathbf{F}'$ inasmuch as the transpose of a product is the product of the transposes but in reverse order. Therefore, since the work is $\frac{1}{2}\mathbf{D}'\mathbf{A}$ (see Eq. 1-16), it follows that still another expression for the work is

$$W = \tfrac{1}{2}\mathbf{A}'\mathbf{F}'\mathbf{A} \tag{b}$$

Comparison of Eqs. (a) and (b) shows that the flexibility matrix and its transpose are equal:

$$\mathbf{F} = \mathbf{F}' \tag{1-17}$$

Therefore, it is concluded that the flexibility matrix $\mathbf{F}$ is a symmetrical matrix, and that the cross flexibility coefficients are related by the expression

$$F_{ij} = F_{ji} \tag{1-18}$$

This relation is known as the reciprocal theorem for flexibilities, and shows that the displacement corresponding to the i-th action and caused by a unit value of the j-th action is equal to the displacement corresponding to the j-th action and caused by a unit value of the i-th action.*

An illustration of the reciprocal theorem can be obtained by referring to the example at the end of the preceding article. It can be seen in that example that the flexibilities F_{12} and F_{21} are equal and that the flexibility matrix $\mathbf{F}$ is symmetrical. Many additional examples of the reciprocal relations will be encountered later in connection with the flexibility method of analysis.

Instead of substituting the displacement equation $\mathbf{D} = \mathbf{FA}$ into the expression for the work of the loads on the structure, it is possible to begin by substituting the action equation $\mathbf{A} = \mathbf{SD}$ (see Eq. 1-12). When this relation is substituted into the second expression for work (see Eq. 1-16), the result is

$$W = \tfrac{1}{2}\mathbf{D}'\mathbf{S}\mathbf{D} \tag{c}$$

Also, when the relation $\mathbf{A}' = \mathbf{D}'\mathbf{S}'$, which is obtained by taking the transpose of both sides of the action equation, is substituted into the first expression for work, the following is obtained:

* The reciprocal theorem for flexibilities is also called *Maxwell's theorem*, because it was first presented by J. C. Maxwell in 1864.

$$W = \tfrac{1}{2}\mathbf{D'S'D} \tag{d}$$

From Eqs. (c) and (d) it can be seen that the stiffness matrix is also a symmetrical matrix, since it is the same as its transpose:

$$\mathbf{S} = \mathbf{S'} \tag{1-19}$$

Of course, the fact that **S** is symmetrical could have been concluded directly from the symmetry of the flexibility matrix **F**, inasmuch as the inverse of any symmetric matrix is another symmetric matrix. From Eq. (1-19) it is seen that the reciprocal theorem for stiffnesses is

$$S_{ij} = S_{ji} \tag{1-20}$$

This theorem shows that the i-th action due to a unit value of the j-th displacement is equal to the j-th action due to a unit value of the i-th displacement. An example of the reciprocal theorem for stiffnesses also can be seen in the example at the end of the preceding article, where the stiffnesses S_{12} and S_{21} are equal to one another, and **S** is symmetrical.

Stiffness matrices are an essential feature of the stiffness method of analysis, and many examples of them will be encountered in Chapters 2 and 4. In all cases the reciprocal relation can be seen to exist between the cross stiffness coefficients.

Problems

1.4-1. A prismatic bar is subjected to equal and opposite axial forces P at each end, thereby producing uniform tension in the bar. Define the displacement that corresponds to the two forces P.

1.4-2. Define the actions that correspond to the displacements Δ_C and Θ_C in the plane frame shown in Fig. 1-6.

1.4-3. The simply supported beam shown in the figure has constant flexural rigidity EI and length L. The loads on the beam are two actions A_1 and A_2 as illustrated. Obtain expressions in terms of A_1, A_2, E, I, and L for each of the following: (a) the displacement D_{11} corresponding to action A_1 and caused by action A_1 acting alone; (b) the displacement D_{12} corresponding to action A_1 and caused by A_2 acting alone; and (c) the displacement D_1 corresponding to A_1 and caused by both actions acting simultaneously.

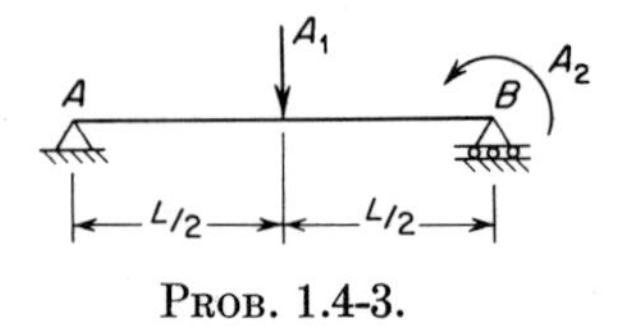

PROB. 1.4-3.

1.4-4. For the cantilever beam shown in the figure, determine the displacements D_{11}, D_{21}, and D_{31}. Assume that the beam has constant flexural rigidity EI.

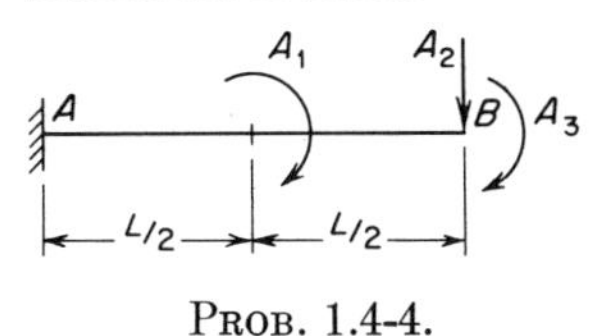

PROB. 1.4-4.

1.4-5. The overhanging beam shown in the figure is subjected to loads A_1, A_2, and A_3. Assuming constant flexural rigidity EI for the beam, determine the displacements D_{11}, D_{23}, and D_{33}.

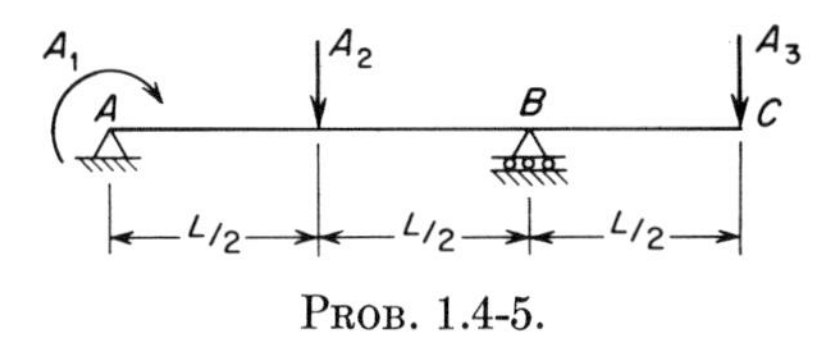

PROB. 1.4-5.

1.4-6. The truss shown in the figure is acted upon by an action A_1 in the form of a horizontal force at joint A and an action A_2 consisting of two equal and opposite forces at joints A and D. All members of the truss are prismatic and have axial rigidity EA. Determine the displacements D_{11}, D_{12}, D_{21}, and D_{22}.

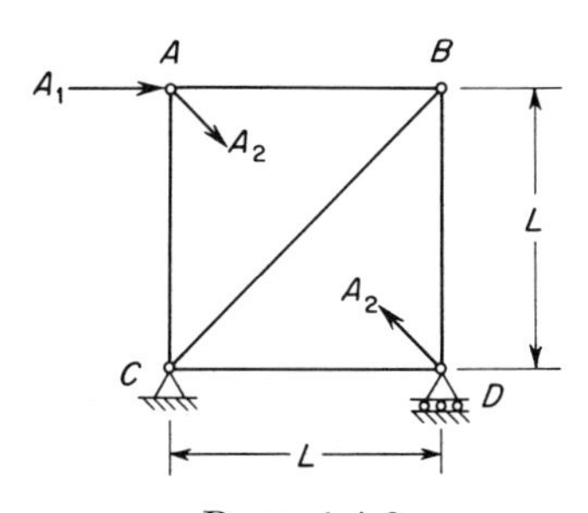

PROB. 1.4-6.

1.7-1. (a) What is the degree of kinematic indeterminacy for a simply supported beam (see figure)? (b) If the effects of axial deformations are neglected, what is the degree of kinematic indeterminacy? (c) and (d) Repeat questions (a) and (b) for a cantilever beam that is fixed at one end and free at the other.

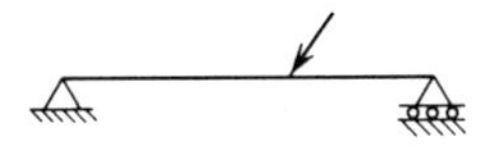

PROB. 1.7-1.

1.7-2. (a) What is the degree of static indeterminacy for the continuous beam shown in Fig. 1-1a? (b) What is the degree of kinematic indeterminacy? (c) If the effects of axial deformations are neglected, what is the degree of kinematic indeterminacy?

1.7-3. (a) Determine the number of degrees of static indeterminateness for the plane truss shown in Fig. 1-lb. (b) Determine the degree of kinematic indeterminacy.

1.7-4. What is the degree of kinematic indeterminacy for the plane truss shown in Fig. 1-9?

1.7-5. Determine the degrees of (a) static indeterminacy and (b) kinematic indeterminacy for the space truss in Fig. 1-1c.

1.7-6. Repeat Prob. 1.7-2 for the plane frame shown in the figure.

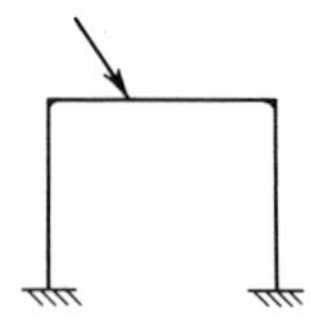

PROB. 1.7-6.

1.7-7. Repeat Prob. 1.7-2 for the plane frame with pinned supports (see figure).

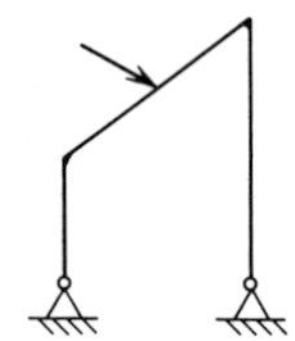

PROB. 1.7-7.

1.7-8. Repeat Prob. 1.7-2 for the plane frame shown in Fig. 1-1d.

1.7-9. For the grid shown in the figure, find (a) the degree of static indeterminacy and (b) the degree of kinematic indeterminacy.

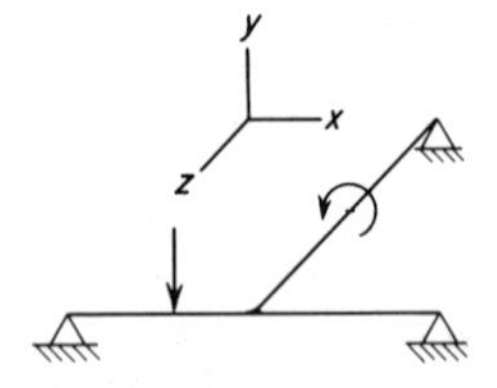

PROB. 1.7-9.

1.7-10. Solve the preceding problem for the grid shown in the figure.

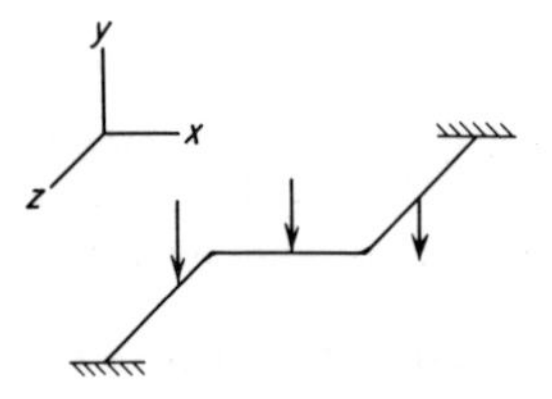

PROB. 1.7-10.

1.7-11. Repeat Prob. 1.7-9 for the grid shown in Fig. 1-1e.

1.7-12. Repeat Prob. 1.7-2 for the space frame pictured in Fig. 1-1f.

1.10-1 to 1.10-10. Illustrate by means of sketches the physical significance of the flexibility and stiffness coefficients corresponding to the actions shown in the figures.

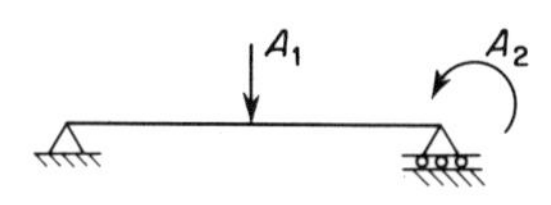

PROB. 1.10-1.

PROB. 1.10-2.

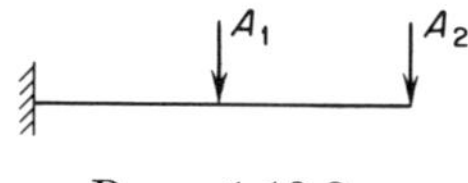

PROB. 1.10-3.

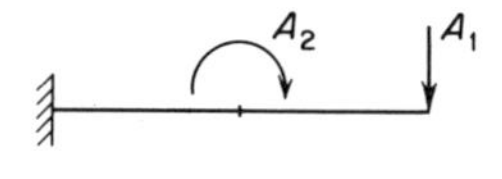

PROB. 1.10-4.

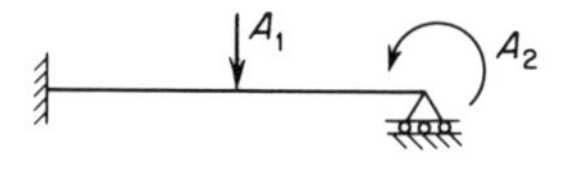

PROB. 1.10-5.

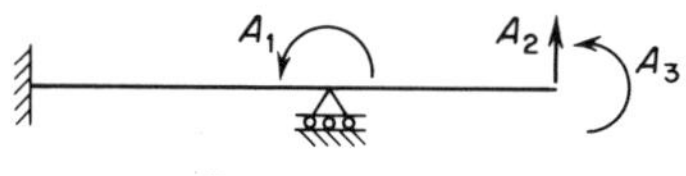

PROB. 1.10-6.

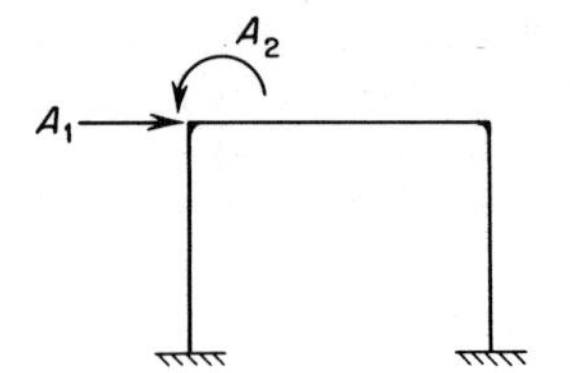

PROB. 1.10-7.

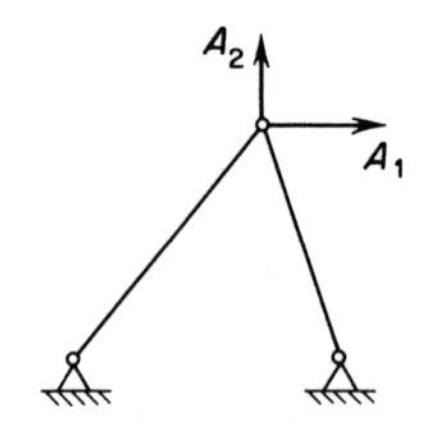

PROB. 1.10-8.

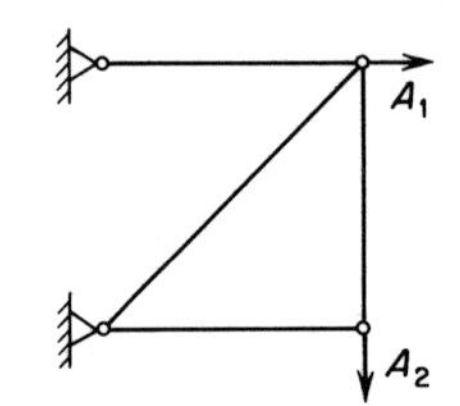

PROB. 1.10-9.

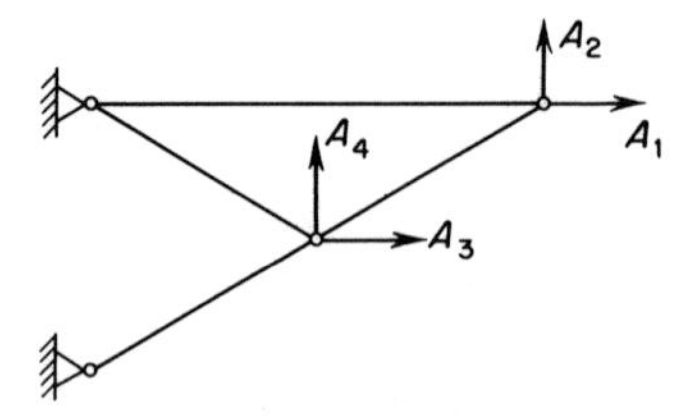

PROB. 1.10-10

Chapter 2

INTRODUCTION TO THE FLEXIBILITY AND STIFFNESS METHODS

2.1 Introduction. In this chapter the basic concepts of the flexibility and stiffness methods (also called the action and displacement methods, respectively) are introduced. These methods are applicable generally to all types of structures, including those constructed of beams, columns, plates, shells, and other structural elements. However, all of the structures to be analyzed in this book are framed structures, as described previously in Art. 1.2. Framed structures are probably the most commonly encountered structures in engineering practice, and they provide excellent examples with which to illustrate the basic ideas of the flexibility and stiffness methods. Both methods are very fundamental in concept and involve similar mathematical formulations, as will be pointed out in later articles.

The formulation of the two methods is made by means of matrix algebra, since this makes it possible to deal in general terms from the beginning, even though the first problems to be solved are very simple and are devised solely to illustrate the basic concepts. However, the expression of the methods in matrix terms permits immediate generalization to very complicated structures, and this is one of the principal advantages of the matrix notation. Also, the use of matrices casts the structural problem in a form which is ideally suited for programming on a digital computer. This fact probably represents the primary motivation for using the flexibility and stiffness methods in the form to be described.

Finally, it should be realized that the flexibility and stiffness methods can be organized into a highly systematic procedure for the analysis of a structure. Once the basic concepts incorporated in the procedure are understood, the methods may be applied to structures of any degree of complexity. Thus, the objective of the later chapters is to develop systematic procedures for analyzing each class of framed structures, including beams, trusses, grids, and frames. These procedures can then be programmed for a digital computer. After this is done, any structure within the class of structures being considered can be analyzed by automatic means.

2.2 Flexibility Method. The first of the two methods to be described is the flexibility method, which may be used to analyze any statically indeterminate structure. In order to illustrate the fundamental ideas of the method, consider the example shown in Fig. 2-1a. The beam ABC in the figure has two spans of equal length and is subjected to a uniform load of intensity w. The beam is statically indeterminate to the first degree, since there are four possible reactions (two reactions at A, one at B, and one at C) and three equations of static equilibrium for actions in a plane. The reaction R_B at the middle support will be taken as the statical redundant, although other possibilities also exist. If this redundant is released (or removed), a statically determinate *released structure* is obtained. In this case, the released structure is the simply supported beam shown in Fig. 2-1b.

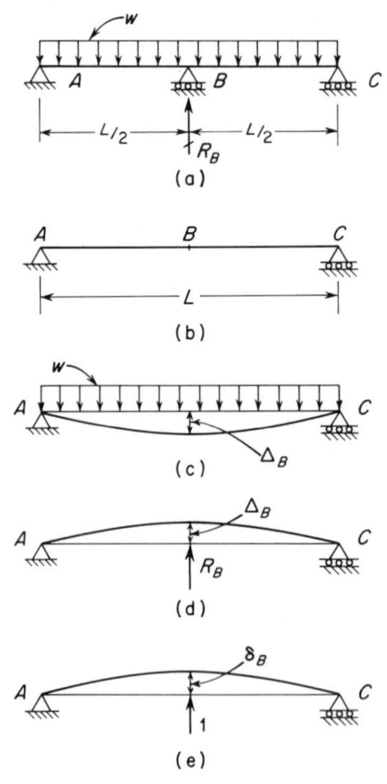

FIG. 2-1. Illustration of flexibility method.

Under the action of the uniform load w, the released structure will deflect as illustrated in Fig. 2-1c. The displacement of the beam at point B is denoted by Δ_B and is given by the expression (see Table A-3 of Appendix A):

$$\Delta_B = \frac{5wL^4}{384EI}$$

in which EI is the flexural rigidity of the beam. However, the actual beam is assumed to have no translation (that is, no settlement) at point B. Therefore, the redundant reaction R_B must be such as to produce in the released structure an upward displacement equal to Δ_B (see Fig. 2-1d). According to the principle of superposition, the final displacement at point B in the released structure is the resultant of the displacements caused by the load w and the redundant R_B. The upward displacement due to R_B is

$$\Delta_B = \frac{R_B L^3}{48EI}$$

as obtained from Table A-3 of the Appendix. Equating the two expressions obtained above for Δ_B gives

$$\frac{5wL^4}{384EI} = \frac{R_B L^3}{48EI} \tag{2-1}$$

The unknown reaction R_B may be obtained by solving Eq. (2-1):

$$R_B = \tfrac{5}{8}wL$$

After R_B has been found, the remaining reactions for the two-span beam can be found from static equilibrium equations.

In the preceding example, the procedure was to calculate displacements in the released structure caused by both the loads and the redundant action and then to formulate an equation pertaining to these displacements. Equation (2-1) expressed the fact that the downward displacement due to the load was equal to the upward displacement due to the redundant. In general, an equation of this type can be called an *equation of compatibility,* because it expresses a condition pertaining to the displacements of the structure. It may also be called an *equation of superposition,* since it is based upon the superposition of displacements caused by more than one action. Still another name would be *equation of geometry,* because the equation expresses a condition that pertains to the geometry of the structure.

A more general approach that can be used in solving the two-span beam of Fig. 2-1a consists of finding the displacement produced by a unit value of R_B and then multiplying this displacement by R_B in order to obtain the displacement caused by R_B. Also, it is more general and systematic to use a consistent sign convention for the actions and displacements at B. For example, it may be assumed that both the displacement at B and the reaction at B are positive when in the upward direction. Then the application of a unit force (corresponding to R_B) to the released structure, as shown in Fig. 2-1e, results in a positive displacement δ_B. This displacement is given by the expression

$$\delta_B = \frac{L^3}{48EI} \tag{2-2}$$

The displacement caused by R_B acting alone on the released structure is $\delta_B R_B$. The displacement caused by the uniform load w acting alone on the released structure is

$$\Delta_B = -\frac{5wL^4}{384EI} \tag{2-3}$$

This displacement is negative because Δ_B is assumed to be positive when in the upward direction. Superposition of the displacements due to the load w and the reaction R_B must produce zero displacement of the beam at point B. Thus, the compatibility equation is

$$\Delta_B + \delta_B R_B = 0 \tag{2-4}$$

from which

$$R_B = -\frac{\Delta_B}{\delta_B} \tag{2-5}$$

When the expressions given above for Δ_B and δ_B (see Eqs. 2-2 and 2-3) are substituted into Eq. (2-5), the result is

$$R_B = \tfrac{5}{8}wL$$

which is the same result as obtained before. The positive sign in the result denotes the fact that R_B is in the upward direction.

An important part of the preceding solution consists of writing the superposition equation (Eq. 2-4) which expresses the geometrical fact that the beam does not deflect at the support B. Included in this equation are the effects of the load and the redundant reaction. The displacement caused by the reaction has been conveniently expressed as the product of the reaction itself and the displacement caused by a unit value of the reaction. The latter is a flexibility influence coefficient, since it is the displacement due to a unit action. If all terms in the equation are expressed with the same sign convention, then the sign of the final result will denote the true direction of the redundant action.

If a structure is statically indeterminate to more than one degree, the approach used in the preceding example must be further organized, and a more generalized notation must be introduced. To illustrate these features, another example of a prismatic beam will be considered (see Fig. 2-2a). The beam shown in the figure is statically indeterminate to the second degree; hence, a statically determinate released structure may be obtained by removing two redundant actions from the beam. Several choices for the redundants and the corresponding released structures are possible. Four such possibilities for the released structure are shown in Figs. 2-2b, 2-2c, 2-2d, and 2-2e. In the first of these figures, the reactive moment at A and the force at B are taken as the redundants; thus, the rotational restraint at A and the translational restraint at B are removed from the original beam in order to obtain the released structure. In the next case (Fig. 2-2c) the reactive moment at A and the internal bending moment at B are released. Therefore, the released structure has no rotational restraint at A and no restraint against bending moment at B. The latter condition is represented by the presence of a hinge in the beam at point B. The released structure shown in Fig. 2-2d is obtained by releasing both the reaction and bending moment at point B. Lastly, the released structure shown in Fig. 2-2e is obtained by selecting the reactions at joints B and C as the redundants. This particular released structure is selected for the analysis which follows, although any of the others would be suitable. All of the released structures shown in Fig. 2-2 are statically determinate and immobile. In general, only structures of this type will be used in the discussions of the flexibility method.

The redundant actions that are selected for the analysis are denoted as Q_1 and Q_2 in Fig. 2-2a. These actions are the reactive forces at joints

B and C. In Fig. 2-2f is shown the released structure acted upon by the various loads on the original beam, which in this example are assumed to be three concentrated forces P_1, P_2, and P_3, and a couple M. These loads produce displacements in the released structure, and, in particular,

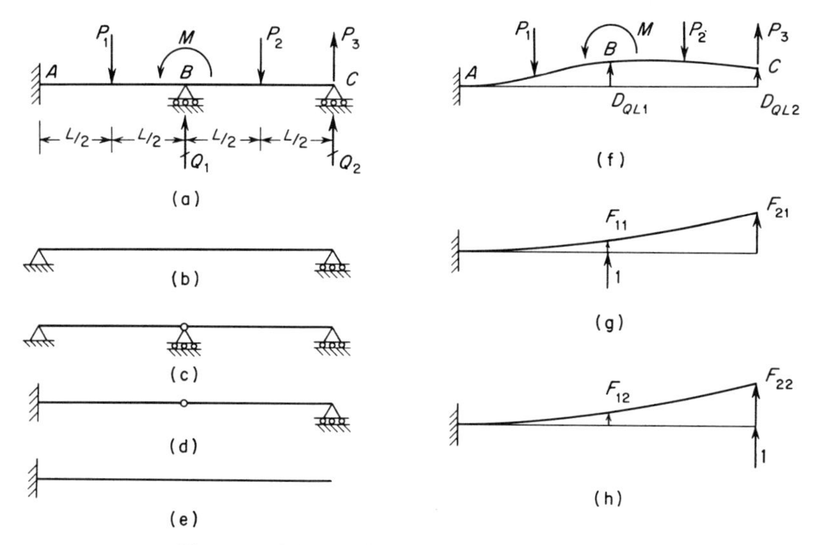

FIG. 2-2. Illustration of flexibility method.

displacements that correspond to Q_1 and Q_2 are produced at joints B and C. These displacements are denoted D_{QL1} and D_{QL2} in the figure. In this notation the symbol D_{QL} is used to represent a displacement corresponding to a redundant Q and caused by the loads on the structure. The numerical subscripts that follow the symbol denote the redundant to which the displacement corresponds. Thus, it is convenient to envisage D_{QL} as one symbol, while the numbers that follow the symbol are the subscripts.* In Fig. 2-2f the displacements D_{QL} are shown in their assumed positive direction, which is upward. The positive directions for the displacements must always be the same as the positive directions of the redundants to which the displacements correspond. Since the redundants are assumed to be positive when in the upward direction, the displacements are also positive upward.

In order to obtain the various flexibility influence coefficients that appear in the equations of compatibility, unit values of the redundants Q_1 and Q_2 are applied separately to the released structure. For the con-

* In computer programming, the symbol D_{QL} is written DQL because of the necessity that all characters, whether alphabetic or numeric, be on the same line. When subscripts are present, they can be added to the basic symbol. For example, in a computer program $DQL[1]$ is the usual way of writing D_{QL1}. A similar style can be used for the other symbols that are encountered in subsequent work.

dition $Q_1 = 1$ shown in Fig. 2-2g, the flexibility coefficient F_{11} is the displacement corresponding to Q_1 due to a unit value of Q_1, and the coefficient F_{21} is the displacement corresponding to Q_2 due to a unit value of Q_1. For the condition $Q_2 = 1$ shown in Fig. 2-2h, F_{12} is the displacement corresponding to Q_1 due to a unit value of Q_2, and F_{22} is the displacement corresponding to Q_2 due to a unit value of Q_2. The flexibility coefficients are shown in their positive directions.

The superposition equations expressing the conditions of compatibility at joints B and C of the actual beam may now be written. Since the translational displacements at supports B and C are zero, the equations become

$$\begin{aligned} D_{QL1} + F_{11}Q_1 + F_{12}Q_2 &= 0 \\ D_{QL2} + F_{21}Q_1 + F_{22}Q_2 &= 0 \end{aligned} \tag{2-6}$$

The first of these equations represents the total displacement at B, which consists of three parts: the displacement due to loads, the displacement due to Q_1, and the displacement due to Q_2. The superposition of all three displacements gives the total displacement, which is zero. Similar comments apply to the second equation. The two equations can be solved simultaneously for Q_1 and Q_2, after which all other actions in the beam can be found by statics.

It is desirable at this stage of the discussion to write the superposition equations in a slightly more general form. It is always possible that support movements corresponding to the redundants will occur in the original beam, and these displacements can be included readily in the analysis. Assume that D_{Q1} and D_{Q2} represent the actual displacements in the beam corresponding to Q_1 and Q_2. Thus, D_{Q1} represents the support displacement at B, with the upward direction being positive. Similarly, D_{Q2} is the support displacement at C. The superposition equations express the fact that the final displacements corresponding to Q_1 and Q_2 are equal to the sums of the displacements caused by the loads and the redundants; thus, the equations are:

$$\begin{aligned} D_{Q1} &= D_{QL1} + F_{11}Q_1 + F_{12}Q_2 \\ D_{Q2} &= D_{QL2} + F_{21}Q_1 + F_{22}Q_2 \end{aligned} \tag{2-7}$$

If there are no support displacements, as assumed in this problem, then D_{Q1} and D_{Q2} are both zero, and Eqs. (2-7) reduce to Eqs. (2-6). If there are support displacements that do not correspond to a redundant, such as a displacement at joint A, they must be handled by the methods described later in Art. 2.4.

The superposition equations (2-7) can be written in matrix form as

$$\mathbf{D}_\mathrm{Q} = \mathbf{D}_\mathrm{QL} + \mathbf{FQ} \tag{2-8}$$

in which $\mathbf{D}_\mathrm{Q}$ is the matrix of actual displacements corresponding to the redundants, $\mathbf{D}_{QL}$ is the matrix of displacements in the released structure corresponding to the redundant actions Q and due to the loads, and $\mathbf{F}$ is the flexibility matrix for the released structure corresponding to the redundant actions Q. For the equations given above, these matrices are:

$$\mathbf{D}_Q = \begin{bmatrix} D_{Q1} \\ D_{Q2} \end{bmatrix} \qquad \mathbf{D}_{QL} = \begin{bmatrix} D_{QL1} \\ D_{QL2} \end{bmatrix} \qquad \mathbf{F} = \begin{bmatrix} F_{11} & F_{12} \\ F_{21} & F_{22} \end{bmatrix} \qquad \mathbf{Q} = \begin{bmatrix} Q_1 \\ Q_2 \end{bmatrix}$$

It should be noted that an alternate (and more consistent) symbol for a flexibility coefficient F would be D_{QQ}, which represents a displacement in the released structure corresponding to a redundant Q and caused by a unit value of a redundant Q. Thus, the matrix $\mathbf{F}$ could also be denoted as $\mathbf{D}_{QQ}$ for consistency in notation, but the use of $\mathbf{F}$ for "flexibility" is more convenient.

The vector $\mathbf{Q}$ of redundants can be obtained by solving Eq. (2-8); thus:

$$\mathbf{Q} = \mathbf{F}^{-1}(\mathbf{D}_Q - \mathbf{D}_{QL}) \tag{2-9}$$

in which $\mathbf{F}^{-1}$ denotes the inverse of the flexibility matrix. From this equation the redundants can be calculated after first obtaining the matrices $\mathbf{D}_Q$, $\mathbf{D}_{QL}$, and $\mathbf{F}$. The first of these matrices will be known from the support conditions that exist in the original structure, while the last two are calculated from the properties of the released structure. The problem can be considered as solved when the matrix $\mathbf{Q}$ is known, since all other actions then can be found by static equilibrium. When the actions throughout the structure have been found, the displacements at any point can also be found. A method for incorporating into the analysis the calculation of actions and displacements at various points in the structure will be given in Art. 2.5.

The matrix $\mathbf{D}_Q$ normally will be a null matrix $\mathbf{O}$ (that is, a matrix with all elements equal to zero), except when one or more of the redundants is a support reaction that has a support displacement corresponding to it. If the matrix $\mathbf{D}_Q$ is null, Eq. (2-9) for the redundant actions $\mathbf{Q}$ becomes

$$\mathbf{Q} = \mathbf{F}^{-1}(\mathbf{O} - \mathbf{D}_{QL}) = -\mathbf{F}^{-1}\mathbf{D}_{QL} \tag{2-10}$$

This equation may be used instead of Eq. (2-9) whenever the displacements D_Q are zero. The method of handling a support displacement when there is no redundant action corresponding to that displacement is described later in Art. 2.4.

To show the use of the matrix equations given above, consider again the beam in Fig. 2-2a. In order to have a specific example, assume that the beam has constant flexural rigidity EI in both spans and that the actions on the beam are as follows:

$$P_1 = 2P \qquad M = PL \qquad P_2 = P \qquad P_3 = P$$

Also, assume that there are no support displacements at any of the supports of the structure.

The matrices to be found first in the analysis are $\mathbf{D}_Q$, $\mathbf{D}_{QL}$, and $\mathbf{F}$, as mentioned previously. Since in the original beam there are no displacements corresponding to Q_1 and Q_2, the matrix $\mathbf{D}_Q$ is a null matrix. The matrix $\mathbf{D}_{QL}$ represents the displacements in the released structure corresponding to the redundants and caused by the loads. These displacements are found by considering Fig. 2-2f, which shows the released structure under the action of the loads. The displacements in this beam corresponding to Q_1 and Q_2 can be found by the methods described in Appendix A (see Example 3, Art. A.2), and the results are:

$$D_{QL1} = \frac{13PL^3}{24EI} \qquad D_{QL2} = \frac{97PL^3}{48EI}$$

The positive signs in these expressions show that both displacements are upward. From the results given above, the vector $\mathbf{D}_{QL}$ is obtained:

$$\mathbf{D}_{QL} = \frac{PL^3}{48EI}\begin{bmatrix} 26 \\ 97 \end{bmatrix}$$

The flexibility matrix $\mathbf{F}$ is found by referring to the beams pictured in Figs. 2-2g and 2-2h. The beam in Fig. 2-2g, which is subjected to a unit load corresponding to Q_1, has displacements given by the expressions

$$F_{11} = \frac{L^3}{3EI} \qquad F_{21} = \frac{5L^3}{6EI}$$

Similarly, the displacements in the beam of Fig. 2-2h are

$$F_{12} = \frac{5L^3}{6EI} \qquad F_{22} = \frac{8L^3}{3EI}$$

From the results listed above, the flexibility matrix can be formed:

$$\mathbf{F} = \frac{L^3}{6EI}\begin{bmatrix} 2 & 5 \\ 5 & 16 \end{bmatrix}$$

The inverse of the flexibility matrix can be found by any one of several standard methods,* the result being

$$\mathbf{F}^{-1} = \frac{6EI}{7L^3}\begin{bmatrix} 16 & -5 \\ -5 & 2 \end{bmatrix}$$

It should be noted that both the flexibility matrix and its inverse are symmetrical matrices.

As the final step in the analysis, Eq. (2-10) may be used to obtain the redundant actions $\mathbf{Q}$, as follows:

* See, for instance, J. M. Gere and W. Weaver, Jr., *Matrix Algebra for Engineers*, D. Van Nostrand Co., Inc., Princeton, N.J., 1965.

$$\mathbf{Q} = \begin{bmatrix} Q_1 \\ Q_2 \end{bmatrix} = -\frac{6EI}{7L^3}\begin{bmatrix} 16 & -5 \\ -5 & 2 \end{bmatrix}\frac{PL^3}{48EI}\begin{bmatrix} 26 \\ 97 \end{bmatrix} = \frac{P}{56}\begin{bmatrix} 69 \\ -64 \end{bmatrix}$$

From this expression it is seen that the vertical reactions at supports B and C of the beam in Fig. 2-2a are

$$Q_1 = \frac{69P}{56} \qquad Q_2 = -\frac{8P}{7}$$

The negative sign for Q_2 indicates that this reaction is downward.

The redundant actions having been obtained in the manner shown in this example, the remaining actions in the beam can be found from static equilibrium equations. Also, the displacements at any point in the beam can now be obtained without difficulty, inasmuch as all actions can be considered as known. For example, one method of finding displacements is to subdivide the beam into two simple beams. Then each simple beam is acted upon by known moments at the ends as well as by the loads, and the displacements can be calculated.

While the matrix equations of the flexibility method (Eqs. 2-8, 2-9, and 2-10) were derived by discussion of the two-span beam of Fig. 2-2, they are actually completely general. They can be applied to any statically indeterminate framed structure having any number of degrees of indeterminacy. Of course, in this event the matrices appearing in the equations will be of a different order than for the two-span beam. In general, if there are n degrees of statical indeterminacy, the order of the flexibility matrix $\mathbf{F}$ will be $n \times n$, and the order of all the other matrices will be $n \times 1$. In the next article several examples of finding redundant actions by the flexibility method will be given.

2.3 Examples. In order to illustrate the application of the flexibility method to structures of various types, several examples are given in this article. In each example it is assumed that the object of the analysis is to calculate the values of certain selected redundants; hence, the problem is considered to be solved when the matrix $\mathbf{Q}$ is determined. The redundants are selected in each example primarily for illustrative purposes; however, many other choices of redundants are possible.

Example 1. The beam AB shown in Fig. 2-3a is fixed at both ends and is subjected to a concentrated load P and a couple M at the midpoint of the beam. It is assumed that the beam has constant flexural rigidity EI.

In order to begin the analysis by the flexibility method, two redundant actions must be selected. In this example the vertical reaction and the reactive moment at end B of the beam are selected as redundants, and are denoted in Fig. 2-3a as Q_1 and Q_2, respectively. The redundant Q_1 is assumed to be positive in the upward direction, and Q_2 is assumed positive in the clockwise direction. Other possible choices for the redundants include the reactive moments at both ends and the bending moment and shearing force at any section along the beam.

For the redundants shown in Fig. 2-3a the released structure consists of a

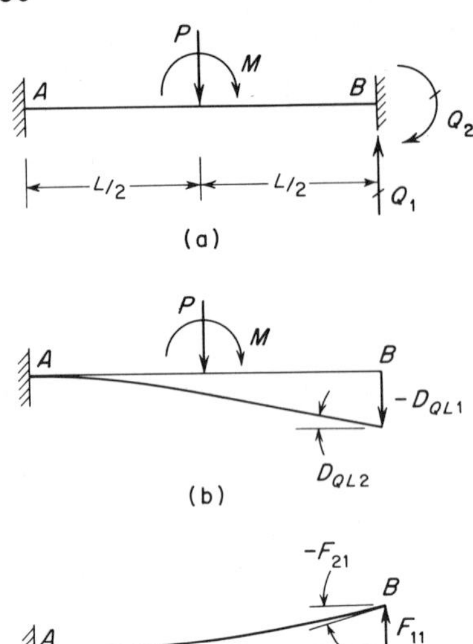

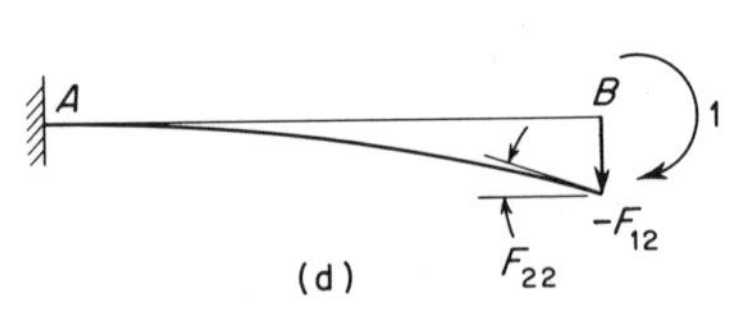

FIG. 2-3. Example 1: Fixed-end beam.

cantilever beam (see Fig. 2-3b). The displacements in this beam corresponding to Q_1 and Q_2 and caused by the loads P and M are denoted by D_{QL1} and D_{QL2} in the figure. The positive directions for these displacements are the same as those for Q_1 and Q_2. Since the displacement D_{QL1} is pictured in the negative direction, it is labeled with a minus sign. However, D_{QL2} is shown in its positive sense (clockwise). The displacements can be found with the aid of Table A-3 in the Appendix, and are as follows:

$$D_{QL1} = -\frac{5PL^3}{48EI} - \frac{3ML^2}{8EI}$$

$$D_{QL2} = \frac{PL^2}{8EI} + \frac{ML}{2EI}$$

Therefore, the vector $\mathbf{D}_{QL}$ can be written as

$$\mathbf{D}_{QL} = \frac{L}{48EI}\begin{bmatrix}(-5PL^2 - 18ML) \\ (6PL + 24M)\end{bmatrix}$$

The flexibility influence coefficients are the displacements of the released structure caused by unit values of Q_1 and Q_2, as shown in Figs. 2-3c and 2-3d. These coefficients are as follows:

$$F_{11} = \frac{L^3}{3EI} \qquad F_{12} = F_{21} = -\frac{L^2}{2EI} \qquad F_{22} = \frac{L}{EI}$$

The flexibility matrix $\mathbf{F}$ can now be written, after which its inverse can be determined; these matrices are:

$$\mathbf{F} = \frac{L}{6EI}\begin{bmatrix} 2L^2 & -3L \\ -3L & 6 \end{bmatrix} \qquad \mathbf{F}^{-1} = \frac{2EI}{L^3}\begin{bmatrix} 6 & 3L \\ 3L & 2L^2 \end{bmatrix}$$

The displacements in the fixed-end beam (Fig. 2-3a) corresponding to Q_1 and Q_2 are both zero, since there is no vertical translation and no rotation at support B. Therefore, the matrix $\mathbf{D}_Q$ is a null matrix, and the redundants can be found from Eq. (2-10). Substituting into that equation gives the following:

$$\mathbf{Q} = -\frac{2EI}{L^3}\begin{bmatrix} 6 & 3L \\ 3L & 2L^2 \end{bmatrix}\frac{L}{48EI}\begin{bmatrix}(-5PL^2 - 18ML) \\ (6PL + 24M)\end{bmatrix} = \frac{1}{8L}\begin{bmatrix}(4PL + 12M) \\ (PL^2 + 2ML)\end{bmatrix}$$

Therefore, the reactive force and moment at end B of the beam (see Fig. 2-3a) are given by the following expressions:

$$Q_1 = \frac{P}{2} + \frac{3M}{2L} \qquad Q_2 = \frac{PL}{8} + \frac{M}{4}$$

These results can be confirmed by comparison with the formulas given in Appendix B (see Cases 1 and 2 of Table B-1).

Example 2. The three-span continuous beam shown in Fig. 2-4a has constant flexural rigidity EI and is acted upon by a uniform load w in span AB and concentrated loads P at the midpoints of spans BC and CD. Since the structure is statically indeterminate to the second degree, two redundant actions must be selected. In this example, the bending moments at joints B and C are chosen. When these moments are removed from the beam by inserting hinges at B and C, the released structure is seen to consist of three simple beams (Fig. 2-4b). The redundant moments Q_1 and Q_2 are shown acting in their positive directions in Fig. 2-4b. Each redundant consists of two couples, one acting on each adjoining span of the structure. For example, the left-hand couple in Q_1 acts on beam AB in the counterclockwise direction, while the right-hand couple in Q_1 acts on span BC in the clockwise direction. The positive direction of each Q corresponds to that of a bending moment which produces compression at the top of the beam. Therefore, a positive sign in the final solution for either Q_1 or Q_2 means that the redundant moment produces compression at the top of the beam; if negative, the redundant moment produces tension at the top of the beam.

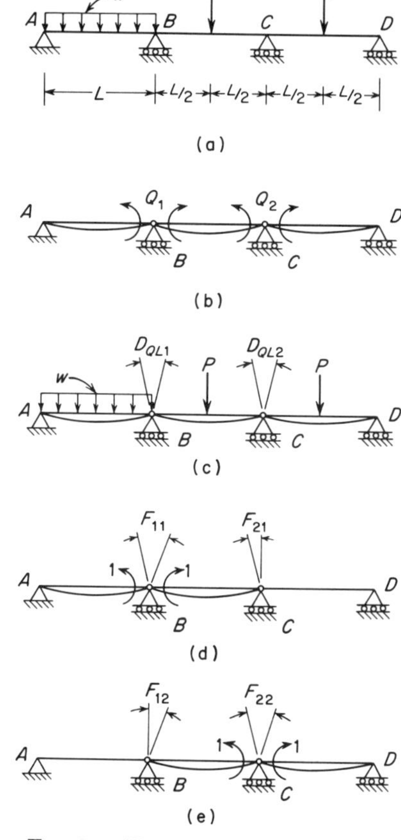

FIG. 2-4. Example 2: Continuous beam.

The displacement corresponding to one of the redundant moments consists of the sum of two rotations, one in each adjoining span. For example, the displacement corresponding to Q_1 consists of the counterclockwise rotation at joint B of the right-hand end of member AB plus the clockwise rotation at joint B of the left-hand end of member BC. In a similar manner, the displacement corresponding to Q_2 is the sum of the two rotations at joint C.

The displacements D_{QL1} and D_{QL2} corresponding to Q_1 and Q_2, respectively, and caused by the loads acting on the released structure, are shown in Fig. 2-4c. Since the counterclockwise rotation at end B of member AB due to the uniform load w is

$$\frac{wL^3}{24EI}$$

and since the clockwise rotation of end B of member BC due to the load P is

$$\frac{PL^2}{16EI}$$

it follows that the displacement D_{QL1} is

$$D_{QL1} = \frac{wL^3}{24EI} + \frac{PL^2}{16EI}$$

In a similar manner, the displacement D_{QL2} is seen to be

$$D_{QL2} = \frac{PL^2}{16EI} + \frac{PL^2}{16EI} = \frac{PL^2}{8EI}$$

Thus, the vector $\mathbf{D}_{QL}$ can be written as

$$\mathbf{D}_{QL} = \frac{L^2}{48EI}\begin{bmatrix} (2wL + 3P) \\ 6P \end{bmatrix}$$

The flexibility matrix must be found next. For this purpose, unit values of Q_1 and Q_2 are shown acting on the beams in Figs. 2-4d and 2-4e. The flexibility coefficient F_{11} (see Fig. 2-4d) is the sum of two rotations at joint B; one rotation is in span AB and the other is in span BC. Similarly, the coefficient F_{21} is the sum of the rotations at joint C. In Fig. 2-4d, however, the rotation in span CD is zero. Therefore, F_{21} is equal to the rotation in span BC alone. Similar comments apply to the flexibility coefficients shown in Fig. 2-4e. The rotations produced at the ends of a simply supported beam by a couple of unit value applied at one end of the beam are

$$\frac{L}{3EI} \quad \text{and} \quad \frac{L}{6EI}$$

at the near and far ends of the span, respectively, as given in Case 5, Table A-3. From these formulas the flexibility coefficients pictured in Figs. 2-4d and 2-4e can be obtained, as follows:

$$F_{11} = \frac{2L}{3EI} \qquad F_{12} = \frac{L}{6EI}$$

$$F_{21} = \frac{L}{6EI} \qquad F_{22} = \frac{2L}{3EI}$$

Therefore, the flexibility matrix is

$$\mathbf{F} = \frac{L}{6EI}\begin{bmatrix} 4 & 1 \\ 1 & 4 \end{bmatrix}$$

and its inverse becomes

$$\mathbf{F}^{-1} = \frac{2EI}{5L}\begin{bmatrix} 4 & -1 \\ -1 & 4 \end{bmatrix}$$

The displacements D_{Q1} and D_{Q2} in the original beam (Fig. 2-4a) corresponding to Q_1 and Q_2, respectively, are both zero because the beam is continuous across supports B and C. Thus, the matrix $\mathbf{D}_Q$ is null and Eq. (2-10) can be used to determine the redundants, as shown:

$$\mathbf{Q} = -\frac{2EI}{5L}\begin{bmatrix} 4 & -1 \\ -1 & 4 \end{bmatrix}\frac{L^2}{48EI}\begin{bmatrix} (2wL + 3P) \\ 6P \end{bmatrix} = -\frac{L}{120}\begin{bmatrix} (8wL + 6P) \\ (-2wL + 21P) \end{bmatrix}$$

Thus, the redundant bending moments Q_1 and Q_2 are given by the following formulas:

$$Q_1 = -\frac{wL^2}{15} - \frac{PL}{20} \qquad Q_2 = \frac{wL^2}{60} - \frac{7PL}{40}$$

Having obtained these moments at the supports, the remaining bending moments in the beam, as well as shearing forces and reactions, can be found by statics.

Example 3. The plane truss shown in Fig. 2-5a is statically indeterminate to the second degree. The horizontal reaction at the support B (positive to the right) and the axial force in bar AD (positive when tension) are selected as the redundants, resulting in the released structure shown in Fig. 2-5b. In the released structure the support at B has been removed, thereby releasing the reactive force Q_1, and the bar AD has been cut at some intermediate location in order to release the axial force Q_2 in the bar. Many other combinations of bar forces and reactions could be taken as redundants in this example, and each would give a different released structure. In all cases in which the force in a bar is selected as a redundant, the bar must be cut to obtain the released structure. The cut bar then remains as part of the released structure, since its deformations must be included in the calculation of displacements in the released structure.

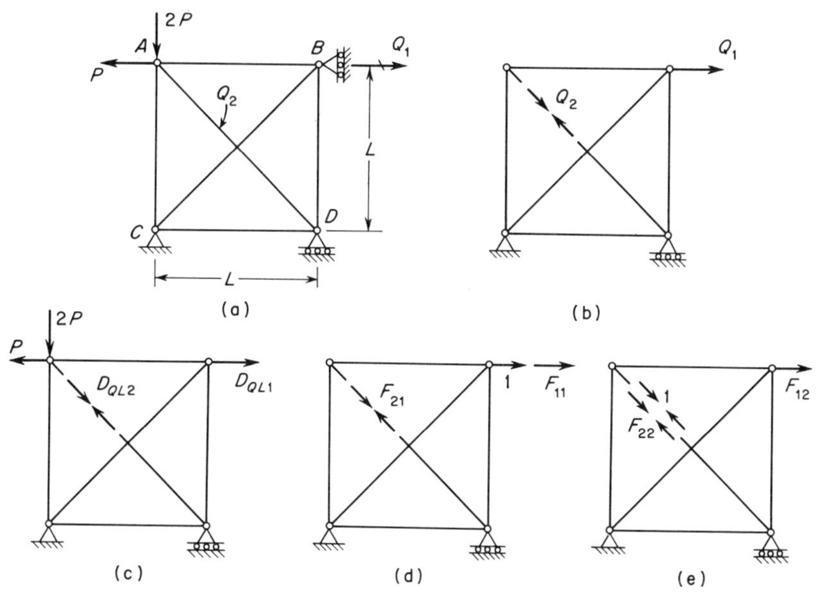

FIG. 2-5. Example 3: Plane truss.

It should be observed that Q_2 consists of a pair of forces acting on the released structure. This type of redundant is analogous to those described in Example 2, where a redundant bending moment was considered as a pair of couples acting on the released structure (see Fig. 2-4b). From this discussion it follows that a displacement corresponding to Q_2 consists of a relative translation of the cut ends of bar AD. When the cut ends are displaced toward one another, the displacement is in the same direction as Q_2 and hence is positive. When the cut ends are displaced apart from one another, the displacement is negative.

The first step in the analysis consists of determining the displacements in the released structure corresponding to Q_1 and Q_2 and due to the loads P and $2P$ acting at joint A. These displacements are denoted as D_{QL1} and D_{QL2} and are represented by vectors in Fig. 2-5c. This manner of representing the displacements is used in lieu of drawing the structure in its displaced configuration, which may become difficult for complicated structures. The determination of these displacements may be carried out by the unit-load method, as illustrated in Example 1 of Art. A.2 of the Appendix. Assuming that all members of the truss have the same axial rigidity EA, it is found that the displacements due to the loads P and $2P$ are

$$D_{QL1} = -\frac{PL}{EA}(1 + 2\sqrt{2}) = -3.828\frac{PL}{EA}$$

$$D_{QL2} = -2\frac{PL}{EA}$$

Therefore, the vector $\mathbf{D}_{QL}$ is

$$\mathbf{D}_{QL} = -\frac{PL}{EA}\begin{bmatrix} 3.828 \\ 2 \end{bmatrix}$$

The minus sign for D_{QL1} means that it is a displacement to the left, and the minus sign for D_{QL2} means that the cut ends of the bars are moved apart from one another.

The next step in the analysis involves the determination of the displacements in the released structure corresponding to Q_1 and Q_2 and caused by unit values of Q_1 and Q_2. Such displacements will constitute the flexibility matrix $\mathbf{F}$ and can be found by the unit-load method. When using the unit-load method for the released structure shown in Fig. 2-5, it is essential that all members of the released structure, including bar AD, be included in the calculations. The flexibility coefficient F_{11} is the displacement corresponding to Q_1 and caused by a unit value of Q_1, and is shown as a vector displacement in Fig. 2-5d. This displacement is

$$F_{11} = \frac{L}{EA}(1 + 2\sqrt{2}) = 3.828\frac{L}{EA}$$

The flexibility coefficient F_{21} is the displacement corresponding to Q_2 and due to a unit value of Q_1 (see Fig. 2-5d). Similarly, the coefficients F_{12} and F_{22} represent displacements in the released structure of Fig. 2-5e. When all of these displacements are obtained, the results are as follows:

$$F_{12} = F_{21} = \frac{L}{2EA}(4 + \sqrt{2}) = 2.707\frac{L}{EA}$$

$$F_{22} = \frac{2L}{EA}(1 + \sqrt{2}) = 4.828\frac{L}{EA}$$

Finally, the flexibility matrix can be formed and its inverse determined:

$$\mathbf{F} = \frac{L}{EA}\begin{bmatrix} 3.828 & 2.707 \\ 2.707 & 4.828 \end{bmatrix}$$

$$\mathbf{F}^{-1} = \frac{EA}{L}\begin{bmatrix} 0.4328 & -0.2426 \\ -0.2426 & 0.3431 \end{bmatrix}$$

Assuming that there are no support displacements in the truss, the redundants Q can be found by means of Eq. (2-10):

$$\mathbf{Q} = \frac{EA}{L}\begin{bmatrix} 0.4328 & -0.2426 \\ -0.2426 & 0.3431 \end{bmatrix}\frac{PL}{EA}\begin{bmatrix} 3.828 \\ 2 \end{bmatrix} = P\begin{bmatrix} 1.172 \\ -0.243 \end{bmatrix}$$

Therefore, the horizontal reactive force at B (Fig. 2-5a) is

$$Q_1 = 1.172P$$

and the axial force in bar AD is

$$Q_2 = -0.243P$$

The minus sign for Q_2 shows that the member is in compression. From the above results, one can calculate the remaining reactions and bar forces by statics.

Now assume that when the loads P and $2P$ act on the truss, support B moves a small distance s horizontally to the left. Therefore, the displacement D_{Q1} (see Eq. 2-7) is equal to minus s. The displacement D_{Q2} is zero, since it represents the relative displacement of the cut ends of bar AD in the original truss. The redundants Q can now be found from Eq. (2-9), as follows:

$$\mathbf{Q} = \frac{EA}{L}\begin{bmatrix} 0.4328 & -0.2426 \\ -0.2426 & 0.3431 \end{bmatrix}\left\{\begin{bmatrix} -s \\ 0 \end{bmatrix} + \frac{PL}{EA}\begin{bmatrix} 3.828 \\ 2 \end{bmatrix}\right\}$$

$$= \begin{bmatrix} -0.433\dfrac{sEA}{L} + 1.172P \\ \\ 0.243\dfrac{sEA}{L} - 0.243P \end{bmatrix}$$

Therefore, the horizontal reaction at B is

$$Q_1 = -0.433\frac{sEA}{L} + 1.172P$$

and the force in bar AD is

$$Q_2 = 0.243\frac{sEA}{L} - 0.243P$$

In this example there was a redundant action (Q_1) corresponding to the support displacement at joint B. Of course, if other redundants had been selected, such as the forces in the two diagonal bars, the displacement at support B could not have been included in the vector $\mathbf{D}_Q$. Then it would be necessary to account for the support displacement by another means, such as that described later in Art. 2.4.

Example 4. The plane frame shown in Fig. 2-6a has fixed supports at A and C and is acted upon by the vertical load P at the midpoint of member AB. It is desired to analyze the frame taking into account the effects of both flexural and axial deformations. The inclusion of axial effects is for illustrative purposes only; normally, in a frame of this type, only flexural effects would be considered, and the analysis would be simplified slightly. The members of the frame have constant flexural rigidity EI and constant axial rigidity EA. The structure is statically indeterminate to the third degree, and a suitable released structure (see Fig. 2-6b) is obtained by cutting the frame at joint B, thereby releasing two forces and a bending moment. These released actions are the redundants Q_1, Q_2, and Q_3, as shown in Fig. 2-6b. A displacement in the released structure corresponding to Q_1 consists of the horizontal translation of end B of member AB (taken positive to the right) plus the horizontal translation of end B of member BC (taken positive to the left). In other words, a displacement corresponding to Q_1 consists of the sum of two translations and represents a relative displacement between the two points labeled B in Fig. 2-6b. In a similar manner the displacements corresponding to Q_2 and Q_3 can be defined as the sum of two vertical translations and two rotations, respectively, at joint B.

The displacements in the released structure caused by the load P and corresponding to Q_1, Q_2, and Q_3 are shown in Fig. 2-6c. For example, the displacement D_{QL1} is shown in the figure to consist of two horizontal translations, as described in the preceding paragraph. Similarly, the displacements D_{QL2} and D_{QL3} are shown as two vertical translations and two rotations, respectively. It is not difficult to calculate these displacements due to the force P, inasmuch as the released structure consists of two cantilever beams. First, taking the beam AB in Fig. 2-6c, it is seen that the displacements at end B are as follows:

$$(D_{QL1})_{AB} = 0 \qquad (D_{QL2})_{AB} = -\frac{5PL^3}{48EI} \qquad (D_{QL3})_{AB} = -\frac{PL^2}{8EI}$$

These expressions are based upon only the flexural deformations of beam AB because there are no axial deformations. Secondly, the beam BC in Fig. 2-6c must be considered. In this example there is no load on member BC and hence no displacements at end B; therefore

$$(D_{QL1})_{BC} = (D_{QL2})_{BC} = (D_{QL3})_{BC} = 0$$

The final displacements caused by the load P can now be obtained by combining the above results:

$$D_{QL1} = 0 \qquad D_{QL2} = -\frac{5PL^3}{48EI} \qquad D_{QL3} = -\frac{PL^2}{8EI}$$

Therefore, the matrix $\mathbf{D}_{QL}$ is:

$$\mathbf{D}_{QL} = \frac{PL^2}{48EI}\begin{bmatrix} 0 \\ -5L \\ -6 \end{bmatrix}$$

The flexibility matrix must be determined next. Consider first the released structure with the action $Q_1 = 1$ applied to it, as shown in Fig. 2-6d. The displacements corresponding to Q_1, Q_2, and Q_3 are shown in the figure as the flexibility coefficients F_{11}, F_{21}, and F_{31}. If both axial and flexural deformations are considered, the displacements at end B of member AB are

$$(F_{11})_{AB} = \frac{L}{EA} \qquad (F_{21})_{AB} = 0 \qquad (F_{31})_{AB} = 0$$

Also, the displacements at end B of member BC are

$$(F_{11})_{BC} = \frac{H^3}{3EI} \qquad (F_{21})_{BC} = 0 \qquad (F_{31})_{BC} = -\frac{H^2}{2EI}$$

Therefore, the final values of the three flexibility coefficients shown in Fig. 2-6d are

$$F_{11} = \frac{L}{EA} + \frac{H^3}{3EI} \qquad F_{21} = 0 \qquad F_{31} = -\frac{H^2}{2EI}$$

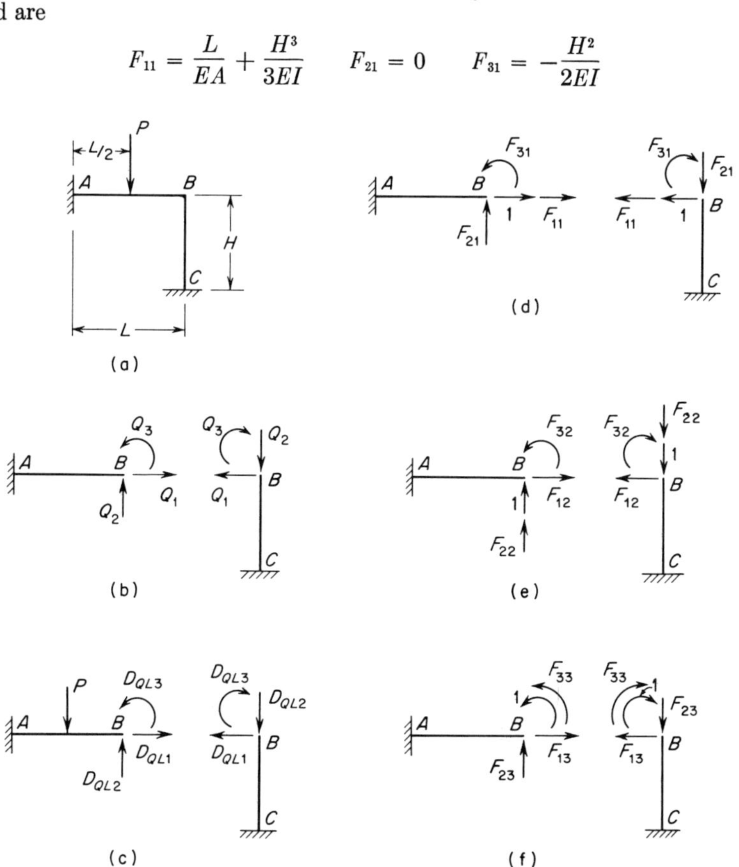

FIG. 2-6. Example 4: Plane frame.

Next, the same kind of analysis must be made for actions $Q_2 = 1$ and $Q_3 = 1$ acting on the released structure. These conditions are shown in Figs. 2-6e and 2-6f, and the various flexibility coefficients are found to be as follows:

$$F_{12} = 0 \qquad F_{22} = \frac{L^3}{3EI} + \frac{H}{EA} \qquad F_{32} = \frac{L^2}{2EI}$$

$$F_{13} = -\frac{H^2}{2EI} \qquad F_{23} = \frac{L^2}{2EI} \qquad F_{33} = \frac{L}{EI} + \frac{H}{EI}$$

Finally, the flexibility matrix can be assembled:

$$\mathbf{F} = \begin{bmatrix} \frac{L}{EA} + \frac{H^3}{3EI} & 0 & -\frac{H^2}{2EI} \\ 0 & \frac{L^3}{3EI} + \frac{H}{EA} & \frac{L^2}{2EI} \\ -\frac{H^2}{2EI} & \frac{L^2}{2EI} & \frac{L}{EI} + \frac{H}{EI} \end{bmatrix} \quad \text{(a)}$$

If axial effects were to be omitted from the analysis, the two fractions in **F** containing the axial rigidity EA in the denominators would be omitted.

The next step in the solution is to obtain the inverse of the flexibility matrix and then to substitute it, as well as the matrix $\mathbf{D}_{QL}$, into Eq. (2-10). It is possible to use Eq. (2-10) in this example because the displacements D_Q are all equal to zero, inasmuch as joint B in the original frame (Fig. 2-6a) is a rigid connection. However, the inverse of the flexibility matrix **F** given above cannot be obtained conveniently in literal form, as was done in the preceding examples where **F** was a 2×2 matrix. Therefore, this solution will be continued by substituting numerical data.

Assume now that the following values are given for the frame shown in Fig. 2-6a:

$$P = 10\text{ k} \qquad L = H = 12\text{ ft} \qquad E = 30{,}000\text{ ksi} \qquad I = 200\text{ in.}^4 \qquad A = 10\text{ in.}^2$$

When these numerical values are substituted into the matrices $\mathbf{D}_{QL}$ and **F**, they become:

$$\mathbf{D}_{QL} = \begin{bmatrix} 0 \\ -0.5184 \\ -0.00432 \end{bmatrix}$$

$$\mathbf{F} = \begin{bmatrix} 0.1664 & 0 & -0.001728 \\ 0 & 0.1664 & 0.001728 \\ -0.001728 & 0.001728 & 0.000048 \end{bmatrix}$$

The numerical values appearing in $\mathbf{D}_{QL}$ and **F** are in units of kips, inches, and radians; thus, D_{QL2} is -0.5184 inches, D_{QL3} is -0.00432 radians, F_{11} is 0.1664 inches per kip, F_{13} is -0.001728 inches per inch-kip, F_{32} is 0.001728 radians per kip, and so on. The inverse of **F** can be found by standard numerical methods, and is equal to

$$\mathbf{F}^{-1} = \begin{bmatrix} 14.92 & -8.913 & 858.1 \\ -8.913 & 14.92 & -858.1 \\ 858.1 & -858.1 & 82{,}620 \end{bmatrix}$$

Finally, the matrices $\mathbf{F}^{-1}$ and $\mathbf{D}_{QL}$ can be substituted into Eq. (2-10) in order to obtain the vector of redundants:

$$\mathbf{Q} = \begin{bmatrix} 14.92 & -8.913 & 858.1 \\ -8.913 & 14.92 & -858.1 \\ 858.1 & -858.1 & 82{,}620 \end{bmatrix} \begin{bmatrix} 0 \\ 0.5184 \\ 0.00432 \end{bmatrix} = \begin{bmatrix} -0.913 \\ 4.03 \\ -87.9 \end{bmatrix}$$

Thus, the redundants are

$$Q_1 = -0.913\text{ k} \qquad Q_2 = 4.03\text{ k} \qquad Q_3 = -87.9\text{ in.-k}$$

The minus signs for Q_1 and Q_3 show that these actions are opposite in direction to the positive directions assumed in Fig. 2-6b.

If the effects of axial deformations are omitted from the analysis, the flexibility matrix becomes

$$\mathbf{F} = \begin{bmatrix} 0.1659 & 0 & -0.001728 \\ 0 & 0.1659 & 0.001728 \\ -0.001728 & 0.001728 & 0.000048 \end{bmatrix}$$

and the following values for the redundants are obtained:

$$Q_1 = -0.938 \text{ k} \qquad Q_2 = 4.06 \text{ k} \qquad Q_3 = -90.0 \text{ in.-k}$$

These results differ by less than 3 per cent from the earlier ones, which is frequently the situation when analyzing plane frames. In such cases only bending deformations need to be considered in the analysis.

Example 5. The grid structure shown in Fig. 2-7a is in a horizontal plane (x-z plane) and carries a load P acting in the vertical direction. The supports of the grid at A and C are fixed, and the members AB and BC have length L, flexural rigidity EI, and torsional rigidity GJ (see Appendix A for definition of torsional rigidity). The redundants that are chosen for this example are released by cutting the grid at joint B (Fig. 2-7b), thereby giving a released structure in the form of two cantilever beams. Each redundant consists of a pair of actions, and these actions are shown in their assumed positive directions in Fig 2-7b. There are no other internal actions at joint B, because in a grid with only vertical loads there are no horizontal forces between members and no couples about a vertical axis.

When the load P is applied to the released structure (Fig. 2-7c), displacements D_{QL} are produced. These displacements are shown in their positive directions in the figure. Note that a translational displacement is denoted by a single-headed arrow, while a rotation is denoted by a double-headed arrow. This representation is analogous to the convention that is used when representing forces and couples. The displacements shown in Fig. 2-7c can be obtained without difficulty, and then the matrix $\mathbf{D}_{QL}$ can be formed:

$$\mathbf{D}_{QL} = \frac{PL^2}{48EI}\begin{bmatrix} 5L \\ 0 \\ -6 \end{bmatrix}$$

The flexibility coefficients are the displacements shown in Figs. 2-7d, 2-7e, and 2-7f. For these figures, it is assumed that unit values of the redundants Q_1, Q_2, and Q_3 act on the released structure. By referring to the figures, the various coefficients can be obtained, and the flexibility matrix can be formed:

$$\mathbf{F} = \begin{bmatrix} \dfrac{2L^3}{3EI} & \dfrac{L^2}{2EI} & -\dfrac{L^2}{2EI} \\ \dfrac{L^2}{2EI} & \dfrac{L}{EI} + \dfrac{L}{GJ} & 0 \\ -\dfrac{L^2}{2EI} & 0 & \dfrac{L}{EI} + \dfrac{L}{GJ} \end{bmatrix}$$

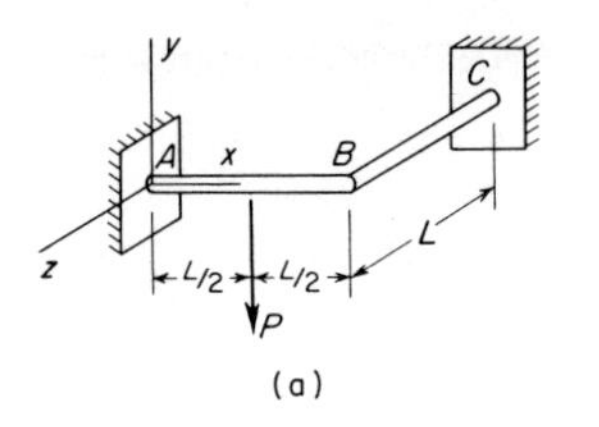

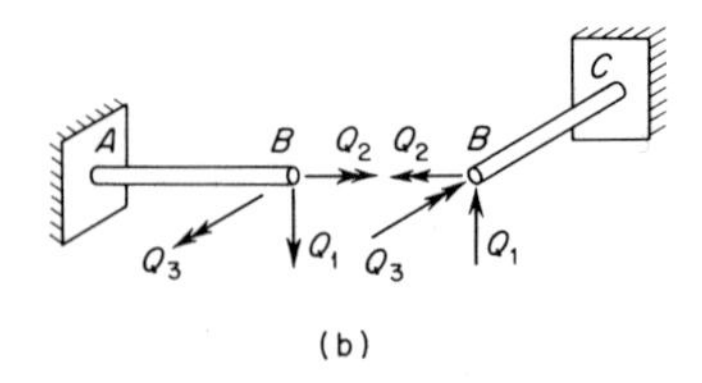

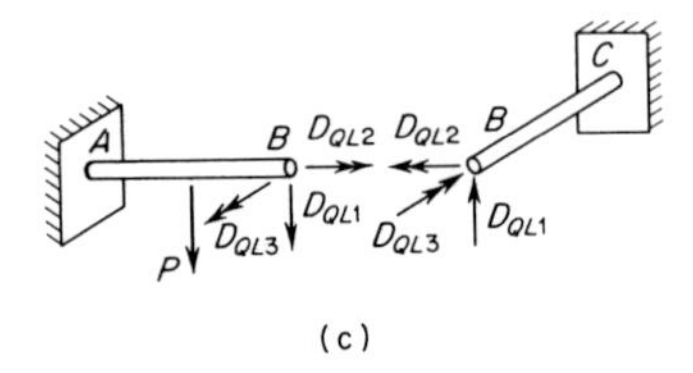

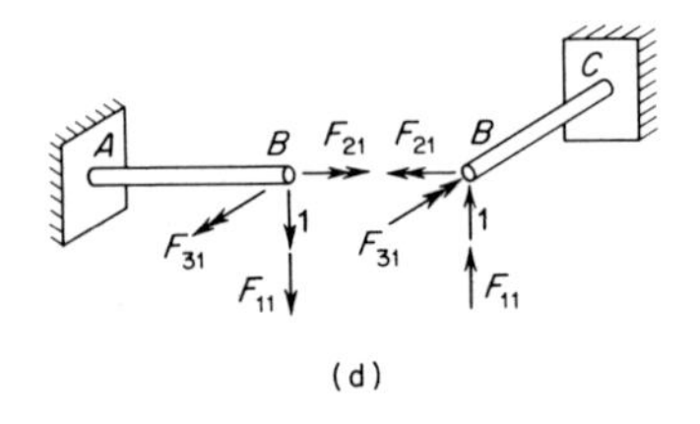

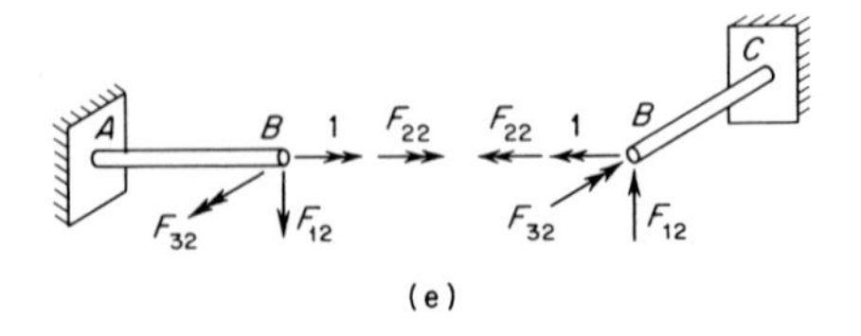

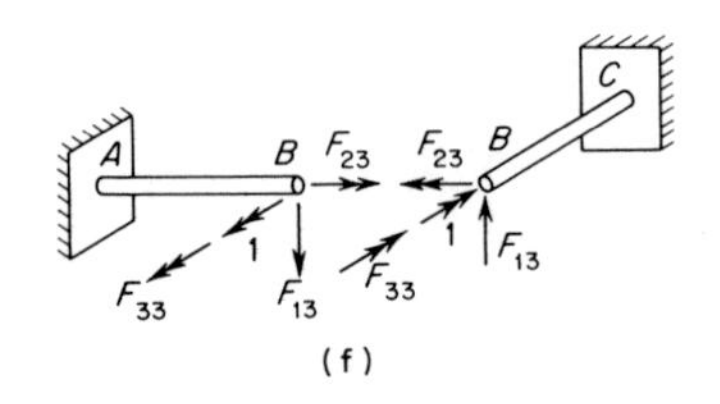

FIG. 2-7. Example 5: Grid.

This matrix can be written in a simpler form by introducing a nondimensional parameter ρ defined as the ratio of the flexural and torsional rigidities, thus:

$$\rho = \frac{EI}{GJ}$$

With this notation the flexibility matrix can be put into the following form:

$$\mathbf{F} = \frac{L}{6EI}\begin{bmatrix} 4L^2 & 3L & -3L \\ 3L & 6(1+\rho) & 0 \\ -3L & 0 & 6(1+\rho) \end{bmatrix}$$

The inverse of $\mathbf{F}$ is

$$\mathbf{F}^{-1} = \frac{EI}{2L^3b_1b_2}\begin{bmatrix} 12b_1^2 & -6Lb_1 & 6Lb_1 \\ -6Lb_1 & L^2b_3 & -3L^2 \\ 6Lb_1 & -3L^2 & L^2b_3 \end{bmatrix}$$

in which the following additional nondimensional parameters are used:

$$b_1 = 1 + \rho \qquad b_2 = 1 + 4\rho \qquad b_3 = 5 + 8\rho$$

Finally, substitution into Eq. (2-10) yields the vector of redundants:

$$\mathbf{Q} = \frac{P}{16(1+\rho)(1+4\rho)}\begin{bmatrix} -2(1+\rho)(2+5\rho) \\ L(2+5\rho) \\ 3L\rho \end{bmatrix}$$

Thus, the redundants are as follows:

$$Q_1 = -\frac{P}{8}\frac{2+5\rho}{1+4\rho}$$

$$Q_2 = \frac{PL}{16}\frac{2+5\rho}{(1+\rho)(1+4\rho)}$$

$$Q_3 = \frac{3PL}{16}\frac{\rho}{(1+\rho)(1+4\rho)}$$

If the members AB and BC are torsionally very weak, then ρ can be considered as infinitely large. The above formulas then give the following values for the redundants (after dividing the numerators and denominators by ρ):

$$Q_1 = -\frac{5P}{32} \qquad Q_2 = Q_3 = 0$$

This result could also be obtained by assuming that a spherical hinge exists at B which transmits vertical force but not moment. Such a grid would be statically indeterminate to the first degree.

2.4 Effects of Temperature, Prestrain, and Support Displacement. It is frequently necessary to include in the analysis of a structure not only the effects of loads but also the effects of temperature changes, prestrain of members, and displacements of one or more supports. In most cases these effects can be incorporated into the analysis by including them in the calculation of displacements in the released structure. As an example, consider the effects of temperature changes. If the changes are assumed to occur in the released structure, there will be displacements in the released structure corresponding to the redundant actions Q. These displacements can be identified by the symbol D_{QT}, which is analogous to the symbol D_{QL} used previously to represent displacements in the released structure corresponding to the redundants and caused by the loads. The temperature displacements D_{QT} in the released structure may be due to either uniform changes in temperature or to differential changes in temperature. A uniform change refers to a change in temperature that is constant throughout the member, and causes the member to increase or decrease in length. A differential change means that the top and bottom of the member are subjected to different temperatures, while the average temperature remains unchanged; thus, the member will not change in length but will undergo a curvature of its longitudinal axis. The calculation of displacements due to temperature changes is discussed in Appendix A.

When the matrix $\mathbf{D}_{\mathrm{QT}}$ of displacements due to temperature changes has been obtained, it can be added to the matrix $\mathbf{D}_{\mathrm{QL}}$ of displacements due to loads to give the sum of all displacements in the released structure. Then these total displacements can be used in the equation of superposition (see Eq. 2-8) in place of $\mathbf{D}_{\mathrm{QL}}$ alone. Therefore, the superposition equation becomes

$$\mathbf{D}_{\mathrm{Q}} = \mathbf{D}_{\mathrm{QL}} + \mathbf{D}_{\mathrm{QT}} + \mathbf{FQ}$$

which can be solved for the vector $\mathbf{Q}$ of redundants as before. An illustration of the calculation of the vector $\mathbf{D}_{\mathrm{QT}}$ to be used in this equation is given in a later example.

The effects of prestrain in any member of the structure can be handled in a manner analogous to that for temperature changes. By prestrain of a member is meant an initial deformation of the member due to any of various causes. For example, a truss member may be fabricated with a length greater or less than the theoretical length of the member, or a beam may be fabricated with an initial curvature. It can be seen intuitively that a truss bar having a prestrain consisting of an initial elongation will produce the same effects in the truss as if the bar had been heated uniformly to a temperature that would produce the same increase in length. Thus, the method of analysis for prestrain effects is similar to that for temperature. The first step is to assume that the prestrain occurs in the released structure. Then the displacements in the released structure corresponding to the redundants must be found. These displacements are denoted D_{QP}, signifying that they correspond to the redundants and are due to prestrain effects. The matrix $\mathbf{D}_{\mathrm{QP}}$ for the prestrain displacements can then be added to the matrices $\mathbf{D}_{QL}$ and $\mathbf{D}_{QT}$ to give the sum of all displacements in the released structure. Finally, the sum of all displacements is included in the equation of superposition as follows:

$$\mathbf{D}_{Q} = \mathbf{D}_{QL} + \mathbf{D}_{QT} + \mathbf{D}_{QP} + \mathbf{FQ}$$

As before, the superposition equation can be solved for the matrix of redundants $\mathbf{Q}$. An example involving the calculation of prestrain displacements will be given later.

Lastly, consider the possibility of known displacements occurring at the restraints (or supports) of a structure. There are two possibilities to be considered, depending upon whether or not the restraint displacement corresponds to one of the redundant actions Q. If the restraint displacement does correspond to a redundant, its effects can be taken into account by including the displacement in the vector $\mathbf{D}_{Q}$ of actual displacements in the structure. This procedure was discussed before in Art. 2.2 and was illustrated in Art. 2.3 in an example of a statically indeterminate truss (see Example 3). In a more general situation, however, there will be restraint displacements that do not correspond to any of the selected redundants. In that event, the effects of such restraint displacements must be incorporated in the analysis of the released structure in the same manner as in the case of temperature and prestrain effects. When the restraint displacements are assumed to occur in the released structure, there will be displacements D_{QR} corresponding to the redundants Q. When these displacements have been found, the matrix $\mathbf{D}_{QR}$ can be obtained. Then this matrix can be added to the other matrices representing displacements in the released structure.

The sum of all matrices representing displacements in the released structure will be denoted $\mathbf{D}_{QS}$ in future discussions, and can be expressed as follows:

$$\mathbf{D}_{QS} = \mathbf{D}_{QL} + \mathbf{D}_{QT} + \mathbf{D}_{QP} + \mathbf{D}_{QR} \qquad (2\text{-}11)$$

Thus, the matrix $\mathbf{D}_{QS}$ represents the sum of displacements due to all causes, including loads, temperature changes, prestrain effects, and restraint displacements. Using the notation from Eq. (2-11), a generalized form of the superposition equation for the flexibility method becomes

$$\mathbf{D}_{Q} = \mathbf{D}_{QS} + \mathbf{FQ} \qquad (2\text{-}12)$$

The superposition equation used previously (Eq. 2-8) can be considered as a special case of Eq. (2-12). When the latter equation is solved for $\mathbf{Q}$, the result is

$$\mathbf{Q} = \mathbf{F}^{-1}(\mathbf{D}_{Q} - \mathbf{D}_{QS}) \qquad (2\text{-}13)$$

and this equation may be used in place of Eq. (2-9) when causes other than loads must be considered. Of course, in any particular analysis it is not likely that all of the matrices given in Eq. (2-11) will be of interest. Some examples of the use of the above equations will now be given.

Example 1. In order to illustrate the analysis of a structure when temperature changes are present, consider the two-span beam ABC in Fig. 2-8a. The beam is assumed to be subjected to a temperature differential such that the top surface of the beam is at temperature T_2 while the lower surface is at temperature T_1. Previously, this same beam was analyzed for the effects of loads (see Fig. 2-2), and in that solution the reactions at supports B and C were taken as the redundants Q_1 and Q_2. These same redundants will be used again in solving for the thermal effects; therefore, the inverse of the flexibility matrix may be taken from the former solution (see Art. 2.2):

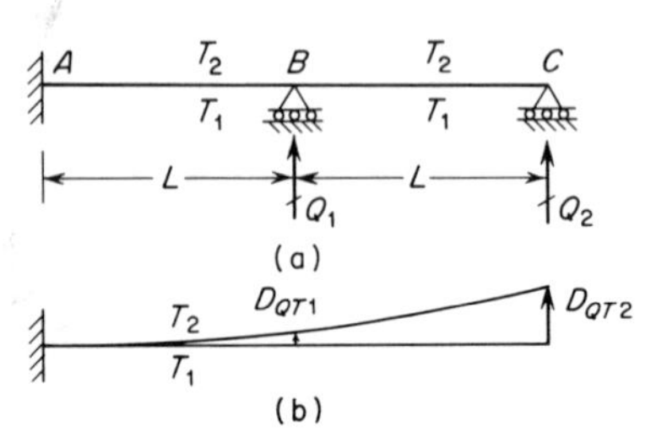

FIG. 2-8. Example 1: Continuous beam of Fig. 2-2.

$$\mathbf{F}^{-1} = \frac{6EI}{7L^3}\begin{bmatrix} 16 & -5 \\ -5 & 2 \end{bmatrix}$$

Also, the vector $\mathbf{D}_Q$ (see Eq. 2-13) giving the actual displacements in the beam is a null vector ($\mathbf{D}_Q = \mathbf{O}$).

The released structure for the beam ABC is the cantilever beam shown in Fig. 2-8b. If the temperature T_1 is greater than T_2, the released beam will deflect upward as indicated in the figure. The displacements corresponding to the redundants are denoted D_{QT1} and D_{QT2}. These displacements can be calculated by the unit-load method described in Appendix A (see Example 4 in Art. A.2). Thus, the displacements D_{QT1} and D_{QT2} are found to be

$$\frac{\alpha(T_1 - T_2)L^2}{2d} \quad \text{and} \quad \frac{2\alpha(T_1 - T_2)L^2}{d}$$

respectively. In these expressions α is the coefficient of thermal expansion for the material of the beam, and d is the depth of the beam. The vector $\mathbf{D}_{QT}$ is

$$\mathbf{D}_{QT} = \frac{\alpha(T_1 - T_2)L^2}{2d}\begin{bmatrix} 1 \\ 4 \end{bmatrix}$$

If only the effects of temperature are considered in the analysis of the beam, the vector $\mathbf{D}_{QT}$ becomes the vector $\mathbf{D}_{QS}$ in Eq. (2-13). Then the redundants as found from that equation are:

$$\mathbf{Q} = \frac{3EI\,\alpha(T_1 - T_2)}{7Ld}\begin{bmatrix} 4 \\ -3 \end{bmatrix}$$

The signs on the elements of $\mathbf{Q}$ show that the redundant reaction Q_1 is upward when T_1 is greater than T_2, while the redundant Q_2 is downward. If T_1 is less than T_2, the directions of the redundants will be reversed. Of course, if T_1 is equal to T_2, the redundants are zero.

If the combined effects of both loads and temperature are desired, the matrix $\mathbf{D}_{QS}$ in Eq. (2-13) is taken as the sum of $\mathbf{D}_{QT}$ (obtained above) and $\mathbf{D}_{QL}$ (obtained in Art. 2.2). The results obtained for the redundants under the combined conditions will be the sum of the results obtained for temperature changes and loads taken separately.

Example 2. As a second example illustrating the effects of a temperature change, consider the plane truss shown previously in Fig. 2-5a and assume that bar BD has its temperature increased uniformly by an amount T. The resulting elongation will produce displacements D_{QT} in the released structure corresponding to the redundants. These displacements, denoted D_{QT1} and D_{QT2}, are shown in Fig. 2-9 in their positive directions (compare with Fig. 2-5c which shows the displacements in the released structure due to loads). The displacements D_{QT} can be readily found by the unit-load method, as illustrated in Example 2 of Art. A.2. By this means the vector $\mathbf{D}_{\mathrm{QT}}$ is found to be:

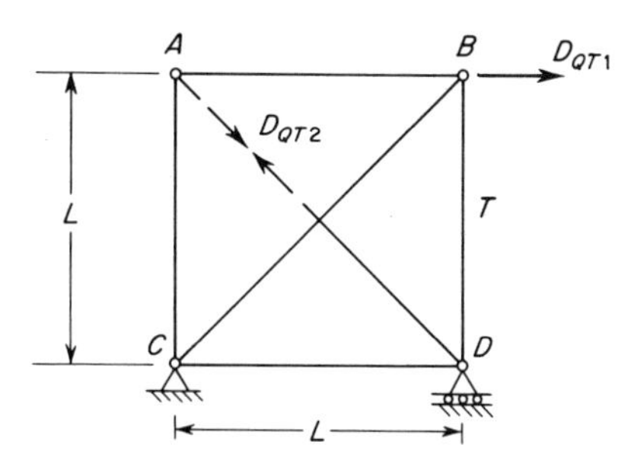

FIG. 2-9. Example 2: Plane truss of Fig. 2-5.

$$\mathbf{D}_{\mathrm{QT}} = \alpha LT \begin{bmatrix} -1 \\ -\dfrac{1}{\sqrt{2}} \end{bmatrix}$$

The temperature effects can now be incorporated into the analysis of the truss by including the matrix $\mathbf{D}_{\mathrm{QT}}$ in the calculation of $\mathbf{D}_{\mathrm{QS}}$ (see Eq. 2-11).

If the member BD of the truss in Fig. 2-5a were fabricated with a length $L + e$, instead of L, the analysis could be handled in the same manner as shown above for a temperature change in the bar. The only difference is that the prestrain elongation e would replace the temperature elongation αLT for bar BD. Therefore, the vector of prestrain displacements in the released structure would be (see Example 2, Art. A.2):

$$\mathbf{D}_{\mathrm{QP}} = e \begin{bmatrix} -1 \\ -\dfrac{1}{\sqrt{2}} \end{bmatrix}$$

A prestrain of any other bar could be handled in a similar fashion.

Example 3. This example illustrates how restraint displacements can be taken into account. Refer again to the two-span beam shown in Fig. 2-2a and assume that the following two support displacements occur. Joint A undergoes a known rotation in the clockwise direction of β radians, and joint B is displaced downward by a distance s. The displacement at B corresponds to one of the redundants and therefore is accounted for in the vector $\mathbf{D}_{\mathrm{Q}}$, which now becomes

$$\mathbf{D}_{\mathrm{Q}} = \begin{bmatrix} -s \\ 0 \end{bmatrix}$$

The minus sign is required in the first element inasmuch as Q_1 is positive when upward. The restraint displacement at support A is included in the analysis by means of the matrix $\mathbf{D}_{\mathrm{QR}}$ (see Eq. 2-11). This matrix consists of the displacements in the released structure (see Fig. 2-10) when joint A is rotated through the clockwise angle β. Therefore, the displacements D_{QR1} and D_{QR2} corresponding to Q_1 and Q_2 are

$$-\beta L \quad \text{and} \quad -2\beta L$$

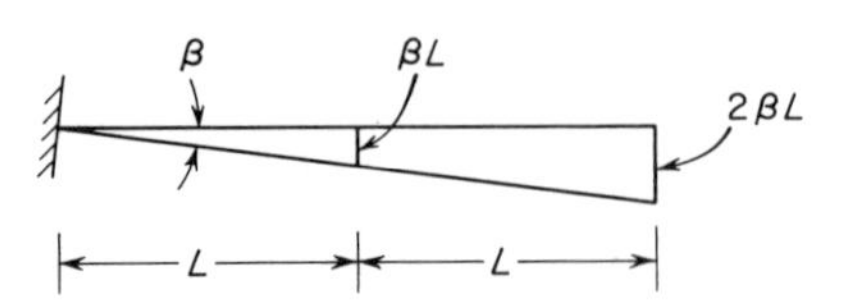

FIG. 2-10. Example 3: Support rotation in released structure.

respectively. These terms are negative because the actual displacements are downward. Thus, the matrix $\mathbf{D}_{QR}$ is

$$\mathbf{D}_{QR} = \beta L \begin{bmatrix} -1 \\ -2 \end{bmatrix}$$

Finally, this matrix can be included in the calculation of $\mathbf{D}_{QS}$, after which Eq. (2-13) can be solved for the redundants. As in the other illustrations, the values of the redundants due to the combined effects will be equal to the sum of the values obtained separately.

2.5 Joint Displacements, Member End-Actions, and Reactions. In the preceding articles the emphasis was on finding redundant actions by the flexibility method. The redundant actions may be either internal stress resultants (such as axial forces and bending moments) or external reactions at points of support. In all cases it is possible to find other actions in the structure by using principles of static equilibrium after the redundants have been calculated. Such calculations would normally include the reactions for the structure and the actions at the ends of each member (member end-actions). Furthermore, when all actions in the structure are available, it is possible to calculate all displacements. This can be done, for example, by isolating individual members of the structure and then finding displacements in the member from action-displacement relationships. Usually the displacements of primary interest are the translations and rotations of the joints.

Instead of following the procedure described above, it is more systematic to incorporate the calculations for the joint displacements, member end-actions, and reactions directly into the basic computations for the flexibility method. That is, the task of finding the various actions and displacements that are of interest can be performed in parallel with the computations for finding the redundants instead of postponing them as separate calculations to be performed at the end of the analysis.

In order to demonstrate the procedure for making a complete analysis of a structure, the two-span beam example given previously in Fig. 2-2 will now be extended. The beam to be analyzed is shown again in Fig. 2-11a. Now assume that it is desired to calculate not only the redundants Q_1 and Q_2 for the beam, but also the joint displacements, member end-actions, and reactions. The joint displacements in a structure will be denoted by the general symbol D_J, and numerical subscripts will be used to identify the individual joint displacements. For example, in the beam of Fig. 2-11a the two joint displacements to be found are the rotations at joints B and C. These displacements will be denoted D_{J1} and D_{J2}, respectively, and are assumed to be positive when counterclockwise, as shown in the figure.

The member end-actions are the couples and forces that act at the ends of a member when it is considered to be isolated from the remainder of the structure. For the beam under consideration, the end-actions are the bending couples and shearing forces at the ends of the members, as shown in Fig. 2-11b. These end-actions must be evaluated according to a specified sign convention, which may be either a deformation sign convention (related to how the member is deformed) or a statical sign convention (related to the direction of the action in space). The positive directions shown in Fig. 2-11b are based upon a statical sign convention that upward forces and counterclockwise moments are positive. In general, the end-actions are denoted by the symbol A_M and are distinguished from one another by numerical subscripts. In the example of Fig. 2-11 there is a total of eight end-actions. However, it will be assumed arbitrarily that only the four actions labeled A_{M1}, A_{M2}, A_{M3}, and A_{M4} in Fig. 2-11b are to be calculated. The end-actions A_{M1} and A_{M2} are the shearing force and bending moment at the right-hand end of member AB, while A_{M3} and A_{M4} are the shearing force and bending moment at the left-hand end of member BC. Thus, the first two end-actions are located just to the left of joint B, and the last two are located just to the right of joint B. In this particular example, the sum of the shearing forces A_{M1} and A_{M3} must be equal to the redundant reaction Q_1, because there is no vertical load on the beam at joint B. Also, the sum of the bending couples A_{M2} and A_{M4} must be equal to the couple M acting as a load at joint B.

Finally, consider the reactions for the beam in Fig. 2-11a. The two reactions at supports B and C will be determined automatically since they are the redundants Q_1 and Q_2. The remaining reactions, denoted generally by the symbol A_R, consist of a vertical force and a couple at support A. These reactions are labeled A_{R1} and A_{R2} in Fig. 2-11a and are assumed positive in the directions shown. While the sign convention for reactions may be selected arbitrarily in each particular case, the positive directions shown in the figure will customarily be used in this book.

In order to find the displacements D_J, end-actions A_M, and reactions A_R for the beam in Fig. 2-11a, the principle of superposition will be used. Previously, this principle was applied to the released structures shown in Figs. 2-11c, 2-11d, and 2-11e in order to obtain an equation for the redundants Q (see Eq. 2-8). In a similar manner, the principle of superposition may be used to obtain the joint displacements D_J in the beam of Fig. 2-11a. In order to accomplish this result, it is necessary to evaluate the displacements in the released structures (Figs. 2-11c, 2-11d, and 2-11e) corresponding to the displacements D_J. In the released structure subjected to the loads, these displacements are denoted by the general symbol D_{JL} and, in particular, the rotations at joints B and C are

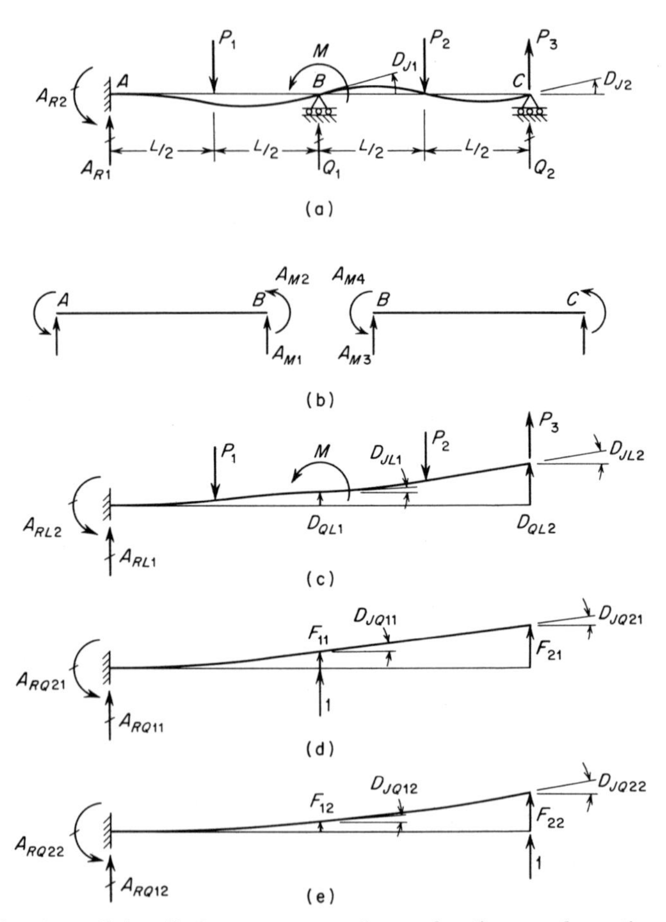

FIG. 2-11. Joint displacements, member end-actions, and reactions.

denoted D_{JL1} and D_{JL2}, respectively. Both of these quantities can be found from an analysis of the cantilever beam in Fig. 2-11c.

Next, the released structure subjected to unit values of the redundants must be considered (Figs. 2-11d and 2-11e). The displacements corresponding to D_J are denoted as D_{JQ}, in which the letter Q is used to indicate that these joint displacements are caused by unit values of the redundants. Consider, for example, the released structure subjected to a unit value of the redundant Q_1 (Fig. 2-11d). The joint displacements caused by this load are denoted D_{JQ11} and D_{JQ21}, in which the first numerical subscript identifies the particular displacement being considered and the second subscript denotes the redundant which is causing the displacement. In the same manner, joint displacements D_{JQ12} and D_{JQ22} caused by a unit value of Q_2 are shown in Fig. 2-11e.

The principle of superposition may now be used to obtain the displacements D_J in the actual beam. Superimposing the displacements from the beams in Figs. 2-11c, 2-11d, and 2-11e gives the displacements in the beam of Fig. 2-11a:

$$D_{J1} = D_{JL1} + D_{JQ11}Q_1 + D_{JQ12}Q_2$$
$$D_{J2} = D_{JL2} + D_{JQ21}Q_1 + D_{JQ22}Q_2$$

These equations can be expressed in simpler form by the following matrix equation:

$$\mathbf{D}_{\mathrm{J}} = \mathbf{D}_{\mathrm{JL}} + \mathbf{D}_{\mathrm{JQ}}\mathbf{Q} \tag{2-14}$$

in which the various matrices are:

$$\mathbf{D}_{\mathrm{J}} = \begin{bmatrix} D_{J1} \\ D_{J2} \end{bmatrix} \qquad \mathbf{D}_{\mathrm{JL}} = \begin{bmatrix} D_{JL1} \\ D_{JL2} \end{bmatrix}$$

$$\mathbf{D}_{\mathrm{JQ}} = \begin{bmatrix} D_{JQ11} & D_{JQ12} \\ D_{JQ21} & D_{JQ22} \end{bmatrix} \qquad \mathbf{Q} = \begin{bmatrix} Q_1 \\ Q_2 \end{bmatrix}$$

Of course, in a more general situation the matrices may be of greater order than in this example. If there are j joint displacements to be obtained, the order of the vectors $\mathbf{D}_{\mathrm{J}}$ and $\mathbf{D}_{\mathrm{JL}}$ will be $j \times 1$. If the number of redundants is q, so that the $\mathbf{Q}$ matrix is of order $q \times 1$, then the matrix $\mathbf{D}_{\mathrm{JQ}}$ will be rectangular and of order $j \times q$. Equation (2-14) can be used to calculate the displacements $\mathbf{D}_{\mathrm{J}}$ by matrix operations only, after the matrices $\mathbf{D}_{\mathrm{JL}}$, $\mathbf{D}_{\mathrm{JQ}}$, and $\mathbf{Q}$ have been obtained.

In a manner similar to that used in obtaining Eq. (2-14), the principle of superposition may be used to obtain the member end-actions A_M and the reactions A_R. In these cases the superposition equations are

$$\mathbf{A}_{\mathrm{M}} = \mathbf{A}_{\mathrm{ML}} + \mathbf{A}_{\mathrm{MQ}}\mathbf{Q} \tag{2-15}$$

$$\mathbf{A}_{\mathrm{R}} = \mathbf{A}_{\mathrm{RL}} + \mathbf{A}_{\mathrm{RQ}}\mathbf{Q} \tag{2-16}$$

in which $\mathbf{A}_{\mathrm{M}}$ and $\mathbf{A}_{\mathrm{R}}$ are vectors of member end-actions and reactions in the actual beam (Fig. 2-11a); $\mathbf{A}_{\mathrm{ML}}$ and $\mathbf{A}_{\mathrm{RL}}$ are vectors of member end-actions and reactions in the released structure due to loads; and $\mathbf{A}_{\mathrm{MQ}}$ and $\mathbf{A}_{\mathrm{RQ}}$ are matrices of end-actions and reactions in the released structure due to unit values of the redundants. In the example of Fig. 2-11 the matrices $\mathbf{A}_{\mathrm{M}}$ and $\mathbf{A}_{\mathrm{ML}}$ are of order 4×1 since there are four end-actions being considered; the matrices $\mathbf{A}_{\mathrm{R}}$ and $\mathbf{A}_{\mathrm{RL}}$ are of order 2×1 because there are two reactions being considered; and the matrices $\mathbf{A}_{\mathrm{MQ}}$ and $\mathbf{A}_{\mathrm{RQ}}$ are of order 4×2 and 2×2, respectively. In the general case in which there are m member end-actions, r reactions, and q redundants, the matrices $\mathbf{A}_M$ and $\mathbf{A}_{ML}$ are of order $m \times 1$, $\mathbf{A}_{MQ}$ is of order $m \times q$, $\mathbf{A}_R$ and $\mathbf{A}_{RL}$ are of order $r \times 1$, and $\mathbf{A}_{RQ}$ is of order $r \times q$.

From the above discussion it is seen that the steps to be followed in analyzing a structure by the flexibility method include rather extensive analyses of the released structure. With the loads on the released structure, it is necessary to find the actions and displacements that constitute the matrices $\mathbf{D}_{QL}$, $\mathbf{D}_{JL}$, $\mathbf{A}_{ML}$, and $\mathbf{A}_{RL}$. With the unit values of the redundants acting on the released structure it is necessary to determine the matrices $\mathbf{F}$, $\mathbf{D}_{JQ}$, $\mathbf{A}_{MQ}$, and $\mathbf{A}_{RQ}$. Then Eq. (2-8) is solved first, yielding the vector $\mathbf{Q}$ of redundants, after which Eqs. (2-14) through (2-16) can be evaluated for the vectors $\mathbf{D}_J$, $\mathbf{A}_M$, and $\mathbf{A}_R$. By this means all actions and displacements of interest in the actual structure can be found.

When the effects of temperature changes, prestrain, and support displacement must be taken into account, the only changes in the superposition equations are in the first terms on the right-hand sides of the equal signs. These terms represent the actions and displacements in the released structure and must include the effects of temperature, prestrain, and restraint displacements. This situation was described in the preceding article for the displacements corresponding to the redundants, and Eq. (2-12) was derived as a generalized form of Eq. (2-8). Using the same approach, a more general form for Eq. (2-14) becomes the following:

$$\mathbf{D}_J = \mathbf{D}_{JS} + \mathbf{D}_{JQ}\mathbf{Q} \tag{2-17}$$

in which the vector $\mathbf{D}_{JS}$ represents the sum of all effects in the released structure and is given by the following expression:

$$\mathbf{D}_{JS} = \mathbf{D}_{JL} + \mathbf{D}_{JT} + \mathbf{D}_{JP} + \mathbf{D}_{JR} \tag{2-18}$$

In Eq. (2-18) the matrices $\mathbf{D}_{JT}$, $\mathbf{D}_{JP}$, and $\mathbf{D}_{JR}$ represent joint displacements due to temperature, prestrain, and restraint displacements, respectively. The restraint displacements that are considered in obtaining $\mathbf{D}_{JR}$ are those that do not correspond to redundants. Those that do correspond to redundants are represented in the matrix $\mathbf{D}_Q$.

There is no need to generalize Eqs. (2-15) and (2-16) to account for temperature changes, prestrain, and support displacements. None of these influences will produce any actions or reactions in a statically determinate released structure; instead, the structure will merely change its configuration to accommodate these effects. The effects of these influences are propagated into the matrices $\mathbf{A}_M$ and $\mathbf{A}_R$ through the values of the redundants Q, which are obtained by solving Eq. (2-12).

In summary, the procedure in any particular example is to analyze the released structure for all causes, and then to sum the appropriate matrices as shown in Eqs. (2-11) and (2-18). Then the redundants are found from Eq. (2-13), after which the various actions and displacements are found from Eqs. (2-15), (2-16), and (2-17).

Example. An extended solution for the two-span beam shown in Fig. 2-11 will now be given for the case of loads only. It is assumed that the object of the analysis is to calculate the various joint displacements D_J, member end-actions A_M, and reactions A_R that are shown in Figs. 2-11a and 2-11b. The beam has constant flexural rigidity EI and is acted upon by the loads P_1, M, P_2, and P_3, which are assumed to have the following values:

$$P_1 = 2P \qquad M = PL \qquad P_2 = P \qquad P_3 = P$$

When these loads act upon the released structure (Fig. 2-11c), the joint displacements D_{JL1} and D_{JL2} are found to be

$$D_{JL1} = \frac{5PL^2}{4EI} \qquad D_{JL2} = \frac{13PL^2}{8EI}$$

and, therefore, the vector $\mathbf{D_{JL}}$ is

$$\mathbf{D_{JL}} = \frac{PL^2}{8EI}\begin{bmatrix} 10 \\ 13 \end{bmatrix}$$

The member end-actions in the beam of Fig. 2-11c can be found by static equilibrium. For instance, A_{ML1} and A_{ML2} are the shearing force and bending moment just to the left of the point where the couple M is applied. These quantities are equal to

$$-P_2 + P_3 \quad \text{and} \quad M - \frac{P_2L}{2} + P_3L$$

respectively, or

$$A_{ML1} = 0 \qquad A_{ML2} = \frac{3PL}{2}$$

Similarly, the shearing force and bending moment just to the right of point B in the beam of Fig. 2-11c are

$$A_{ML3} = P_2 - P_3 = 0 \qquad A_{ML4} = \frac{P_2L}{2} - P_3L = -\frac{PL}{2}$$

Thus, the matrix $\mathbf{A_{ML}}$ is

$$\mathbf{A_{ML}} = \begin{bmatrix} 0 \\ \frac{3PL}{2} \\ 0 \\ -\frac{PL}{2} \end{bmatrix}$$

Also, the reactions in the beam of Fig. 2-11c are

$$\mathbf{A_{RL}} = \begin{bmatrix} 2P \\ -\frac{PL}{2} \end{bmatrix}$$

The joint displacements due to unit values of the redundants are shown in Figs. 2-11d and 2-11e. These displacements can be readily calculated for the released structure and are given in the matrix $\mathbf{D}_{\mathrm{JQ}}$:

$$\mathbf{D}_{\mathrm{JQ}} = \frac{L^2}{2EI}\begin{bmatrix} 1 & 3 \\ 1 & 4 \end{bmatrix}$$

The member end-actions and reactions in the beams of Figs. 2-11d and 2-11e are found by statics and constitute the matrices $\mathbf{A}_{\mathrm{MQ}}$ and $\mathbf{A}_{\mathrm{RQ}}$:

$$\mathbf{A}_{\mathrm{MQ}} = \begin{bmatrix} 1 & 1 \\ 0 & L \\ 0 & -1 \\ 0 & -L \end{bmatrix} \qquad \mathbf{A}_{\mathrm{RQ}} = \begin{bmatrix} -1 & -1 \\ -L & -2L \end{bmatrix}$$

The matrices $\mathbf{D}_{\mathrm{QL}}$ and $\mathbf{F}$ which appear in Eq. (2-8) were determined previously for the beam in Fig. 2-11 (see Art. 2.2) and the vector $\mathbf{Q}$ of redundants was found to be

$$\mathbf{Q} = \frac{P}{56}\begin{bmatrix} 69 \\ -64 \end{bmatrix}$$

Using this matrix as well as the matrices $\mathbf{D}_{\mathrm{JL}}$ and $\mathbf{D}_{\mathrm{JQ}}$ in Eq. (2-14) gives the joint displacements in the actual beam:

$$\mathbf{D}_{\mathrm{J}} = \frac{PL^2}{112EI}\begin{bmatrix} 17 \\ -5 \end{bmatrix}$$

This result shows that the rotation D_{J1} at joint B is in the counterclockwise sense and equal to

$$D_{J1} = \frac{17PL^2}{112EI}$$

while the rotation at joint C is

$$D_{J2} = -\frac{5PL^2}{112EI}$$

and is clockwise as indicated by the negative sign.

The member end-actions A_M and reactions A_R are obtained by substituting the appropriate matrices given above into Eqs. (2-15) and (2-16); the results are

$$\mathbf{A}_{\mathrm{M}} = \frac{P}{56}\begin{bmatrix} 5 \\ 20L \\ 64 \\ 36L \end{bmatrix} \qquad \mathbf{A}_{\mathrm{R}} = \frac{P}{56}\begin{bmatrix} 107 \\ 31L \end{bmatrix}$$

The results obtained for $\mathbf{A}_{\mathrm{M}}$ and $\mathbf{A}_{\mathrm{R}}$ can be readily verified by static equilibrium. With the redundants Q_1 and Q_2 known, the equilibrium of the entire beam can be used to verify the reactions at support A; also, the member end-actions can be obtained by considering the equilibrium of the individual members.

2.6 Inverse of the Flexibility Matrix. The analysis of a structure by the flexibility method requires the determination of the flexibility matrix $\mathbf{F}$. This matrix is obtained by calculating displacements in the released structure, after which **F** is inverted and substituted into Eq. (2-13). Then the equation is solved for the vector **Q** of redundants. Since the inverse of the flexibility matrix is a stiffness matrix, it is natural to inquire into the possibility of ascertaining $\mathbf{F}^{-1}$ directly from a consideration of the structure. Such a procedure is quite possible but, as shown later, the inversion of the flexibility matrix is still an essential step in the analysis. In order to understand why this is so, the following discussion dealing with the physical meaning of $\mathbf{F}^{-1}$ is presented.

As discussed in Art. 1.11, every flexibility and stiffness matrix can be associated with a set of corresponding actions and displacements (see Eqs. 1-10 and 1-12). In the case of the matrix **F** used in the flexibility method, the actions are redundants acting on the released structure. The flexibility matrix itself consists of the displacements corresponding to the redundants and due to unit values of the redundants. Therefore, it follows that the inverse of the flexibility matrix consists of the actions corresponding to the redundants and due to unit values of the displacements corresponding to the redundants. Each element of the matrix is an action in the structure due to a unit displacement corresponding to a particular redundant, while all other displacements corresponding to the redundants are zero. This stiffness matrix, equal to $\mathbf{F}^{-1}$, will be denoted by the symbol **K**.

In order to show the physical meaning of the matrix **K**, refer to the two-span beam in Fig. 2-12a. This is the same beam that has been considered in previous examples (see Figs. 2-2 and 2-11), and the redundants are the reactions Q_1 and Q_2 at supports B and C. In Fig. 2-12b the beam is shown with a unit displacement corresponding to Q_1 while the displacement corresponding to Q_2 is equal to zero. The actions corresponding to Q_1 and Q_2 are shown as K_{11} and K_{21} in the figure; these actions are the elements of the first column of the stiffness matrix **K**,

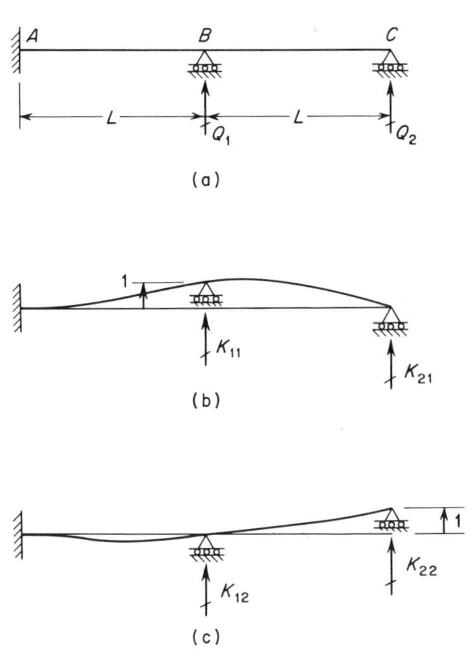

FIG. 2-12. Inverse of the flexibility matrix.

which is the inverse of the flexibility matrix **F** for this beam. Similarly, in Fig. 2-12c the beam is subjected to a unit displacement corresponding to Q_2, and the resulting actions corresponding to Q_1 and Q_2 are denoted K_{12} and K_{22}. If these actions are evaluated for the beams shown in the figure, the results are found to be

$$K_{11} = \frac{96EI}{7L^3} \qquad K_{12} = K_{21} = -\frac{30EI}{7L^3} \qquad K_{22} = \frac{12EI}{7L^3}$$

Therefore, the inverse matrix **K** is

$$\mathbf{K} = \begin{bmatrix} K_{11} & K_{12} \\ K_{21} & K_{22} \end{bmatrix} = \frac{6EI}{7L^3}\begin{bmatrix} 16 & -5 \\ -5 & 2 \end{bmatrix}$$

This matrix is the same as the matrix $\mathbf{F}^{-1}$ found previously in Art. 2.2 for the same beam.

If the four actions K_{11}, K_{12}, K_{21}, and K_{22} (given above) are obtained by the flexibility method using Q_1 and Q_2 as redundants, the calculation of the flexibility matrix **F** and its inverse are required. Therefore, the task of determining the matrix **K** directly from a consideration of the structure (see Figs. 2-12b and 2-12c) involves the very same operations as the usual procedure. In general, the most convenient mode of solution by the flexibility method consists of determining the flexibility matrix from an analysis of the released structure, inverting this matrix, and substituting into Eq. (2-13) to find the redundants.

2.7 Summary of the Flexibility Method. As shown by the examples in the preceding articles, the flexibility method of analysis is quite general and may be applied to any type of framed structure. Of course, the structures chosen for illustrative purposes in the preceding articles are relatively simple in order to keep the calculations to a minimum. However, the method may be applied in theory to structures of any degree of complexity. To facilitate the analysis of more complicated structures, further development of the method with emphasis on making the calculations as systematic as possible will be given in Chapter 3. For reference purposes when performing calculations by hand methods, the essential features of the flexibility method, as well as the notation, are summarized in this article.

The analysis of a structure by the flexibility method may be described by the following steps:

1. Problem Statement. The problem to be solved must be clearly defined by describing the structure and the loads, temperature changes, prestrains, and support displacements to which it is subjected. The description of the structure includes the type of structure, the locations of joints, positions of members, and locations and types of supports. It is also necessary to state the types of deformations to be considered in the analysis, such as flexural deformations, axial deformations, etc. Depending upon the types of deformations to be considered, the appropriate rigidities of the members must be given. For example, if flexural deformations are considered, the flexural rigidity EI must be given for each member; if axial deformations are considered, the axial rigidity EA must be given; and so forth.

2. Selection of Released Structure. The degree of statical indeterminacy of the structure must be determined, and a corresponding number of unknown redundant actions Q must be selected. The released structure is obtained by releasing the selected redundants. The redundants should be carefully chosen in order to obtain a statically determinate released structure that is immobile and easy to analyze. In some cases, the selection of redundants requires a considerable amount of judgment on the part of the analyst, if the analysis is to be made as simple as possible.

3. Analysis of Released Structure Under Loads. Various actions and displacements must be evaluated in the released structure due to the loads. The most important displacements to be determined are the displacements D_{QL} which correspond to the redundants. Other displacements of interest are the displacements D_{JL} at the joints of the structure. The actions to be determined include the end-actions A_{ML} for the members and the reactions A_{RL} at the supports.

4. Analysis of Released Structure for Other Causes. If there are temperature changes, prestrain effects, or support displacements to be included in the analysis, their effects must be evaluated in the released structure. The displacements to be found are those that correspond to the redundants (D_{QT}, D_{QP}, D_{QR}) and also the joint displacements (D_{JT}, D_{JP}, D_{JR}). The actions at the ends of the members and the support reactions are not affected.

5. Analysis of Released Structure for Unit Values of Redundants. Actions and displacements in the released structure due to unit values of the redundants must be determined. The displacements to be calculated are those that correspond to the redundants (the flexibility influence coefficients F) and the displacements of the joints (D_{JQ}). The actions to be evaluated are the member end-actions and reactions (A_{MQ} and A_{RQ}).

6. Determination of Redundants. The superposition equation for the displacements D_Q corresponding to the redundants in the actual structure is Eq. (2-12):

$$\mathbf{D}_{\mathrm{Q}} = \mathbf{D}_{\mathrm{QS}} + \mathbf{FQ} \tag{2-12 repeated}$$

In this equation the vector $\mathbf{D}_{\mathrm{QS}}$ includes the effects of loads, temperature, prestrain, and support displacements (other than those already accounted for in $\mathbf{D}_{\mathrm{Q}}$), as follows:

$$\mathbf{D}_{\mathrm{QS}} = \mathbf{D}_{\mathrm{QL}} + \mathbf{D}_{\mathrm{QT}} + \mathbf{D}_{\mathrm{QP}} + \mathbf{D}_{\mathrm{QR}} \tag{2-11 repeated}$$

When Eq. (2-12) is solved for the redundants, the result is

$$\mathbf{Q} = \mathbf{F}^{-1}(\mathbf{D}_{\mathrm{Q}} - \mathbf{D}_{\mathrm{QS}}) \qquad \text{(2-13) repeated}$$

7. Determination of Other Displacements and Actions. The vectors $\mathbf{D}_{\mathrm{J}}$, $\mathbf{A}_{\mathrm{M}}$, and $\mathbf{A}_{\mathrm{R}}$ for the joint displacements, member end-actions, and reactions, respectively, in the actual structure are obtained from the following superposition equations:

$$\mathbf{D}_{\mathrm{J}} = \mathbf{D}_{\mathrm{JS}} + \mathbf{D}_{\mathrm{JQ}}\mathbf{Q} \qquad \text{(2-17) repeated}$$

$$\mathbf{A}_{\mathrm{M}} = \mathbf{A}_{\mathrm{ML}} + \mathbf{A}_{\mathrm{MQ}}\mathbf{Q} \qquad \text{(2-15) repeated}$$

$$\mathbf{A}_{\mathrm{R}} = \mathbf{A}_{\mathrm{RL}} + \mathbf{A}_{\mathrm{RQ}}\mathbf{Q} \qquad \text{(2-16) repeated}$$

In Eq. (2-17) the vector $\mathbf{D}_{\mathrm{JS}}$ represents displacements in the released structure due to all causes, as follows:

$$\mathbf{D}_{\mathrm{JS}} = \mathbf{D}_{\mathrm{JL}} + \mathbf{D}_{\mathrm{JT}} + \mathbf{D}_{\mathrm{JP}} + \mathbf{D}_{\mathrm{JR}} \qquad \text{(2-18) repeated}$$

When the vectors $\mathbf{D}_{\mathrm{J}}$, $\mathbf{A}_{\mathrm{M}}$, and $\mathbf{A}_{\mathrm{R}}$ have been obtained, the analysis can be considered as completed.

All of the matrices used in the flexibility method are summarized in Table 2-1.

2.8 Stiffness Method. The stiffness method differs from the flexibility method in the physical concepts that are involved, although the methods are similar in their mathematical formulation. In both methods the fundamental equations are derived by using the principle of superposition. In the flexibility method the unknown quantities are redundant actions, but in the stiffness method the unknowns are the joint displacements in the structure. Therefore, in the stiffness method the number of unknowns to be calculated is the same as the degree of kinematic indeterminacy. The stiffness method involves extensive use of actions and displacements in members having fixed ends, and hence reference will be made frequently to the material presented in Appendix B.

In order to illustrate the concepts of the stiffness method in their simplest form, consider the analysis of the beam in Fig. 2-13a. This beam has a fixed support at A, a roller support at B, and is subjected to a uniform load of intensity w. The beam is kinematically indeterminate to the first degree (if axial deformations are neglected), because the only unknown joint displacement is the rotation θ_B at joint B. The first phase of the analysis is to determine this rotation. After this rotation is found, the various actions and displacements throughout the beam can be determined, as will be shown later.

TABLE 2-1

Matrices Used in the Flexibility Method

Matrix	*Order*	*Definition*
$\mathbf{Q}$	$q \times 1$	Unknown redundant actions (q = number of redundants)
$\mathbf{D}_Q$	$q \times 1$	Displacements in the actual structure corresponding to the redundants
$\mathbf{D}_{QL}$	$q \times 1$	Displacements in the released structure corresponding to the redundants and due to loads
$\mathbf{F}$ (or $\mathbf{D}_{QQ}$)	$q \times q$	Displacements in the released structure corresponding to the redundants and due to unit values of the redundants (flexibility coefficients)
$\mathbf{D}_{QT}$, $\mathbf{D}_{QP}$, $\mathbf{D}_{QR}$	$q \times 1$	Displacements in the released structure corresponding to the redundants and due to temperature, prestrain, and restraint displacements (other than those in $\mathbf{D}_Q$)
$\mathbf{D}_{QS}$	$q \times 1$	$\mathbf{D}_{QS} = \mathbf{D}_{QL} + \mathbf{D}_{QT} + \mathbf{D}_{QP} + \mathbf{D}_{QR}$
$\mathbf{D}_J$	$j \times 1$	Joint displacements in the actual structure (j = number of joint displacements)
$\mathbf{D}_{JL}$	$j \times 1$	Joint displacements in the released structure due to loads
$\mathbf{D}_{JQ}$	$j \times q$	Joint displacements in the released structure due to unit values of the redundants
$\mathbf{D}_{JT}$, $\mathbf{D}_{JP}$, $\mathbf{D}_{JR}$	$j \times 1$	Joint displacements in the released structure due to temperature, prestrain, and restraint displacements (other than those in $\mathbf{D}_Q$)
$\mathbf{D}_{JS}$	$j \times 1$	$\mathbf{D}_{JS} = \mathbf{D}_{JL} + \mathbf{D}_{JT} + \mathbf{D}_{JP} + \mathbf{D}_{JR}$
$\mathbf{A}_M$	$m \times 1$	Member end-actions in the actual structure (m = number of end-actions)
$\mathbf{A}_{ML}$	$m \times 1$	Member end-actions in the released structure due to loads
$\mathbf{A}_{MQ}$	$m \times q$	Member end-actions in the released structure due to unit values of the redundants
$\mathbf{A}_R$	$r \times 1$	Reactions in the actual structure (r = number of reactions)
$\mathbf{A}_{RL}$	$r \times 1$	Reactions in the released structure due to loads
$\mathbf{A}_{RQ}$	$r \times q$	Reactions in the released structure due to unit values of the redundants

In the flexibility method a statically determinate released structure is obtained by altering the actual structure in such a manner that the selected redundant actions are zero. The analogous operation in the stiffness method is to obtain a kinematically determinate structure by altering the actual structure in such a manner that all unknown displacements are zero. Since the unknown displacements are the translations and rotations of the joints, they can be made equal to zero by restraining the joints of the structure against displacements of any kind. The structure obtained by restraining all joints of the actual structure is called the *restrained structure*. For the beam in Fig. 2-13a the restrained structure is obtained by restraining joint B against rotation. Thus, the restrained structure is the fixed-end beam shown in Fig. 2-13b.

When the loads act on the restrained beam (see Fig. 2-13b), there will be a couple M_B developed at support B. This reactive couple is in the clockwise direction and is given by the expression

$$M_B = \frac{wL^2}{12} \tag{2-19}$$

which can be found from the table of fixed-end moments given in Appendix B (see Table B-1). Note that the couple M_B is an action corresponding to the rotation Θ_B, which is the unknown quantity in the analysis. Because there is no couple at joint B in the actual beam of Fig. 2-13a, it is necessary to consider next that the restrained beam is subjected to a couple equal and opposite to the couple M_B. Such a couple is shown acting on the beam in Fig. 2-13c. When the actions acting on the two beams in (b) and (c) are superimposed, they produce the actions on the actual beam. Thus, the analysis of the beam in Fig. 2-13a can be considered as the superposition of the analyses shown in Figs. 2-13b and 2-13c. It follows, therefore, that the rotation produced by the couple M_B in Fig. 2-13c is equal to Θ_B, the unknown rotation in the actual beam.

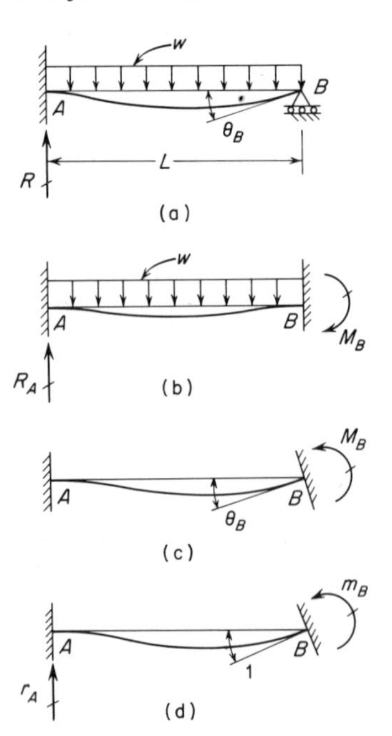

FIG. 2-13. Illustration of stiffness method.

The relation between the moment M_B and the rotation Θ_B in the beam of Fig. 2-13c is

$$M_B = \frac{4EI}{L}\Theta_B \tag{2-20}$$

in which EI is the flexural rigidity of the beam. Equation (2-20) is obtained from Case 3 of Table B-4. Equating the two expressions for the moment M_B from Eqs. (2-19) and (2-20) gives the equation

$$\frac{wL^2}{12} = \frac{4EI}{L}\Theta_B$$

from which

$$\Theta_B = \frac{wL^3}{48EI}$$

Thus, the rotation at joint B of the beam has been determined.

In a manner analogous to that used in the flexibility method, it is convenient in the above example to consider the restrained structure under the effect of a unit value of the unknown rotation. It is also more systematic to formulate the equation for the rotation as an equation of superposition and to use a consistent sign convention for all terms in the equation. This procedure will now be followed for the beam in Fig. 2-13.

The effect of a unit value of the unknown rotation is shown in Fig. 2-13d, where the restrained beam is acted upon by a couple m_B that produces a unit value of the rotation Θ_B at the right-hand end. Since the moment m_B is an action corresponding to the rotation Θ_B and caused by a unit value of that rotation (while all other joint displacements are zero), it is recognized that m_B is a *stiffness coefficient* for the restrained structure (see Art. 1.10). The value of the couple m_B (see Eq. 2-20) is

$$m_B = \frac{4EI}{L}$$

In formulating the equation of superposition the couples at joint B will be superimposed as follows. The couple in the restrained beam subjected to the load (Fig. 2-13b) will be added to the couple m_B (corresponding to a unit value of Θ_B) multiplied by Θ_B itself. The sum of these two terms must give the couple at joint B in the actual beam, which is zero in this example. All terms in the superposition equation will be expressed in the same sign convention, namely, that couples and rotations at joint B are positive when counterclockwise. According to this convention, the couple M_B in the beam of Fig. 2-13b is negative:

$$M_B = -\frac{wL^2}{12}$$

The equation for the superposition of moments at support B now becomes

$$M_B + m_B\Theta_B = 0 \tag{2-21}$$

or

$$-\frac{wL^2}{12} + \frac{4EI}{L}\Theta_B = 0$$

Solving this equation yields

$$\Theta_B = \frac{wL^3}{48EI}$$

which is the same result as before. The positive sign for the result means that the rotation is counterclockwise.

The most essential part of the preceding solution consists of writing the superposition equation (2-21), which expresses the fact that the moment at B in the actual beam is zero. Included in this equation are the moment caused by the loads on the restrained structure and the moment caused by rotating the end B of the restrained structure. The latter term in the equation was expressed conveniently as the product of the moment caused by a unit value of the unknown displacement (stiffness coefficient) times the unknown displacement itself. The two effects are summed algebraically, using the same sign convention for all terms in the equation. When the equation is solved for the unknown displacement, the sign of the result will give the true direction of the displacement. The equation may be referred to either as an *equation of superposition* or as an *equation of joint equilibrium.* The latter name is used because the equation may be considered to express the equilibrium of moments at joint B.

Having obtained the unknown rotation Θ_B for the beam, it is now possible to calculate other quantities, such as member end-actions and reactions. As an example, assume that the reactive force R acting at support A of the beam (Fig. 2-13a) is to be found. This force is the sum of the corresponding reactive force R_A at support A in Fig. 2-13b and Θ_B times the force r_A in Fig. 2-13d, as shown in the following superposition equation:

$$R = R_A + \Theta_B r_A$$

The forces R_A and r_A can be readily calculated for the restrained beam (see Case 6, Table B-1, and Case 3, Table B-4):

$$R_A = \frac{wL}{2} \qquad r_A = \frac{6EI}{L^2}$$

When these values, as well as the previously found value for Θ_B, are substituted into the equation above, the result is

$$R = \frac{5wL}{8}$$

The same concepts can be used to calculate any other actions or displacements for the beam. However, in all cases the unknown joint displacements must be found first.

If a structure is kinematically indeterminate to more than one degree, a more organized approach to the solution, as well as a generalized notation, must be introduced. For this purpose, the same two-span beam used previously as an example in the flexibility method will be analyzed now by the stiffness method (see Fig. 2-14a). The beam has constant flexural rigidity EI, and is subjected to the loads P_1, M, P_2, and P_3.

Since rotations can occur at joints B and C, the structure is kinematically indeterminate to the second degree when axial deformations are neglected. Let the unknown rotations at these joints be D_1 and D_2, respectively, and assume that counterclockwise rotations are positive. These unknown displacements may be determined by solving equations of superposition for the actions at joints B and C, as described in the following discussion.

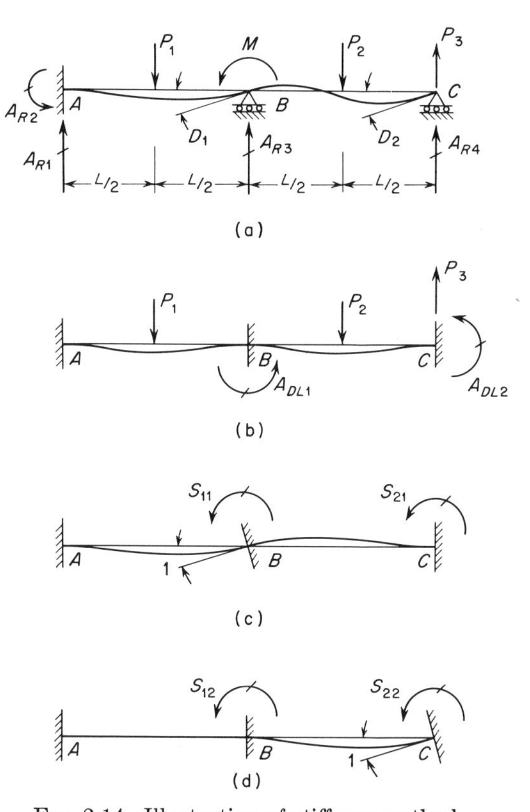

FIG. 2-14. Illustration of stiffness method.

The first step in the analysis consists of applying imaginary restraints at the joints to prevent all joint displacements. The restrained structure which is obtained by this means is shown in Fig. 2-14b and consists of two fixed-end beams. The restrained structure is assumed to be acted upon by all of the loads except those that correspond to the unknown displacements. Thus, only the loads P_1, P_2, and P_3 are shown in Fig. 2-14b. All loads that correspond to the unknown joint displacements, such as the couple M in this example, are taken into account later. The moments A_{DL1} and A_{DL2} (Fig. 2-14b) are the actions of the restraints (against the restrained structure) corresponding to D_1 and D_2, respectively, and caused by loads acting on the structure. For example, the restraint action A_{DL1} is the sum of the reactive moment at B due to the load P_1 acting on member AB and the reactive moment at B due to the load P_2 acting on member BC. These actions can be found with the aid of formulas for fixed-end moments in beams (see Appendix B), as illustrated later.

In order to generate the stiffness coefficients at joints B and C, unit values of the unknown displacements D_1 and D_2 are induced separately in the restrained structure. A unit displacement corresponding to D_1 consists of a unit rotation of joint B, as shown in Fig. 2-14c. The displacement D_2 remains equal to zero in this beam. Thus, the actions corresponding to D_1 and D_2 are the stiffness coefficients S_{11} and S_{21}, respectively. These stiffnesses consist of the couples exerted by the restraints on the beam at joints B and C, respectively. The calculation of these actions is not difficult when formulas for fixed-end moments in beams are available. Their determination in this example will be described later. The condition that D_2 is equal to unity while D_1 is equal to zero is shown in Fig. 2-14d. In the figure the stiffness S_{12} is the action corresponding to D_1 while the stiffness S_{22} is the action corresponding to D_2. Note that in each case the stiffness coefficient is the action that the artificial restraint exerts upon the structure.

Two superposition equations expressing the conditions pertaining to the moments acting on the original structure (Fig. 2-14a) at joints B and C may now be written. Let the actions in the actual structure corresponding to D_1 and D_2 be denoted A_{D1} and A_{D2}, respectively. These actions will be zero in all cases except when a concentrated external action is applied at a joint corresponding to a degree of freedom. In the example of Fig. 2-14, the action A_{D1} is equal to the couple M while the action A_{D2} is zero. The superposition equations express the fact that the actions in the original structure (Fig. 2-14a) are equal to the corresponding actions in the restrained structure due to the loads (Fig. 2-14b) plus the corresponding actions in the restrained structure under the unit displacements (Figs. 2-14c and 2-14d) multiplied by the displacements themselves. Therefore, the superposition equations are

$$\begin{aligned} A_{D1} &= A_{DL1} + S_{11}D_1 + S_{12}D_2 \\ A_{D2} &= A_{DL2} + S_{21}D_1 + S_{22}D_2 \end{aligned} \tag{2-22}$$

The sign convention used throughout these equations is that moments are positive when in the same sense (counterclockwise) as the corresponding unknown displacements.

When Eqs. (2-22) are expressed in matrix form they become

$$\mathbf{A}_{\mathrm{D}} = \mathbf{A}_{\mathrm{DL}} + \mathbf{SD} \tag{2-23}$$

in which the vector $\mathbf{A}_{\mathrm{D}}$ represents the actions in the original beam corresponding to the unknown joint displacements $\mathbf{D}$, the vector $\mathbf{A}_{DL}$ represents actions in the restrained structure corresponding to the unknown joint displacements and caused by the loads (that is, all loads except those corresponding to the unknown displacements), and $\mathbf{S}$ is the stiffness matrix corresponding to the unknown displacements. The stiffness matrix $\mathbf{S}$ could also be denoted $\mathbf{A}_{DD}$, since it represents actions corresponding to the unknown joint displacements and caused by unit values of those displacements. For the example of Fig. 2-14 the matrices are as follows:

$$\mathbf{A}_D = \begin{bmatrix} A_{D1} \\ A_{D2} \end{bmatrix} \qquad \mathbf{A}_{DL} = \begin{bmatrix} A_{DL1} \\ A_{DL2} \end{bmatrix} \qquad \mathbf{S} = \begin{bmatrix} S_{11} & S_{12} \\ S_{21} & S_{22} \end{bmatrix} \qquad \mathbf{D} = \begin{bmatrix} D_1 \\ D_2 \end{bmatrix}$$

In general, these matrices will have as many rows as there are unknown joint displacements. Thus, if d represents the number of unknown displacements, the order of the stiffness matrix $\mathbf{S}$ is $d \times d$, while $\mathbf{A}_D$, $\mathbf{A}_{DL}$, and $\mathbf{D}$ are vectors of order $d \times 1$.

Subtracting $\mathbf{A}_{DL}$ from both sides of Eq. (2-23) and then premultiplying by $\mathbf{S}^{-1}$ gives the following equation for the unknown displacements:

$$\mathbf{D} = \mathbf{S}^{-1}(\mathbf{A}_D - \mathbf{A}_{DL}) \tag{2-24}$$

This equation represents the solution for the displacements in matrix terms because the elements of $\mathbf{A}_D$, $\mathbf{A}_{DL}$, and $\mathbf{S}$ are either known or may be obtained from the restrained structure. Moreover, the member end-actions and reactions for the structure may be found after the joint displacements are known. The procedure for performing such calculations will be illustrated later.

In order to demonstrate the use of Eq. (2-24), the beam in Fig. 2-14a will be analyzed for the values of the loads previously given:

$$P_1 = 2P \qquad M = PL \qquad P_2 = P \qquad P_3 = P$$

When the loads P_1, P_2, and P_3 act upon the restrained structure (Fig. 2-14b) the actions A_{DL1} and A_{DL2}, corresponding to D_1 and D_2, respectively, are developed at the supports B and C. Since the couple M corresponds to one of the unknown displacements, it is taken into account later by means of the matrix $\mathbf{A}_D$. The actions A_{DL1} and A_{DL2} are found from the formulas for fixed-end moments (see Case 1, Table B-1):

$$A_{DL1} = -\frac{P_1 L}{8} + \frac{P_2 L}{8} = -\frac{PL}{8}$$

$$A_{DL2} = -\frac{P_2 L}{8} = -\frac{PL}{8}$$

Therefore, the matrix $\mathbf{A}_{DL}$ is

$$\mathbf{A}_{DL} = \frac{PL}{8}\begin{bmatrix} -1 \\ -1 \end{bmatrix}$$

It may be observed from these calculations that the load P_3 does not enter into the matrix $\mathbf{A}_{DL}$, hence, it does not affect the calculations for the joint displacements. However, this load does affect the calculations for the reactions of the actual beam, which are given later.

The stiffness matrix **S** consists of the stiffness coefficients shown in Figs. 2-14c and 2-14d. Each coefficient is a couple corresponding to one of the unknown displacements and due to a unit value of one of the displacements. Elements of the first column of the stiffness matrix are shown in Fig. 2-14c, and elements of the second column in Fig. 2-14d. In order to find these coefficients, consider first the fixed-end beam shown in Fig. 2-15. This beam is subjected to a unit rotation at end B, and, as a result, the moment developed at end B is $4EI/L$ while the moment at the opposite end is $2EI/L$ (see Case 3, Table B-4). The reactive forces at the ends of the beam are each equal to $6EI/L^2$, and are also indicated in the figure. All of the actions shown in Fig. 2-15 are called *member stiffnesses*, because they are actions at the ends of the member due to a unit displacement of one end. The end of the beam which undergoes the unit displacement is sometimes called the *near end* of the beam, and the opposite end is called the *far end*. Thus, the member stiffnesses at ends A and B are sometimes referred to as the stiffnesses at the far and near ends of the beam. The subject of member stiffnesses is dealt with extensively in Chapter 4 in conjunction with a more detailed discussion of the stiffness method.

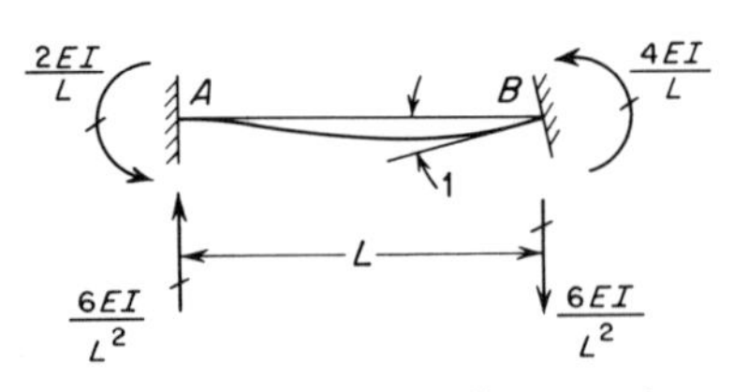

FIG. 2-15. Member stiffnesses for a beam member.

The task of calculating the joint stiffnesses S_{11} and S_{21} in Fig. 2-14c may now be performed through the use of member stiffnesses. When the beam is rotated through a unit angle at joint B, a moment equal to $4EI/L$ is developed at B because of the rotation of the end of member AB. Also, a moment equal to $4EI/L$ is developed at B because of the rotation of the end of member BC. Thus, the total moment at B, equal to S_{11}, is

$$S_{11} = \frac{4EI}{L} + \frac{4EI}{L} = \frac{8EI}{L} \tag{a}$$

The stiffness S_{21} is the moment developed at joint C when joint B is rotated through a unit angle. Since joint C is at the far end of the member, the stiffness coefficient is

$$S_{21} = \frac{2EI}{L}$$

Both S_{11} and S_{21} are positive because they act in the counterclockwise sense. The stiffness coefficients S_{12} and S_{22} are shown in Fig. 2-14d. The first of these is equal to $2EI/L$, since it is an action at the far end of the member BC, while the latter is equal to $4EI/L$ since it is at the near end of the member.

The stiffness matrix $\mathbf{S}$ can be formed from the stiffness coefficients described above:

$$\mathbf{S} = \frac{EI}{L}\begin{bmatrix} 8 & 2 \\ 2 & 4 \end{bmatrix}$$

Each of the elements in $\mathbf{S}$ is a *joint stiffness*, inasmuch as it represents the action at one of the joints of the structure due to a unit value of a displacement at one of the joints. In this example, the joint stiffness S_{11} (see Fig. 2-14c) is the sum of the near-end member stiffnesses (see Eq. a) for the two members meeting at the joint. Similarly, the stiffness S_{22} is a near-end member stiffness. On the other hand, the stiffnesses S_{12} and S_{21} consist of far-end member stiffnesses for members which connect to a joint that is rotated. In a more general example, it will be found that stiffness elements on the principal diagonal are always composed of near-end stiffnesses while those off the diagonal may consist of either far-end or near-end stiffnesses, as will be seen in later examples. After the stiffness matrix $\mathbf{S}$ is determined, its inverse can be found:

$$\mathbf{S}^{-1} = \frac{L}{14EI}\begin{bmatrix} 2 & -1 \\ -1 & 4 \end{bmatrix}$$

The next matrix to be determined is the matrix $\mathbf{A}_\mathrm{D}$, representing the actions in the actual structure corresponding to the unknown displacements. In this example the external load which corresponds to the rotation D_1 is the couple M (equal to PL) at joint B. There is no moment at joint C corresponding to D_2, and therefore the matrix $\mathbf{A}_\mathrm{D}$ is

$$\mathbf{A}_\mathrm{D} = \begin{bmatrix} PL \\ 0 \end{bmatrix}$$

Now that the matrices $\mathbf{A}_\mathrm{D}$, $\mathbf{S}^{-1}$, and $\mathbf{A}_\mathrm{DL}$ have been obtained, the matrix of displacements $\mathbf{D}$ in the actual structure can be found by substituting into Eq. (2-24) and solving:

$$\mathbf{D} = \frac{L}{14EI}\begin{bmatrix} 2 & -1 \\ -1 & 4 \end{bmatrix}\left\{\begin{bmatrix} PL \\ 0 \end{bmatrix} - \frac{PL}{8}\begin{bmatrix} -1 \\ -1 \end{bmatrix}\right\} = \frac{PL^2}{112EI}\begin{bmatrix} 17 \\ -5 \end{bmatrix}$$

Thus, the rotations D_1 and D_2 at joints B and C are, respectively,

$$D_1 = \frac{17PL^2}{112EI} \qquad D_2 = -\frac{5PL^2}{112EI} \qquad \text{(b)}$$

These results agree with the joint displacements found by the flexibility method in the example of Art. 2.5.

The next step after finding the joint displacements is to determine the member end-actions and the reactions for the structure. As in the flexibility method, there are two approaches that can be followed when performing the calculations by hand. One approach is to obtain the end-actions and reactions by making separate calculations after the joint displacements have been found. The other approach is to perform the calculations in a systematic manner simultaneously with the calculations for finding the displacements. In using the first of the two methods, the end-actions are obtained by considering consecutively each individual member of the structure. For each member the effects of the loads on the restrained member and the effects of the displacements at the ends of the member are determined. Then these results are combined to give the end-actions in the original beam. (This method is quite suitable for computer programming, and is discussed in Chapter 4.) After all such end-actions for a structure are determined, the reactions can be found by static equilibrium.

The second approach to finding member end-actions and reactions is suitable, however, for hand calculations because it is systematic and can be easily generalized. In order to show how the calculations are performed, consider again the two-span beam shown in Fig. 2-14. As in the flexibility method, the matrices of member end-actions and reactions in the actual structure (Fig. 2-14a) will be denoted $\mathbf{A}_{\mathrm{M}}$ and $\mathbf{A}_{\mathrm{R}}$, respectively. In the restrained structure subjected to the loads (Fig. 2-14b), the matrices of end-actions and reactions corresponding to $\mathbf{A}_{\mathrm{M}}$ and $\mathbf{A}_{\mathrm{R}}$ will be denoted $\mathbf{A}_{\mathrm{ML}}$ and $\mathbf{A}_{\mathrm{RL}}$, respectively. It should be noted again that when any reference is made to the loads on the restrained structure, it is assumed that all of the actual loads are taken into account except those that correspond to an unknown displacement. Thus, the joint load M shown in Fig. 2-14a does not appear on the restrained structure in Fig. 2-14b. However, all other loads are considered to act on the restrained beam in Fig. 2-14b, including the load P_3. This load does not affect the end-actions A_{ML} in the restrained structure, but the reactions A_{RL} are affected. Each of the matrices $\mathbf{A}_{\mathrm{M}}$ and $\mathbf{A}_{\mathrm{ML}}$ is of order $m \times 1$, assuming that m represents the number of member end-actions; similarly, the matrices $\mathbf{A}_{\mathrm{R}}$ and $\mathbf{A}_{\mathrm{RL}}$ are of order $r \times 1$, in which r denotes the number of reactions.

In the restrained structure subjected to unit displacements (Figs. 2-14c and 2-14d), the matrices of end-actions and reactions will be denoted $\mathbf{A}_{\mathrm{MD}}$ and $\mathbf{A}_{\mathrm{RD}}$, respectively. The first column of each of the matrices will contain the actions obtained from the restrained beam in Fig. 2-14c, while the second column is made up of actions obtained from the beam in Fig. 2-14d. In the general case the matrices $\mathbf{A}_{\mathrm{MD}}$ and $\mathbf{A}_{\mathrm{RD}}$ are of order $m \times d$ and $r \times d$, respectively, in which d represents the number of unknown displacements.

The superposition equations for the end-actions and reactions in the actual structure may now be expressed in matrix form:

$$\mathbf{A}_{\mathrm{M}} = \mathbf{A}_{\mathrm{ML}} + \mathbf{A}_{\mathrm{MD}}\mathbf{D} \qquad (2\text{-}25)$$

$$\mathbf{A}_{\mathrm{R}} = \mathbf{A}_{\mathrm{RL}} + \mathbf{A}_{\mathrm{RD}}\mathbf{D} \qquad (2\text{-}26)$$

The above two equations and Eq. (2-23) together constitute the three superposition equations of the stiffness method. The complete solution of a structure consists of solving for the matrix $\mathbf{D}$ of displacements from Eq. (2-24) and then substituting into Eqs. (2-25) and (2-26) to determine $\mathbf{A}_{\mathrm{M}}$ and $\mathbf{A}_{\mathrm{R}}$. When this is done, all joint displacements, member end-actions, and reactions for the structure will be known.

Consider now the use of Eqs. (2-25) and (2-26) in the solution of the two-span beam shown in Fig. 2-14. The unknown displacements D have already been found (see Eqs. b) and all that remains is the determination of the matrices $\mathbf{A}_{\mathrm{ML}}$, $\mathbf{A}_{\mathrm{RL}}$, $\mathbf{A}_{\mathrm{MD}}$, and $\mathbf{A}_{\mathrm{RD}}$. Assume that the member end-actions to be calculated are the shearing force A_{M1} and moment A_{M2} at end B of member AB, and the shearing force A_{M3} and moment A_{M4} at end B of member BC. These are the same end-actions considered previously in the solution by the flexibility method (see Fig. 2-11b), and are selected solely for illustrative purposes. Also, assume that the reactions to be calculated are the force A_{R1} and couple A_{R2} at support A, and the forces A_{R3} and A_{R4} at supports B and C (see Fig. 2-14a). The first two of these reactions are the same as in the earlier solution, and the last two are the redundants from the earlier solution. All of these actions are assumed positive either when upward or counterclockwise.

In the restrained structure subjected to the loads (Fig. 2-14b), the end-moments and reactions in terms of the loads P_1, P_2, and P_3 are seen to be as follows:

$$A_{ML1} = \frac{P_1}{2} \qquad A_{ML2} = -\frac{P_1L}{8} \qquad A_{ML3} = \frac{P_2}{2} \qquad A_{ML4} = \frac{P_2L}{8}$$

$$A_{RL1} = \frac{P_1}{2} \qquad A_{RL2} = \frac{P_1L}{8} \qquad A_{RL3} = \frac{P_1}{2} + \frac{P_2}{2} \qquad A_{RL4} = \frac{P_2}{2} - P_3$$

The values of the loads ($P_1 = 2P$, $P_2 = P$, $P_3 = P$) can now be substituted into these expressions, after which the matrices $\mathbf{A}_{\mathrm{ML}}$ and $\mathbf{A}_{\mathrm{RL}}$ can be formed:

$$\mathbf{A}_{\mathrm{ML}} = \frac{P}{8}\begin{bmatrix} 8 \\ -2L \\ 4 \\ L \end{bmatrix} \qquad \mathbf{A}_{\mathrm{RL}} = \frac{P}{4}\begin{bmatrix} 4 \\ L \\ 6 \\ -2 \end{bmatrix}$$

The matrices $\mathbf{A}_{\mathrm{MD}}$ and $\mathbf{A}_{\mathrm{RD}}$ are obtained from an analysis of the beams shown in Figs. 2-14c and 2-14d. For example, the member end-action A_{MD11} is the shearing force at end B of member AB due to a unit displacement corresponding to D_1 (Fig. 2-14c). Thus, this end-action is

$$A_{MD11} = -\frac{6EI}{L^2}$$

as can be seen from Fig. 2-15. The reaction A_{RD11} is the vertical force at support A in the beam of Fig. 2-14c, and is

$$A_{RD11} = \frac{6EI}{L^2}$$

In a similar manner, the other member end-actions and reactions can be found for the beam shown in Fig. 2-14c. These quantities constitute the first columns of the matrices $\mathbf{A}_{\mathrm{MD}}$ and $\mathbf{A}_{\mathrm{RD}}$. The terms in the second columns are found by similar analyses that are made for the beam shown in Fig. 2-14d. The results are as follows:

$$\mathbf{A}_{\mathrm{MD}} = \frac{EI}{L^2}\begin{bmatrix} -6 & 0 \\ 4L & 0 \\ 6 & 6 \\ 4L & 2L \end{bmatrix} \qquad \mathbf{A}_{\mathrm{RD}} = \frac{EI}{L^2}\begin{bmatrix} 6 & 0 \\ 2L & 0 \\ 0 & 6 \\ -6 & -6 \end{bmatrix}$$

Substituting the matrices $\mathbf{A}_{\mathrm{ML}}$ and $\mathbf{A}_{\mathrm{MD}}$ given above, as well as the matrix $\mathbf{D}$ obtained earlier, into Eq. (2-25) gives the following:

$$\mathbf{A}_{\mathrm{M}} = \frac{P}{8}\begin{bmatrix} 8 \\ -2L \\ 4 \\ L \end{bmatrix} + \frac{EI}{L^2}\begin{bmatrix} -6 & 0 \\ 4L & 0 \\ 6 & 6 \\ 4L & 2L \end{bmatrix}\frac{PL^2}{112EI}\begin{bmatrix} 17 \\ -5 \end{bmatrix} = \frac{P}{56}\begin{bmatrix} 5 \\ 20L \\ 64 \\ 36L \end{bmatrix}$$

This result agrees with that found previously by the flexibility method (see Art. 2.5). By substituting the matrices $\mathbf{A}_{\mathrm{RL}}$, $\mathbf{A}_{\mathrm{RD}}$, and $\mathbf{D}$ into Eq. (2-26) the reactions can be found also:

$$\mathbf{A}_{\mathrm{R}} = \frac{P}{56}\begin{bmatrix} 107 \\ 31L \\ 69 \\ -64 \end{bmatrix}$$

These results also agree with those found previously by the flexibility method.

The method of solution described above for the two-span beam in Fig. 2-14 is quite general in its basic concepts, and the matrix equations (2-23) to (2-26) may be used in the solution of any type of framed structure. Also, the equations apply to structures having any number of degrees of kinematic indeterminacy. Several examples illustrating the stiffness method are given in the following article.

2.9 Examples. The examples given in this article illustrate the application of the stiffness method to several types of structures. In each example the object of the calculations is to determine the unknown joint displacements and certain selected member end-actions and reactions.

Since the number of degrees of kinematic indeterminacy is small, the problems are suitable for solution by hand. Most of the examples are solved in literal form since this illustrates more clearly how the various terms in the matrices are obtained.

Example 1. The three-span continuous beam shown in Fig. 2-16a has fixed supports at A and D and roller supports at B and C; the length of the middle span is equal to 1.5 times the length of the end spans. The loads on the beam are assumed to be two concentrated forces acting at the positions shown, a uniform load of intensity w acting on spans BC and CD, and a couple M applied at joint C. All members of the beam are assumed to have the same flexural rigidity EI.

The unknown joint displacements for the beam are the rotations at supports B and C, denoted D_1 and D_2, respectively, as shown in Fig. 2-16b. For illustrative purposes in this example, it will be assumed that the only member end-actions to be determined are the shearing force A_{M1} and the moment A_{M2} at the left-hand end of member AB, and the shearing force A_{M3} and moment A_{M4} at the left-hand end of member BC, as shown in Fig. 2-16b. The reactions to be found in this example are the vertical forces A_{R1} and A_{R2} at supports B and C, respectively. Other reactions could also be obtained if desired. All of the end-actions, reactions, and joint displacements are assumed to be positive when upward or counterclockwise.

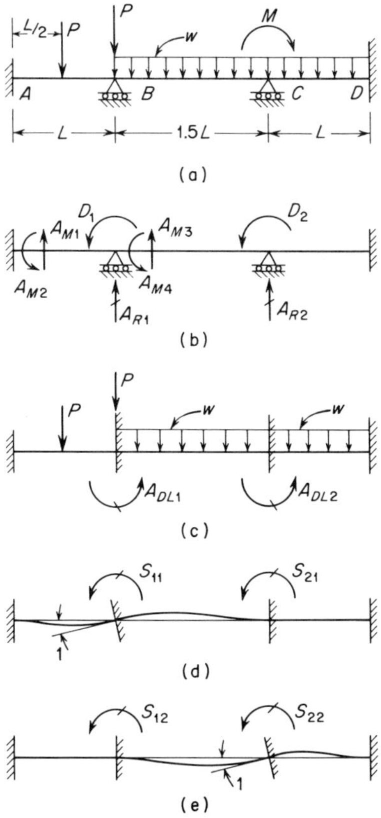

The only load on the structure that corresponds to one of the unknown joint displacements is the couple M, which corresponds to the rotation D_2 (except that it is in the opposite sense). Therefore, the vector $\mathbf{A}_D$ of actions corresponding to the unknown displacements is

FIG. 2-16. Example 1: Continuous beam.

$$\mathbf{A}_D = \begin{bmatrix} 0 \\ -M \end{bmatrix}$$

The remaining loads on the beam are taken into account by considering them to act on the restrained structure shown in Fig. 2-16c. This structure consists of fixed-end beams and is obtained by preventing joints B and C from rotating. The actions A_{DL1} and A_{DL2} exerted on the beam by the artificial restraints are the couples at supports B and C. Each of the couples is an action corresponding to a displacement D and caused by loads on the beam. The couples can be evaluated without difficulty by referring to the formulas for fixed-end moments given in Table B-1 of the Appendix (see Cases 1 and 6):

$$A_{DL1} = -\frac{PL}{8} + \frac{w(1.5L)^2}{12} = -\frac{PL}{8} + \frac{3wL^2}{16}$$

$$A_{DL2} = -\frac{w(1.5L)^2}{12} + \frac{wL^2}{12} = -\frac{5wL^2}{48}$$

Thus, the vector $\mathbf{A}_{\mathrm{DL}}$ becomes

$$\mathbf{A}_{\mathrm{DL}} = \frac{L}{48}\begin{bmatrix} -6P + 9wL \\ -5wL \end{bmatrix}$$

The end-actions A_{ML} and the reactions A_{RL} for the restrained beam of Fig. 2-16c can be determined also by referring to the table of fixed-end actions. For example, the end-action A_{ML1} is the shearing force at the left-hand end of member AB and is equal to $P/2$; similarly, the reaction A_{RL1} is the vertical reaction at support B, obtained as follows:

$$A_{RL1} = \frac{P}{2} + P + \frac{w(1.5L)}{2} = \frac{3P}{2} + \frac{3wL}{4}$$

By continuing in the same manner, all of the required actions in the restrained structure can be found. The resulting vectors $\mathbf{A}_{\mathrm{ML}}$ and $\mathbf{A}_{\mathrm{RL}}$ are:

$$\mathbf{A}_{\mathrm{ML}} = \begin{bmatrix} \dfrac{P}{2} \\ \dfrac{PL}{8} \\ \dfrac{3wL}{4} \\ \dfrac{3wL^2}{16} \end{bmatrix} \qquad \mathbf{A}_{\mathrm{RL}} = \begin{bmatrix} \dfrac{3P}{2} + \dfrac{3wL}{4} \\ \dfrac{5wL}{4} \end{bmatrix}$$

Note that the load P acting downward at joint B affects the reactions in the restrained beam but not the member end-actions.

In order to simplify the subsequent calculations, assume now that the following relationships exist between the various loads on the beam:

$$wL = P \qquad M = PL$$

Substituting these relations into the matrices given in the preceding paragraphs yields the following matrices:

$$\mathbf{A}_{\mathrm{D}} = PL\begin{bmatrix} 0 \\ -1 \end{bmatrix} \qquad \mathbf{A}_{\mathrm{DL}} = \frac{PL}{48}\begin{bmatrix} 3 \\ -5 \end{bmatrix} \qquad \mathbf{A}_{\mathrm{ML}} = \frac{P}{16}\begin{bmatrix} 8 \\ 2L \\ 12 \\ 3L \end{bmatrix} \qquad \mathbf{A}_{\mathrm{RL}} = \frac{P}{4}\begin{bmatrix} 9 \\ 5 \end{bmatrix}$$

The next step in the solution is the analysis of the restrained beam for the effects of unit displacements corresponding to the unknowns. The two conditions to be taken into account are unit rotations at joints B and C, as illustrated in

Figs. 2-16d and 2-16e. The four couples acting at joints B and C in these figures represent the elements of the stiffness matrix **S**. Using the formulas given in Fig. 2-15, each of these stiffnesses can be found without difficulty, as shown in the following calculations:

$$S_{11} = S_{22} = \frac{4EI}{L} + \frac{4EI}{1.5L} = \frac{20EI}{3L}$$

$$S_{12} = S_{21} = \frac{2EI}{1.5L} = \frac{4EI}{3L}$$

Therefore, the stiffness matrix **S** becomes

$$\mathbf{S} = \frac{4EI}{3L}\begin{bmatrix} 5 & 1 \\ 1 & 5 \end{bmatrix}$$

and the inverse matrix is

$$\mathbf{S}^{-1} = \frac{L}{32EI}\begin{bmatrix} 5 & -1 \\ -1 & 5 \end{bmatrix}$$

The joint displacements may now be found by substituting the matrices $\mathbf{S}^{-1}$, $\mathbf{A}_D$, and $\mathbf{A}_{DL}$ into Eq. (2-24) and solving for **D**, as shown:

$$\mathbf{D} = \frac{L}{32EI}\begin{bmatrix} 5 & -1 \\ -1 & 5 \end{bmatrix}\left\{PL\begin{bmatrix} 0 \\ -1 \end{bmatrix} - \frac{PL}{48}\begin{bmatrix} 3 \\ -5 \end{bmatrix}\right\} = \frac{PL^2}{384EI}\begin{bmatrix} 7 \\ -53 \end{bmatrix}$$

This matrix gives the rotations at joints B and C of the continuous beam shown in Fig. 2-16a.

As the next step in the determination of the member end-actions and the reactions, it is necessary to consider again the restrained beams shown in Figs. 2-16d and 2-16e. In each of these beams there are end-actions and reactions that correspond to the end-actions and reactions selected previously and shown in Fig. 2-16b. These quantities for the beams with unit displacements are denoted A_{MD} and A_{RD}, respectively. For example, the end-action A_{MD11} is the shearing force at the left-hand end of member AB due to a unit value of D_1 (see Fig. 2-16d). The end-action A_{MD21} is the moment at the same location. In all cases the first subscript identifies the end-action itself and the second signifies the unit displacement that produces the action. The reactions in the beams of Figs. 2-16d and 2-16e follow a similar pattern, with A_{RD11} and A_{RD21} being the reactions at supports B and C, respectively, due to a unit value of the displacement D_1 (Fig. 2-16d). With this identification scheme in mind, and also using the formulas given in Fig. 2-15, it is not difficult to calculate the various end-actions and reactions. For example, the end-actions in the beam of Fig. 2-16d are the following:

$$A_{MD11} = \frac{6EI}{L^2} \qquad A_{MD21} = \frac{2EI}{L} \qquad A_{MD31} = \frac{6EI}{(1.5L)^2} = \frac{8EI}{3L^2} \qquad A_{MD41} = \frac{4EI}{1.5L} = \frac{8EI}{3L}$$

Also, the reactions in the same beam are

$$A_{RD11} = -\frac{6EI}{L^2} + \frac{6EI}{(1.5L)^2} = -\frac{10EI}{3L^2} \qquad A_{RD21} = -\frac{6EI}{(1.5L)^2} = -\frac{8EI}{3L^2}$$

Similarly, the end-actions and reactions for the beam in Fig. 2-16e can be found, after which the matrices $\mathbf{A}_{MD}$ and $\mathbf{A}_{RD}$ are constructed. These matrices are

$$\mathbf{A}_{MD} = \frac{2EI}{3L^2}\begin{bmatrix} 9 & 0 \\ 3L & 0 \\ 4 & 4 \\ 4L & 2L \end{bmatrix} \qquad \mathbf{A}_{RD} = \frac{2EI}{3L^2}\begin{bmatrix} -5 & 4 \\ -4 & 5 \end{bmatrix}$$

The final steps in the solution consist of calculating the matrices $\mathbf{A}_M$ and $\mathbf{A}_R$ for the end-actions and reactions in the original beam of Fig. 2-16a. These matrices are obtained by substituting into Eqs. (2-25) and (2-26) the matrices $\mathbf{D}$, $\mathbf{A}_{ML}$, $\mathbf{A}_{MD}$, $\mathbf{A}_{RL}$, and $\mathbf{A}_{RD}$, all of which have been determined above. The following results are thereby obtained:

$$\mathbf{A}_M = \frac{P}{576}\begin{bmatrix} 351 \\ 93L \\ 248 \\ 30L \end{bmatrix} \qquad \mathbf{A}_R = \frac{P}{576}\begin{bmatrix} 1049 \\ 427 \end{bmatrix}$$

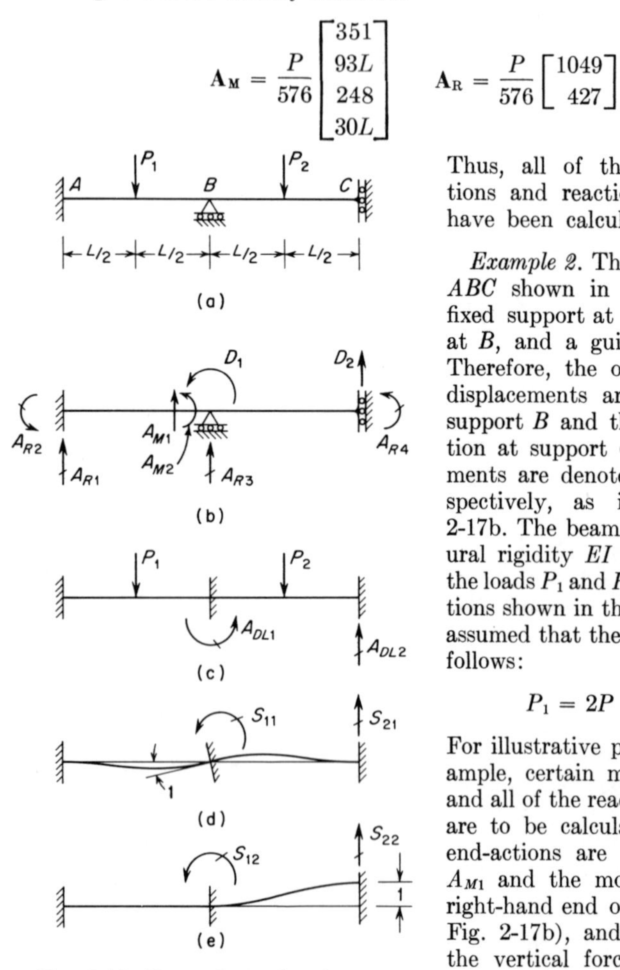

FIG. 2-17. Example 2: Continuous beam.

Thus, all of the selected end-actions and reactions for the beam have been calculated.

Example 2. The continuous beam ABC shown in Fig. 2-17a has a fixed support at A, a roller support at B, and a guided support at C. Therefore, the only unknown joint displacements are the rotation at support B and the vertical translation at support C. These displacements are denoted D_1 and D_2, respectively, as identified in Fig. 2-17b. The beam has constant flexural rigidity EI and is subjected to the loads P_1 and P_2 acting at the positions shown in the figure. It will be assumed that the loads are given as follows:

$$P_1 = 2P \qquad P_2 = P$$

For illustrative purposes in this example, certain member end-actions and all of the reactions for the beam are to be calculated. The selected end-actions are the shearing force A_{M1} and the moment A_{M2} at the right-hand end of member AB (see Fig. 2-17b), and the reactions are the vertical force A_{R1} and couple A_{R2} at support A, the vertical force A_{R3} at support B, and the couple A_{R4} at support C (see Fig. 2-17b). All actions and displacements are shown in their positive directions in the figure.

The restrained structure is obtained by preventing rotation at joint B and translation at joint C, thereby giving the two fixed-end beams shown in Fig.

2-17c. Due to the loads on this restrained structure, the actions corresponding to the unknown displacements D are

$$A_{DL1} = -\frac{P_1L}{8} + \frac{P_2L}{8} = -\frac{PL}{8} \qquad A_{DL2} = \frac{P_2}{2} = \frac{P}{2}$$

Also, the member end-actions in the same beam are

$$A_{ML1} = \frac{P_1}{2} = P \qquad A_{ML2} = -\frac{P_1L}{8} = -\frac{PL}{4}$$

and the reactions are

$$A_{RL1} = \frac{P_1}{2} = P \qquad A_{RL2} = \frac{P_1L}{8} = \frac{PL}{4}$$

$$A_{RL3} = \frac{P_1}{2} + \frac{P_2}{2} = \frac{3P}{2} \qquad A_{RL4} = -\frac{P_2L}{8} = -\frac{PL}{8}$$

From the values given above, the following matrices that are required in the solution can be formed:

$$\mathbf{A}_{\mathrm{DL}} = \frac{P}{8}\begin{bmatrix} -L \\ 4 \end{bmatrix} \qquad \mathbf{A}_{\mathrm{ML}} = \frac{P}{4}\begin{bmatrix} 4 \\ -L \end{bmatrix} \qquad \mathbf{A}_{\mathrm{RL}} = \frac{P}{8}\begin{bmatrix} 8 \\ 2L \\ 12 \\ -L \end{bmatrix}$$

After obtaining the matrices of actions due to loads, the next step is to analyze the restrained beam for unit values of the unknown displacements, as shown in Figs. 2-17d and 2-17e. The stiffnesses S_{11} and S_{21} caused by a unit rotation at joint B are readily obtained from the formulas given in Fig. 2-15, as follows:

$$S_{11} = \frac{4EI}{L} + \frac{4EI}{L} = \frac{8EI}{L} \qquad S_{21} = -\frac{6EI}{L^2}$$

In the case of a unit displacement corresponding to D_2 (Fig. 2-17e) it is necessary to have formulas for the forces and couples at the ends of a fixed-end beam subjected to a translation of one end relative to the other. The required formulas can be obtained from Table B-4 of Appendix B (see Case 2). When the translation is equal to unity, the couples at the ends are equal to $6EI/L^2$, and the forces are equal to $12EI/L^3$, as shown in Fig. 2-18. Using these values, the stiffnesses S_{12} and S_{22} for the beam in Fig. 2-17e are seen to be as follows:

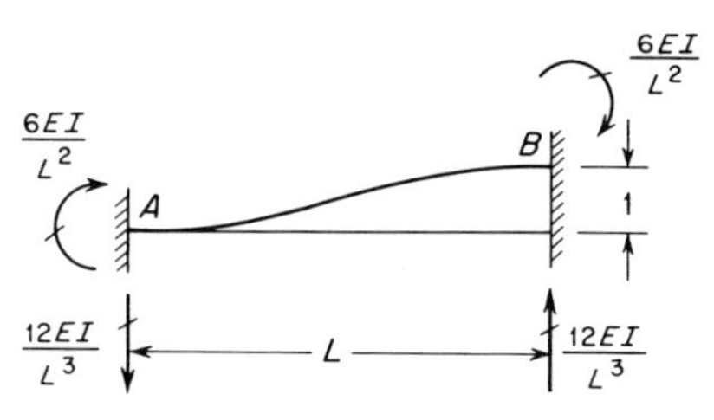

FIG. 2-18. Member stiffnesses for a beam member.

$$S_{12} = -\frac{6EI}{L^2} \qquad S_{22} = \frac{12EI}{L^3}$$

Therefore, the stiffness matrix can be constructed, and its inverse obtained:

$$\mathbf{S} = \frac{2EI}{L^3}\begin{bmatrix} 4L^2 & -3L \\ -3L & 6 \end{bmatrix} \qquad \mathbf{S}^{-1} = \frac{L}{30EI}\begin{bmatrix} 6 & 3L \\ 3L & 4L^2 \end{bmatrix}$$

The inverse matrix, as well as the matrix $\mathbf{A}_{DL}$ determined previously, can now be substituted into Eq. (2-24) in order to obtain the matrix $\mathbf{D}$ of unknown displacements. The matrix $\mathbf{A}_D$ appearing in the equation is a null matrix since there are no loads on the original beam corresponding to either D_1 or D_2. The solution for $\mathbf{D}$ is found to be

$$\mathbf{D} = \frac{PL^2}{240EI}\begin{bmatrix} -6 \\ -13L \end{bmatrix}$$

The matrices $\mathbf{A}_{MD}$ and $\mathbf{A}_{RD}$ which appear in Eqs. (2-25) and (2-26) represent the end-actions and reactions, respectively, in the restrained beams of Figs. 2-17d and 2-17e. The first column of each matrix is associated with a unit value of the displacement D_1 (Fig. 2-17d), and the second column with a unit value of D_2 (Fig. 2-17e). All of the elements in these matrices can be obtained with the aid of the formulas given in Figs. 2-15 and 2-18, and the results are as follows:

$$\mathbf{A}_{MD} = \begin{bmatrix} -\frac{6EI}{L^2} & 0 \\ \frac{4EI}{L} & 0 \end{bmatrix} \qquad \mathbf{A}_{RD} = \begin{bmatrix} \frac{6EI}{L^2} & 0 \\ \frac{2EI}{L} & 0 \\ 0 & -\frac{12EI}{L^3} \\ \frac{2EI}{L} & -\frac{6EI}{L^2} \end{bmatrix}$$

or, in simpler form, the matrices are

$$\mathbf{A}_{MD} = \frac{2EI}{L^2}\begin{bmatrix} -3 & 0 \\ 2L & 0 \end{bmatrix} \qquad \mathbf{A}_{RD} = \frac{2EI}{L^3}\begin{bmatrix} 3L & 0 \\ L^2 & 0 \\ 0 & -6 \\ L^2 & -3L \end{bmatrix}$$

As a final step, the matrices $\mathbf{A}_M$ and $\mathbf{A}_R$ can be found by substituting the matrices $\mathbf{A}_{ML}$, $\mathbf{A}_{MD}$, $\mathbf{A}_{RL}$, $\mathbf{A}_{RD}$, and $\mathbf{D}$ into Eqs. (2-25) and (2-26) and solving. The results are

$$\mathbf{A}_M = \frac{P}{20}\begin{bmatrix} 23 \\ -7L \end{bmatrix} \qquad \mathbf{A}_R = \frac{P}{20}\begin{bmatrix} 17 \\ 4L \\ 43 \\ 3L \end{bmatrix}$$

Thus, all of the desired actions in the beam, as well as the joint displacements have been calculated.

Example 3. The purpose of this example is to illustrate the analysis of a plane truss by the stiffness method. The truss to be solved is shown in Fig. 2-19a and consists of four bars meeting at a common joint E. This particular truss is selected because it has only two degrees of freedom for joint displacement, namely, the horizontal and vertical translations at joint E. However, most of the ensuing discussion pertaining to the solution of this truss is also applicable to more complicated trusses.

It is a convenience in the analysis to identify the bars of the truss numerically. Therefore, the bars are numbered from 1 to 4 as shown by the numbers in circles in Fig. 2-19a. Also, for the purposes of general discussion it will be assumed that the four bars have lengths L_1, L_2, L_3, and L_4, and axial rigidities EA_1, EA_2, EA_3, and EA_4, respectively. Later, all of these quantities will be given specific values in order that the solution may be carried to completion.

The loads on the truss consist of the two concentrated forces P_1 and P_2 acting at joint E, as well as the weights of the members. The weights act as uniformly distributed loads along the bars and are assumed to be of intensity w_1, w_2, w_3, and w_4, respectively, for each of the four bars. In all cases the intensity w is the weight of the bar per unit distance measured along the axis of the bar. For example, the total weight of bar 1 is w_1L_1.

The unknown displacements at joint E, denoted D_1 and D_2 in Fig. 2-19b, are taken as the horizontal and vertical translations of the joint. These displacements, as well as the applied loads at joint E, will be assumed positive when directed toward the right or upward. The member end-actions to be calculated are selected as the axial forces in the four bars at the ends A, B, C, and D, respectively. These actions are shown in Fig. 2-19b and are denoted A_{M1}, A_{M2},

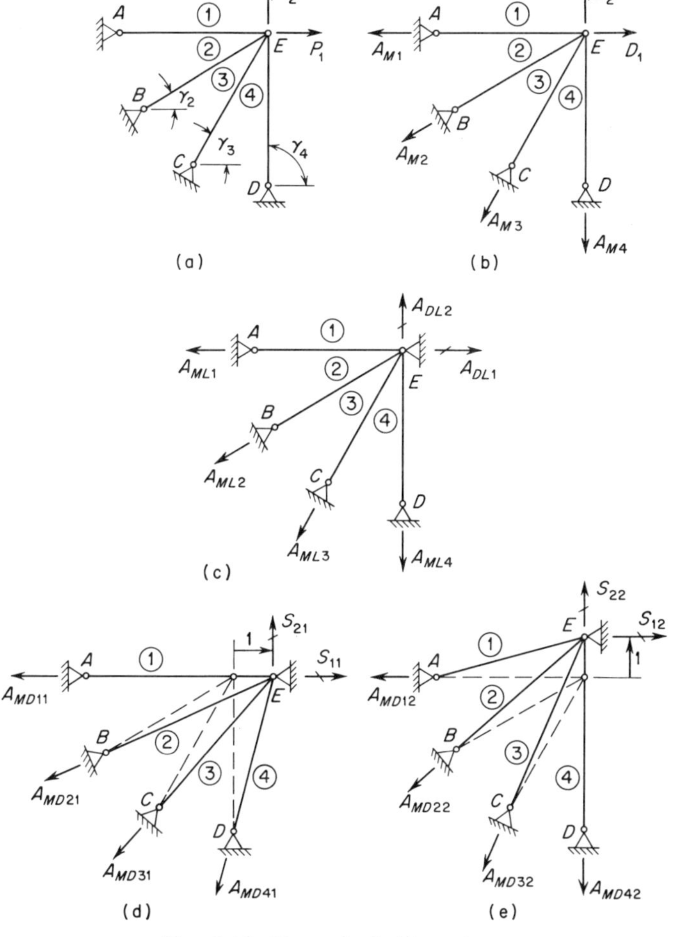

FIG. 2-19. Example 3: Plane truss.

A_{M3}, and A_{M4}. Because of the weights of the bars, the axial forces at the other ends of the bars (that is, at joint E) will have different values from those at ends A, B, C, and D. The axial forces at end E of the bars could be included in the calculations if desired; also, the shearing forces at the ends of the bars could be included. However, the axial forces at end E and the shearing forces are omitted in this example for simplicity. The end-actions are assumed positive when they produce tension in the bars. It is superfluous to calculate reactions for the truss in Fig. 2-19 inasmuch as the reactions are the same as the end-actions. However, there are other situations, such as when several bars of a truss are joined at one point of support, in which the reactions should be determined separately.

The loads P_1 and P_2 acting at joint E of the structure (Fig. 2-19a) are loads that correspond to the unknown displacements D_1 and D_2, respectively. Therefore, these loads appear in the vector $\mathbf{A}_D$ as follows:

$$\mathbf{A}_D = \begin{bmatrix} P_1 \\ P_2 \end{bmatrix}$$

The remaining loads on the truss are considered to act on the restrained structure, which is obtained by preventing translation of joint E (Fig. 2-19c). Each member of the truss is, therefore, in the condition indicated in Fig. 2-20, which shows an inclined truss member that is supported at both ends by immovable pin supports. For purposes of general discussion, the points of support are denoted A and B in the figure. The weight of the bar itself is represented by the uniform load of intensity w. The reactions for the bar are two vertical forces, each of which is equal to one-half the weight of the bar (see Appendix B, Table B-5). Therefore, the actions A_{DL} for the restrained truss shown in Fig. 2-19c can be readily calculated. The actions to be determined at joint E are the horizontal force A_{DL1} and the vertical force A_{DL2}, which correspond to the displacements D_1 and D_2, respectively. Since the weights of the bars produce no horizontal reactions, the action A_{DL1} must be zero. However, the action A_{DL2} will be equal to one-half the weight of all bars meeting at joint E. Therefore, the vector $\mathbf{A}_{DL}$ becomes

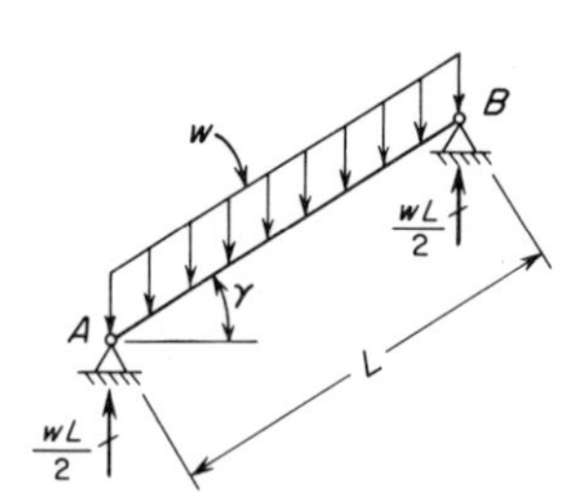

FIG. 2-20. End-actions for restrained truss member.

$$\mathbf{A}_{DL} = \begin{bmatrix} 0 \\ \dfrac{w_1 L_1}{2} + \dfrac{w_2 L_2}{2} + \dfrac{w_3 L_3}{2} + \dfrac{w_4 L_4}{2} \end{bmatrix} = \begin{bmatrix} 0 \\ \dfrac{W}{2} \end{bmatrix}$$

The quantity W is the total weight of all bars meeting at joint E, which in this example is also the total weight of the truss.

For the purpose of calculating the end-actions for the members, it is also necessary to obtain the vector $\mathbf{A}_{ML}$ from a consideration of the restrained structure shown in Fig. 2-19c. This vector consists of the end-actions A_{ML1}, A_{ML2}, A_{ML3}, and A_{ML4}, which are shown in the figure in the positive directions (tension in the members). Each such quantity is given by the general formula

$$-\frac{wL}{2} \sin \gamma$$

in which γ is the angle between the axis of the bar and the horizontal, and the minus sign indicates that the force is compression. To apply this formula to the truss in Fig. 2-19, it is necessary first to identify the angles between the members and the horizontal. Let these angles be denoted γ_1, γ_2, γ_3, and γ_4 for the four bars 1, 2, 3, and 4, respectively. These angles are shown in Fig. 2-19a for all bars except bar 1, which is assumed to be horizontal in this particular example. Thus, the end-action A_{ML2}, for example, is given by the expression

$$A_{ML2} = -\frac{w_2 L_2}{2} \sin \gamma_2$$

and it can be seen that the vector $\mathbf{A}_{ML}$ is the following:

$$\mathbf{A}_{ML} = -\frac{1}{2} \begin{bmatrix} w_1 L_1 \sin \gamma_1 \\ w_2 L_2 \sin \gamma_2 \\ w_3 L_3 \sin \gamma_3 \\ w_4 L_4 \sin \gamma_4 \end{bmatrix}$$

This vector can be evaluated in any particular case by substituting the appropriate values for each bar of the truss. If it is assumed that bar 1 is horizontal and bar 4 is vertical, then the first and last elements of the vector can be simplified to zero and $-w_4L_4/2$, respectively.

To obtain the stiffness matrix $\mathbf{S}$, it is necessary to impose unit displacements corresponding to D_1 and D_2 on the restrained structure, as shown in Figs. 2-19d and 2-19e, respectively. The actions corresponding to the joint displacements are the four joint stiffnesses S_{11}, S_{21}, S_{12}, and S_{22}. Each of these stiffnesses is shown in the figure acting in the positive direction. The joint stiffnesses are composed of contributions from each member of the truss; that is, from the stiffnesses of the members themselves. For example, S_{11} is the total force in the horizontal direction when a unit displacement in that direction is imposed upon the truss (see Fig. 2-19d) and consists of the sum of the horizontal components of the forces in all bars of the truss. Thus, to obtain the joint stiffnesses shown in the figure it is first necessary to find the forces acting on the individual members when joint E is displaced.

The forces acting on a typical truss member due to a unit horizontal displacement of one end are shown in Fig. 2-21a. For purposes of general discussion, the lower and upper ends of the bar are denoted as ends A and B, respectively. The upper end of the bar is assumed to be moved a unit distance to the right, while all other end displacements are zero. As a result, the bar becomes lengthened, and restraint forces are required at each end. The elongation of the bar is determined from the displacements occurring at end B, as shown in Fig. 2-21b. From the triangle in the figure it is seen that the elongation of the bar is $\cos \gamma$, in which γ is the angle of inclination of the bar. Therefore, the tensile force in the bar is (see Fig. 2-21a)

$$\frac{EA}{L} \cos \gamma$$

in which EA is the axial rigidity and L is the length of the bar. The components of this axial force in the horizontal and vertical directions are also shown in the figure. These components, which are member stiffnesses, can be readily determined from the geometry of the figure.

In the case of a unit vertical displacement of end B of the bar (Fig. 2-21c), the axial force in the bar is

$$\frac{EA}{L} \sin \gamma$$

from which the components in the horizontal and vertical directions can be easily found. All of the forces for this case are shown in Fig. 2-21c.

If the lower end A of the bar in Fig. 2-21a is displaced a unit distance to the right while the upper end B remains fixed, all of the actions shown in the figure

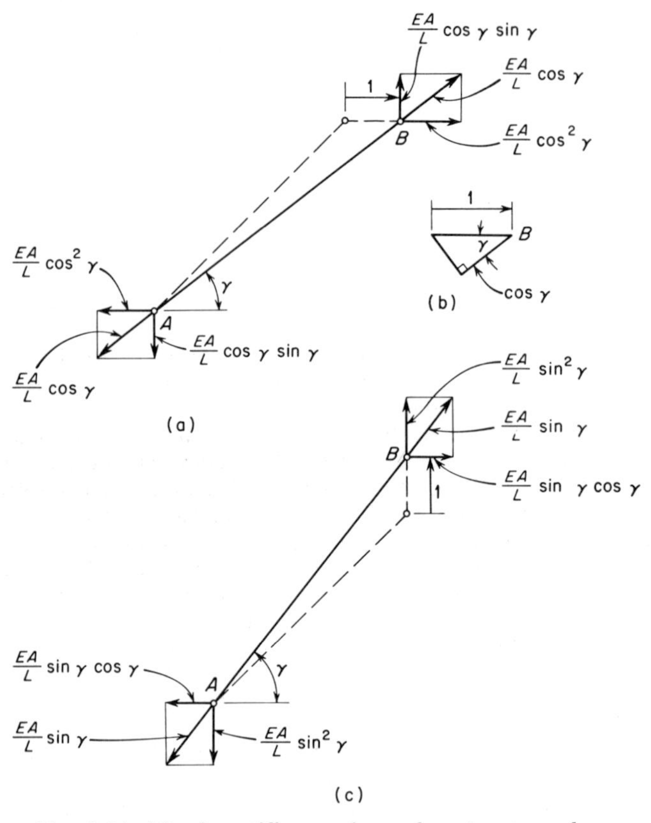

FIG. 2-21. Member stiffnesses for a plane truss member.

will have their directions reversed. The same thing happens to the bar shown in Fig. 2-21c if end A is displaced upward a unit distance while end B remains fixed. All of the formulas given in Fig. 2-21 are suitable for use in analyzing plane trusses by the stiffness method when the calculations are being performed by hand. Later, a more systematic approach to the use of member stiffnesses will be given, not only for trusses but also for other types of structures (see Chapter 4).

Now consider again the calculation of the joint stiffnesses for the restrained truss shown in Figs. 2-19d and 2-19e. The stiffness S_{11} is composed of contributions from the various members of the truss. For example, the contribution to S_{11} from member 3 is (see Fig. 2-21a):

$$\frac{EA_3}{L_3} \cos^2 \gamma_3$$

Similarly, the contribution of member 3 to the stiffness S_{21} is (see Fig. 2-21a):

$$\frac{EA_3}{L_3} \cos \gamma_3 \sin \gamma_3$$

Expressions of the same form as the two expressions given above can be written for all four bars of the truss. The sums of such expressions yield the stiffnesses S_{11} and S_{21}. Inasmuch as bar 1 is horizontal ($\gamma_1 = 0$) and bar 4 is vertical ($\gamma_4 = 90°$), the expressions for the stiffnesses are

$$S_{11} = \frac{EA_1}{L_1} + \frac{EA_2}{L_2} \cos^2 \gamma_2 + \frac{EA_3}{L_3} \cos^2 \gamma_3$$

$$S_{21} = \frac{EA_2}{L_2} \cos \gamma_2 \sin \gamma_2 + \frac{EA_3}{L_3} \cos \gamma_3 \sin \gamma_3$$

By following an analogous procedure but using the formulas in Fig. 2-21c, the stiffnesses S_{12} and S_{22} (see Fig. 2-19e) are obtained:

$$S_{12} = \frac{EA_2}{L_2} \sin \gamma_2 \cos \gamma_2 + \frac{EA_3}{L_3} \sin \gamma_3 \cos \gamma_3$$

$$S_{22} = \frac{EA_2}{L_2} \sin^2 \gamma_2 + \frac{EA_3}{L_3} \sin^2 \gamma_3 + \frac{EA_4}{L_4}$$

The above four expressions constitute the elements of the stiffness matrix **S**. The remaining steps in the calculation of the joint displacements D consist of inverting **S** and substituting into Eq. (2-24). This part of the solution will be performed later using specific values for the quantities EA, L, and γ. Before proceeding to that stage of the solution, however, the matrix $\mathbf{A_{MD}}$ of member end-actions in the restrained structures of Figs. 2-19d and 2-19e will be obtained in general terms.

All of the member end-actions due to the unit joint displacements are shown in Figs. 2-19d and 2-19e. For example, the axial force A_{MD31} is the force in bar 3 due to a unit value of the displacement D_1. Each such force is obtained by referring to Fig. 2-21a or 2-21c. The force A_{MD31}, for instance, is given by the expression

$$\frac{EA_3}{L_3} \cos \gamma_3$$

as found from Fig. 2-21a. If one proceeds in this manner, all of the end-actions due to unit displacements can be found, and the matrix $\mathbf{A_{MD}}$ formed:

$$\mathbf{A_{MD}} = \begin{bmatrix} \frac{EA_1}{L_1} & 0 \\ \frac{EA_2}{L_2} \cos \gamma_2 & \frac{EA_2}{L_2} \sin \gamma_2 \\ \frac{EA_3}{L_3} \cos \gamma_3 & \frac{EA_3}{L_3} \sin \gamma_3 \\ 0 & \frac{EA_4}{L_4} \end{bmatrix}$$

As mentioned previously, other end-actions, as well as reactions, may also be included in the analysis. Under such conditions, the appropriate formulas from Fig. 2-21 can be used in finding the values of the various actions due to the unit displacements. When all of the required matrices have been formed, Eqs. (2-24) through (2-26) are used in finding the resultant effects.

The preceding discussion has served to illustrate in general terms how the various matrices are obtained in a truss analysis. In order to complete the solution of the truss in Fig. 2-19, it is necessary to assume specific values for the quantities appearing in the matrices. Therefore, it will be assumed that all bars have the same length L, the same axial rigidity EA, and the same weight w per unit length. Furthermore, it is assumed that the angles between adjoining bars are 30 degrees, so that the angles between the bars and the horizontal are

$$\gamma_1 = 0 \qquad \gamma_2 = 30° \qquad \gamma_3 = 60° \qquad \gamma_4 = 90°$$

Also, the total weight W of the truss is

$$W = 4wL$$

When these substitutions are made, the matrices described above simplify to the following:

$$\mathbf{A}_{\mathrm{DL}} = wL\begin{bmatrix} 0 \\ 2 \end{bmatrix} \qquad \mathbf{S} = \frac{EA}{2L}\begin{bmatrix} 4 & \sqrt{3} \\ \sqrt{3} & 4 \end{bmatrix}$$

$$\mathbf{A}_{\mathrm{ML}} = -\frac{wL}{4}\begin{bmatrix} 0 \\ 1 \\ \sqrt{3} \\ 2 \end{bmatrix} \qquad \mathbf{A}_{\mathrm{MD}} = \frac{EA}{2L}\begin{bmatrix} 2 & 0 \\ \sqrt{3} & 1 \\ 1 & \sqrt{3} \\ 0 & 2 \end{bmatrix}$$

The inverse of $\mathbf{S}$ is

$$\mathbf{S}^{-1} = \frac{2L}{13EA}\begin{bmatrix} 4 & -\sqrt{3} \\ -\sqrt{3} & 4 \end{bmatrix}$$

and the vector $\mathbf{D}$, found from Eq. (2-24), is

$$\mathbf{D} = \frac{2L}{13EA}\begin{bmatrix} 4P_1 - \sqrt{3}\,P_2 + 2\sqrt{3}\,wL \\ -\sqrt{3}\,P_1 + 4P_2 - 8wL \end{bmatrix}$$

Into this vector can be substituted any particular values of the loads P_1 and P_2, as well as the weight w per unit length. Then the vector $\mathbf{A}_{\mathrm{M}}$ of end-actions (or bar forces) can be found by substituting the matrices $\mathbf{A}_{\mathrm{ML}}$, $\mathbf{A}_{\mathrm{MD}}$, and $\mathbf{D}$ into Eq. (2-25) and solving. If the weight of the truss is not included in the analysis, the matrices $\mathbf{A}_{\mathrm{DL}}$ and $\mathbf{A}_{\mathrm{ML}}$ become null.

As a particular case, assume that $P_1 = 0$, $P_2 = -P$, and $wL = P/10$. Then the vectors $\mathbf{D}$ and $\mathbf{A}_{\mathrm{M}}$ are found to be

$$\mathbf{D} = \frac{12PL}{65EA}\begin{bmatrix} \sqrt{3} \\ -4 \end{bmatrix} \qquad \mathbf{A}_{\mathrm{M}} = \frac{P}{520}\begin{bmatrix} 96\sqrt{3} \\ -61 \\ -157\sqrt{3} \\ -410 \end{bmatrix}$$

These results show that, under the assumed loading, joint E is displaced to the right and downward, and all bars of the truss are subjected to compression except bar 1.

Example 4. The analysis of the plane frame ABC shown in Fig. 2-22a is described in this example. This frame is the same one that was analyzed previously by the flexibility method in Example 4 of Art. 2.3. When using the stiffness method, the first step is to determine the number of unknown joint displacements. In this example, the number of unknowns depends upon whether both flexural and axial deformations are considered in the analysis. If both types of deformation are included, joint B can translate as well as rotate. An analysis of this kind is made later in the article. However, in a plane frame it is normally not necessary to include the axial deformations. If they are omitted, joint B will rotate but not translate, and there will be only one unknown displacement D (see Fig. 2-22b). In addition to finding this joint displacement, it will be assumed

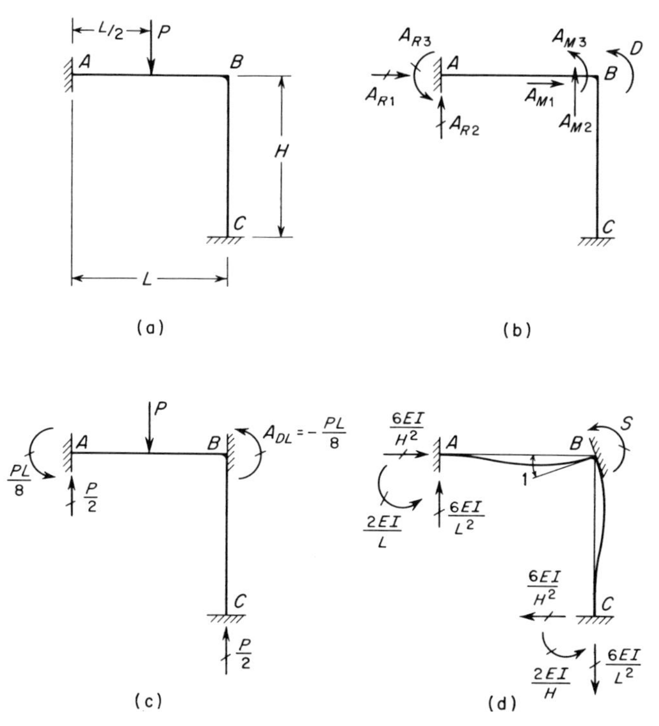

FIG. 2-22. Example 4: Plane frame (flexural effects only).

arbitrarily that the following end-actions and reactions are to be found as part of the solution: the axial force A_{M1}, shearing force A_{M2}, and moment A_{M3} at the right-hand end of member AB (see Fig. 2-22b) and the three reactive actions A_{R1}, A_{R2}, and A_{R3} at support A. Both members of the frame are assumed to have the same flexural rigidity EI.

The restrained structure (Fig. 2-22c) is obtained by preventing rotation of joint B. The action A_{DL} corresponding to the unknown displacement D and caused by the load P is equal to

$$-\frac{PL}{8}$$

The end-actions and reactions in the restrained structure due to the load P are

$$A_{ML1} = 0 \qquad A_{ML2} = \frac{P}{2} \qquad A_{ML3} = -\frac{PL}{8}$$

$$A_{RL1} = 0 \qquad A_{RL2} = \frac{P}{2} \qquad A_{RL3} = \frac{PL}{8}$$

All of the reactions in the restrained structure due to the load P are shown in Fig. 2-22c. It should be recalled at this stage of the calculations that the restraint at joint B is a rotational restraint only, and hence it offers no restraint against translation. Therefore, horizontal and vertical forces can be transmitted through joint B from one member to the other. As a result, the vertical reactive force $P/2$ at the right-hand end of member AB is sustained at joint C rather than at joint B. The vectors $\mathbf{A}_{DL}$, $\mathbf{A}_{ML}$, and $\mathbf{A}_{RL}$ become

$$\mathbf{A}_{DL} = \frac{P}{8}[-L] \qquad \mathbf{A}_{ML} = \frac{P}{8}\begin{bmatrix} 0 \\ 4 \\ -L \end{bmatrix} \qquad \mathbf{A}_{RL} = \frac{P}{8}\begin{bmatrix} 0 \\ 4 \\ L \end{bmatrix}$$

In this problem the vector $\mathbf{A}_D$ is a null vector containing one zero element because there is no couple applied as a load at joint B of the frame.

The stiffness matrix for the frame is found by imposing a unit displacement corresponding to D on the restrained structure (see Fig. 2-22d). From the figure it is seen that the stiffness S is

$$\frac{4EI}{L} + \frac{4EI}{H}$$

Therefore, the stiffness matrix $\mathbf{S}$ and its inverse are

$$\mathbf{S} = \left[4EI\left(\frac{1}{L} + \frac{1}{H}\right)\right] \qquad \mathbf{S}^{-1} = \left[\frac{LH}{4EI(L+H)}\right]$$

The end-actions and reactions for the restrained frame of Fig. 2-22d are evaluated next. Again recalling that there is only a rotational restraint at joint B, it is seen that the reactions for the restrained frame have the values shown in Fig. 2-22d. Therefore, the matrices of end-actions and reactions are as follows:

$$\mathbf{A}_{MD} = \begin{bmatrix} -\dfrac{6EI}{H^2} \\ -\dfrac{6EI}{L^2} \\ \dfrac{4EI}{L} \end{bmatrix} \qquad \mathbf{A}_{RD} = \begin{bmatrix} \dfrac{6EI}{H^2} \\ \dfrac{6EI}{L^2} \\ \dfrac{2EI}{L} \end{bmatrix}$$

All of the above matrices are now substituted into Eqs. (2-24), (2-25), and (2-26), which are then solved for the vectors $\mathbf{D}$, $\mathbf{A}_M$, and $\mathbf{A}_R$. The results are

$$\mathbf{D} = \frac{PL^2}{32EI}\left[\frac{H}{L+H}\right]$$

$$\mathbf{A}_M = \frac{P}{16H(L+H)}\begin{bmatrix} -3L^2 \\ H(8L+5H) \\ -2HL^2 \end{bmatrix} \qquad \mathbf{A}_R = \frac{P}{16H(L+H)}\begin{bmatrix} 3L^2 \\ H(8L+11H) \\ HL(2L+3H) \end{bmatrix}$$

In the special case when $H = L$, these results simplify to

$$\mathbf{D} = \frac{PL^2}{64EI}[1] \qquad \mathbf{A}_{\mathrm{M}} = \frac{P}{32}\begin{bmatrix} -3 \\ 13 \\ -2L \end{bmatrix} \qquad \mathbf{A}_{\mathrm{R}} = \frac{P}{32}\begin{bmatrix} 3 \\ 19 \\ 5L \end{bmatrix}$$

In order to obtain numerical values for these quantities, the data given previously in Example 4 of Art. 2.3 will be used:

$$P = 10 \text{ k} \quad L = H = 12 \text{ ft} \quad E = 30{,}000 \text{ ksi} \quad I = 200 \text{ in.}^4 \quad A = 10 \text{ in.}^2$$

When these values are substituted into the expressions given above, the vectors become

$$\mathbf{D} = [0.00054] \qquad \mathbf{A}_{\mathrm{M}} = \begin{bmatrix} -0.938 \\ 4.06 \\ -90.0 \end{bmatrix} \qquad \mathbf{A}_{\mathrm{R}} = \begin{bmatrix} 0.938 \\ 5.94 \\ 225 \end{bmatrix}$$

in which D is in radians and kip and inch units are used for all other terms. The three elements of the vector $\mathbf{A}_{\mathrm{M}}$ have the same physical significance as the actions Q_1, Q_2, and Q_3 of Example 4, Art. 2.3 (compare Figs. 2-22b and 2-6b). Therefore, the numerical values given above for the end-actions agree with the values of Q_1, Q_2, and Q_3 obtained in the earlier example for the case when axial deformations are neglected.

When axial deformations are included in the analysis of the plane frame in Fig. 2-22a, three degrees of freedom must be recognized at joint B instead of only one. The displacements D_1, D_2, and D_3 for this case are shown in Fig. 2-23a. Also shown in the figure are the end-actions and reactions that are to be determined. The restrained structure for this new analysis is obtained by preventing horizontal translation, vertical translation, and rotation at joint B (see Fig. 2-23b). The actions in this restrained structure corresponding to the unknown displacements and caused by the load P are shown in the figure as A_{DL1}, A_{DL2}, and A_{DL3}. These actions are

$$A_{DL1} = 0 \qquad A_{DL2} = \frac{P}{2} \qquad A_{DL3} = -\frac{PL}{8}$$

The end-actions and reactions in the restrained structure due to the load P are the same as found previously. Therefore, the vectors $\mathbf{A}_{\mathrm{DL}}$, $\mathbf{A}_{\mathrm{ML}}$, and $\mathbf{A}_{\mathrm{RL}}$ are as follows:

$$\mathbf{A}_{\mathrm{DL}} = \frac{P}{8}\begin{bmatrix} 0 \\ 4 \\ -L \end{bmatrix} \qquad \mathbf{A}_{\mathrm{ML}} = \frac{P}{8}\begin{bmatrix} 0 \\ 4 \\ -L \end{bmatrix} \qquad \mathbf{A}_{\mathrm{RL}} = \frac{P}{8}\begin{bmatrix} 0 \\ 4 \\ L \end{bmatrix}$$

As before, the vector $\mathbf{A}_{\mathrm{D}}$ is a null vector because there are no forces or couples applied as loads at joint B.

The elements of the stiffness matrix for the frame are pictured in Figs. 2-23c, 2-23d, and 2-23e. These figures show the actions developed at the restrained joint B when unit displacements corresponding to D_1, D_2, and D_3 are introduced. Each action can be found from the expressions for end-actions due to a unit displacement (see Figs. 2-15, 2-18, and 2-21). Thus, the stiffness matrix is

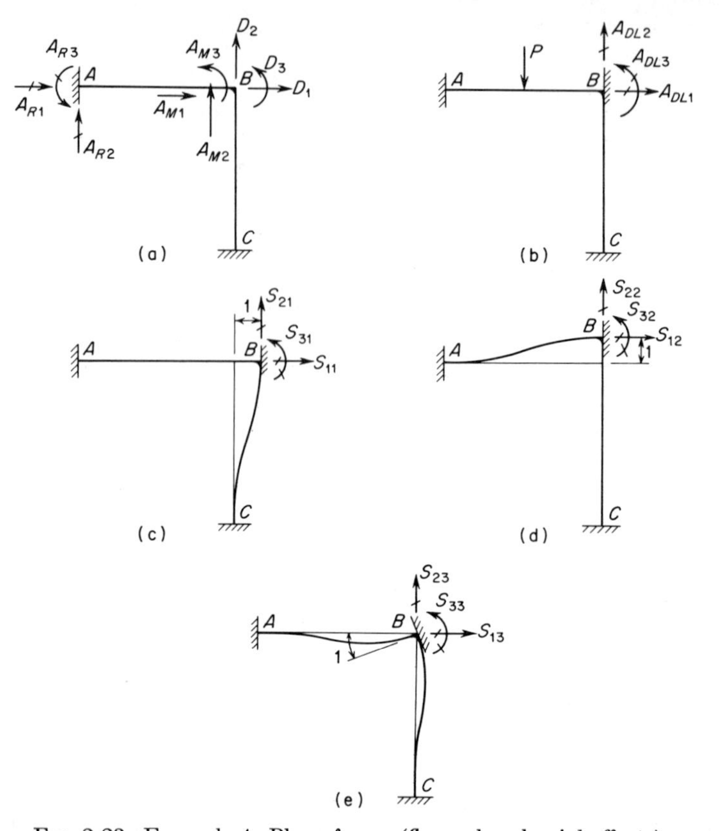

FIG. 2-23. Example 4: Plane frame (flexural and axial effects).

$$\mathbf{S} = \begin{bmatrix} \frac{EA}{L} + \frac{12EI}{H^3} & 0 & \frac{6EI}{H^2} \\ 0 & \frac{12EI}{L^3} + \frac{EA}{H} & -\frac{6EI}{L^2} \\ \frac{6EI}{H^2} & -\frac{6EI}{L^2} & \frac{4EI}{L} + \frac{4EI}{H} \end{bmatrix}$$

The end-actions and reactions for the restrained frames in Figs. 2-23c, 2-23d, and 2-23e are evaluated next, thereby yielding the matrices $\mathbf{A}_{\mathrm{MD}}$ and $\mathbf{A}_{\mathrm{RD}}$:

$$\mathbf{A}_{\mathrm{MD}} = \begin{bmatrix} \frac{EA}{L} & 0 & 0 \\ 0 & \frac{12EI}{L^3} & -\frac{6EI}{L^2} \\ 0 & -\frac{6EI}{L^2} & \frac{4EI}{L} \end{bmatrix} \qquad \mathbf{A}_{\mathrm{RD}} = \begin{bmatrix} -\frac{EA}{L} & 0 & 0 \\ 0 & -\frac{12EI}{L^3} & \frac{6EI}{L^2} \\ 0 & -\frac{6EI}{L^2} & \frac{2EI}{L} \end{bmatrix}$$

All of the above matrices are now substituted into Eqs. (2-24), (2-25), and

(2-26), which then are solved for the vectors $\mathbf{D}$, $\mathbf{A}_{\mathrm{M}}$, and $\mathbf{A}_{\mathrm{R}}$. In order to carry out these steps, the numerical values given previously will be used. When those values are substituted into the vectors $\mathbf{A}_{\mathrm{DL}}$, $\mathbf{A}_{\mathrm{ML}}$, and $\mathbf{A}_{\mathrm{RL}}$, they become

$$\mathbf{A}_{\mathrm{DL}} = \mathbf{A}_{\mathrm{ML}} = \begin{bmatrix} 0 \\ 5 \\ -180 \end{bmatrix} \qquad \mathbf{A}_{\mathrm{RL}} = \begin{bmatrix} 0 \\ 5 \\ 180 \end{bmatrix}$$

The units used in these and subsequent matrices are kips, inches, and radians. The stiffness matrix $\mathbf{S}$ and its inverse become

$$\mathbf{S} = \begin{bmatrix} 2107 & 0 & 1736 \\ 0 & 2107 & -1736 \\ 1736 & -1736 & 333{,}333 \end{bmatrix} \qquad \mathbf{S}^{-1} = 10^{-6} \begin{bmatrix} 476.6 & -2.054 & -2.493 \\ -2.054 & 476.6 & 2.493 \\ -2.493 & 2.493 & 3.026 \end{bmatrix}$$

and the matrices $\mathbf{A}_{\mathrm{MD}}$ and $\mathbf{A}_{\mathrm{RD}}$ are

$$\mathbf{A}_{\mathrm{MD}} = \begin{bmatrix} 2083 & 0 & 0 \\ 0 & 24.11 & -1736 \\ 0 & -1736 & 166{,}667 \end{bmatrix} \qquad \mathbf{A}_{\mathrm{RD}} = \begin{bmatrix} -2083 & 0 & 0 \\ 0 & -24.11 & 1736 \\ 0 & -1736 & 83{,}333 \end{bmatrix}$$

The vector $\mathbf{D}$ of joint displacements is found by substituting into Eq. (2-24) and solving; the result is:

$$\mathbf{D} = 10^{-6} \begin{bmatrix} -438.4 \\ -1934 \\ 532.2 \end{bmatrix}$$

Thus, the joint displacements at B are

$$D_1 = -0.000438 \text{ in.} \qquad D_2 = -0.00193 \text{ in.} \qquad D_3 = 0.000532 \text{ radians}$$

The minus signs for D_1 and D_2 show that these translations are to the left and downward, respectively.

The vectors $\mathbf{A}_{\mathrm{M}}$ and $\mathbf{A}_{\mathrm{R}}$ are found from Eqs. (2-25) and (2-26), as follows:

$$\mathbf{A}_{\mathrm{M}} = \begin{bmatrix} -0.913 \\ 4.03 \\ -87.9 \end{bmatrix} \qquad \mathbf{A}_{\mathrm{R}} = \begin{bmatrix} 0.913 \\ 5.97 \\ 228 \end{bmatrix}$$

These vectors give the end-actions at end B of member AB and the reactions at support A. The former agree with the values calculated previously for the redundants Q_1, Q_2, and Q_3 (see Example 4 of Art. 2.3).

The results obtained above for $\mathbf{A}_{\mathrm{M}}$ and $\mathbf{A}_{\mathrm{R}}$ differ by less than 3 per cent from those obtained earlier when axial deformations were neglected. This small difference is to be expected in plane frames where bending is the principal effect.

Example 5. The grid shown in Fig. 2-24a consists of two members (AB and BC) that are rigidly joined at B. The load on the grid is a concentrated force P acting at the midpoint of member AB. Each member is assumed to have flexural rigidity EI and torsional rigidity GJ. Because the supports at A and C are fixed, the only unknown joint displacements for the structure occur at joint B. These are a translation D_1 in the y direction, a rotation D_2 about the x axis, and a rotation D_3 about the z axis, as indicated in Fig. 2-24b. In the figure the positive direction of each displacement vector is assumed to be the same as the positive direction of one of the coordinate axes. As in earlier articles, a double-headed arrow is used to distinguish a rotation from a translation. The number of degrees of freedom of the grid happens to be the same as the degree of statical indeterminateness (see Example 5, Art. 2.3).

When analyzing the grid by the stiffness method, an artificial restraint is supplied at joint B to prevent displacements corresponding to D_1, D_2, and D_3 (see Fig. 2-24c). When the loads act upon this restrained structure, the actions A_{DL} (corresponding to the displacements D) are developed at the restraint. These actions are evaluated readily by making use of the formulas for fixed-end actions in beams (see Appendix B). In the case of the load P acting on member AB, the actions are

$$A_{DL1} = \frac{P}{2} \qquad A_{DL2} = 0 \qquad A_{DL3} = -\frac{PL}{8}$$

and, hence, the vector $\mathbf{A}_{\mathrm{DL}}$ is

$$\mathbf{A}_{\mathrm{DL}} = \frac{P}{8}\begin{bmatrix} 4 \\ 0 \\ -L \end{bmatrix}$$

The vector $\mathbf{A}_{\mathrm{D}}$ representing actions in the actual beam (Fig. 2-24a) corresponding to the unknown displacements is a null vector because there are no concentrated forces or couples at joint B.

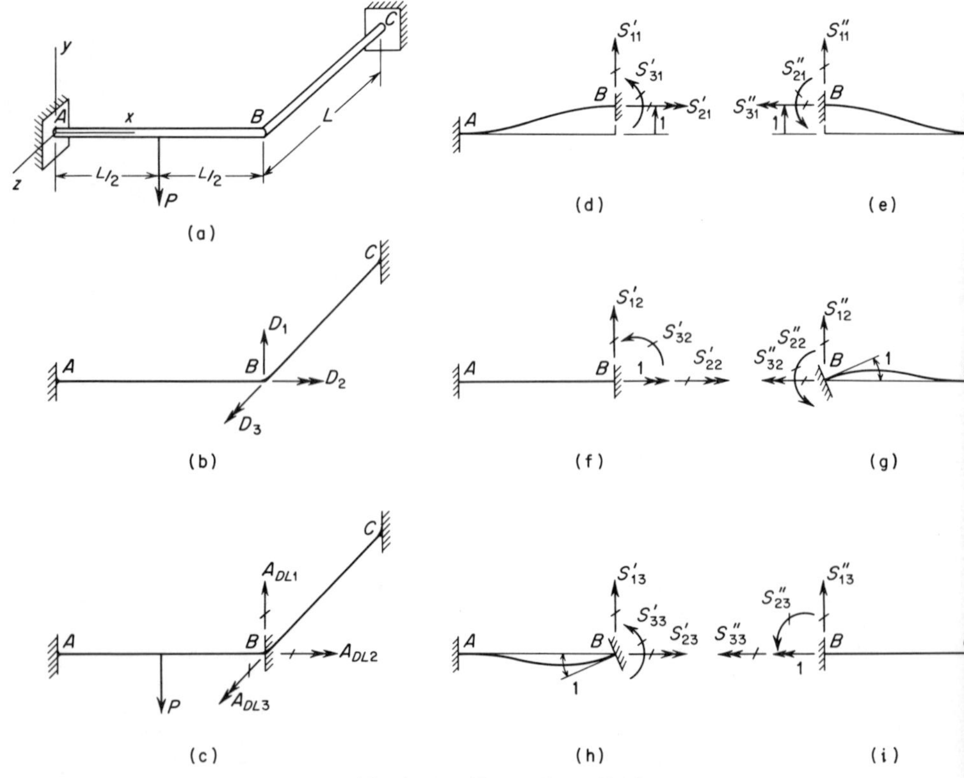

FIG. 2-24. Example 5: Grid.

The stiffness matrix for the grid is found by analyzing the restrained structure for the effects of unit values of the unknown displacements. In the case of a unit value of D_1, joint B is displaced upward by a unit distance without rotating. Then the actions developed at the restraint corresponding to D_1, D_2, and D_3 are the joint stiffnesses S_{11}, S_{21}, and S_{31}, respectively. The effects of this unit translation on members AB and BC are shown separately in Figs. 2-24d and 2-24e. In Fig. 2-24d the contributions of member AB to the joint stiffnesses are shown with a single prime, while in Fig. 2-24e the contributions from member BC are shown with a double prime. From the figures it can be seen that the stiffness terms are as follows:

$$S'_{11} = \frac{12EI}{L^3} \qquad S'_{21} = 0 \qquad S'_{31} = -\frac{6EI}{L^2}$$

$$S''_{11} = \frac{12EI}{L^3} \qquad S''_{21} = \frac{6EI}{L^2} \qquad S''_{31} = 0$$

Therefore, the joint stiffnesses, found by summing the effects from both members, are

$$S_{11} = \frac{24EI}{L^3} \qquad S_{21} = \frac{6EI}{L^2} \qquad S_{31} = -\frac{6EI}{L^2}$$

In a similar manner the stiffnesses resulting from unit rotations corresponding to D_2 and D_3 can be found. The contributions from the individual members are pictured in Figs. 2-24f through 2-24i, and can be evaluated by inspection. In the case of a unit value of D_2 (see Figs. 2-24f and 2-24g) the contributions are

$$S'_{12} = 0 \qquad S'_{22} = \frac{GJ}{L} \qquad S'_{32} = 0$$

$$S''_{12} = \frac{6EI}{L^2} \qquad S''_{22} = \frac{4EI}{L} \qquad S''_{32} = 0$$

For a unit value of D_3 the contributions are (see Figs. 2-24h and 2-24i):

$$S'_{13} = -\frac{6EI}{L^2} \qquad S'_{23} = 0 \qquad S'_{33} = \frac{4EI}{L}$$

$$S''_{13} = 0 \qquad S''_{23} = 0 \qquad S''_{33} = \frac{GJ}{L}$$

Summing the individual terms given above yields the joint stiffnesses:

$$S_{12} = \frac{6EI}{L^2} \qquad S_{22} = \frac{4EI}{L} + \frac{GJ}{L} \qquad S_{32} = 0$$

$$S_{13} = -\frac{6EI}{L^2} \qquad S_{23} = 0 \qquad S_{33} = \frac{4EI}{L} + \frac{GJ}{L}$$

Finally, the stiffness matrix $\mathbf{S}$ can be expressed in the following form:

$$\mathbf{S} = \frac{EI}{L^3}\begin{bmatrix} 24 & 6L & -6L \\ 6L & (4+\eta)L^2 & 0 \\ -6L & 0 & (4+\eta)L^2 \end{bmatrix}$$

in which the dimensionless parameter η is

$$\eta = \frac{GJ}{EI}$$

The quantity η is the reciprocal of the parameter ρ used in solving this same grid by the flexibility method. The inverse matrix is

$$\mathbf{S}^{-1} = \frac{L}{24EIc_1c_2}\begin{bmatrix} L^2c_1^2 & -6Lc_1 & 6Lc_1 \\ -6Lc_1 & 12c_3 & -36 \\ 6Lc_1 & -36 & 12c_3 \end{bmatrix}$$

in which the following additional parameters are used:

$$c_1 = 4 + \eta \qquad c_2 = 1 + \eta \qquad c_3 = 5 + 2\eta$$

The quantities η, c_1, c_2, and c_3 depend only on the ratio of the torsional rigidity GJ to the bending rigidity EI.

The displacements at joint B can be found by substituting the matrices $\mathbf{A}_D$, $\mathbf{A}_{DL}$, and $\mathbf{S}^{-1}$ into Eq. (2-24) and solving for $\mathbf{D}$. This yields the result

$$\mathbf{D} = \frac{PL^2}{96EI(1+\eta)(4+\eta)}\begin{bmatrix} -L(4+\eta)(5+2\eta) \\ 6(5+2\eta) \\ -18 \end{bmatrix}$$

from which

$$D_1 = -\frac{PL^3}{96EI}\frac{5+2\eta}{1+\eta}$$

$$D_2 = \frac{PL^2}{16EI}\frac{5+2\eta}{(1+\eta)(4+\eta)}$$

$$D_3 = -\frac{3PL^2}{16EI}\frac{1}{(1+\eta)(4+\eta)}$$

If the members AB and BC are torsionally very weak, the grid can be considered to consist of two members joined at B by a hinge that is capable of transmitting a vertical force but not a couple. In the analysis of such a grid by the stiffness method, the only joint displacement to be treated as an unknown in the analysis is the translation in the y direction. Its value can be found from the result given above for D_1 by letting η become zero; thus:

$$D_1 = -\frac{5PL^3}{96EI}$$

In general, the analysis of a grid consisting of beams that cross one another (thereby transmitting a vertical force but no moment at each crossing point) will have one unknown joint displacement at each such point.

End-actions and reactions for the grid can also be calculated, following the general techniques illustrated in the preceding examples. Such calculations are given as problems at the end of the chapter.

2.10 Effects of Temperature, Prestrain, and Support Displacement. The effects of temperature changes, prestrain of members, and support displacements can be readily incorporated into the analysis of a structure by the stiffness method. A convenient procedure is to consider all such effects to occur in the restrained structure and to add the resulting actions to the actions produced by the loads. For example, in the restrained structure subjected to loads only, it is necessary to calculate the actions A_{DL} corresponding to the unknown displacements (see Eq. 2-23). When temperature changes are assumed to occur in the same restrained structure, additional actions corresponding to the unknown displacements may occur. Such actions will be denoted by the symbol A_{DT}, which is consistent with the symbol A_{DL} except that the cause is temperature rather than loads. The same idea can be applied to prestrains and restraint displacements, which will produce in the restrained structure the actions A_{DP} and A_{DR}, respectively. When all such actions have been determined for the restrained structure, the vectors $\mathbf{A}_{\mathrm{DL}}$, $\mathbf{A}_{\mathrm{DT}}$, $\mathbf{A}_{\mathrm{DP}}$, and $\mathbf{A}_{\mathrm{DR}}$ can be formed. These vectors are of order $d \times 1$, where d denotes the number of unknown displacements. The sum of these vectors contains the sum of all actions corresponding to the unknown displacements, and is denoted $\mathbf{A}_{\mathrm{DS}}$. Hence, this vector is

$$\mathbf{A}_{\mathrm{DS}} = \mathbf{A}_{\mathrm{DL}} + \mathbf{A}_{\mathrm{DT}} + \mathbf{A}_{\mathrm{DP}} + \mathbf{A}_{\mathrm{DR}} \tag{2-27}$$

The vector $\mathbf{A}_{\mathrm{DS}}$ is included in the first of the three equations of superposition, in place of the matrix $\mathbf{A}_{\mathrm{DL}}$ alone (see Eq. 2-23):

$$\mathbf{A}_{\mathrm{D}} = \mathbf{A}_{\mathrm{DS}} + \mathbf{SD} \tag{2-28}$$

This equation is a more general equation of the stiffness method and should be used instead of Eq. (2-23) whenever effects other than loads are to be included in the calculations. The equation can be solved for the displacements as follows:

$$\mathbf{D} = \mathbf{S}^{-1}(\mathbf{A}_{\mathrm{D}} - \mathbf{A}_{\mathrm{DS}}) \tag{2-29}$$

The techniques used in calculating the vector $\mathbf{A}_{\mathrm{DS}}$ will be discussed later in this article.

The presence of temperature changes, prestrains, and restraint displacements also affects the determination of member end-actions and reactions in a structure. The vectors of end-actions in the restrained structure due to these causes are denoted $\mathbf{A}_{\mathrm{MT}}$, $\mathbf{A}_{\mathrm{MP}}$, and $\mathbf{A}_{\mathrm{MR}}$, respectively, and the sum of all such actions, including loads, is the vector

$$\mathbf{A}_{\mathrm{MS}} = \mathbf{A}_{\mathrm{ML}} + \mathbf{A}_{\mathrm{MT}} + \mathbf{A}_{\mathrm{MP}} + \mathbf{A}_{\mathrm{MR}} \tag{2-30}$$

In an analogous manner the sum of all reactions due to these causes in the restrained structure gives the vector $\mathbf{A}_{\mathrm{RS}}$:

$$\mathbf{A}_{\mathrm{RS}} = \mathbf{A}_{\mathrm{RL}} + \mathbf{A}_{\mathrm{RT}} + \mathbf{A}_{\mathrm{RP}} + \mathbf{A}_{\mathrm{RR}} \tag{2-31}$$

in which the four matrices on the right-hand side of the equation represent reactions due to loads, temperature changes, prestrains, and restraint displacements, respectively. The equations of superposition for the member end-actions and the reactions now become

$$\mathbf{A}_{\mathrm{M}} = \mathbf{A}_{\mathrm{MS}} + \mathbf{A}_{\mathrm{MD}}\mathbf{D} \tag{2-32}$$

$$\mathbf{A}_{\mathrm{R}} = \mathbf{A}_{\mathrm{RS}} + \mathbf{A}_{\mathrm{RD}}\mathbf{D} \tag{2-33}$$

These equations may be considered as generalized forms of Eqs. (2-25) and (2-26), which were used previously when only loads acted on the structure. The use of the above equations will now be illustrated by extending the two-span beam example given in Art. 2.8 (see Fig. 2-14).

The beam to be analyzed is pictured again in Fig. 2-25a. The loads are not shown in the figure because their effects were considered in the earlier example. The unknown joint displacements D_1 and D_2 are indicated in Fig. 2-25a by the curved arrows. For illustrative purposes, consider first the effects of temperature changes, and assume that member BC is subjected to a temperature differential such that the lower side of the beam is at temperature T_1 while the upper side is at temperature T_2. Member AB is assumed to remain at a constant temperature. When these temperature effects are assumed to occur in the restrained structure (see Fig. 2-25b), there will be moments developed at the ends of the member. Therefore, a restraint action A_{DT1} will be developed at joint B corresponding to the displacement D_1, and an action A_{DT2} will be developed at joint C corresponding to D_2. These actions are shown in their positive directions in the figure and can be evaluated from the expressions for fixed-end actions due to temperature changes given in Table B-2 of Appendix B, as follows:

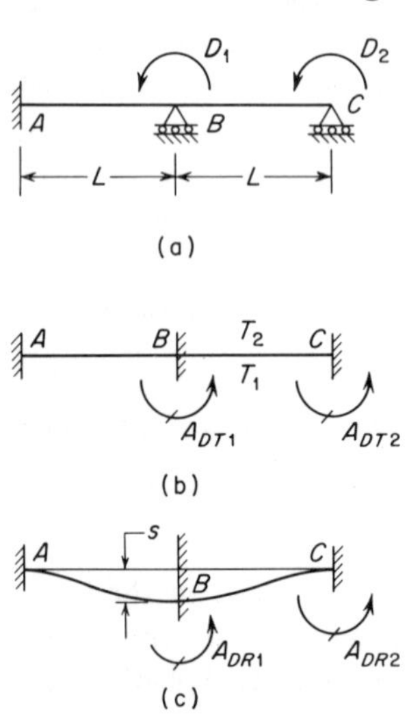

FIG. 2-25. Continuous beam with temperature differential and support displacement.

$$A_{DT1} = -A_{DT2} = \frac{\alpha EI(T_1 - T_2)}{d}$$

in which α is the coefficient of thermal expansion, EI is the flexural rigidity of the beam, and d is the depth of the beam. The vector $\mathbf{A}_{\mathrm{DT}}$ can now be expressed as

$$\mathbf{A}_{\mathrm{DT}} = \frac{\alpha EI(T_1 - T_2)}{d}\begin{bmatrix} 1 \\ -1 \end{bmatrix}$$

This vector can be added to the vector $\mathbf{A}_{\mathrm{DL}}$ obtained in the example of Art. 2.8 to give the sum $\mathbf{A}_{\mathrm{DS}}$ for use in Eq. (2-28), assuming that the beam is to be analyzed for the combined effects of the loads (see Fig. 2-14a) and the temperature differential in member BC. Of course, the displacements D calculated for both effects together will be the sum of the displacements obtained when the loads and the temperature effects are considered separately.

In a similar manner the member end-actions and the reactions due to the temperature change in member BC can be determined from the restrained beam in Fig. 2-25b. In this particular example the member end-actions are zero in span AB, and in member BC are the same as the fixed-end actions. Also, in this example all the reactions are zero inasmuch as the temperature differential does not produce any vertical force at joints B and C. However, in a more general situation there will be values for both the reactions and the end-actions, and these two sets of values are placed in the vectors $\mathbf{A}_{\mathrm{RT}}$ and $\mathbf{A}_{\mathrm{MT}}$, respectively. Then these vectors are added to the corresponding quantities caused by the loads, and the summation vectors are substituted into Eqs. (2-32) and (2-33).

The procedure is similar for prestrain effects and support displacements. In both cases three vectors of actions in the restrained structure are to be determined ($\mathbf{A}_{\mathrm{DP}}$, $\mathbf{A}_{\mathrm{MP}}$, $\mathbf{A}_{\mathrm{RP}}$, and $\mathbf{A}_{\mathrm{DR}}$, $\mathbf{A}_{\mathrm{MR}}$, $\mathbf{A}_{\mathrm{RR}}$). All of these vectors can be evaluated readily when the fixed-end actions are known. Such fixed-end actions are given in Tables B-3 and B-4 of Appendix B. Suppose, for example, that support B of the beam is known to undergo a downward displacement equal to s (see Fig. 2-25c). This results in actions A_{DR1} and A_{DR2}, which are evaluated as follows (see Table B-4):

$$A_{DR1} = \frac{6EIs}{L^2} - \frac{6EIs}{L^2} = 0 \qquad A_{DR2} = -\frac{6EIs}{L^2}$$

The vector $\mathbf{A}_{\mathrm{DR}}$ is formed from these expressions and then included in the calculation of $\mathbf{A}_{\mathrm{DS}}$ (see Eq. 2-27). In a similar manner the vectors of end-actions and reactions ($\mathbf{A}_{\mathrm{MR}}$ and $\mathbf{A}_{\mathrm{RR}}$) due to the restraint displacement shown in Fig. 2-25c can be found. These vectors are included in the calculations for $\mathbf{A}_{\mathrm{MS}}$ and $\mathbf{A}_{\mathrm{RS}}$ (see Eqs. 2-30 and 2-31).

As an example in which prestrain effects are to be determined, consider the analysis of the plane truss shown in Fig. 2-19. Previously this truss was analyzed for the effects of the loads acting at joint E and for the weights of the members. Now assume that one of the members, such as bar 3, is constructed with a length $L_3 + e$, instead of the theoretical length L_3. The effect of the additional increment of length is assumed to take place in the restrained truss shown in Fig. 2-19c. The result is that actions A_{DP1} and A_{DP2} are developed at joint E, corresponding to D_1 and D_2, respectively, and an axial force A_{MP3} is developed in bar 3, corresponding to the action A_{M3}. All of these actions can be found without difficulty. For instance, the axial force is

$$A_{MP3} = -\frac{EA_3e}{L_3}$$

and the vector $\mathbf{A}_{\mathrm{MP}}$ becomes

$$\mathbf{A}_{\mathrm{MP}} = \begin{bmatrix} 0 \\ 0 \\ -\dfrac{EA_3e}{L_3} \\ 0 \end{bmatrix}$$

The horizontal and vertical components of the axial force in the member are used to obtain the actions at joint E in the restrained truss:

$$A_{DP1} = -\frac{EA_3e}{L_3}\cos\gamma_3 \qquad A_{DP2} = -\frac{EA_3e}{L_3}\sin\gamma_3$$

Therefore, the vector $\mathbf{A}_{\mathrm{DP}}$ is

$$\mathbf{A}_{\mathrm{DP}} = \begin{bmatrix} -\dfrac{EA_3e}{L_3}\cos\gamma_3 \\ -\dfrac{EA_3e}{L_3}\sin\gamma_3 \end{bmatrix}$$

The vectors $\mathbf{A}_{\mathrm{MP}}$ and $\mathbf{A}_{\mathrm{DP}}$, found above, are added to the vectors $\mathbf{A}_{\mathrm{ML}}$ and $\mathbf{A}_{\mathrm{DL}}$, found in the earlier example, to give the sums $\mathbf{A}_{\mathrm{MS}}$ and $\mathbf{A}_{\mathrm{DS}}$. Then these vectors are used in Eqs. (2-29) and (2-32) when solving for the joint displacements and the member end-actions. The final results for the combined effects of loads and prestrain will be the sum of the results obtained for loads and prestrain occurring separately.

2.11 Inverse of the Stiffness Matrix. The stiffness matrix $\mathbf{S}$ used in the stiffness method of analysis is inverted during the process of solution (see Eq. 2-29). The inverse matrix is, of course, a flexibility matrix, but it is not the flexibility matrix $\mathbf{F}$ used in analyzing the same structure by the flexibility method. In order for them to be the same, the redundant actions in the flexibility solution would have to correspond to the unknown displacements in the stiffness method, which is an impossible condition. This lack of correspondence can be shown by referring to the two-span beam used as an example in the preceding articles (see Fig. 2-26a). In the stiffness method the unknown displacements for this beam are the rotations at joints B and C. In the flexibility method the redundant actions may be support reactions or internal stress resultants, depending upon the selections that are made. It is apparent, however, that a set of redundants which corresponds to the unknown joint displacements does not exist.

The physical significance of the inverse of the stiffness matrix $\mathbf{S}$ may be ascertained without difficulty in any particular structure. Since the matrix $\mathbf{S}$ consists of actions corresponding to the joint displacements due to unit values of those displacements, the inverse matrix must consist of the joint displacements due to unit values of the actions corresponding to the displacements. For the two-span beam shown in Fig. 2-26 unit values of the actions corresponding to the joint displacements are shown in parts (b) and (c). Each of the unit actions is a couple since the joint displacements are rotations. The rotations at the joints due to these unit couples are the elements of the inverse matrix $\mathbf{S}^{-1}$, which is denoted by the symbol $\mathbf{R}$. Hence, in the figure the joint rotations are denoted R_{11}, R_{21}, R_{12}, and R_{22}, in which the first subscript denotes the displacement and the second subscript denotes the unit load causing the displacement. All rotations are assumed to be positive in the counterclockwise sense. If these joint

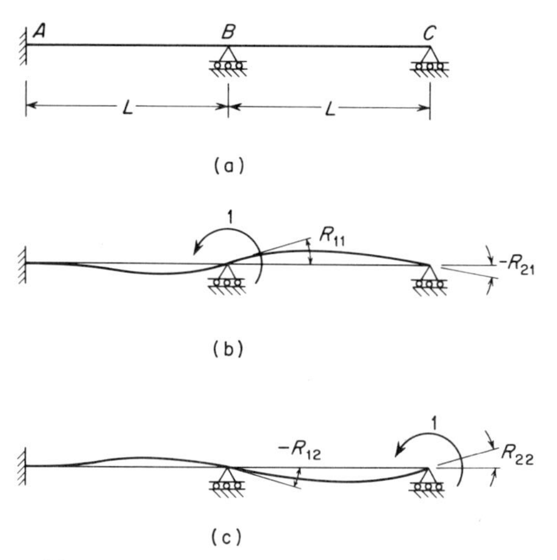

FIG. 2-26. Inverse of the stiffness matrix.

rotations are evaluated for the beams shown in the figure, the results are found to be as follows:

$$R_{11} = \frac{L}{7EI} \qquad R_{12} = R_{21} = -\frac{L}{14EI} \qquad R_{22} = \frac{2L}{7EI}$$

Therefore, the inverse matrix **R** is

$$\mathbf{R} = \begin{bmatrix} R_{11} & R_{12} \\ R_{21} & R_{22} \end{bmatrix} = \frac{L}{14EI}\begin{bmatrix} 2 & -1 \\ -1 & 4 \end{bmatrix}$$

which is the same as the matrix $\mathbf{S}^{-1}$ found previously in Art. 2.8 for the same beam.

The direct determination of the stiffness matrix **S**, as carried out in Art. 2.8, is generally the preferred method of solution. It involves an analysis of the restrained structure for unit values of each joint displacement, which is not difficult since stiffness coefficients for each member are known. On the other hand, the direct determination of the inverse matrix **R** requires an analysis of the beams shown in Figs. 2-26b and 2-26c. Such an analysis may be performed by the stiffness method, in which case it becomes necessary to obtain **S** in the usual way. Then **S** must be inverted in order to complete the analysis and find the rotations at the joints of the two beams. Thus, any attempt to find directly the rotations R shown in the figure involves essentially the same steps as those in the usual procedure.

In general, the most convenient mode of solution by the stiffness method consists of determining the stiffness matrix from an analysis of the restrained structure, inverting this matrix, and then substituting into Eq. (2-29) to find the unknown displacements.

2.12 Summary of the Stiffness Method. The stiffness method of analysis is a very general method that may be applied to the solution of any type of framed structure. The examples in the preceding articles serve to illustrate the fundamental ideas of the method, which are the same regardless of the complexity of the structure. Further development of the method, with emphasis on making the calculations as systematic as possible, will be given in Chapter 4. Also, descriptions of computer programs for the stiffness method are given in Chapter 5, and various modifications to the method are given in Chapter 6. However, the techniques described in this chapter are suitable for hand calculations, and are summarized in this article for reference purposes.

The analysis of a structure by the stiffness method may be described by the following steps:

1. Problem Statement. The problem to be solved must be clearly defined by describing the structure and the loads, temperature changes, prestrains, and support displacements to which it is subjected. The description of the structure includes the type of structure, the locations of joints, positions of members, and locations and types of supports. It is also necessary to state the types of deformations to be considered in the analysis, such as flexural deformations and axial deformations. Depending upon the types of deformations to be considered, the appropriate rigidities of the members must be given. For example, if flexural deformations are considered, the flexural rigidity EI must be given for each member; if axial deformations are considered, the axial rigidity EA must be given; and so forth.

2. Restrained Structure. The number of unknown joint displacements (or degrees of freedom) in the structure must be determined. A corresponding number of artificial restraints must be supplied to produce the restrained structure, in which all joint displacements are zero. There is only one possible restrained structure, and its determination is made automatically.

3. Analysis of Restrained Structure Under Loads. All loads except those that correspond to an unknown joint displacement are considered to be applied to the restrained structure, and various actions in the structure are evaluated. The most important actions to be determined are the actions A_{DL} which correspond to the unknown displacements. Other actions of interest are the end-actions A_{ML} for the members and the reactions A_{RL} at the supports. All of these actions can be found readily with the aid of tables of fixed-end actions (Appendix B).

4. Analysis of Restrained Structure for Other Causes. If there are temperature changes, prestrain effects, or support displacements to be included in the analysis, their effects must be evaluated in the restrained structure. The actions to be found are those that correspond to the unknown displacements (A_{DT}, A_{DP}, A_{DR}), the member end-actions (A_{MT}, A_{MP}, A_{MR}), and the reactions (A_{RT}, A_{RP}, A_{RR}). All of these actions can be found with the aid of tables of fixed-end actions.

5. Analysis of Restrained Structure for Unit Values of the Displacements. Various actions in the restrained structure due to unit values of the unknown joint displacements must be determined. The most important actions to be found are those that correspond to the unknown displacements (the stiffness coefficients S). The other actions to be evaluated are the end-actions and the reactions (A_{MD} and A_{RD}, respectively).

6. Determination of Displacements. The superposition equation for the actions A_D corresponding to the displacements in the actual structure is Eq. (2-28):

$$\mathbf{A_D} = \mathbf{A_{DS}} + \mathbf{SD} \tag{2-28}$$
repeated

In this equation the vector $\mathbf{A_{DS}}$ includes the effects of loads, temperature changes, prestrains, and support displacements, as follows:

$$\mathbf{A_{DS}} = \mathbf{A_{DL}} + \mathbf{A_{DT}} + \mathbf{A_{DP}} + \mathbf{A_{DR}} \tag{2-27}$$
repeated

When Eq. (2-28) is solved for the displacements, the result is

$$\mathbf{D} = \mathbf{S}^{-1}(\mathbf{A_D} - \mathbf{A_{DS}}) \tag{2-29}$$
repeated

7. Determination of End-Actions and Reactions. The vectors $\mathbf{A_M}$ and $\mathbf{A_R}$ for the member end-actions and the reactions, respectively, in the actual structure are obtained from the following superposition equations:

$$\mathbf{A_M} = \mathbf{A_{MS}} + \mathbf{A_{MD}D} \tag{2-32}$$
repeated

$$\mathbf{A_R} = \mathbf{A_{RS}} + \mathbf{A_{RD}D} \tag{2-33}$$
repeated

In these equations the vectors $\mathbf{A_{MS}}$ and $\mathbf{A_{RS}}$ represent actions in the restrained structure due to all causes, as follows:

$$\mathbf{A_{MS}} = \mathbf{A_{ML}} + \mathbf{A_{MT}} + \mathbf{A_{MP}} + \mathbf{A_{MR}} \tag{2-30}$$
repeated

$$\mathbf{A_{RS}} = \mathbf{A_{RL}} + \mathbf{A_{RT}} + \mathbf{A_{RP}} + \mathbf{A_{RR}} \tag{2-31}$$
repeated

When the vectors $\mathbf{D}$, $\mathbf{A_M}$, and $\mathbf{A_R}$ have been obtained, the analysis can be considered to be completed.

All of the matrices used in the stiffness method are summarized in Table 2-2.

TABLE 2-2

Matrices Used in the Stiffness Method

Matrix	*Order*	*Definition*
$\mathbf{D}$	$d \times 1$	Unknown joint displacements (d = number of displacements)
$\mathbf{A}_{D}$	$d \times 1$	Actions in the actual structure corresponding to the unknown displacements
$\mathbf{A}_{DL}$	$d \times 1$	Actions in the restrained structure corresponding to the unknown displacements and due to all loads except those that correspond to the unknown displacements
$\mathbf{S}$ (or $\mathbf{A}_{DD}$)	$d \times d$	Actions in the restrained structure corresponding to the unknown displacements and due to unit values of the displacements (stiffness coefficients)
$\mathbf{A}_{DT}$, $\mathbf{A}_{DP}$, $\mathbf{A}_{DR}$	$d \times 1$	Actions in the restrained structure corresponding to the unknown displacements and due to temperature, prestrain, and restraint displacement
$\mathbf{A}_{DS}$	$d \times 1$	$\mathbf{A}_{DS} = \mathbf{A}_{DL} + \mathbf{A}_{DT} + \mathbf{A}_{DP} + \mathbf{A}_{DR}$
$\mathbf{A}_{M}$	$m \times 1$	Member end-actions in the actual structure (m = number of end-actions)
$\mathbf{A}_{ML}$	$m \times 1$	Member end-actions in the restrained structure due to all loads except those that correspond to the unknown displacements
$\mathbf{A}_{MD}$	$m \times d$	Member end-actions in the restrained structure due to unit values of the displacements
$\mathbf{A}_{MT}$, $\mathbf{A}_{MP}$, $\mathbf{A}_{MR}$	$m \times 1$	Member end-actions in the restrained structure due to temperature, prestrain, and restraint displacement
$\mathbf{A}_{MS}$	$m \times 1$	$\mathbf{A}_{MS} = \mathbf{A}_{ML} + \mathbf{A}_{MT} + \mathbf{A}_{MP} + \mathbf{A}_{MR}$
$\mathbf{A}_{R}$	$r \times 1$	Reactions in the actual structure (r = number of reactions)
$\mathbf{A}_{RL}$	$r \times 1$	Reactions in the restrained structure due to all loads except those that correspond to the unknown displacements
$\mathbf{A}_{RD}$	$r \times d$	Reactions in the restrained structure due to unit values of the displacements
$\mathbf{A}_{RT}$, $\mathbf{A}_{RP}$, $\mathbf{A}_{RR}$	$r \times 1$	Reactions in the restrained structure due to temperature, prestrain, and restraint displacement
$\mathbf{A}_{RS}$	$r \times 1$	$\mathbf{A}_{RS} = \mathbf{A}_{RL} + \mathbf{A}_{RT} + \mathbf{A}_{RP} + \mathbf{A}_{RR}$

2.13 Comparison of Methods. The flexibility and stiffness methods are very similar in their mathematical formulation, and both require the principle of superposition in order to obtain the fundamental equations. The similarities between the two approaches, as well as the differences, are readily seen when the two methods are outlined in parallel, as in Table 2-3. The table shows all of the principal steps in the solution of a structure by both methods.

In the flexibility method the choice of redundants can have a significant effect on the amount of calculating effort required. For example, in continuous beams the bending moments at the supports should usually be selected as redundants because the released structure then consists of a series of simple beams. This released structure is easy to analyze both for the effects of loads and for the effects of unit values of the redundants. The application of a unit value of each redundant influences only the adjacent spans of the beam. Other choices for the redundants do not give this advantageous localization of effects, and, instead, the effects of a unit redundant may be propagated throughout the structure. In the case of structures other than continuous beams, it is normally not possible to localize the effects when using the flexibility method.

In the stiffness method there is never any question about the selection of the restrained structure since there is only one possibility. The analysis of the restrained structure is usually not difficult because all the effects are localized. For example, the effect of a unit displacement at a joint is limited to the members framing into the joint.

In general, both methods of analysis serve useful purposes for hand calculations. The preferred method of solution will usually be the one that involves the smaller number of unknowns. For computer programming the stiffness method is normally much more suitable than the flexibility method. The advantage of the stiffness method arises from the automatic determination of the restrained structure and from the fact that all effects are localized. The suitability of one method or the other is indicated in general terms in Table 2-4. Of course, it must be realized that exceptions to the general rule occasionally will be encountered.

2.14 Multiple Load Systems. In many cases it is necessary to analyze a structure for several different systems of loading. One method is to repeat the analysis of the structure for each new set of loads. In so doing, only those matrices that depend upon the loads must be recalculated. Included in this category are the vectors $\mathbf{D}_{QL}$, $\mathbf{D}_{JL}$, $\mathbf{A}_{ML}$, and $\mathbf{A}_{RL}$ in the flexibility method, and the vectors $\mathbf{A}_{D}$, $\mathbf{A}_{DL}$, $\mathbf{A}_{ML}$, and $\mathbf{A}_{RL}$ in the stiffness method. The matrices that depend only on the properties of the structure will remain the same when a new loading system is considered.

TABLE 2-3

Comparison of Methods

Flexibility Method	Stiffness Method
(a) Problem Statement	
(b) Selection of Unknowns	
Redundants Q	Joint Displacements D
(c) Set Unknowns Equal to Zero	
Released Structure	Restrained Structure
(d) Effects of Loads	
D_{QL}, D_{JL}, A_{ML}, A_{RL}	A_D, A_{DL}, A_{ML}, A_{RL}
(e) Effects of Other Causes	
D_Q, D_{QT}, D_{QP}, D_{QR} D_{JT}, D_{JP}, D_{JR}	A_{DT}, A_{DP}, A_{DR} A_{MT}, A_{MP}, A_{MR} A_{RT}, A_{RP}, A_{RR}
(f) Sums of All Effects	
$D_{QS} = D_{QL} + D_{QT} + D_{QP} + D_{QR}$ $D_{JS} = D_{JL} + D_{JT} + D_{JP} + D_{JR}$	$A_{DS} = A_{DL} + A_{DT} + A_{DP} + A_{DR}$ $A_{MS} = A_{ML} + A_{MT} + A_{MP} + A_{MR}$ $A_{RS} = A_{RL} + A_{RT} + A_{RP} + A_{RR}$
(g) Effects of Unit Values of Unknowns	
F, D_{JQ}, A_{MQ}, A_{RQ}	S, A_{MD}, A_{RD}
(h) Inversion of Matrices	
F^{-1}	S^{-1}
(i) Superposition Equations	
$D_Q = D_{QS} + FQ$ or $Q = F^{-1}(D_Q - D_{QS})$	$A_D = A_{DS} + SD$ or $D = S^{-1}(A_D - A_{DS})$
$D_J = D_{JS} + D_{JQ}Q$ $A_M = A_{ML} + A_{MQ}Q$ $A_R = A_{RL} + A_{RQ}Q$	$A_M = A_{MS} + A_{MD}D$ $A_R = A_{RS} + A_{RD}D$

TABLE 2-4

Suitability of Methods

Degree of Indeterminacy		Suitable Method	
Static	Kinematic	By Hand	By Computer
Low	Low	Either	Stiffness
Low	High	Flexibility	Stiffness
High	Low	Stiffness	Stiffness
High	High	Neither	Stiffness

These matrices are $\mathbf{F}$, $\mathbf{D}_{JQ}$, $\mathbf{A}_{MQ}$, and $\mathbf{A}_{RQ}$ in the flexibility method, and $\mathbf{S}$, $\mathbf{A}_{MD}$, and $\mathbf{A}_{RD}$ in the stiffness method. Of course, all of the superposition equations must be solved for each new loading system, but this involves only matrix addition and multiplication after the inverse matrix (either $\mathbf{F}^{-1}$ or $\mathbf{S}^{-1}$) has been determined for the structure. This simplicity is one of the advantages of formulating the solution in matrix terms.

A more systematic method of handling multiple load systems is to generalize the matrix equations that have been derived previously. Every column vector in the equations of the flexibility and stiffness methods may be changed to a rectangular matrix having one column for each loading system. Otherwise, the matrix equations remain unchanged symbolically.

As an example, consider Eq. (2-29) of the stiffness method, which is repeated here:

$$\mathbf{D} = \mathbf{S}^{-1}(\mathbf{A}_D - \mathbf{A}_{DS}) \tag{2-29}$$

repeated

If there are p load systems that may act on the structure, the matrices $\mathbf{A}_D$ and $\mathbf{A}_{DS}$ will be of order $d \times p$, in which d is the number of unknown joint displacements. The first column of each of these matrices corresponds to the first loading system, the second column corresponds to the second loading system, and so forth. When the term $\mathbf{A}_D - \mathbf{A}_{DS}$ is premultiplied by $\mathbf{S}^{-1}$, which is of order $d \times d$, the result is the matrix $\mathbf{D}$ of joint displacements. This matrix is also of order $d \times p$, and each column represents the joint displacements due to the corresponding load system. Similar comments apply to all of the matrix equations used in the flexibility and stiffness methods.

The method described above for multiple-load systems also can be adapted to handle separately the effects of loads, temperature changes, prestrains, and support displacements. Instead of combining these effects into summation vectors (such as $\mathbf{A}_{DS}$ and $\mathbf{D}_{QS}$), they can be treated as multiple loading systems and handled as separate columns in the matrices.

Many other methods for organizing the matrix equations also are possible. For example, sometimes it is convenient to analyze the structure for unit values of all possible loads that may act on the structure, and then to multiply the final results (the matrices $\mathbf{D}$, $\mathbf{A}_M$, $\mathbf{A}_R$) by a *load vector*, consisting of the values of the particular set of loads being considered.

Problems

The problems for Art. 2.3 are to be solved by the flexibility method using Eq. (2-9). In each problem the redundant actions are to be obtained, unless stated otherwise.

2.3-1. Determine the reactive moments at each end of the fixed-end beam shown in Fig. 2-3a, due to the force P and couple M acting at the middle of the span. The beam has constant flexural rigidity EI and length L. Select the reactive moments themselves as the redundant actions, and assume these moments are positive when they produce compression on the bottom of the beam. Take the first redundant at end A of the beam and the second at end B.

2.3-2. Analyze the two-span beam shown in Fig. 2-2a by taking the reactive moment at support A and the bending moment just to the left of support B as the redundants Q_1 and Q_2, respectively. Assume that these moments are positive when they produce compression on the top of the beam. Also, assume that the loads on the beam are $P_1 = 2P$, $M = PL$, $P_2 = P$, $P_3 = P$, and the flexural rigidity EI is constant.

2.3-3. Analyze the two-span beam of Fig. 2-2a if support B is displaced downward by a small distance s. Select the redundants to be the vertical reactions at supports B and C, as shown in Fig. 2-2a, and omit the effects of the loads in the analysis. Assume that EI is constant for both spans.

2.3-4. Find the redundant actions for the two-span beam of Fig. 2-2a using the released structure shown in Fig. 2-2b. Assume that EI is constant for the beam and that the loads are $P_1 = P$, $M = 0$, $P_2 = P$, $P_3 = P$. Number the redundants from left to right along the beam; also, assume that the redundant couple is positive when counterclockwise, and that the redundant force is positive when upward.

2.3-5. Determine the bending moments at supports B and C of the continuous beam shown in the figure, using these moments as the redundants Q_1 and Q_2, respectively. Assume that the redundants are positive when they produce compression on the top of the beam. The beam has constant flexural rigidity EI.

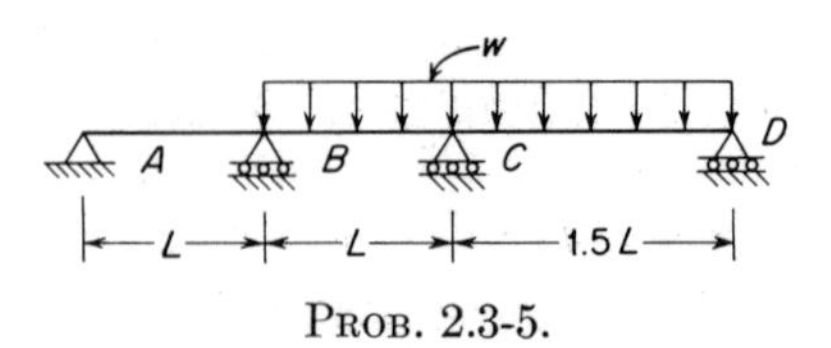

PROB. 2.3-5.

2.3-6. Find the bending moments at supports B and C of the continuous beam (see figure), using these moments as the redundants Q_1 and Q_2, respectively.

Assume that Q_1 and Q_2 are positive when they produce compression on the top of the beam. The flexural rigidity of the beam is EI.

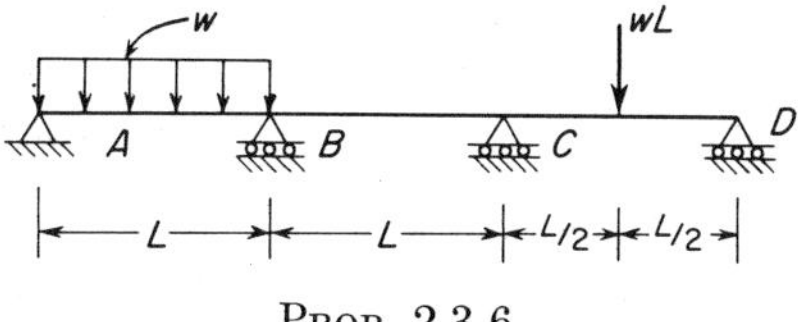

PROB. 2.3-6.

2.3-7. Analyze the plane truss shown in Fig. 2-5a by taking the forces in the two diagonal members AD and BC as the redundants Q_1 and Q_2, respectively. Assume that tension in a member is positive, and assume that there are no support displacements. All members have the same axial rigidity EA.

2.3-8. Solve the preceding problem using the force in bar AD and the reaction at D as the redundants Q_1 and Q_2, respectively.

2.3-9. Find the forces in the bars AE and CE of the truss shown in the figure by taking these bar forces as the redundants Q_1 and Q_2, respectively. Consider tension in a member to be positive. The axial rigidity for the vertical and horizontal members is EA and for the diagonal members is $2EA$.

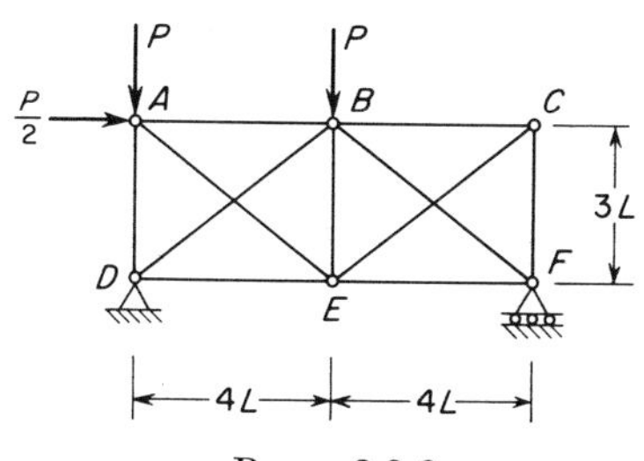

PROB. 2.3-9.

2.3-10. Find the forces in the bars AB and BC of the truss in the figure, using these forces as the redundants Q_1 and Q_2, respectively. Assume that tensile force is positive. Use the following numerical data in the solution: $L = 40$ in., $H = 30$ in., $P_1 = 250$ lb, $P_2 = 150$ lb, $E = 12{,}000{,}000$ psi, and $A = 1.5$ in.^{2} for all members. Perform all calculations using units of inches and pounds.

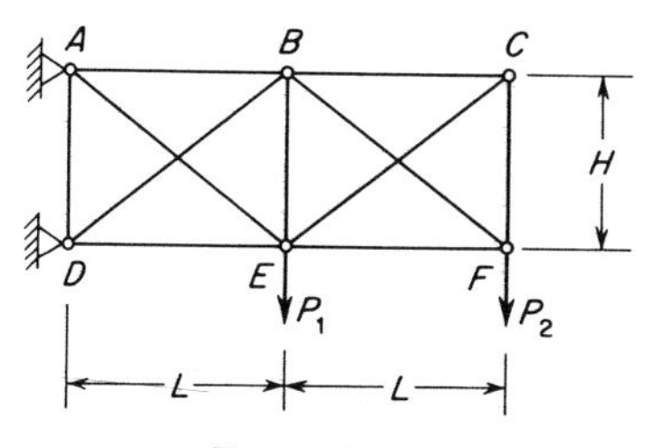

PROB. 2.3-10.

2.3-11. Analyze the plane frame shown in Fig. 2-6a due to a uniform load of intensity $w = 0.4$ k/ft acting downward on member AB. Omit the load P

from the analysis. Use the redundants Q_1, Q_2, and Q_3 shown in Fig. 2-6b, and consider only the effects of flexural deformations. Use the following numerical data for both members of the frame: $L = H = 12$ ft, $E = 30{,}000$ ksi, and $I = 200$ in.[4] Perform all calculations using units of inches and kips.

2.3-12. Obtain the flexibility matrix **F** for the plane frame in Fig. 2-6a, corresponding to the redundants shown in Fig. 2-6b, by considering flexural, axial, and shearing deformations. Both members have flexural rigidity EI, axial rigidity EA, and shearing rigidity GA/f (see Art. A.1).

2.3-13. For the plane frame shown in the figure, find the flexibility matrix **F** for the following conditions: (a) considering flexural deformations only and (b) considering flexural, axial, and shearing deformations. Select the redundants Q_1, Q_2, and Q_3 as the axial force, shearing force, and bending moment, respectively, at the midpoint of member BC. Take these quantities as positive when in the same directions as the actions Q_1, Q_2, and Q_3 shown in Fig. 2-6b. Assume that all members of the frame have flexural rigidity EI, axial rigidity EA, and shearing rigidity GA/f.

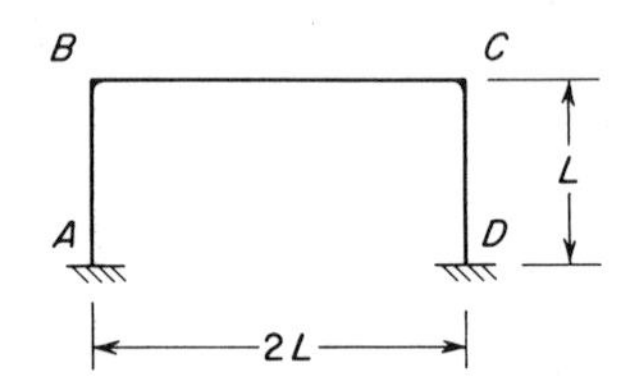

PROB. 2.3-13 and PROB. 2.3-14.

2.3-14. Obtain the flexibility matrix **F** for the plane frame in the preceding problem if the redundants Q_1, Q_2, and Q_3 are the horizontal force (positive to the right), vertical force (positive upward), and couple (positive when counterclockwise), respectively, at support D. Consider only the effects of flexural deformations and assume that each member has constant EI.

2.3-15. Find the redundants Q_1 and Q_2 for the plane frame shown in the figure, considering only flexural deformations. The flexural rigidity EI is the same for all members.

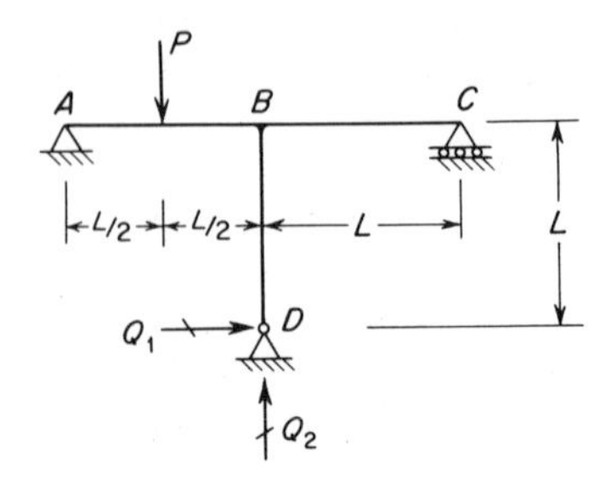

PROB. 2.3-15.

2.3-16. Find the reactions at the support D of the plane frame shown in the figure by taking those reactions as the redundants. Assume that Q_1 is the horizontal reaction (positive to the right), Q_2 is the vertical reaction (positive upward), and Q_3 is the couple (positive when counterclockwise). Consider only the effects of flexural deformations in the analysis. Use the following numerical data: $P = 16$ k, $L = 60$ ft, $H = 20$ ft, $E = 30{,}000$ ksi, I for member $BC = 1100$ in.^{4}, and I for members AB and $CD = 650$ in.^{4}

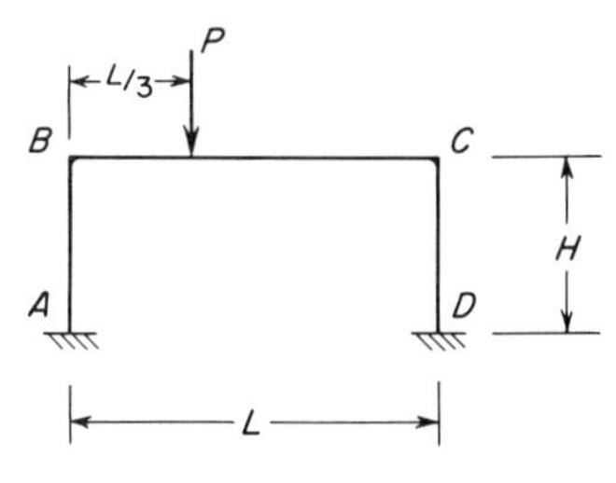

PROB. 2.3-16.

2.3-17. Obtain the flexibility matrix **F** for the grid shown in Fig. 2-7a, considering both flexural and torsional deformations, if the reactions at support C are taken as the redundants. Assume that Q_1 is the force in the positive y direction, Q_2 is the positive couple about the x axis, and Q_3 is the positive couple about the z axis. The flexural and torsional rigidities of the members are EI and GJ, respectively.

2.3-18. Calculate the redundant reaction Q at support D for the horizontal grid shown in the figure. The grid is constructed of three members (AB, BC, and CD) that are rigidly joined together at right angles and supported by simple supports at A, B, C, and D. Each member has flexural rigidity EI, torsional rigidity GJ, and length L. Assume that the loads P act at the midpoints of members AB and BC.

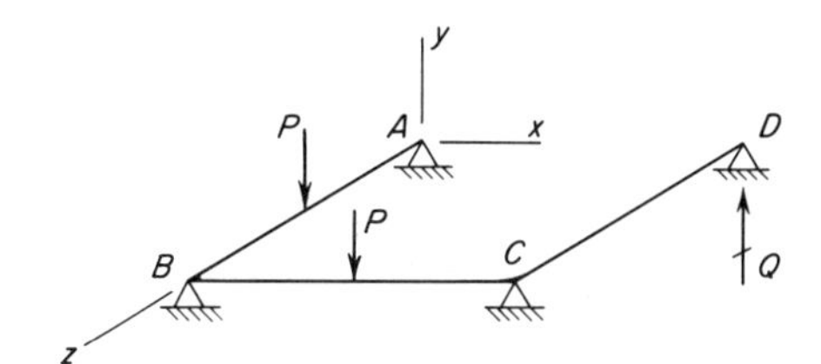

PROB. 2.3-18.

2.3-19. Determine the redundant moments Q_1 and Q_2 at support D of the grid shown in the figure. The supports at A and C are simple supports, and the support at D is a fixed support. The members of the grid are rigidly connected at joint B; also, members AB, BC, and BD each have flexural rigidity EI and torsional rigidity GJ. The load P acts at the midpoint of member BC. Use the following numerical data in the calculations: $P = 300$ lb, $L = 80$ in., $L_1 = 40$ in., $E = 30{,}000$ ksi, $G = 12{,}000$ ksi, $I = 28.2$ in.^{4}, and $J = 56.4$ in.^{4} Express all results in units of pounds and inches.

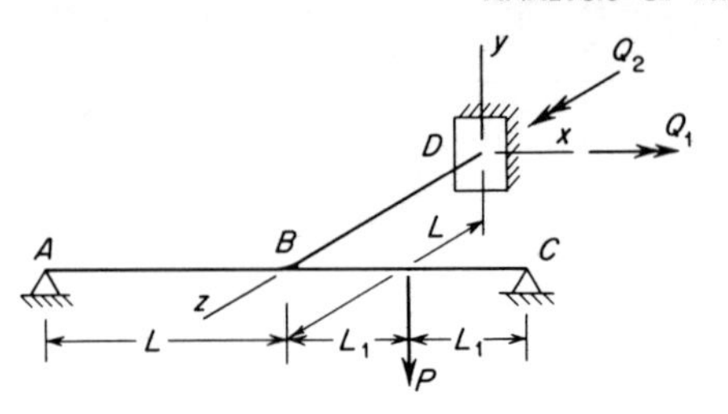

PROB. 2.3-19 and PROB. 2.3-20.

2.3-20. Solve the preceding problem if $L = L_1 = 60$ in.

2.3-21. The space frame $ABCD$ has pin supports at A, C, and D; thus, each support is capable of resisting a force in any direction, but is not capable of transmitting a couple. The members are rigidly connected at joint B. Each member of the frame is of tubular cross-section with length L, flexural rigidity EI, and axial rigidity EA. The effects of both flexural and axial deformations are to be considered. Obtain in literal form the flexibility matrix $\mathbf{F}$ assuming that the redundants Q_1, Q_2, and Q_3 are the reactions at D in the x direction, at C in the x direction, and at C in the z direction, respectively, as shown in the figure.

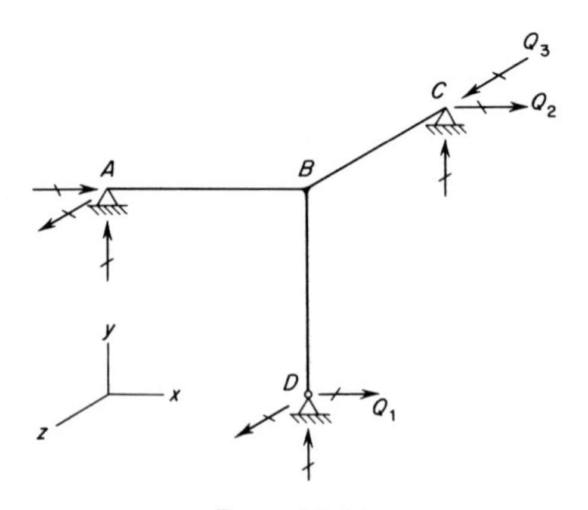

PROB. 2.3-21.

2.3-22. Determine the matrices $\mathbf{D}_{QL}$ and $\mathbf{F}$ for the space frame shown in the figure, considering flexural and torsional deformations. The load on the frame is a vertical force P acting at joint B. The frame has a fixed support at A and a pin support at D, and the members are rigidly connected at right angles to one another at joints B and C. The redundants are selected as the reactions at joint D, as shown in the figure. Each member of the frame has flexural rigidity EI and torsional rigidity GJ.

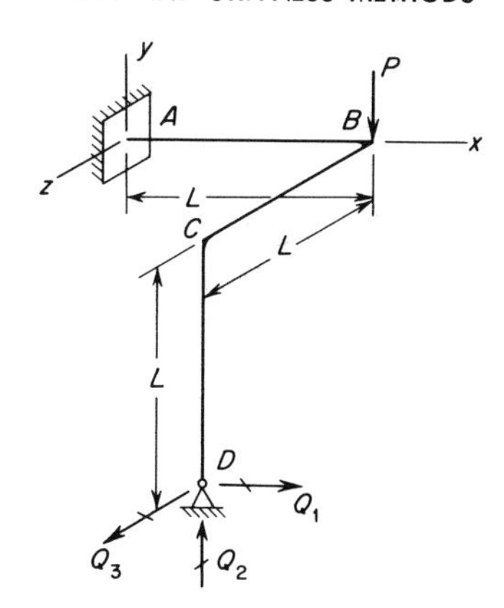

PROB. 2.3-22.

2.4-1. Find the redundant reactions Q_1 and Q_2 at support B of the fixed-end beam shown in Fig. 2-3a, assuming that the beam is subjected to a temperature differential such that the top of the beam is at temperature T_2 and the bottom of the beam is at temperature T_1. The coefficient of thermal expansion for the beam is α, the depth of the beam is d, and the flexural rigidity is EI. Omit the effects of the loads in the analysis.

2.4-2. Obtain the matrix $\mathbf{D}_{QT}$ for the continuous beam shown in Fig. 2-4a, assuming that members BC and CD are heated to a temperature T_1 on the lower surface while the upper surface is at temperature T_2. Let α denote the coefficient of thermal expansion and let d denote the depth of the beam. (Use the redundants Q_1 and Q_2 shown in Fig. 2-4b.)

2.4-3. Obtain the matrix $\mathbf{D}_{QT}$ for the plane truss shown in Fig. 2-5a, assuming that the entire truss has its temperature increased uniformly by an amount T. Use the redundants Q_1 and Q_2 shown in Fig. 2-5b, and let α denote the coefficient of thermal expansion.

2.4-4. Find the redundant reactions Q_1 and Q_2 at support B of the fixed-end beam shown in Fig. 2-3a, assuming that the beam is constructed initially of two straight bars rigidly joined together but slightly out of alignment (see figure). The angle between the two halves of the beam is β, and the flexural rigidity of the beam is EI. Do not include the effects of the loads in the analysis.

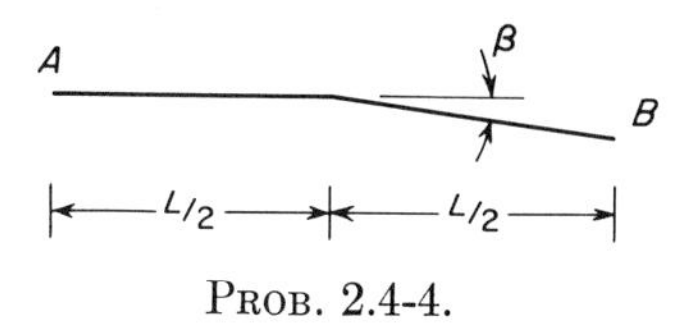

PROB. 2.4-4.

2.4-5. Obtain the matrix $\mathbf{D}_{QP}$ for the plane truss shown in Fig. 2-5a assuming that bars AB and CD are constructed with lengths $L + e$ instead of L. Take the redundants Q_1 and Q_2 as shown in Fig. 2-5b.

2.4-6. Find the redundant reactions Q_1 and Q_2 at support B of the fixed-end beam shown in Fig. 2-3a, assuming that support A rotates β radians in the clockwise direction and support B is displaced downward by a distance s. The flexural rigidity of the beam is EI. Omit the effects of the loads in the analysis.

2.4-7. Obtain the matrix $\mathbf{D}_{QR}$ for the plane truss of Fig. 2-5a if support C is displaced downward a distance s. Use the redundants Q_1 and Q_2 shown in Fig. 2-5b.

2.4-8. Find the matrix $\mathbf{D}_{QR}$ for the continuous beam shown in Fig. 2-4a if support B is displaced downward a distance s_1 and support C is displaced downward a distance s_2. The redundants Q_1 and Q_2 are to be taken as shown in Fig. 2-4b.

2.4-9. Obtain the matrix $\mathbf{D}_{QS}$, representing the sum of all effects in the released structure, for the plane frame shown in Fig. 2-6a if, in addition to the load P, the frame has its temperature increased uniformly by an amount T, support A is displaced downward by an amount s, and support C rotates clockwise by an amount β. The members of the frame have flexural rigidity EI and coefficient of thermal expansion α. Take the redundants Q_1, Q_2, and Q_3 as shown in Fig. 2-6b.

Problems 2.5-1 to 2.5-3 are to be solved using Eqs. (2-14), (2-15), and (2-16). Assume that reactions and joint displacements are positive to the right, upward, and counterclockwise.

2.5-1. Find the joint displacements and reactions for the continuous beam shown in Fig. 2-4a, assuming that the intensity w of the distributed load is such that $wL = P$. The beam has constant flexural rigidity EI. The four joint displacements and four reactions are to be numbered consecutively from left to right in the figure. Use the solution given in Example 2, Art. 2.3, for the redundants.

2.5-2. For the plane truss shown in Fig. 2-5a, obtain the horizontal and vertical displacements of joint A, the forces in bars AB, AC, and BD, and the horizontal and vertical reactions at support C. Assume that there are no support displacements and consider only the effects of the loads in the analysis. Assume that all bars have axial rigidity EA and that tension in a member is positive. Number the displacements, bar forces, and reactions in the order stated above. Use the solution given in Example 3, Art. 2.3, for the redundants.

2.5-3. For the plane frame in Fig. 2-6a, obtain the displacements of joint B (translations in the horizontal and vertical directions, and rotation) and the reactions at support C. Consider both flexural and axial deformations in the analysis, and use the numerical data of Example 4, Art. 2.3, in the solution.

The problems for Art. 2.9 are to be solved by the stiffness method using Eqs. (2-24) to (2-26). Assume all actions and displacements are positive either to the right, upward, or counterclockwise, unless stated otherwise.

2.9-1. Find the reactions for the beam AB shown in Fig. 2-13a. The beam carries a uniform load of intensity w and has constant flexural rigidity EI. The reactions are to be taken in the following order: (1) the force at support A, (2) the couple at support A, and (3) the force at support B.

2.9-2. Find the end-actions for member AB of the beam shown in Fig. 2-13a if, instead of the uniform load, the beam is subjected to a vertical, downward, concentrated force P at the midpoint. The beam has constant flexural rigidity EI, and the end-actions are to be taken in the following order: (1) the shearing force at the left-hand end, (2) the couple at the left-hand end, and (3) the shearing force at the right-hand end.

2.9-3. Find the reactions for the beam AB shown in the figure. The beam has a fixed support at end A and a guided support at end B, and is subjected to two concentrated loads acting at the positions shown. Assume that the beam has constant flexural rigidity EI. The reactions are to be taken in the following order:

(1) and (2), the vertical force and couple, respectively, at support A; and (3), the couple at support B.

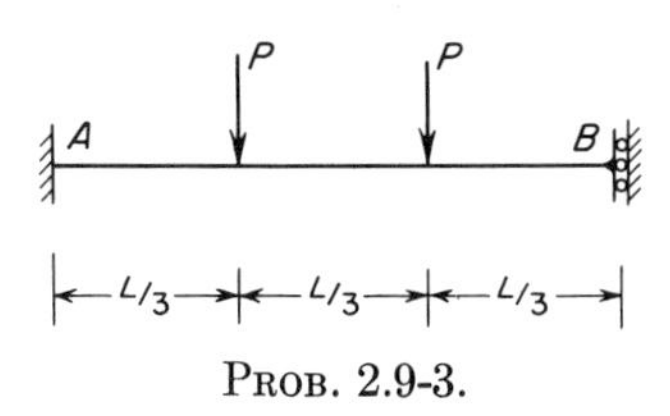

PROB. 2.9-3.

2.9-4. Analyze the two-span beam shown in Fig. 2-14a if $P_1 = P$, $M = PL$, $P_2 = 0$, and $P_3 = 0$. Assume that members AB and BC have lengths L and $1.5L$, respectively, and that the flexural rigidity EI is constant for both spans. Determine end-actions for the members as follows: (1) the shearing force at end A of member AB, (2) the moment at end A of member AB, (3) the shearing force at end B of member BC, and (4) the moment at end B of member BC. Determine reactions for the structure as follows: (1) and (2), the forces at supports B and C, respectively. The unknown displacements are to be numbered from left to right along the beam.

2.9-5. Analyze the beam ABC shown in Fig. 2-17a if $P_1 = 3P_2 = P$. Assume that the lengths of members AB and BC are L and $2L$, respectively, and that the loads P_1 and P_2 act at the midpoints of the members. Also assume that the flexural rigidity EI is constant for both spans. Determine the following end-actions for the beam: (1) the shearing force at the left-hand end of member AB and (2) the moment at the left-hand end of member AB. Determine the following reactions: (1) the force at support B and (2) the couple at support C. The unknown displacements are to be numbered from left to right along the beam.

2.9-6. Analyze the three-span beam shown in the figure if $L_1 = L_2 = L_3 = L$, $P_1 = P$, $P_2 = P_3 = 0$, $M = 0$, and $wL = P$. The flexural rigidity EI is constant for all members. Determine the following end-actions: (1) and (2), the shearing force and moment, respectively, at the left-hand end of member AB; (3) and (4), the shearing force and moment, respectively, at the right-hand end of member AB. Determine the following reactions: (1) and (2), the forces at supports B and C, respectively. The unknown displacements are to be numbered from left to right along the beam.

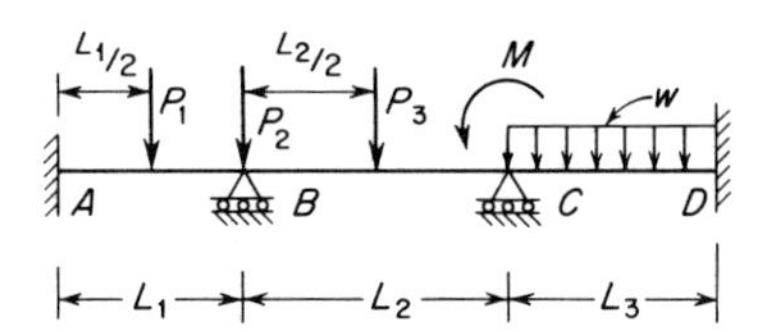

PROB. 2.9-6 and PROB. 2.9-7.

2.9-7. Analyze the three-span beam in the preceding problem if $L_1 = L_3 = L$, $L_2 = 2L$, $P_1 = P_2 = P_3 = P$, $M = PL$, and $wL = P$. The flexural rigidity for members AB and CD is EI and for member BC is $2EI$. Determine the end-moments for all members, numbering the six actions consecutively from left to right. Also, determine reactions as follows: (1) and (2), the forces at supports B and C, respectively. The unknown displacements are to be numbered from left to right along the beam.

2.9-8. Obtain the stiffness matrix **S** for the continuous beam shown in the figure, assuming that the beam has constant flexural rigidity EI. The unknown displacements are to be numbered consecutively from left to right along the beam.

PROB. 2.9-8.

2.9-9. Solve the preceding problem for the beam of Prob. 2.3-5.

2.9-10. Solve Prob. 2.9-8 for a continuous beam on simple supports having five equal spans.

2.9-11. Obtain the stiffness matrix **S** for the beam shown in the figure, assuming that the beam has constant flexural rigidity EI. The unknown displacements are to be numbered from left to right along the beam with translations preceding rotations when both occur at the same joint.

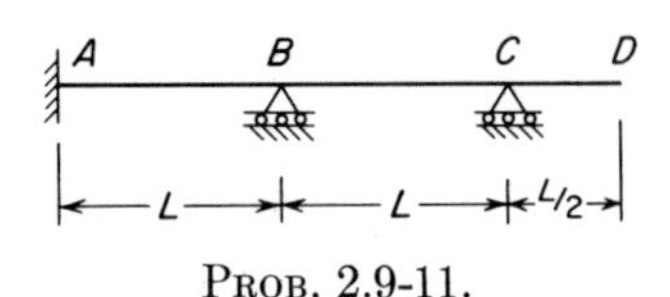

PROB. 2.9-11.

2.9-12. Obtain the stiffness matrix **S** for the beam shown in the figure, assuming that the flexural rigidity of the middle span is twice that of the end spans. The unknown displacements are to be numbered from left to right in the figure.

PROB. 2.9-12.

2.9-13. Obtain the stiffness matrix **S** for the four-span continuous beam shown in the figure if $L_1 = L_2 = L_4 = 24$ ft, $L_3 = 20$ ft, $EI_1 = EI_4 = 12{,}000{,}000$ k-in.^{2}, and $EI_2 = EI_3 = 9{,}000{,}000$ k-in.^{2} Express the results in units of inches and kips. The unknown displacements are to be numbered from left to right along the beam.

PROB. 2.9-13.

2.9-14. Find the reactions at the supports A and C for the beam with fixed ends shown in the figure. Assume that $EI_1 = 2EI$ and $EI_2 = EI$. Number the reactions in the following order: vertical force at A, moment at A, vertical force at C, and moment at C. (Hint: Perform the calculations by considering AB and BC as separate members of the structure.)

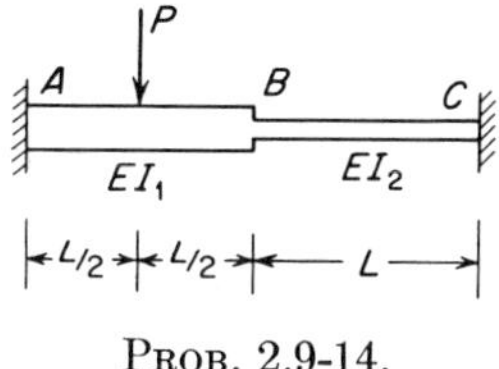

PROB. 2.9-14.

2.9-15. Find the axial forces in all bars of the truss shown in the figure. The truss is subjected to a horizontal force P at joint A. Omit the weights of the bars from the analysis. Each bar has length L and axial rigidity EA. Assume that tensile force in a member is positive, and use the bar-numbering system shown in the figure.

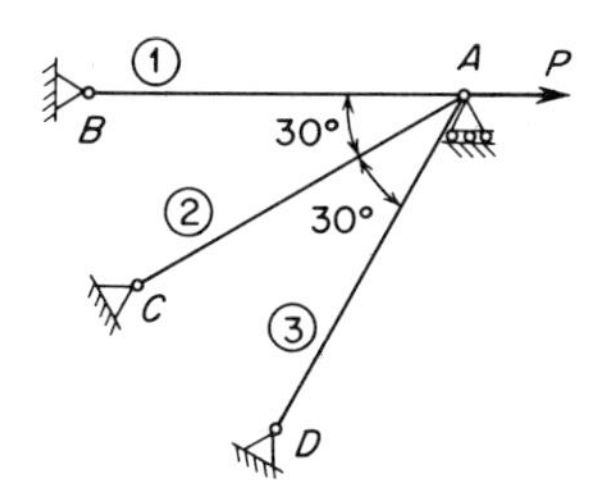

PROB. 2.9-15.

2.9-16. Calculate the axial forces in all bars of the truss due to the force P only (see figure), assuming that EA is the same for all bars. Assume that tensile forces are positive, and use the bar-numbering system shown in the figure. (Hint: Because of the symmetry of the structure and loading, there is only one unknown joint displacement.)

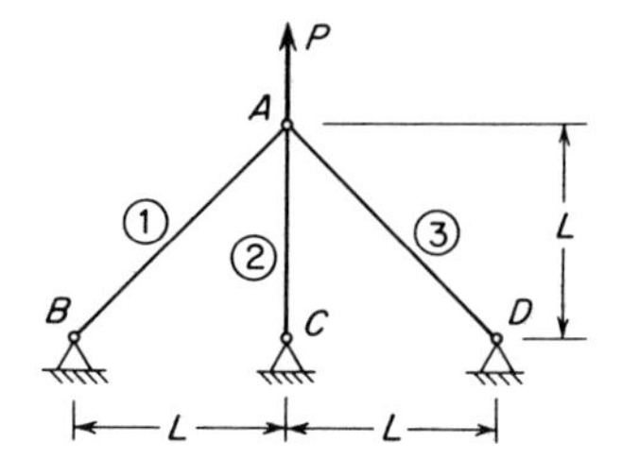

PROB. 2.9-16.

2.9-17. Solve the preceding problem if, in addition to the force P, the weights of all members are included in the analysis. Assume that each bar has weight w per unit length. Calculate the axial forces at the upper ends of the members.

2.9-18. Find the axial forces in all bars of the truss for Prob. 2.9-16 if there is a horizontal force P acting to the right at joint A (in addition to the upward force P). Assume that EA is the same for all bars, and consider tension in a bar to be positive. Omit the effects of the weights of the members.

2.9-19. Find the axial forces in the bars of the truss shown in the figure if all bars have the same length L and the same axial rigidity EA. The angles between the bars are 45 degrees, and the load P makes an angle of 45 degrees with bar 5, which is horizontal. Assume that tension in a bar is positive. Omit the effects of the weights of the members.

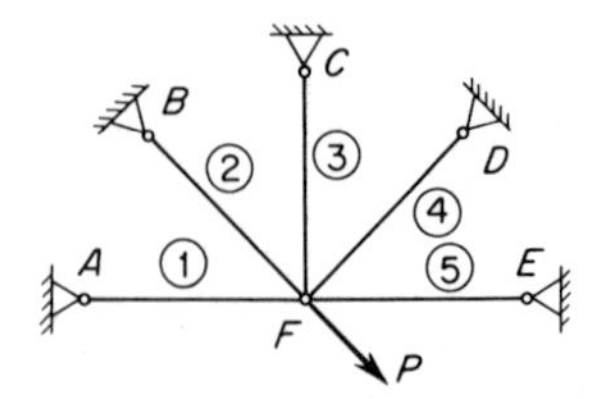

Prob. 2.9-19.

2.9-20. Calculate the horizontal and vertical reactions at the supports of the truss shown in the figure due to the weights of the bars. Number the six reactions in a counterclockwise order around the truss beginning at joint A, and take the horizontal reaction before the vertical reaction when both occur at the same support. Each bar has the same axial rigidity EA and the same weight w per unit length.

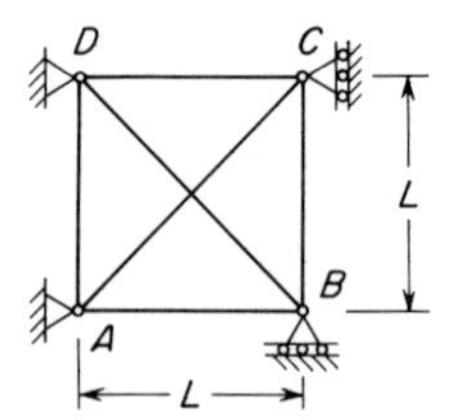

Prob. 2.9-20.

2.9-21. Construct the stiffness matrix $\mathbf{S}$ for the truss shown in the figure. All bars have the same axial rigidity EA. Number the unknown joint displacements in a counterclockwise order around the truss beginning at joint B, and take the horizontal displacement before the vertical displacement when both occur at the same joint.

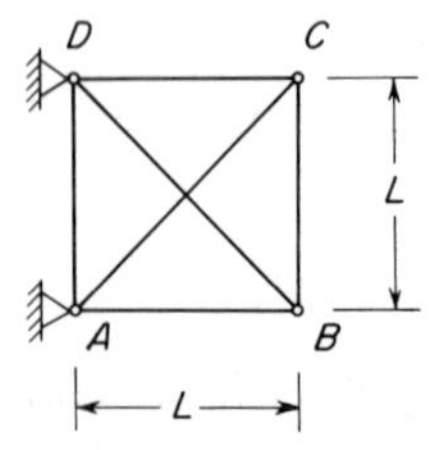

Prob. 2.9-21.

2.9-22. Obtain the stiffness matrix $\mathbf{S}$ for the truss shown in Fig. 2-5a, assuming that all bars have the same axial rigidity EA. Number the joint displacements in a counterclockwise order around the truss beginning at joint D, and take the horizontal displacement before the vertical displacement when both occur at the same joint.

2.9-23. Obtain the stiffness matrix **S** for the truss of Prob. 2.3-9 if the axial rigidity for the vertical and horizontal members is EA and for the diagonal members is $2EA$. Number the joint displacements as described in the preceding problem except begin with joint E.

2.9-24. Analyze the plane frame shown in Fig. 2-22a if a clockwise couple M acts at joint B. Omit the load P from the analysis, and consider only flexural deformations. Determine the end-actions and reactions shown in Fig. 2-22b.

2.9-25. Find the reactions at joints A and D for the plane frame shown in the figure, considering only flexural deformations. Assume that all members have flexural rigidity EI and that $L = 1.5H$. Number the reactions in the following order, first for joint A and then for joint D: horizontal force, vertical force, and couple.

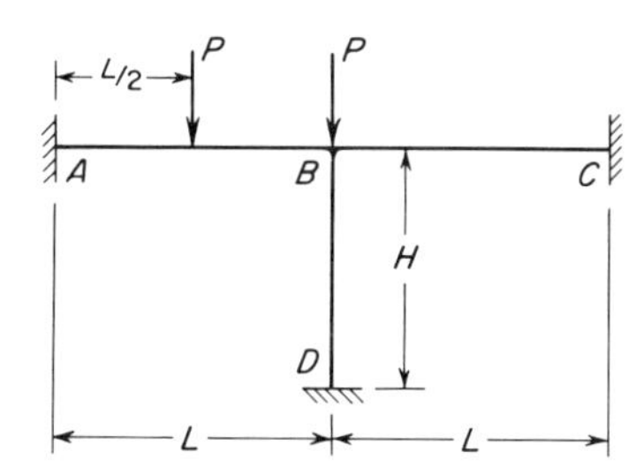

PROB. 2.9-25 and PROB. 2.9-27.

2.9-26. Analyze the plane frame shown in the figure considering only the effects of flexural deformations. Assume that $M = 2wL^2$, $H = L$, and that both members have flexural rigidity EI. Determine the following end-actions: (1) the axial force, (2) the shearing force, and (3) the moment at end B of member BC. Also determine the reactions at support C, taking the horizontal force before the vertical force.

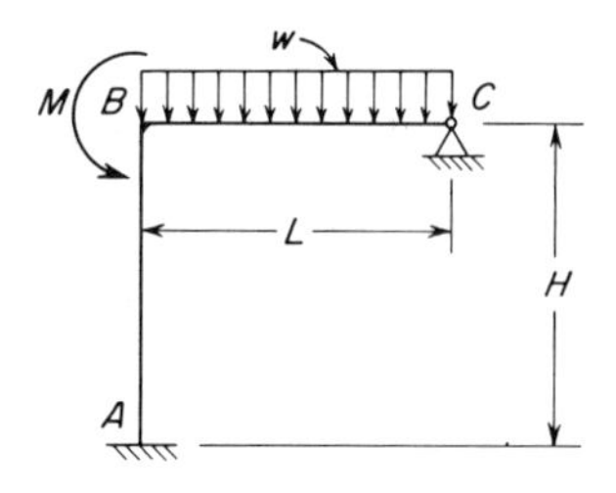

PROB. 2.9-26.

2.9-27. Solve Prob. 2.9-25 if both flexural and axial deformations are taken into account for member BD, and only flexural deformations are considered for members AB and BC. Use numerical data as follows: $P = 6$ k, $L = 24$ ft, $H = 16$ ft, $E = 30{,}000$ ksi, $I = 350$ in.^{4}, $A = 16$ in.^{2} Express the results in units of kips and inches.

2.9-28. Determine the reactions at support D for the plane frame of Prob. 2.3-16, considering only flexural deformations. Number the reactions in the following order: horizontal force, vertical force, and couple. Use the following numerical values: $P = 16$ k, $L = 60$ ft, $H = 20$ ft, $E = 30{,}000$ ksi, I for member $BC = 1100$ in.^{4}, and I for members AB and $CD = 650$ in.^{4}

2.9-29. Obtain the stiffness matrix **S** for the plane frame of Prob. 2.3-13, considering (a) flexural deformations only and (b) flexural and axial deformations. Number the unknown displacements for part (a) as follows: horizontal displacement of B, rotation of B, and rotation of C. For part (b), number the displacements by taking joint B before joint C, and by taking the displacements at a joint in the following order: horizontal translation, vertical translation, and rotation. Assume that all members have flexural rigidity EI and axial rigidity EA.

2.9-30. Obtain the stiffness matrix **S** for the plane frame of Prob. 2.3-15 considering only flexural deformations. Number the unknown displacements in the same sequence that the joints are labeled.

2.9-31. Obtain the stiffness matrix **S** for the plane frame shown in the figure if only flexural deformations are taken into account. The flexural rigidities for the columns are EI_1 and for the beams are EI_2. Number the unknown joint displacements in the following order: (1) horizontal translation of beam AB, (2) horizontal translation of beam CD, (3) rotation of joint A, (4) rotation of B, (5) rotation of C, and (6) rotation of D.

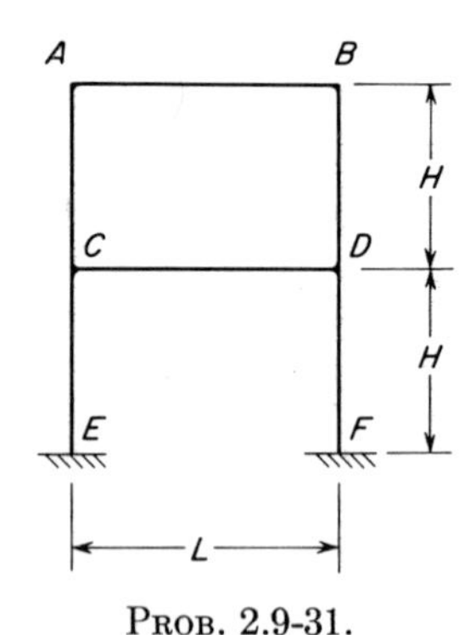

PROB. 2.9-31.

2.9-32. Find the reactions at support A for the grid in Fig. 2-24a. The load on the grid consists of the concentrated force P shown in the figure, and both members have the same flexural rigidity EI and torsional rigidity GJ. Number the reactions in the following order: (1) force in the y direction, (2) couple about the x axis, and (3) couple about the z axis. Assume all actions and displacements are positive when their vectors are in the positive directions of the coordinate axes.

2.9-33. Find the member end-actions at end C of member BC for the grid in Fig. 2-24a. The only load on the grid is the concentrated force P shown in the figure, and both members have flexural rigidity EI and torsional rigidity GJ. Number the end-actions as follows: (1) shearing force in the y direction, (2) bending couple about the x axis, and (3) twisting couple about the z axis. Use the sign convention described in the preceding problem.

2.9-34. Determine the displacements D_1, D_2, and D_3 at joint B of the grid shown in Fig. 2-24 due only to the weight of the members. Assume that each member has weight w per unit length, and that the rigidities EI and GJ are the same for both members.

2.9-35. Analyze the grid in Fig. 2-24a if there is a simple support at joint B. The support prevents translation in the y direction but does not offer any restraint against rotation of the joint. The load on the grid consists of the force P shown in the figure, and both members have the same flexural rigidity EI and torsional rigidity GJ. Number the unknown displacements in the following order: (1) rotation about the x axis and (2) rotation about the z axis. Determine the reactions at support A, using the numbering system and sign convention described in Prob. 2.9-32.

2.9-36. Obtain the stiffness matrix $\mathbf{S}$ for the grid of Prob. 2.3-18. All members of the grid have the same flexural rigidity EI and torsional rigidity GJ. Number the unknown joint displacements in the same sequence that the joints are labeled, taking at each joint the rotation about the x axis before the rotation about the z axis. Assume all displacements are positive when their vectors are in the positive directions of the axes.

2.9-37. Determine the stiffness matrix $\mathbf{S}$ for the grid pictured in Prob. 2.3-19. Assume that the members AB, BC, and BD each have length L, flexural rigidity EI, and torsional rigidity GJ. Number the unknown joint displacements in the same sequence that the joints are labeled, taking the displacements at each joint (when they exist) in the following order: translation in the y direction, rotation about the x axis, and rotation about the z axis. Assume all displacements are positive when their vectors are in the positive directions of the axes.

In solving the problems for Art. 2.10, assume that all actions and displacements are positive either to the right, upward, or counterclockwise.

2.10-1. Find the reactions for the beam of Fig. 2-14a due to a temperature differential such that the lower surface of member AB is at temperature T_1 while the upper surface is at temperature T_2. Omit the loads from the analysis, and assume that both members have the same flexural rigidity EI. The depth of member AB is d and the coefficient of thermal expression is α. Take the reactions in the following order: vertical force at A, couple at A, force at B, and force at C.

2.10-2. Find the reactions for the beam of Fig. 2-14a if support B is displaced downward a distance s_1 and support C is displaced downward a distance s_2. Omit the loads from the analysis, and assume that both members have flexural rigidity EI. Take the reactions in the order described in the preceding problem.

2.10-3. Determine member end-actions and reactions for the beam of Fig. 2-16 if member BC initially has a sharp bend at the position shown in the accompanying figure (the angle β is a small angle). Omit the loads from the analysis, and assume all members have the same flexural rigidity EI. Determine the end-actions and reactions shown in Fig. 2-16b.

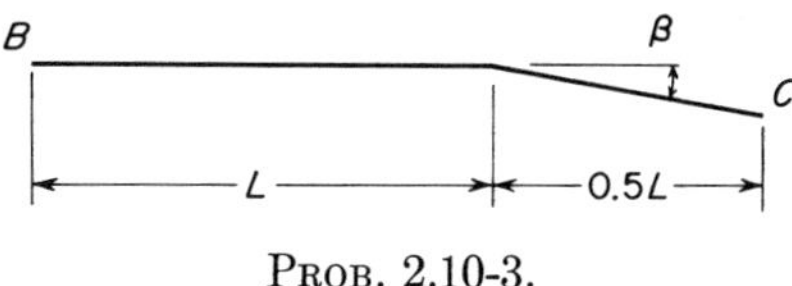

PROB. 2.10-3.

2.10-4. Find the end-actions and reactions shown for the beam in Fig. 2-17b if the entire beam has a temperature differential such that the lower surface is at temperature T_1 while the upper surface is at temperature T_2. Omit the loads from the analysis. Assume that both members have flexural rigidity EI, depth d, and thermal coefficient α.

2.10-5. Solve Prob. 2.9-3 if support A rotates clockwise by an amount β, in which β is a small angle. Omit the loads from the analysis.

2.10-6. Determine the vectors $\mathbf{A}_{DS}$ and $\mathbf{A}_{RS}$ for the beam of Prob. 2.9-6 if, in addition to the loads, the following effects occur: support A is displaced downward a distance s, support D rotates counterclockwise through an angle β, and the entire beam has a temperature differential (T_1 is the temperature on the bottom and T_2 is the temperature on the top).

2.10-7. Assume that the plane truss of Fig. 2-19 has its temperature increased uniformly by an amount T. Omitting the loads from the analysis, find the end-actions shown in Fig. 2-19b. The coefficient of thermal expression is α, all bars have the same axial rigidity EA and length L, and the angles between the bars and the horizontal are $\gamma_1 = 0$, $\gamma_2 = 30°$, $\gamma_3 = 60°$, and $\gamma_4 = 90°$.

2.10-8. Solve Prob. 2.9-15 if bar 2 initially has length $L + e$ instead of L. Omit the loads from the analysis.

2.10-9. Determine the vectors $\mathbf{A}_{DS}$ and $\mathbf{A}_{MS}$ for the truss of Prob. 2.9-16 if the following effects occur: support C is displaced upward a distance s, and bars 1 and 3 have their temperature raised by an amount T. The coefficient of thermal expression is α.

2.10-10. Determine the vector $\mathbf{A}_{DS}$ for the truss of Prob. 2.9-21, considering the following effects: support A is displaced to the left a distance s, member BC has an initial length $L + e$ instead of L, and the entire truss has its temperature increased by T. The coefficient of thermal expression is α.

2.10-11. Find the vectors $\mathbf{A}_{DS}$ and $\mathbf{A}_{MS}$ for the plane frame in Fig. 2-23 if the following effects occur: member AB has its temperature increased uniformly by an amount T, member BC has a temperature differential (T_1 is the temperature on the left and T_2 the temperature on the right), support C is displaced downward a distance s_1, and support A is displaced downward a distance s_2. Assume that both members have the same flexural rigidity EI and axial rigidity EA. Also, assume $H = L$, the coefficient of thermal expression is α, and the depth of the members is d. The unknown displacements D and member end-actions A_M are to be taken as shown in Fig. 2-23a. Omit the load P from the analysis.

Chapter 3

FLEXIBILITY METHOD

3.1 Introduction. The fundamentals of the flexibility method of analysis were described in the preceding chapter, and several examples were given in order to illustrate the use of the matrix equations. It was assumed that all calculations were to be performed by hand methods, using whatever techniques were most appropriate for the particular analysis under consideration. In this chapter the flexibility method is developed further, primarily for the purpose of including the calculation of displacements in the matrix formulation of the problem. In carrying out this objective, the calculations become more organized and formalized. Furthermore, the analysis tends to divide into two distinct phases: (1) there is a "setting up" phase, which must be made by the structural analyst himself; and (2) there is a mathematical phase, which is entirely routine in nature because it involves only formalized matrix calculations. Thus, the latter phase can be performed by anyone who is familiar with matrix algebra, or it can be programmed for a computer.

In general, the flexibility method is not as suitable as the stiffness method for the preparation of computer programs for analyzing a broad class of structures. For example, a computer program which is sufficiently general to handle any plane frame structure should be prepared by the stiffness method (as described in Chapters 4 and 5). One of the reasons for this choice is that in the stiffness method the restrained structure is determined in a very definite way, whereas in the flexibility method there are many choices for the released structure. However, the flexibility method can be used for programming purposes if a more limited class of structures is to be treated (for example, single-story plane frames). Under these special conditions the released structure can be selected according to some particular rule, and therefore the solution can be programmed in a definite manner.

If the analysis is to be performed by hand methods, the most important factor in deciding between the two methods is the size of the matrix to be inverted. There are many structures that will have fewer degrees of statical indeterminacy than kinematic indeterminacy, and in such cases the methods described in this chapter afford a systematic and complete solution of the structure.

From the discussion of the flexibility method in Chapter 2 it is apparent that the calculation of displacements in the released structure is an important part of the solution. Previously, it was left to the reader to calculate the displacements by any convenient methods, such as those described in Appendix A. In this chapter, however, the displacement calculations are performed systematically by matrix methods (see Arts. 3.3 to 3.7). One of the requirements of the matrix approach presented herein is that the structure be subjected to loads at the joints only. Therefore, as a preliminary matter, the conversion of the actual loads on the structure to equivalent joint loads is discussed in the next article. The same conversion is required in the formalized approach to the stiffness method given in Chapter 4.

Throughout most of this chapter, it is assumed that only the effects of loads on the structure are being considered. The effects of temperature changes, prestrain effects, and support displacements are discussed in Art. 3.10.

3.2 Equivalent Joint Loads. The calculation of displacements in a structure by means of the matrix equations derived later in this chapter requires that the structure be subjected to loads acting only at the joints. In general, however, the actual loads on a structure do not meet this requirement. Instead, the loads may be divided into two types: loads acting at the joints and loads acting on the members. Loads of the latter type must be replaced by statically equivalent loads acting at the joints if the requirement stated above is to be fulfilled. The joint loads that are determined from the loads on the members are called *equivalent joint loads.* When these loads are added to the actual joint loads, the total loads which result are called *combined joint loads.* Thereafter, the structure can be analyzed by matrix methods for the effects of the combined joint loads.

It is advantageous in the analysis if the combined joint loads are evaluated in such a manner that the resulting displacements of the structure are the same as the displacements produced by the actual loads. This result can be achieved if the equivalent loads are obtained through the use of fixed-end actions, as demonstrated by the example in Fig. 3-1. Part (a) of this figure shows a beam ABC supported at joints A and B and subjected to several loads. Some of these loads are actual joint loads (see Fig. 3-1b) while the remaining loads act on the members (see Fig. 3-1c). To accomplish the replacement of the member loads by equivalent joint loads, the joints of the structure are restrained against all displacements. For the beam in the figure, this procedure results in two fixed-end beams (Fig. 3-1d). When these fixed-end beams are subjected to the member loads, a set of fixed-end actions is produced. The end-actions can be obtained by means of the formulas in Appendix B and are shown in Fig. 3-1d for the particular loads in this example. The same fixed-end actions are also shown in Fig. 3-1e, where they are represented as restraint actions for the restrained structure. If these restraint actions are reversed in direction, they constitute a set of forces and couples that is statically equivalent to the member loads. Such equivalent joint loads, when added to the actual joint loads (Fig. 3-1b),

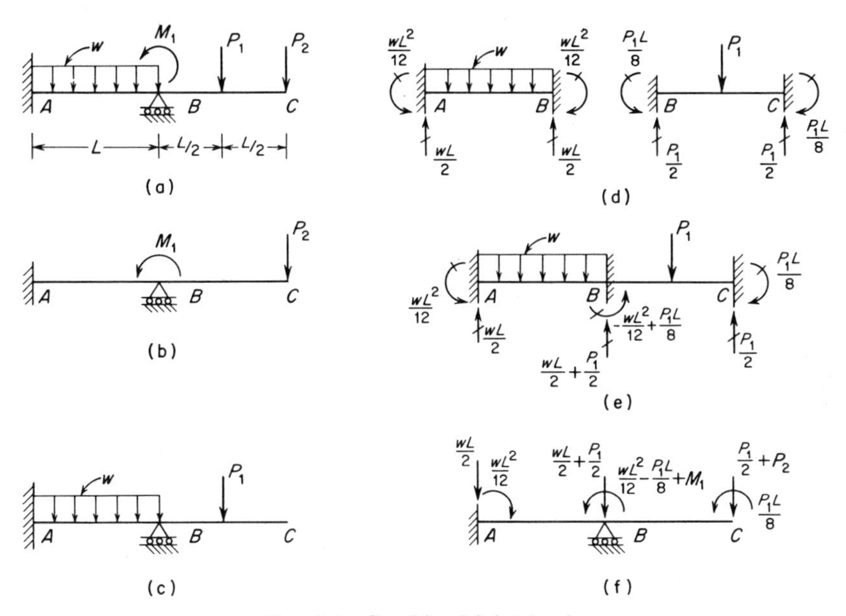

FIG. 3-1. Combined joint loads.

produce the combined joint loads shown in Fig. 3-1f. Then the combined loads are used in carrying out the structural analysis, which is described later.

In general, the combined joint loads for any type of structure can be found by the procedure illustrated in Fig. 3-1. The first step is to separate the actual joint loads from the member loads. Then the structure is restrained against joint displacement by introducing appropriate joint restraints, following the same techniques as those described in Chapter 2 for the stiffness method. Next, the restraint actions produced by the member loads on the restrained structure are calculated, using the formulas given in Table B-1. The negatives of these actions are the equivalent joint loads. Such loads are added to the actual joint loads to give the combined loads.

It was stated earlier that the displacements of the structure under the action of the combined loads should be the same as those produced by the actual loads. In order to observe that this requirement is satisfied, consider again the beam pictured in Fig. 3-1. It is apparent from the figure that the superposition of the combined loads (Fig. 3-1f) and the actions on the restrained structure (Fig. 3-1e) will give the actual loads on the beam (Fig. 3-1a). It follows, therefore, that the superposition of the joint displacements in the beams of Figs. 3-1e and 3-1f must produce the joint displacements in the actual beam. But, since all joint displacements for the restrained beam are zero, it can be concluded that the joint displacements in the beam under the actual loads and the combined loads are the same.

Furthermore, the support reactions for the structure subjected to the combined loads are the same as the support reactions caused by the actual loads. This conclusion also can be verified by superposition of the actions in the beams of Figs. 3-1e and 3-1f. All restraint actions in the beam of Fig. 3-1e are negated by the equal and opposite equivalent joint loads acting on the beam of Fig. 3-1f. Hence, the reactions for the beam with the combined loads are the same as for the beam with the actual loads (Fig. 3-1a). This conclusion, as well as the one in the preceding paragraph, applies to all types of framed structures.

In contrast to the preceding conclusions, the member end-actions caused by the combined joint loads acting on the structure are not usually the same as those caused by the actual loads. Instead, the end-actions due to the actual loads must be obtained by adding the end-actions in the restrained structure to those caused by the combined loads. For example, in the case of the beam shown in Fig. 3-1, the actual end-actions (Fig. 3-1a) are found by superimposing the end-actions from the beams in Figs. 3-1e and 3-1f. The latter will be obtained as a result of the structural analysis itself, and the former are known from the calculations for the fixed-end actions.

The remaining quantities of interest in a structural analysis performed by the flexibility method are the redundant actions themselves. Whether or not the redundant actions are the same in the actual structure and the structure with the combined joint loads depends upon the particular situation. If the redundant action is a reaction for the structure, it will be the same in both cases. If it is a member end-action, it must be treated in the manner described above for end-actions.

Examples illustrating the ideas mentioned above are given in Art. 3.9. However, in the following articles (Arts. 3.3 to 3.7), which deal with the determination of joint displacements, it is assumed usually that the structure is subjected to combined joint loads.

3.3 Joint Displacements in Trusses. The trusses to be considered in this article may be plane or space trusses; each truss is assumed to be statically determinate and subjected to loads acting only at the joints. The methods for calculating displacements that are developed in the following discussion will be incorporated later (see Art. 3.8) into the analysis of statically indeterminate trusses.

The unit-load method can be used conveniently for finding truss displacements, and it constitutes the basic method to be described. The unit-load formula for a displacement D at a particular joint of a truss is (see Eq. A-32, Appendix A):

$$D = \sum \frac{N_U N_L L}{EA} \tag{a}$$

In this equation D represents the displacement that is to be calculated, N_U represents the axial force in any bar of the truss due to a unit load corresponding to the displacement, N_L represents the axial force in the same bar due to the loads that cause the displacement, L is the length of the member, and EA is the axial rigidity of the member. The summation sign in the above equation indicates that the expression on the right-hand side must be evaluated for every member of the structure, and the results summed to obtain the displacement D.

The sign conventions that are used in connection with Eq. (a) are the following. The axial forces N_U and N_L must be evaluated according to the same sign convention for any particular member; in this book both N_U and N_L will be assumed positive when the member is in tension. Also, the positive sense for the displacement D will be the same as the positive sense of the unit load. Thus, the unit load is an action corresponding to the displacement D. For convenience, it will be assumed that the unit load and the displacement D are positive when in the positive directions of the coordinate axes used as a reference for the structure.

Equation (a) is expressed in terms of the axial forces N_U and N_L in the members, as already mentioned. However, in order to make the equations which are derived in this article consistent with those derived later for other types of structures, it is desirable to rewrite Eq. (a) in terms of member end-actions. In the case of either a plane or space truss member, there are only two possible member end-actions, namely, the axial forces acting at the two ends of the member. For the purposes of the following development, either one of the two end-actions can be used. However, it is desirable for uniformity to adopt a specific convention that can be followed consistently.

The conventions to be used in this chapter are shown in Figs. 3-2a and 3-2b, which show a typical member i from a plane and a space truss, respectively. The plane truss member lies in the x-y plane, which is assumed to contain the truss itself, while the space truss member may have any direction with respect to the x, y, z axes. In each case, the member is assumed to be connected at its ends to joints of the structure that are denoted as joints j and k. The axes x, y, z in the figures are assumed to be *structure-oriented axes*, that is, axes oriented in some convenient manner to the structure as a whole. On the other hand, the axes x_M, y_M, z_M are associated with the particular member i under consideration, and are called *member-oriented axes*. The member axes will always have their origin at the j end of the member, with the x_M axis always taken along the axis of the member from end j to end k. The

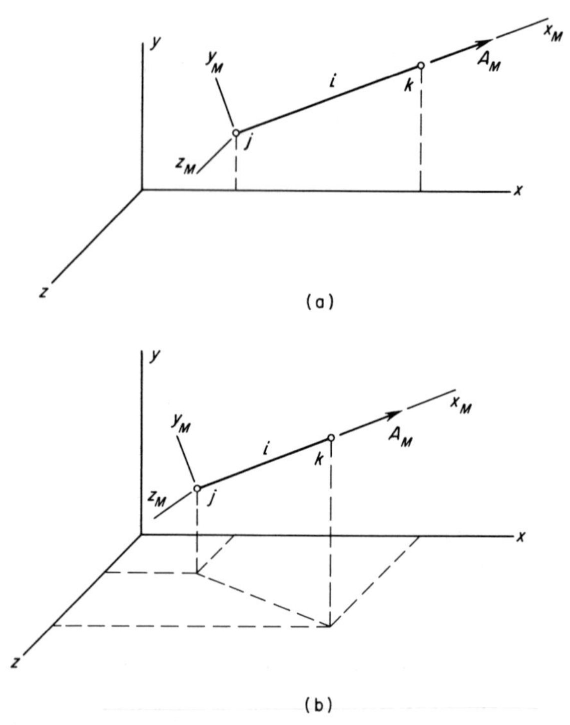

FIG. 3-2. End-actions for (a) plane truss member and (b) space truss member.

position of the y_M and z_M axes at end j is not material to a truss analysis, since all end-actions are in the x_M direction. However, for other types of structures the y_M and z_M axes must be selected as principal axes of the cross-section of the member.

The member end-action to be used in subsequent calculations for trusses is the axial force at the k end of the member, denoted as A_M in Fig. 3-2. Note that a positive value of the end-action at the k end corresponds to a tension force in the member. The end-action at the j end of the member is omitted from the figure, since it is not used in the analysis. If the end-action A_M is caused by the unit load acting on the structure, it is denoted A_{MU}; if caused by the actual loads, it is denoted A_{ML}. Thus, when Eq. (a) for the displacement D is rewritten in terms of end-actions it becomes

$$D = \sum \frac{A_{MU}A_{ML}L}{EA} \tag{b}$$

An example illustrating the calculation of the end-actions A_{ML} and A_{MU} for a truss will be given later.

A further change in Eq. (b) can be made by introducing the notion of a *member flexibility*, denoted as F_M. In the case of an axially loaded member, the flexibility F_M may be visualized as the displacement of the end of the member caused by a unit axial force, as shown for the member i in Fig. 3-3. Therefore, the flexibility for a truss member is defined as

$$F_M = \frac{L}{EA} \tag{3-1}$$

assuming that the member is prismatic. When Eq. (3-1) is substituted into Eq. (b), and the order of the terms is changed slightly, the equation for the displacement D becomes

$$D = \sum A_{MU} F_M A_{ML} \tag{c}$$

The preceding equation for the displacement in a truss can be expressed in matrix form by introducing three matrices. One of these is a

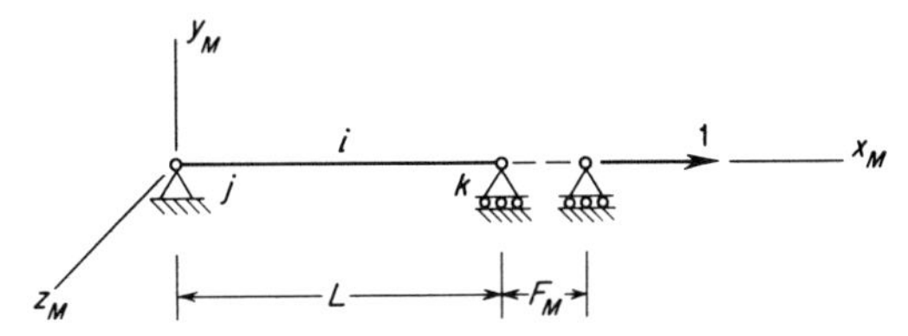

FIG. 3-3. Flexibility F_M for truss member.

column vector $\mathbf{A_{MU}}$ consisting of the member end-actions A_{MU} due to the unit load, and another is a column vector $\mathbf{A_{ML}}$ consisting of the member end-actions A_{ML} due to the actual loads. Each of these column vectors has as many elements as there are members in the truss. Hence, assuming there are m members in the truss, they have the following form:*

$$\mathbf{A_{MU}} = \{A_{MU1}, A_{MU2}, \ldots, A_{MUm}\} \tag{3-2}$$

$$\mathbf{A_{ML}} = \{A_{ML1}, A_{ML2}, \ldots, A_{MLm}\} \tag{3-3}$$

The numerical subscripts in these expressions identify the members of the truss.

The third matrix to be introduced is a diagonal matrix of the member flexibilities, denoted as $\mathbf{F_M}$:

$$\mathbf{F_M} = \begin{bmatrix} F_{M1} & 0 & \cdots & 0 \\ 0 & F_{M2} & \cdots & 0 \\ \cdots & \cdots & \cdots & \cdots \\ 0 & 0 & \cdots & F_{Mm} \end{bmatrix} \tag{3-4}$$

* In this book, whenever a column vector is written in a row, braces { } are used to enclose the vector.

This square matrix is of order $m \times m$ and contains the member flexibilities along its principal diagonal.

In order to express Eq. (c) in matrix form, it is only necessary to observe the result of carrying out the matrix product $\mathbf{A}'_{\mathbf{MU}}\mathbf{F}_{\mathbf{M}}\mathbf{A}_{\mathbf{ML}}$, in which $\mathbf{A}'_{\mathbf{MU}}$ is the transpose of the vector $\mathbf{A}_{\mathbf{MU}}$. If this product is evaluated, it is found to give the same expression as the one obtained by expanding the right-hand side of Eq. (c), namely,

$$A_{MU1}F_{M1}A_{ML1} + A_{MU2}F_{M2}A_{ML2} + \cdots + A_{MUm}F_{Mm}A_{MLm}$$

Therefore, the matrix equation for the displacement D becomes

$$D = \mathbf{A}'_{\mathbf{MU}}\mathbf{F}_{\mathbf{M}}\mathbf{A}_{\mathbf{ML}} \tag{3-5}$$

Equation (3-5) may be used to find any joint displacement D in a plane or space truss.

The matrices $\mathbf{A}_{\mathbf{MU}}$ and $\mathbf{A}_{\mathbf{ML}}$ appearing in Eq. (3-5) represent member end-actions due to the unit load and the actual loads, respectively. All such end-actions are obtained by static equilibrium analyses of the truss. The matrices $\mathbf{A}_{\mathbf{MU}}$ and $\mathbf{A}_{\mathbf{ML}}$ are called *transfer matrices,* because they represent the result of a transfer (by means of static equilibrium relationships) from one set of actions to another.

Example 1. As an illustration of the use of Eq. (3-5) for finding displacements, consider the plane truss shown in Fig. 3-4a and assume that the vertical displacement of joint B is to be determined. This displacement is assumed to be positive when in the positive direction of the y axis. Note that the truss is statically determinate and subjected to a load P in the negative y direction at

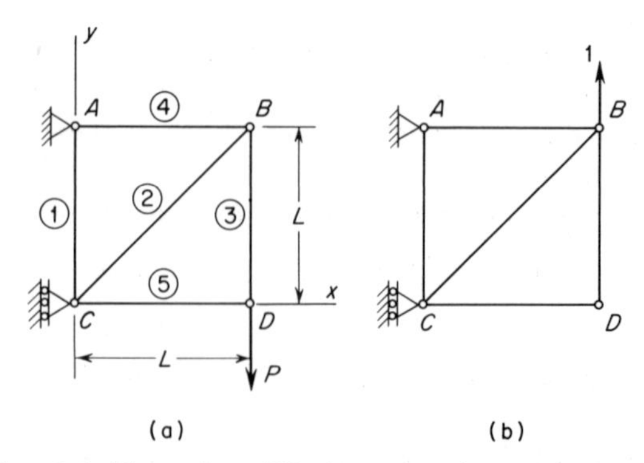

FIG. 3-4. Example 1: Displacements in a plane truss.

joint D. All members of the truss are assumed to have the same axial rigidity EA. The members are numbered consecutively from 1 to 5 as shown by the numbers enclosed in circles; the order in which the members are numbered is selected arbitrarily.

The matrix $\mathbf{F}_{\mathbf{M}}$ of member flexibilities may be constructed without difficulty, as follows:

$$\mathbf{F}_{M} = \frac{L}{EA}\begin{bmatrix} 1 & 0 & 0 & 0 & 0 \\ 0 & \sqrt{2} & 0 & 0 & 0 \\ 0 & 0 & 1 & 0 & 0 \\ 0 & 0 & 0 & 1 & 0 \\ 0 & 0 & 0 & 0 & 1 \end{bmatrix} \tag{d}$$

The transfer matrices $\mathbf{A}_{ML}$ and $\mathbf{A}_{MU}$ must be obtained next. The first of these matrices consists of the member end-actions due to the load P (see Fig. 3-4a), and each term in the matrix can be found readily by statics; thus, the matrix $\mathbf{A}_{ML}$ is

$$\mathbf{A}_{ML} = P\{1, -\sqrt{2}, 1, 1, 0\} \tag{e}$$

The transfer matrix $\mathbf{A}_{MU}$ consists of the end-actions due to a unit load corresponding to the desired displacement D, which means that the unit load must be taken in the positive y direction at joint B (see Fig. 3-4b). The member end-actions caused by the unit load can be found also by statics, and the results are as follows:

$$\mathbf{A}_{MU} = \{-1, \sqrt{2}, 0, -1, 0\} \tag{f}$$

The vertical displacement D at joint B is found by substituting the matrices $\mathbf{A}'_{MU}$, $\mathbf{F}_{M}$, and $\mathbf{A}_{ML}$ into Eq. (3-5) and performing the matrix multiplication; the result is

$$D = -2(1 + \sqrt{2})\frac{PL}{EA} = -4.83\frac{PL}{EA}$$

The minus sign in this result means that the displacement is in the negative y direction (downward).

Calculation of Several Displacements. It is normally necessary in the flexibility method to calculate several joint displacements in the structure; hence, it is desirable to extend the techniques illustrated by the preceding example to include this possibility. All that is necessary is to generalize the definition of the transfer matrix $\mathbf{A}_{MU}$. In the above discussion, this matrix was assumed to consist of one column, representing the member end-actions caused by the unit load (see Eq. 3-2). If there are several displacements to be calculated, there will be one unit load for each displacement and one set of end-actions for each unit load. Each of these sets of end-actions can be represented as a column in the matrix $\mathbf{A}_{MU}$; thus, $\mathbf{A}_{MU}$ becomes a rectangular matrix of order $m \times n$, in which m is the number of members in the truss and n is the number of displacements to be calculated. Therefore, the general form of this matrix is:

$$\mathbf{A}_{MU} = \begin{bmatrix} A_{MU11} & A_{MU12} & \cdots & A_{MU1n} \\ A_{MU21} & A_{MU22} & \cdots & A_{MU2n} \\ \cdots & \cdots & \cdots & \cdots \\ A_{MUm1} & A_{MUm2} & \cdots & A_{MUmn} \end{bmatrix} \tag{3-6}$$

Each element in the matrix $\mathbf{A}_{MU}$ has two subscripts, the first denoting the member of the truss and the second denoting the unit load. In other words, the element A_{MUij} is the force in the i-th bar due to the j-th unit load. In general terms, the matrix $\mathbf{A}_{MU}$ can be described as the transfer matrix that gives the member end-actions caused by the unit loads.

The various displacements in the truss can be placed in a displacement vector $\mathbf{D}$, as follows:

$$\mathbf{D} = \{D_1, D_2, \ldots, D_n\} \tag{3-7}$$

When the matrices $\mathbf{A}_{MU}$ and $\mathbf{D}$ are defined according to Eqs. (3-6) and (3-7), the equation for the displacements takes the following form:

$$\mathbf{D} = \mathbf{A}'_{MU}\mathbf{F}_M\mathbf{A}_{ML} \tag{3-8}$$

which can be considered as a generalized form of Eq. (3-5). In Eq. (3-8) the matrices $\mathbf{A}_{ML}$ and $\mathbf{F}_M$ are not altered from their previous meanings (see Eqs. 3-3 and 3-4) and are of order $m \times 1$ and $m \times m$, respectively. Note, however, that $\mathbf{A}_{MU}$ is now of order $m \times n$, $\mathbf{A}'_{MU}$ is of order $n \times m$, and $\mathbf{D}$ is of order $n \times 1$. The use of Eq. (3-8) in calculating displacements is illustrated in the next example.

Example 2. The plane truss shown in Fig. 3-5a is repeated from the earlier example of Fig. 3-4. However, the object of the analysis now is to calculate the horizontal and vertical components of the displacements of joints B and D. These displacements are denoted D_1, D_2, D_3, and D_4 (see Fig. 3-5a), and their positive directions correspond to the positive directions of the x and y axes shown in the figure.

The matrix $\mathbf{F}_M$ of member flexibilities and the transfer matrix $\mathbf{A}_{ML}$ of member end-actions caused by the load P are the same as in the preceding example (see Eqs. d and e). However, the transfer matrix $\mathbf{A}_{MU}$ must be determined anew. The first column of $\mathbf{A}_{MU}$ represents the end-actions due to a unit load corresponding to D_1 (Fig. 3-5b), the second column represents end-actions due to a unit load

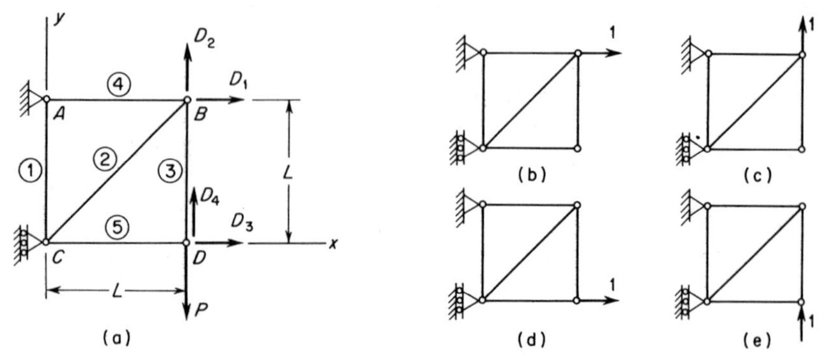

Fig. 3-5. Example 2: Displacements in a plane truss.

corresponding to D_2 (Fig. 3-5c), and so forth, for the remaining unit loads shown in Figs. 3-5d and 3-5e. Thus, the matrix $\mathbf{A}_{MU}$ is as follows:

$$\mathbf{A}_{\mathrm{MU}} = \begin{bmatrix} 0 & -1 & 0 & -1 \\ 0 & \sqrt{2} & 0 & \sqrt{2} \\ 0 & 0 & 0 & -1 \\ 1 & -1 & 0 & -1 \\ 0 & 0 & 1 & 0 \end{bmatrix} \tag{g}$$

Transposing the matrix $\mathbf{A}_{\mathrm{MU}}$ and then substituting $\mathbf{A}'_{\mathrm{MU}}$, $\mathbf{F}_{\mathrm{M}}$, and $\mathbf{A}_{\mathrm{ML}}$ into Eq. (3-8) yields the following result for the displacement vector $\mathbf{D}$:

$$\mathbf{D} = \frac{PL}{EA}\begin{bmatrix} 1 \\ -2(1+\sqrt{2}) \\ 0 \\ -3-2\sqrt{2} \end{bmatrix} = \frac{PL}{EA}\begin{bmatrix} 1 \\ -4.83 \\ 0 \\ -5.83 \end{bmatrix}$$

Thus, the horizontal displacement D_1 of joint B is equal to PL/EA; the vertical displacement D_2 of joint B is $-4.83\ PL/EA$; the horizontal displacement D_3 of joint D is zero; and the vertical displacement D_4 of joint D is $-5.83\ PL/EA$. The negative signs for D_2 and D_4 indicate that these displacements are in the negative y direction (downward).

This example illustrates how the joint displacements for a truss can be calculated by the matrix operations shown in Eq. (3-8). The equation is suitable when only one load system is being considered. However, the procedure can be generalized to include more than one load system, as described in the following discussion.

Several Load Systems. When a truss is to be analyzed for several load systems, it is necessary to have additional columns in the transfer matrix $\mathbf{A}_{\mathrm{ML}}$. Each column of the matrix contains the member end-actions due to one of the load systems. For instance, if there are p load systems, the matrix $\mathbf{A}_{\mathrm{ML}}$ will be of order $m \times p$, in which (as before) m is the number of members; thus:

$$\mathbf{A}_{\mathrm{ML}} = \begin{bmatrix} A_{ML11} & A_{ML12} & \cdots & A_{ML1p} \\ A_{ML21} & A_{ML22} & \cdots & A_{ML2p} \\ \cdots & \cdots & \cdots & \cdots \\ A_{MLm1} & A_{MLm2} & \cdots & A_{MLmp} \end{bmatrix} \tag{3-9}$$

The matrix $\mathbf{F}_{\mathrm{M}}$ of member flexibilities and the transfer matrix $\mathbf{A}_{\mathrm{MU}}$ remain unchanged (see Eqs. 3-4 and 3-6). However, the displacement matrix $\mathbf{D}$ now has one column corresponding to each load system, and therefore is of order $n \times p$:

$$\mathbf{D} = \begin{bmatrix} D_{11} & D_{12} & \cdots & D_{1p} \\ D_{21} & D_{22} & \cdots & D_{2p} \\ \cdots & \cdots & \cdots & \cdots \\ D_{n1} & D_{n2} & \cdots & D_{np} \end{bmatrix} \tag{3-10}$$

Finally, the equation for the displacements has the same form as in earlier equations, namely:

$$\mathbf{D} = \mathbf{A}'_{MU}\mathbf{F}_M\mathbf{A}_{ML} \tag{3-8}$$

repeated

except that the matrices $\mathbf{A}_{ML}$ and $\mathbf{D}$ are now given by Eqs. (3-9) and (3-10), respectively.

Equation (3-8) can be used in calculating any number of joint displacements in either a plane or space truss, due to several different load systems. Example 3 illustrates the use of the equation.

Example 3. The truss shown in Fig. 3-6 is the same truss considered in the earlier examples, except that it is subjected to two load systems. The first load system consists of the force P_1 acting at joint D (Fig. 3-6a), while the second load system consists of the three forces P_2 acting at joints B and D (Fig. 3-6b). The displacements to be calculated are the horizontal and vertical displacements at joints B and D, as shown in Fig. 3-5a.

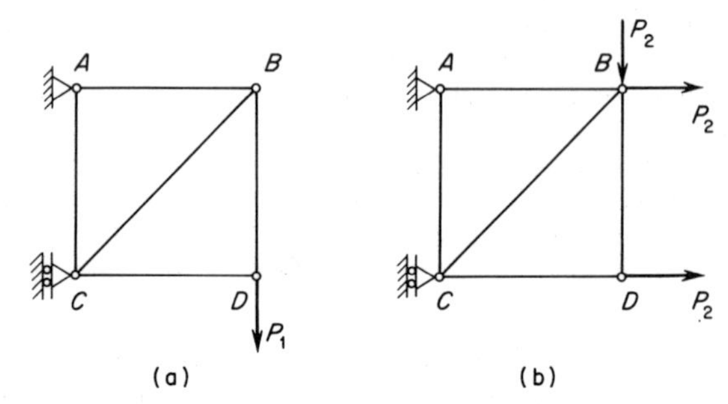

FIG. 3-6. Example 3: Truss with two load systems.

The matrices $\mathbf{A}'_{MU}$ and $\mathbf{F}_M$ to be substituted into Eq. (3-8) are the same as those used in Example 2 (see Eqs. g and d). The transfer matrix $\mathbf{A}_{ML}$ consists of the end-actions caused by the two load systems:

$$\mathbf{A}_{ML} = \begin{bmatrix} P_1 & P_2 \\ -\sqrt{2}P_1 & -\sqrt{2}P_2 \\ P_1 & 0 \\ P_1 & 2P_2 \\ 0 & P_2 \end{bmatrix}$$

When the three matrices $\mathbf{A}'_{MU}$, $\mathbf{F}_M$, and $\mathbf{A}_{ML}$ are substituted into Eq. (3-8), the result is

$$\mathbf{D} = \frac{L}{EA}\begin{bmatrix} P_1 & 2P_2 \\ -2(1+\sqrt{2})P_1 & -(3+2\sqrt{2})P_2 \\ 0 & P_2 \\ -(3+2\sqrt{2})P_1 & -(3+2\sqrt{2})P_2 \end{bmatrix}$$

and thus the desired displacements have been determined for the two load systems.

3.4 Joint Displacements in Beams. The analysis of a continuous beam by the flexibility method requires the calculation of joint displacements in the released structure. Therefore, in this article the matrix equations which are needed for finding joint displacements in a statically determinate beam are developed. For the reasons explained in Art. 3.2, all loads on the beam are assumed to be in the form of combined joint loads.

The basic equation for finding beam displacements is the unit-load equation derived in Appendix A (see Eq. A-33):

$$D = \int \frac{M_U M_L \, dx}{EI} \tag{a}$$

where D is the displacement to be calculated, M_U is the bending moment at any cross-section of the beam due to a unit load corresponding to the displacement, M_L is the bending moment due to the actual loads which cause the displacement, and EI denotes the flexural rigidity of the cross-section of the beam. It should be noted that all members of the structure must be included in the integration of Eq. (a).

In order to put Eq. (a) into matrix form, it is desirable to express the displacement D as the sum of separate terms which are evaluated for each member of the structure. Thus, Eq. (a) can be rewritten in the more explicit form

$$D = \Sigma \int \frac{M_U M_L \, dx}{EI} \tag{b}$$

in which the summation sign is used to indicate that the integral is summed for all members of the structure in order to obtain the displacement. For any particular member it is necessary that both M_U and M_L be evaluated according to the same sign convention for the bending moment.

Now consider the evaluation of the integral in Eq. (b) for a typical member i of a beam (see Fig. 3-7). The joints at the left- and right-hand ends of the member are denoted as joints j and k, respectively. The member-oriented axes have their origin at the j end, and the x_M axis is directed from end j to end k. The y_M axis is chosen so that the x_M-y_M plane is the plane of bending of the beam. The member is subjected to forces and couples at its ends only, since all loads act at the joints. At the k end of the member the end-actions will be a shearing force A_{M1} (positive in the y_M sense) and a bending moment A_{M2} (positive in the z_M sense). There may also be an axial force in the member, but this is omitted when only the effects of flexural deformations are under consideration. The end-actions caused by the loads will be

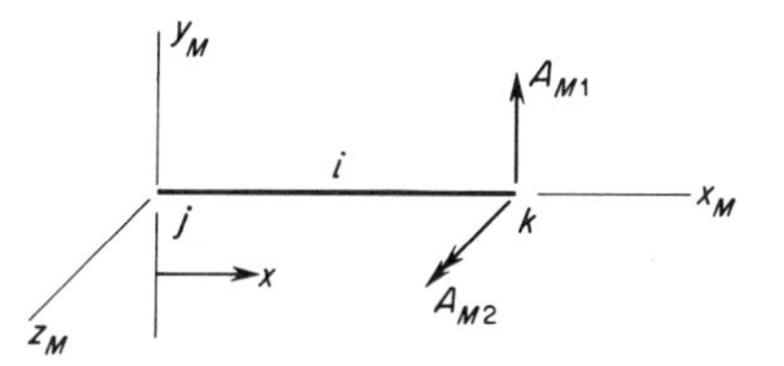

FIG. 3-7. End-actions for beam member.

denoted A_{ML1} and A_{ML2} for the shearing force and bending moment, respectively, and the end-actions due to the unit load will be denoted A_{MU1} and A_{MU2}.

The moments M_U and M_L (see Eq. b) at any section of a member can be readily expressed in terms of the end-actions. Taking a coordinate distance x from the left-hand end of the member (see Fig. 3-7), and also assuming that compression on top of the beam represents positive bending moment, the equations for M_U and M_L become

$$M_U = A_{MU1}(L - x) + A_{MU2}$$

$$M_L = A_{ML1}(L - x) + A_{ML2}$$

in which L is the length of the beam. These expressions for M_U and M_L can be substituted into the integral appearing in Eq. (b). Then, when the integral is evaluated between the limits 0 and L, and if it is assumed that EI is constant, the resulting expression can be written in the form

$$\int \frac{M_U M_L \, dx}{EI} = \frac{1}{EI}\left[A_{MU1}A_{ML1}\frac{L^3}{3} + A_{MU1}A_{ML2}\frac{L^2}{2} + A_{MU2}A_{ML1}\frac{L^2}{2} + A_{MU2}A_{ML2}L\right] \quad \text{(c)}$$

The expression appearing on the right-hand side of Eq. (c) can be written in matrix form. To accomplish this result, two vectors of end-actions for member i are introduced:

$$\mathbf{A}_{\mathbf{MU}i} = \{A_{MU1}, A_{MU2}\} \quad \text{(3-11)}$$

$$\mathbf{A}_{\mathbf{ML}i} = \{A_{ML1}, A_{ML2}\} \quad \text{(3-12)}$$

The first of these vectors contains the shearing force A_{MU1} and bending moment A_{MU2} at the k end of member i due to the unit load; similarly, the second vector contains the end-actions due to the joint loads. Finally, a member flexibility matrix $\mathbf{F}_{\mathbf{M}i}$ is defined:

$$\mathbf{F}_{\mathbf{M}i} = \begin{bmatrix} F_{M11} & F_{M12} \\ F_{M21} & F_{M22} \end{bmatrix} = \begin{bmatrix} \frac{L^3}{3EI} & \frac{L^2}{2EI} \\ \frac{L^2}{2EI} & \frac{L}{EI} \end{bmatrix} \quad \text{(3-13)}$$

The elements of this member flexibility matrix may be visualized as the displacements in a cantilever beam caused by unit values of the end-actions (see Figs. 3-8b and 3-8c). Equation (c) can now be expressed in matrix form as follows:

$$\int \frac{M_U M_L \, dx}{EI} = \mathbf{A}'_{\mathbf{MU}i}\mathbf{F}_{\mathbf{M}i}\mathbf{A}_{\mathbf{ML}i} \quad \text{(d)}$$

The validity of this expression can be checked readily by performing the indicated matrix multiplications.

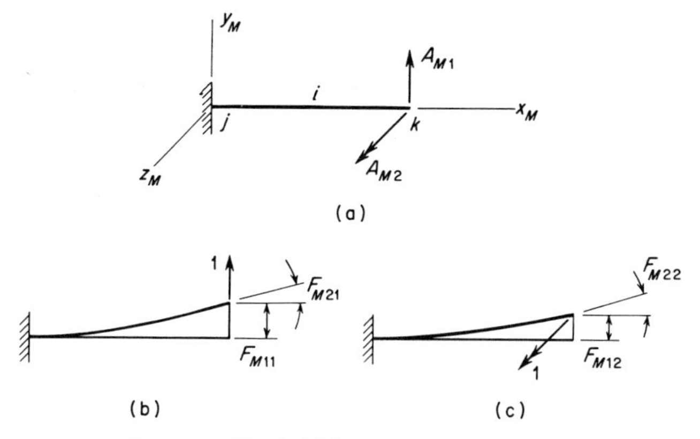

FIG. 3-8. Flexibilities for beam member.

Expressions of the form given by Eq. (d) must now be summed for all members of the structure in order to obtain the displacement D (see Eq. b). Thus, the displacement D is given by the following equation:

$$D = \sum (\mathbf{A}'_{\mathrm{MU}i}\mathbf{F}_{\mathrm{M}i}\mathbf{A}_{\mathrm{ML}i}) = \mathbf{A}'_{\mathrm{MU1}}\mathbf{F}_{\mathrm{M1}}\mathbf{A}_{\mathrm{ML1}} + \mathbf{A}'_{\mathrm{MU2}}\mathbf{F}_{\mathrm{M2}}\mathbf{A}_{\mathrm{ML2}} + \cdots + \mathbf{A}'_{\mathrm{MU}m}\mathbf{F}_{\mathrm{M}m}\mathbf{A}_{\mathrm{ML}m} \qquad \text{(e)}$$

This sum may also be expressed in matrix form. To do so requires the introduction of transfer matrices of end-actions for all of the members, as well as a matrix of flexibilities for all of the members. The matrices of end-actions are

$$\mathbf{A}_{\mathrm{MU}} = \{\mathbf{A}_{\mathrm{MU1}}, \mathbf{A}_{\mathrm{MU2}}, \cdots, \mathbf{A}_{\mathrm{MU}m}\} \qquad \text{(3-14)}$$

$$\mathbf{A}_{\mathrm{ML}} = \{\mathbf{A}_{\mathrm{ML1}}, \mathbf{A}_{\mathrm{ML2}}, \cdots, \mathbf{A}_{\mathrm{ML}m}\} \qquad \text{(3-15)}$$

Each of these matrices is of order $2m \times 1$ and consists of submatrices containing the end-actions for the individual members. In the case of the matrix $\mathbf{A}_{\mathrm{MU}}$ the end-actions are due to the unit load on the structure, while in the case of $\mathbf{A}_{\mathrm{ML}}$ they are due to the joint loads.

The matrix $\mathbf{F}_{\mathrm{M}}$ of member flexibilities is formed by placing the individual flexibility matrices (see Eq. 3-13) on the principal diagonal:

$$\mathbf{F}_{\mathrm{M}} = \begin{bmatrix} \mathbf{F}_{\mathrm{M1}} & \mathbf{O} & \cdots & \mathbf{O} \\ \mathbf{O} & \mathbf{F}_{\mathrm{M2}} & \cdots & \mathbf{O} \\ \cdots & \cdots & \cdots & \cdots \\ \mathbf{O} & \mathbf{O} & \cdots & \mathbf{F}_{\mathrm{M}m} \end{bmatrix} \qquad \text{(3-16)}$$

Thus, the matrix $\mathbf{F}_{\mathrm{M}}$ is of order $2m \times 2m$ since it consists of submatrices of order 2×2. Having defined the transfer matrices $\mathbf{A}_{\mathrm{MU}}$ and $\mathbf{A}_{\mathrm{ML}}$ (Eqs. 3-14 and 3-15) and the matrix $\mathbf{F}_{\mathrm{M}}$ of member flexibilities (Eq. 3-16), it is possible to write Eq. (e) in the form of Eq. (3-5):

$$D = \mathbf{A}'_{\mathrm{MU}}\mathbf{F}_{\mathrm{M}}\mathbf{A}_{\mathrm{ML}} \qquad \text{(3-5)}$$

repeated

Thus, this equation can be used for calculating a beam displacement as well as a truss displacement, provided the matrices on the right-hand side of the equation are formulated in the appropriate manner.

In using Eq. (3-5) to calculate a displacement in a beam, the procedure is to begin by obtaining the flexibility matrices for each member (see Eq. 3-13). These matrices can be placed along the principal diagonal of the matrix $\mathbf{F}_{\mathrm{M}}$ (Eq. 3-16). Next, the end-actions at the k ends of all members due to the unit load and due to the actual loads are calculated by statics. These end-actions are assembled into submatrices, such as $\mathbf{A}_{\mathrm{MU}i}$ and $\mathbf{A}_{\mathrm{ML}i}$ (Eqs. 3-11 and 3-12), and then into the transfer matrices $\mathbf{A}_{\mathrm{MU}}$ and $\mathbf{A}_{\mathrm{ML}}$ (Eqs. 3-14 and 3-15). Finally, the execution of the matrix operations indicated in Eq. (3-5) will lead to the value of the displacement D. These steps are analogous to those performed previously when finding displacements of trusses. An example of the use of Eq. (3-5) when finding beam displacements will now be given.

Example 1. The beam shown in Fig. 3-9a is subjected to a couple M acting at joint B and a concentrated force P acting at joint C. The members AB and BC have lengths L and $L/2$, respectively, and both members have the same flexural rigidity EI. The vertical displacement at joint C (positive in the positive direction of the y axis) is to be calculated.

With the members AB and BC denoted as members 1 and 2, respectively, the member flexibility matrices $\mathbf{F}_{\mathrm{M1}}$ and $\mathbf{F}_{\mathrm{M2}}$ are as follows:

$$\mathbf{F}_{\mathrm{M1}} = \begin{bmatrix} \dfrac{L^3}{3EI} & \dfrac{L^2}{2EI} \\ \dfrac{L^2}{2EI} & \dfrac{L}{EI} \end{bmatrix} \qquad \mathbf{F}_{\mathrm{M2}} = \begin{bmatrix} \dfrac{L^3}{24EI} & \dfrac{L^2}{8EI} \\ \dfrac{L^2}{8EI} & \dfrac{L}{2EI} \end{bmatrix}$$

Next, the matrix $\mathbf{F}_{\mathrm{M}}$ is formed as indicated by Eq. (3-16):

$$\mathbf{F}_{\mathrm{M}} = \frac{L}{24EI} \begin{bmatrix} 8L^2 & 12L & 0 & 0 \\ 12L & 24 & 0 & 0 \\ 0 & 0 & L^2 & 3L \\ 0 & 0 & 3L & 12 \end{bmatrix}$$

With the loads M and P (Fig. 3-9a) acting on the beam, the vectors of end-actions can be calculated by statics. The vectors are denoted $\mathbf{A}_{\mathrm{ML1}}$ and $\mathbf{A}_{\mathrm{ML2}}$ for members 1 and 2, respectively, and are:

$$\mathbf{A}_{\mathrm{ML1}} = \begin{bmatrix} -\dfrac{M}{L} + \dfrac{P}{2} \\ M - \dfrac{PL}{2} \end{bmatrix} \qquad \mathbf{A}_{\mathrm{ML2}} = \begin{bmatrix} -P \\ 0 \end{bmatrix}$$

The first element in each of these vectors represents the shearing force at the right-hand end of the member, and the second represents the bending moment. The positive directions of these quantities are as shown in Fig. 3-7. The matrices $\mathbf{A}_{\mathrm{MU1}}$ and $\mathbf{A}_{\mathrm{MU2}}$ are found in the same manner, except that the shearing force

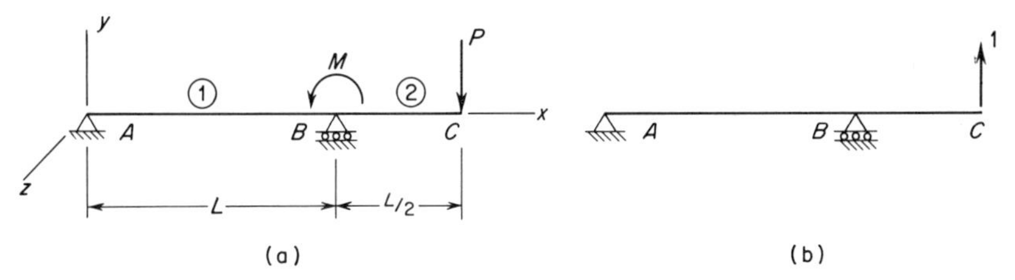

FIG. 3-9. Example 1: Displacements in a beam.

and bending moment are caused by the unit load corresponding to the desired displacement (see Fig. 3-9b):

$$\mathbf{A}_{\mathrm{MU1}} = \begin{bmatrix} -\dfrac{1}{2} \\ \dfrac{L}{2} \end{bmatrix} \qquad \mathbf{A}_{\mathrm{MU2}} = \begin{bmatrix} 1 \\ 0 \end{bmatrix}$$

Finally, the transfer matrices $\mathbf{A}_{\mathrm{ML}}$ and $\mathbf{A}_{\mathrm{MU}}$ are assembled by using the matrices of end-actions as submatrices:

$$\mathbf{A}_{\mathrm{ML}} = \left\{ -\frac{M}{L} + \frac{P}{2}, \quad M - \frac{PL}{2}, \quad -P, \quad 0 \right\}$$

$$\mathbf{A}_{\mathrm{MU}} = \left\{ -\frac{1}{2}, \quad \frac{L}{2}, \quad 1, \quad 0 \right\}$$

The last step in the calculation of the displacement D is to substitute the vectors $\mathbf{A}_{\mathrm{ML}}$ and $\mathbf{A}'_{\mathrm{MU}}$ into Eq. (3-5) and to perform the matrix multiplications. When this is done, the result is

$$D = \frac{ML^2}{6EI} - \frac{PL^3}{8EI}$$

Thus, the vertical displacement at joint C has been obtained.

Calculation of Several Displacements. If it is desired to calculate several displacements in the structure, a generalized form of the matrix equation (3-5) must be used. In the new equation the transfer matrix $\mathbf{A}_{\mathrm{ML}}$ of end-actions will not be changed, nor will the matrix $\mathbf{F}_{\mathrm{M}}$ of member flexibilities. However, additional columns must be added to the matrix $\mathbf{A}_{\mathrm{MU}}$. There will be one column of submatrices of member end-actions for each unit load that must be considered. If there are n joint displacements to be calculated, there will be n columns of submatrices in the matrix $\mathbf{A}_{\mathrm{MU}}$, as shown:

$$\mathbf{A}_{\mathrm{MU}} = \begin{bmatrix} \mathbf{A}_{\mathrm{MU}11} & \mathbf{A}_{\mathrm{MU}12} & \cdots & \mathbf{A}_{\mathrm{MU}1n} \\ \mathbf{A}_{\mathrm{MU}21} & \mathbf{A}_{\mathrm{MU}22} & \cdots & \mathbf{A}_{\mathrm{MU}2n} \\ \cdots & \cdots & \cdots & \cdots \\ \mathbf{A}_{\mathrm{MU}m1} & \mathbf{A}_{\mathrm{MU}m2} & \cdots & \mathbf{A}_{\mathrm{MU}mn} \end{bmatrix} \tag{3-17}$$

The first column of this matrix contains submatrices of member end-actions due to the first unit load, the second column contains end-actions due to the second unit load, and so on for the n unit loads corresponding to the n displacements to be calculated. In general, the submatrix $\mathbf{A}_{\mathrm{MU}ij}$ represents end-actions for the i-th member due to a unit load corresponding to the j-th displacement. Since each submatrix is of order 2×1, the order of the transfer matrix $\mathbf{A}_{\mathrm{MU}}$ is $2m \times n$.

The n joint displacements being calculated are denoted $D_1, D_2, \ldots D_n$ and are the elements of a displacement vector $\mathbf{D}$ having the form shown in Eq. (3-7). This displacement vector is given by the matrix equation (3-8):

$$\mathbf{D} = \mathbf{A}'_{\mathrm{MU}}\mathbf{F}_{\mathrm{M}}\mathbf{A}_{\mathrm{ML}} \tag{3-8}$$

repeated

in which the matrix $\mathbf{A}_{\mathrm{MU}}$ is now defined by Eq. (3-17). The sizes of the matrices in this equation are as follows: $\mathbf{D}$ is of order $n \times 1$, $\mathbf{A}'_{\mathrm{MU}}$ is of order $n \times 2m$; $\mathbf{F}_{\mathrm{M}}$ is of order $2m \times 2m$, and $\mathbf{A}_{\mathrm{ML}}$ is of order $2m \times 1$.

An example illustrating the calculation of several displacements is given later.

Several Load Systems. If the displacements in the structure are to be evaluated for more than one load system, it is necessary to have additional columns in the matrix $\mathbf{A}_{\mathrm{ML}}$. The first column of $\mathbf{A}_{\mathrm{ML}}$ contains member end-actions due to the first load system, the second column contains member end-actions due to the second load system, and so on for as many load systems as necessary. If there are p load systems, the matrix $\mathbf{A}_{\mathrm{ML}}$ will be as follows:

$$\mathbf{A}_{\mathrm{ML}} = \begin{bmatrix} \mathbf{A}_{\mathrm{ML11}} & \mathbf{A}_{\mathrm{ML12}} & \cdots & \mathbf{A}_{\mathrm{ML1}p} \\ \mathbf{A}_{\mathrm{ML21}} & \mathbf{A}_{\mathrm{ML22}} & \cdots & \mathbf{A}_{\mathrm{ML2}p} \\ \cdots & \cdots & \cdots & \cdots \\ \mathbf{A}_{\mathrm{ML}m1} & \mathbf{A}_{\mathrm{ML}m2} & \cdots & \mathbf{A}_{\mathrm{ML}mp} \end{bmatrix} \tag{3-18}$$

The matrix $\mathbf{D}$ of displacements will also have additional columns, one of which corresponds to each load system (see Eq. 3-10).

Therefore, if there are several load systems as well as several displacements to be found, the matrix equation for the displacements has the same form as Eq. (3-8), but the matrices $\mathbf{A}_{\mathrm{MU}}$, $\mathbf{F}_{\mathrm{M}}$, and $\mathbf{A}_{\mathrm{ML}}$ have the forms given by Eqs. (3-17), (3-16), and (3-18), respectively.

The following example illustrates the use of Eq. (3-8) for finding displacements of beams.

Example 2. Figure 3-10a shows the beam of Example 1, and Fig. 3-10b shows the same beam subjected to a second system of loads. Assume that all joint displacements of the beam are to be determined for both load systems. The joint displacements are indicated in Fig. 3-10c with translations positive in the positive y direction and rotations positive in the positive z sense.

In this example the transfer matrix $\mathbf{A}_{\mathrm{ML}}$ has two columns, one for each of the load systems. For instance, under the first load system (Fig. 3-10a) the end-actions for members 1 and 2 are given by the following vectors:

$$\mathbf{A}_{\mathrm{ML11}} = \begin{bmatrix} -\dfrac{M}{L} + \dfrac{P}{2} \\ \\ M - \dfrac{PL}{2} \end{bmatrix} \qquad \mathbf{A}_{\mathrm{ML21}} = \begin{bmatrix} -P \\ 0 \end{bmatrix}$$

Similarly, the end-actions for the second load system (Fig. 3-10b) are:

$$\mathbf{A}_{\mathrm{ML12}} = \begin{bmatrix} \dfrac{P}{2} \\ \\ 2M - \dfrac{PL}{2} \end{bmatrix} \qquad \mathbf{A}_{\mathrm{ML22}} = \begin{bmatrix} -P \\ 0 \end{bmatrix}$$

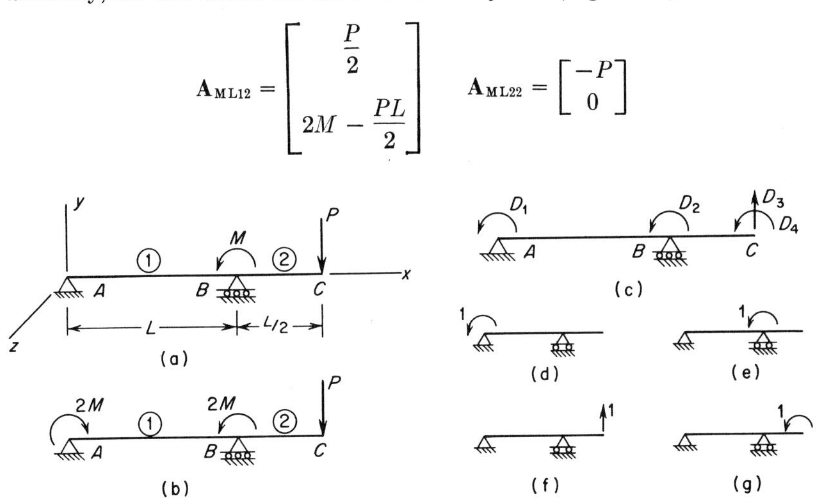

FIG. 3-10. Example 2: Beam with two load systems.

Thus, the transfer matrix $\mathbf{A}_{\mathrm{ML}}$ becomes

$$\mathbf{A}_{\mathrm{ML}} = \begin{bmatrix} -\dfrac{M}{L} + \dfrac{P}{2} & \dfrac{P}{2} \\ \\ M - \dfrac{PL}{2} & 2M - \dfrac{PL}{2} \\ \\ -P & -P \\ 0 & 0 \end{bmatrix}$$

The four unit loads corresponding to the four displacements to be calculated are shown in Figs. 3-10d, 3-10e, 3-10f, and 3-10g. The transfer matrix $\mathbf{A}_{\mathrm{MU}}$ is constructed of the various submatrices representing end-actions due to the unit loads. For example, the end-actions for members 1 and 2, respectively, due to the first unit load are:

$$\mathbf{A}_{\mathrm{MU11}} = \begin{bmatrix} -\dfrac{1}{L} \\ \\ 0 \end{bmatrix} \qquad \mathbf{A}_{\mathrm{MU21}} = \begin{bmatrix} 0 \\ 0 \end{bmatrix}$$

These submatrices constitute the first column of the $\mathbf{A}_{\mathrm{MU}}$ matrix (see Eq. 3-17). The submatrices for the second column of $\mathbf{A}_{\mathrm{MU}}$ are obtained by finding end-actions for the beam of Fig. 3-10e:

$$\mathbf{A}_{MU12} = \begin{bmatrix} -\frac{1}{L} \\ 1 \end{bmatrix} \qquad \mathbf{A}_{MU22} = \begin{bmatrix} 0 \\ 0 \end{bmatrix}$$

Similarly, the submatrices for the third and fourth columns of $\mathbf{A}_{MU}$ are obtained from Figs. 3-10f and 3-10g, respectively. When all of the submatrices have been calculated, they are used in forming the complete transfer matrix of end-actions:

$$\mathbf{A}_{MU} = \begin{bmatrix} -\frac{1}{L} & -\frac{1}{L} & -\frac{1}{2} & -\frac{1}{L} \\ 0 & 1 & \frac{L}{2} & 1 \\ 0 & 0 & 1 & 0 \\ 0 & 0 & 0 & 1 \end{bmatrix}$$

The matrix $\mathbf{F}_M$ of member flexibilities is the same as in Example 1, which dealt with the same beam.

The remaining step in the solution consists of substituting $\mathbf{A}'_{MU}$, $\mathbf{F}_M$, and $\mathbf{A}_{ML}$ into Eq. (3-8), thereby yielding the matrix $\mathbf{D}$:

$$\mathbf{D} = \begin{bmatrix} D_{11} & D_{12} \\ D_{21} & D_{22} \\ D_{31} & D_{32} \\ D_{41} & D_{42} \end{bmatrix} = \frac{L}{24EI} \begin{bmatrix} -4M + 2PL & -24M + 2PL \\ 8M - 4PL & 24M - 4PL \\ 4ML - 3PL^2 & 12ML - 3PL^2 \\ 8M - 7PL & 24M - 7PL \end{bmatrix}$$

Thus, all four joint displacements caused by each of the two load systems have been calculated by means of one matrix equation.

Effects of Shearing Deformations. The effects of shearing deformations can be included readily in the calculation of displacements of beams if desired. All that is required is to incorporate shearing effects in the determination of the member flexibility matrix $\mathbf{F}_{Mi}$ (Eq. 3-13). The only element in this matrix that is affected is F_{M11} (see Fig. 3-8b), which becomes

$$F_{M11} = \frac{L^3}{3EI} + \frac{Lf}{GA} \tag{3-19}$$

in which GA/f is the shearing rigidity for the beam (see Art. A.1, Appendix A). After this change is made in the matrix $\mathbf{F}_{Mi}$, all remaining calculations proceed as previously described. The resulting displacements will include the effects of both flexural and shearing deformations.

3.5 Joint Displacements in Plane Frames. A typical member i from a plane frame is shown in Fig. 3-11. The joints at the ends of the member are denoted as joints j and k, and the member axes x_M, y_M, and z_M have their origin at the j end. The x_M-y_M plane is assumed to be the plane of bending of the member. Normally, the effects of axial deformations in a plane frame are negligible in comparison with the effects of flexural deformations. As a result, the only end-actions that must be considered at end k are the shearing force A_{M1} and bending moment A_{M2} shown in Fig. 3-11. If axial deformations are to be included in the analysis, a third end-action (the axial force) must be taken into account also, as described later.

The end-actions shown in Fig. 3-11 for a plane frame member are the same as described in the preceding article for a beam member (see Fig.

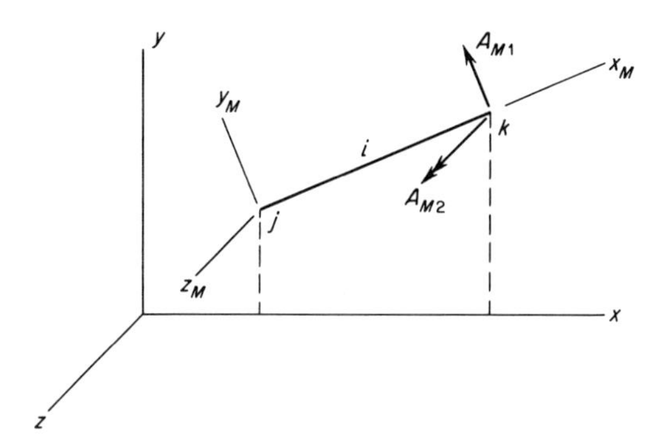

FIG. 3-11. End-actions for plane frame member.

3-7). Therefore, the member flexibility matrix for a plane frame member (when axial deformations are neglected) is the same as for a beam member (see Eq. 3-13). Furthermore, it also follows that all of the matrix equations given in the preceding article can be used in finding displacements of plane frames. The following example illustrates the procedure.

Example 1. The statically determinate plane frame shown in Fig. 3-12a is subjected to loads at joints B and C. The members of the frame are numbered as shown in the figure, and both members have the same flexural rigidity EI. The displacement D_1 of joint C in the x direction is to be determined (see Fig. 3-12b, which shows a unit load corresponding to D_1).

The j and k ends of each member must be specified in order that the positions of the member-oriented axes and the end-actions can be established according to the convention shown in Fig. 3-11. In this example, it is assumed that the j and k ends of member 1 are A and B, respectively, and of member 2 are B and C, respectively.

The member flexibility matrices are found easily (see Eq. 3-13):

$$\mathbf{F}_{M1} = \mathbf{F}_{M2} = \frac{L}{6EI}\begin{bmatrix} 2L^2 & 3L \\ 3L & 6 \end{bmatrix}$$

after which the matrix $\mathbf{F}_M$ can be formed by using $\mathbf{F}_{M1}$ and $\mathbf{F}_{M2}$ as submatrices (see Eq. 3-16).

The end-actions for members 1 and 2 due to the loads are found by statical analysis of the structure shown in Fig. 3-12a; thus, the vectors $\mathbf{A}_{ML1}$ and $\mathbf{A}_{ML2}$ are

$$\mathbf{A}_{ML1} = \frac{3P}{8}\begin{bmatrix} -1 \\ L \end{bmatrix} \qquad \mathbf{A}_{ML2} = \frac{P}{24}\begin{bmatrix} 9 \\ -4L \end{bmatrix}$$

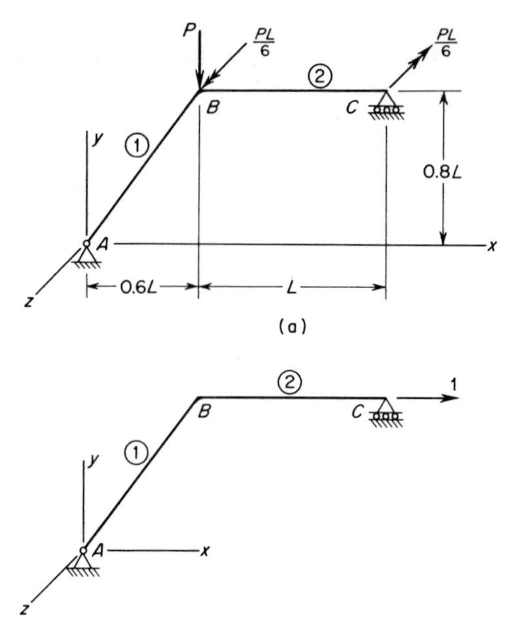

FIG. 3-12. Example 1: Displacements in a plane frame.

In each of these vectors the first element represents the shearing force at the k end of the member, and the second element represents the bending moment. The complete transfer matrix of end-actions is

$$\mathbf{A}_{\mathbf{ML}} = \frac{P}{24}\{-9, 9L, 9, -4L\}$$

The unit load corresponding to the displacement D_1 is a force in the x direction at joint C, as shown in Fig. 3-12b. This unit load produces end-actions for the two members as follows:

$$\mathbf{A}_{\mathbf{MU1}} = \tfrac{1}{2}\begin{bmatrix} -1 \\ L \end{bmatrix} \qquad \mathbf{A}_{\mathbf{MU2}} = \tfrac{1}{2}\begin{bmatrix} 1 \\ 0 \end{bmatrix}$$

From these results the transfer matrix $\mathbf{A}_{\mathbf{MU}}$ is constructed:

$$\mathbf{A}_{\mathbf{MU}} = \tfrac{1}{2}\{-1, L, 1, 0\}$$

The displacement D_1 may now be found by substituting into Eq. (3-5), and solving:

$$D_1 = \mathbf{A}'_{\mathbf{MU}}\mathbf{F}_{\mathbf{M}}\mathbf{A}_{\mathbf{ML}} = \frac{PL^3}{12EI}$$

Thus, the displacement at joint C is in the positive x direction.

If the structure in this example were subjected to several load systems, the matrix $\mathbf{A}_{\mathbf{ML}}$ would become a rectangular matrix with several columns. Also, if more than one displacement were to be calculated, the matrix $\mathbf{A}_{\mathbf{MU}}$ would have one column corresponding to each displacement. The matrix $\mathbf{D}$ for the displacements then would have as many rows as displacements to be found, and as many columns as load systems. Thus, the example can be extended without difficulty to include these more general possibilities, as illustrated previously in Art. 3.4 for a beam.

Effects of Axial Deformations. The preceding example illustrated the techniques for calculating displacements in a plane frame when only bending effects are taken into consideration. The solution was very similar to that for a beam, because the member flexibility matrices for a beam and a plane frame have the same form. However, there are instances in a plane frame analysis when it becomes necessary to take into account the effects of axial deformations. These effects can be incorporated into the analysis by adding suitable terms to the member flexibility matrices and by including the axial forces in the transfer matrices of end-actions. Specifically, the end-actions for a typical member i now become the axial force A_{M1}, the shearing force A_{M2}, and the bending moment A_{M3} shown in Fig. 3-13. These three end-actions will always be taken in the order named. The corresponding member flexibility matrix is

$$\mathbf{F}_{\mathrm{M}i} = \begin{bmatrix} \frac{L}{EA} & 0 & 0 \\ 0 & \frac{L^3}{3EI} & \frac{L^2}{2EI} \\ 0 & \frac{L^2}{2EI} & \frac{L}{EI} \end{bmatrix} \qquad (3\text{-}20)$$

as can be readily deduced from the expressions for member flexibilities given previously in Eqs. (3-1) and (3-13).

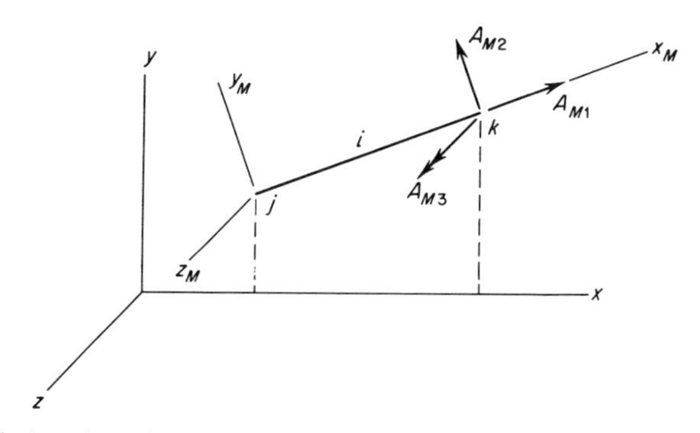

FIG. 3-13. End-actions for plane frame member when axial deformations are considered.

Because there are three end-actions for each member when axial effects are included in the analysis, the sizes of the various matrices in the equations for the displacements will be increased. The matrix $\mathbf{F}_{\mathrm{M}}$ of member flexibilities now is of order $3m \times 3m$, while the sizes of the transfer matrices $\mathbf{A}_{MU}$ and $\mathbf{A}_{ML}$ (see Eqs. 3-17 and 3-18) are $3m \times n$ and $3m \times p$, respectively, in which n is the number of joint displacements and p is the number of load systems.

The following example illustrates the calculation of displacements in a plane frame when the effects of both axial and bending deformations are taken into account.

Example 2. The displacement D_1 at joint C of the plane frame described in Example 1 is to be recalculated with the effects of axial deformations included. The axial rigidity EA is assumed to be the same for both members.

The member flexibility matrices now have the form shown by Eq. (3-20), and are as follows:

$$\mathbf{F}_{M1} = \mathbf{F}_{M2} = \frac{L}{6EI}\begin{bmatrix} 6L^2\psi & 0 & 0 \\ 0 & 2L^2 & 3L \\ 0 & 3L & 6 \end{bmatrix}$$

in which

$$\psi = \frac{I}{AL^2}$$

The matrix $\mathbf{F}_M$ of member flexibilities (see Eq. 3-16) will be of order 6×6.

The vectors of end-actions are augmented by the inclusion of the axial forces in the member; for instance, $\mathbf{A}_{ML1}$ and $\mathbf{A}_{MU1}$ become

$$\mathbf{A}_{ML1} = \frac{P}{8}\begin{bmatrix} -4 \\ -3 \\ 3L \end{bmatrix} \qquad \mathbf{A}_{MU1} = \frac{1}{2}\begin{bmatrix} 2 \\ -1 \\ L \end{bmatrix}$$

The first element in each of these vectors is the axial force at the k end of the member, while the remaining two elements are the shearing force and bending moment, respectively, and are the same as in Example 1. In a similar manner, the vectors $\mathbf{A}_{ML2}$ and $\mathbf{A}_{MU2}$ can be obtained, and then the transfer matrices are formed:

$$\mathbf{A}_{ML} = \frac{P}{24}\{-12, -9, 9L, 0, 9, -4L\}$$

$$\mathbf{A}_{MU} = \tfrac{1}{2}\{2, -1, L, 2, 1, 0\}$$

Lastly, the displacement D_1 is calculated by matrix multiplications:

$$D_1 = \mathbf{A}'_{MU}\mathbf{F}_M\mathbf{A}_{ML} = \frac{PL^3}{12EI}(1 - 6\psi)$$

Note that this expression for D_1 reduces to the value given in Example 1 if axial deformations are neglected (a result that may be obtained by letting the area A approach infinity, causing ψ to approach zero).

3.6 Joint Displacements in Grids. The general pattern of the calculations for finding joint displacements in grids is the same as that described in the preceding articles for trusses, beams, and plane frames. However, the significant end-actions for a grid member are different from those for the other types of structures, and hence the member flexibility matrix has a different form. In order to identify the end-actions for a grid, consider the typical member i shown in Fig. 3-14. Member-oriented axes having their origin at the j end of the member are shown in the figure. The grid itself is assumed to be in the x-z plane, while all concentrated loads acting on the grid have their vectors parallel to the y axis; all couples acting as loads on the grid have their vectors in the x-z plane.

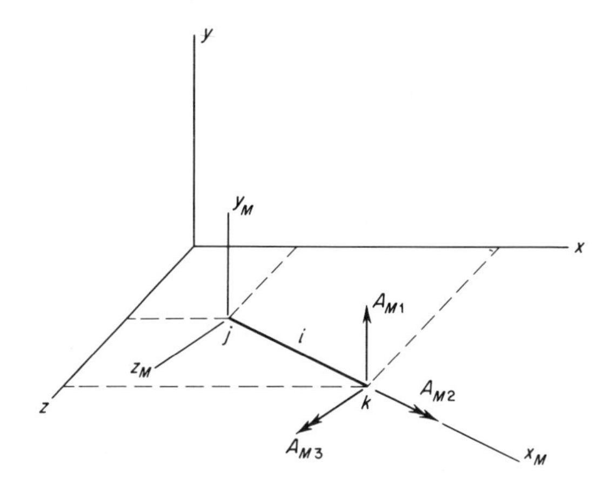

FIG. 3-14. End-actions for grid member.

The end-actions for a grid member are the shearing force in the y_M direction, the twisting couple in the x_M sense, and the bending couple in the z_M sense. These three end-actions are shown in Fig. 3-14 as A_{M1}, A_{M2}, and A_{M3}, respectively. The elements of the member flexibility matrix may be interpreted as the displacements of a cantilever member due to unit values of the end-actions, as shown in Fig. 3-15. Thus, the member flexibility matrix becomes

$$\mathbf{F}_{Mi} = \begin{bmatrix} F_{M11} & F_{M12} & F_{M13} \\ F_{M21} & F_{M22} & F_{M23} \\ F_{M31} & F_{M32} & F_{M33} \end{bmatrix} = \begin{bmatrix} \dfrac{L^3}{3EI} & 0 & \dfrac{L^2}{2EI} \\ 0 & \dfrac{L}{GJ} & 0 \\ \dfrac{L^2}{2EI} & 0 & \dfrac{L}{EI} \end{bmatrix} \tag{3-21}$$

where GJ represents the torsional rigidity of the member. When the member flexibilities are obtained for all members of the grid, the matrix $\mathbf{F}_M$ can be constructed as described previously (see Eq. 3-16).

The transfer matrices $\mathbf{A}_{MU}$ and $\mathbf{A}_{ML}$ for a grid consist of submatrices that are evaluated for each member, as shown by Eqs. (3-17) and (3-18). Each submatrix is of order 3×1, since there are three end-actions per member. When the transfer matrices have been determined, as well as the matrix $\mathbf{F}_M$ of member flexibilities, the displacements can be calculated by one of the matrix equations already discussed (see Eqs. 3-5 and 3-8).

In some grid structures the effects of torsional deformations are not important in determining the behavior of the structure. This situation

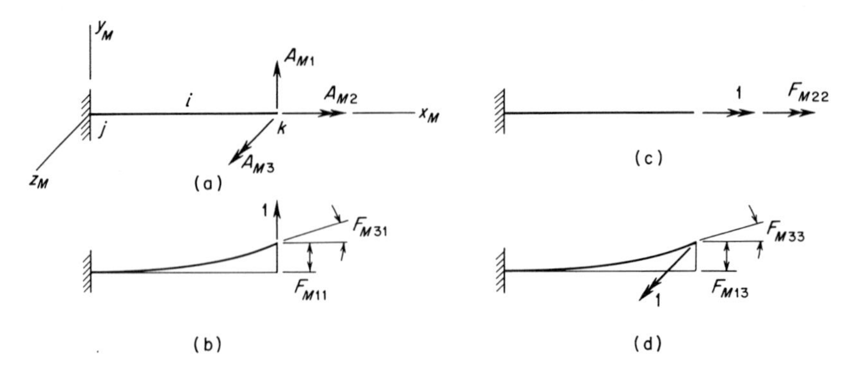

Fig. 3-15. Flexibilities for grid member.

arises either when the members of the grid are very weak in torsion or when the joints of the grid are constructed in such a manner that twisting couples are not developed in the members. An example of the latter type of structure is a grid that consists of several beams crossing one another, and having a connection between the beams that transmits a force but not a moment. The released structure for a grid of this type consists of beam members only.

In either of the situations described above, the effects of torsion can be omitted from the analysis. Therefore, the only end-actions for a member are the shearing force in the y_M direction and the bending moment in the z_M sense. These end-actions are the same as for a beam (see Fig. 3-7), and hence, in such a case, the flexibility matrix for a beam member (Eq. 3-13) can be used also for a grid member.

Example. The grid structure shown in Fig. 3-16 consists of two members (AB and BC) meeting at right angles in the horizontal x-z plane. Each member has length L, flexural rigidity EI, and torsional rigidity GJ. The j and k ends of the two members are assumed to be at A and B, and at B and C, respectively. The loads on the grid consist of forces and couples acting at joints B and C. Assume that the displacements of joint C (Fig. 3-16b) are to be found.

The member flexibility matrices, given by Eq. (3-21), are

$$\mathbf{F}_{M1} = \mathbf{F}_{M2} = \frac{L}{6EI}\begin{bmatrix} 2L^2 & 0 & 3L \\ 0 & 6\rho & 0 \\ 3L & 0 & 6 \end{bmatrix}$$

in which

$$\rho = \frac{EI}{GJ}$$

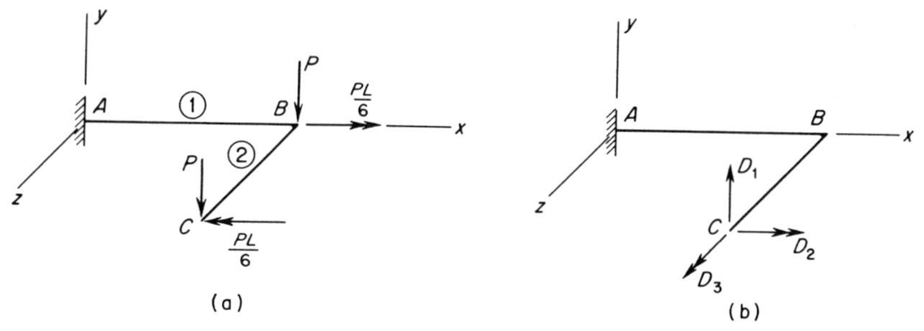

FIG. 3-16. Example: Displacements in a grid.

The end-actions caused by the loads are

$$\mathbf{A}_{\text{ML1}} = P\begin{bmatrix} -2 \\ L \\ 0 \end{bmatrix} \qquad \mathbf{A}_{\text{ML2}} = \frac{P}{6}\begin{bmatrix} -6 \\ 0 \\ L \end{bmatrix}$$

from which the transfer matrix $\mathbf{A}_{\text{ML}}$ is

$$\mathbf{A}_{\text{ML}} = \frac{P}{6}\{-12, 6L, 0, -6, 0, L\}$$

The three displacements to be calculated are shown in Fig. 3-16b. In order to obtain the transfer matrix $\mathbf{A}_{\text{MU}}$, unit loads corresponding to each of these displacements must be assumed to act upon the structure. Under the action of the first such unit load, the following end-actions are developed:

$$\mathbf{A}_{\text{MU11}} = \begin{bmatrix} 1 \\ -L \\ 0 \end{bmatrix} \qquad \mathbf{A}_{\text{MU21}} = \begin{bmatrix} 1 \\ 0 \\ 0 \end{bmatrix}$$

as can be verified by statics. Similarly, the end-actions caused by the other two unit loads can be obtained; then the transfer matrix $\mathbf{A}_{\text{MU}}$ is formed:

$$\mathbf{A}_{\text{MU}} = \begin{bmatrix} 1 & 0 & 0 \\ -L & 1 & 0 \\ 0 & 0 & 1 \\ 1 & 0 & 0 \\ 0 & 0 & 1 \\ 0 & -1 & 0 \end{bmatrix}$$

The final step in the analysis is to solve for the displacements by means of Eq. (3-8):

$$\mathbf{D} = \mathbf{A}'_{\text{MU}}\mathbf{F}_{\text{M}}\mathbf{A}_{\text{ML}} = \frac{PL^2}{12EI}\begin{bmatrix} -L(11 + 12\rho) \\ 4(1 + 3\rho) \\ -12 \end{bmatrix}$$

3.7 Joint Displacements in Space Frames. A space frame is the most general type of framed structure, and a typical member may have as many as six end-actions at each end. The six possible end-actions at the k end of the member are identified in Fig. 3-17, which shows a member i oriented in an arbitrary manner in space. The member-oriented axes are assumed to be so located that the x_M-y_M and x_M-z_M planes are principal planes of bending of the member. The end-actions are numbered in the order x, y, and z, with forces taken before couples.

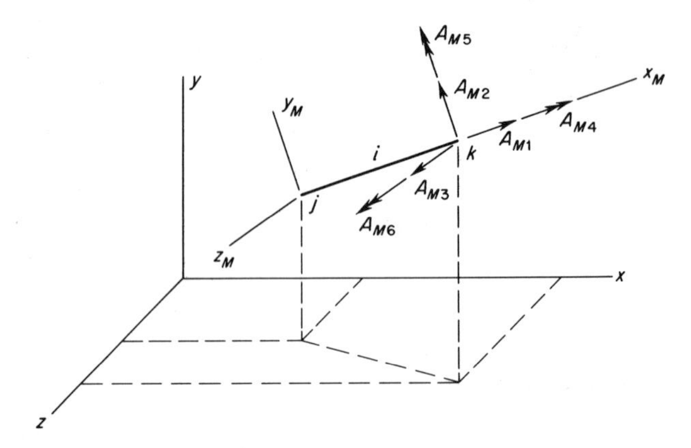

FIG. 3-17. End-actions for space frame member.

The elements of the member flexibility matrix are shown in Fig. 3-18 as the displacements in a cantilever member caused by unit values of the six possible end-actions. For a prismatic member the various displacements indicated in the figure can be found readily and placed in a 6×6 member flexibility matrix as follows:

$$\mathbf{F}_{Mi} = \begin{bmatrix} \frac{L}{EA} & 0 & 0 & 0 & 0 & 0 \\ 0 & \frac{L^3}{3EI_Z} & 0 & 0 & 0 & \frac{L^2}{2EI_Z} \\ 0 & 0 & \frac{L^3}{3EI_Y} & 0 & \frac{-L^2}{2EI_Y} & 0 \\ 0 & 0 & 0 & \frac{L}{GJ} & 0 & 0 \\ 0 & 0 & \frac{-L^2}{2EI_Y} & 0 & \frac{L}{EI_Y} & 0 \\ 0 & \frac{L^2}{2EI_Z} & 0 & 0 & 0 & \frac{L}{EI_Z} \end{bmatrix} \qquad (3\text{-}22)$$

In the above matrix the moments of inertia of the cross-section of the member about the y_M and z_M axes are denoted I_Y and I_Z, respectively.

The transfer matrices $\mathbf{A}_{ML}$ and $\mathbf{A}_{MU}$ in a space frame analysis consist of submatrices of order 6×1, since there are six end-actions for each member. Thus, the sizes of the matrices appearing in the equation for the displacements (Eq. 3-8) are as follows: $\mathbf{D}$ is of order $n \times p$ (see Eq. 3-10); $\mathbf{A}_{\mathrm{MU}}$ is of order $6m \times n$ (see Eq. 3-17); $\mathbf{F}_{\mathrm{M}}$ is of order $6m \times 6m$ (see Eq. 3-16); and $\mathbf{A}_{\mathrm{ML}}$ is of order $6m \times p$ (see Eq. 3-18).

In general, the determination of the end-actions in the transfer

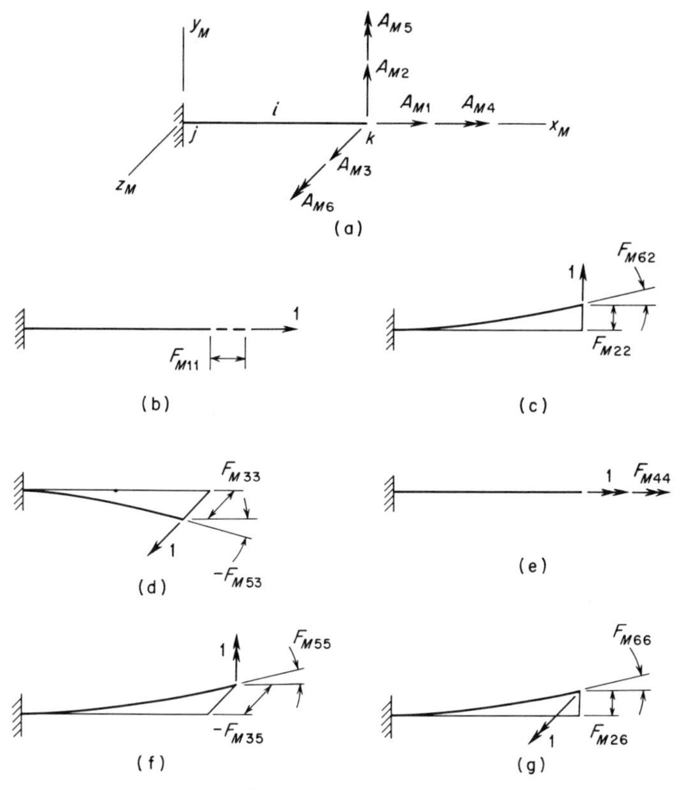

FIG. 3-18. Flexibilities for space frame member.

matrices for a space frame will be more complicated than for a plane structure, primarily because of the necessity to deal with arbitrary directions in three-dimensional space. A systematic approach for handling the three-dimensional aspects of the analysis can be developed by using techniques involving rotation of axes, as described in Chapter 4 for the stiffness method.

3.8 Equations of the Flexibility Method. The preceding articles dealt with matrix methods for finding displacements in statically determinate framed structures, because the determination of displacements in the released structure is an important part of the flexibility method. In this article the basic equations of the flexibility method are reviewed in order to show how the displacement terms in those equations can be expressed in matrix form. The resulting formulation provides a very systematic method for analyzing statically indeterminate structures.

When a structure is to be analyzed only for the effects of loads, the equations of the flexibility method are (see Eqs. 2-8, 2-15, 2-16, and 2-14):

$$\mathbf{D}_{\mathrm{Q}} = \mathbf{D}_{\mathrm{QL}} + \mathbf{FQ} \quad \text{or} \quad \mathbf{Q} = \mathbf{F}^{-1}(\mathbf{D}_{\mathrm{Q}} - \mathbf{D}_{\mathrm{QL}}) \tag{3-23a}$$

$$\mathbf{A}_{\mathrm{M}} = \mathbf{A}_{\mathrm{ML}} + \mathbf{A}_{\mathrm{MQ}}\mathbf{Q} \tag{3-23b}$$

$$\mathbf{A}_{\mathrm{R}} = \mathbf{A}_{\mathrm{RL}} + \mathbf{A}_{\mathrm{RQ}}\mathbf{Q} \tag{3-23c}$$

$$\mathbf{D}_{\mathrm{J}} = \mathbf{D}_{\mathrm{JL}} + \mathbf{D}_{\mathrm{JQ}}\mathbf{Q} \tag{3-23d}$$

It is assumed in this chapter that prior to using the above equations the actual loads on the structure are replaced by combined joint loads. Therefore, all of the quantities appearing in Eqs. (3-23) are assumed henceforth to refer to the structure with combined loads, and these quantities are not necessarily the same as the corresponding quantities for the structure with the actual loads. In particular, the end-actions $\mathbf{A}_{\mathrm{M}}$ found for the structure with combined loads must be added to the corresponding end-actions in the restrained structure in order to obtain the end-actions due to the actual loads (see Art. 3.2). Expressed in symbolic terms, this relationship becomes

$$(\mathbf{A}_{\mathrm{M}})_{\mathrm{A}} = \mathbf{A}_{\mathrm{M}} + (\mathbf{A}_{\mathrm{M}})_{\mathrm{R}} \tag{3-24}$$

in which the matrices $(\mathbf{A}_{\mathrm{M}})_{\mathrm{A}}$ and $(\mathbf{A}_{\mathrm{M}})_{\mathrm{R}}$ represent end-actions in the actual structure and in the restrained structure, respectively. The use of Eq. (3-24) will be illustrated later by examples.

The reactions $\mathbf{A}_{\mathrm{R}}$ and joint displacements $\mathbf{D}_{\mathrm{J}}$ are the same for both the actual loads and the combined loads, as explained in Art. 3.2. Therefore, the matrices $\mathbf{A}_{\mathrm{R}}$ and $\mathbf{D}_{\mathrm{J}}$ as calculated from Eqs. (3-23c) and (3-23d) apply without modification to the structure with the actual loads.

Whether or not the redundants Q that are obtained from Eq. (3-23a) will be valid for the actual loads on the structure depends upon the nature of the redundants themselves. A redundant that is a reaction for the structure will be valid, whereas a redundant that is a member end-action must be treated in the same manner as the end-actions A_M (see Eq. 3-24).

Returning now to Eqs. (3-23), consider how the various matrices in those equations can be calculated with the aid of the displacement equations discussed in Arts. 3.3 to 3.7. The vector $\mathbf{D}_{\mathrm{QL}}$ in Eq. (3-23a) represents the displacements in the released structure corresponding to the redundant actions Q and caused by the combined joint loads, and can be calculated by means of the matrix equation (3-8). Equation (3-8) can be used for any type of framed structure, as shown previously, provided only that the flexibility and transfer matrices ($\mathbf{F}_{\mathrm{M}}$, $\mathbf{A}_{\mathrm{MU}}$, $\mathbf{A}_{\mathrm{ML}}$)

have the correct form. The matrix $\mathbf{F}_M$ can be obtained without difficulty; all that is required is to place the member flexibility matrices along the principal diagonal of $\mathbf{F}_M$ (see Eqs. 3-4 and 3-16).

The transfer matrix $\mathbf{A}_{MU}$ appearing in Eq. (3-8) represents (in general) the member end-actions due to the unit loads corresponding to the desired displacements. In the case of the displacements $\mathbf{D}_{QL}$, the unit loads are unit values of the redundants Q themselves. Therefore, the matrix $\mathbf{A}_{MU}$ is the same as the matrix $\mathbf{A}_{MQ}$ (see Eq. 3-23b), which consists of the member end-actions in the released structure due to unit values of the redundants. (It is assumed in using Eq. 3-23b that the matrix $\mathbf{A}_M$ consists of the end-actions at the k ends of all the members; hence the matrix $\mathbf{A}_{MQ}$ also includes all members of the structure.)

The transfer matrix $\mathbf{A}_{ML}$ appearing in Eq. (3-8) represents (in general) the member end-actions due to the loads causing the displacements. Hence, in the case of the displacements $\mathbf{D}_{QL}$, the matrix $\mathbf{A}_{ML}$ consists of the member end-actions in the released structure due to the combined joint loads. This definition for the matrix $\mathbf{A}_{ML}$ is entirely consistent with the meaning of the matrix $\mathbf{A}_{ML}$ appearing in Eq. (3-23b). Therefore, the vector $\mathbf{D}_{QL}$ can be expressed as follows:

$$\mathbf{D}_{QL} = \mathbf{A}'_{MQ}\mathbf{F}_M\mathbf{A}_{ML} \tag{3-25}$$

Equation (3-25) provides a formalized method for calculating the displacements D_{QL}. In order to use the equation, the transfer matrices of end-actions ($\mathbf{A}_{MQ}$ and $\mathbf{A}_{ML}$) and the matrix $\mathbf{F}_M$ must be obtained first. The determination of the transfer matrices requires that the released structure be analyzed by statics for unit values of the redundants and also for the combined joint loads.

Consider next the calculation of the flexibility matrix $\mathbf{F}$ for the structure. This matrix consists of displacements in the released structure corresponding to the redundants and caused by unit values of the redundants, and may also be determined from Eq. (3-8). The transfer matrix $\mathbf{A}_{MU}$ in Eq. (3-8) becomes the matrix $\mathbf{A}_{MQ}$, inasmuch as the unit loads are unit values of the redundants. Moreover, the matrix $\mathbf{A}_{ML}$ in Eq. (3-8) is the same because the loads causing the displacements are also unit values of the redundants. Thus, the expression for $\mathbf{F}$, which may also be denoted $\mathbf{D}_{QQ}$, is:

$$\mathbf{F} = \mathbf{D}_{QQ} = \mathbf{A}'_{MQ}\mathbf{F}_M\mathbf{A}_{MQ} \tag{3-26}$$

The flexibility matrix $\mathbf{F}$ is a property of the entire released structure, and is dependent upon the arrangement of the members of the structure, whereas the matrix $\mathbf{F}_M$ of member flexibilities depends only on the properties of the isolated members, without regard to how they are assembled into a structure. For this reason the matrix $\mathbf{F}_M$ is sometimes called the flexibility matrix of the *unassembled structure*, while $\mathbf{F}$ is called the flexibility matrix of the *assembled structure*. The use of Eqs. (3-25) and (3-26) makes it possible to find the matrices $\mathbf{D}_{\mathrm{QL}}$ and $\mathbf{F}$ in a formalized manner. Since the vector $\mathbf{D}_{\mathrm{Q}}$ is known from the given conditions of the structure, the redundants $\mathbf{Q}$ now can be obtained by solving Eq. (3-23a).

After finding the vector $\mathbf{Q}$ of redundants, the next step in the analysis is to calculate the member end-actions $\mathbf{A}_{\mathrm{M}}$ from Eq. (3-23b). This step can be performed without difficulty, because the transfer matrices $\mathbf{A}_{\mathrm{ML}}$ and $\mathbf{A}_{\mathrm{MQ}}$ have already been determined for use in Eqs. (3-25) and (3-26).

To find the reactions $\mathbf{A}_{\mathrm{R}}$ from Eq. (3-23c) requires that the transfer matrices $\mathbf{A}_{\mathrm{RL}}$ and $\mathbf{A}_{\mathrm{RQ}}$ be obtained. These matrices represent reactions in the released structure due to the combined loads and due to unit values of the redundants, respectively. Both matrices are found by statical analysis of the released structure.

Lastly, consider the matrices appearing in the equation for the joint displacements (see Eq. 3-23d). The matrices $\mathbf{D}_{\mathrm{JL}}$ and $\mathbf{D}_{\mathrm{JQ}}$ represent displacements in the released structure and can be calculated by matrix operations. For this purpose it is necessary to introduce another transfer matrix, denoted $\mathbf{A}_{\mathrm{MJ}}$, consisting of member end-actions due to unit loads corresponding to the desired joint displacements D_J. The first column of $\mathbf{A}_{\mathrm{MJ}}$ contains the member end-actions due to a unit load corresponding to D_{J1}, the second column contains end-actions due to a unit load corresponding to D_{J2}, etc. In using Eq. (3-8) to calculate $\mathbf{D}_{\mathrm{JL}}$, the matrix $\mathbf{A}_{\mathrm{MU}}$ becomes $\mathbf{A}_{\mathrm{MJ}}$ and the matrix $\mathbf{A}_{\mathrm{ML}}$ remains unchanged. Thus, the equation for $\mathbf{D}_{\mathrm{JL}}$ is

$$\mathbf{D}_{\mathrm{JL}} = \mathbf{A}'_{\mathrm{MJ}}\mathbf{F}_{\mathrm{M}}\mathbf{A}_{\mathrm{ML}} \tag{3-27}$$

In a similar manner, the matrix $\mathbf{D}_{\mathrm{JQ}}$ can be determined by means of Eq. (3-8), as follows:

$$\mathbf{D}_{\mathrm{JQ}} = \mathbf{A}'_{\mathrm{MJ}}\mathbf{F}_{\mathrm{M}}\mathbf{A}_{\mathrm{MQ}} \tag{3-28}$$

In these equations, the matrices $\mathbf{D}_{\mathrm{JL}}$ and $\mathbf{D}_{\mathrm{JQ}}$ are of orders $j \times 1$ and $j \times q$, respectively.

When the two above equations are substituted into Eq. (3-23d), the equation for $\mathbf{D}_{\mathrm{J}}$ becomes

$$\mathbf{D}_{\mathrm{J}} = \mathbf{D}_{\mathrm{JL}} + \mathbf{D}_{\mathrm{JQ}}\mathbf{Q} = \mathbf{A}'_{\mathrm{MJ}}\mathbf{F}_{\mathrm{M}}(\mathbf{A}_{\mathrm{ML}} + \mathbf{A}_{\mathrm{MQ}}\mathbf{Q})$$

The term in parentheses represents the vector $\mathbf{A}_{\mathrm{M}}$ of member end-actions (see Eq. 3-23b), and therefore an alternate expression for the joint displacement vector is

$$\mathbf{D}_{\mathrm{J}} = \mathbf{A}'_{\mathrm{MJ}}\mathbf{F}_{\mathrm{M}}\mathbf{A}_{\mathrm{M}} \tag{3-29}$$

Either Eq. (3-23d) or Eq. (3-29) can be used to find the joint displacements. The latter equation is somewhat simpler to use, provided the end-actions $\mathbf{A}_{\mathrm{M}}$ have been determined previously.

The transfer matrices that appear in Eqs. (3-23) to (3-29) are summarized in Table 3-1. The first three matrices ($\mathbf{A}_{ML}$, $\mathbf{A}_{MQ}$, $\mathbf{A}_{MJ}$) consist of member end-actions in the released structure due to various causes, and the last two matrices ($\mathbf{A}_{RL}$ and $\mathbf{A}_{RQ}$) consist of reactions in the released structure due to the combined loads and due to unit values of the redundants.

TABLE 3-1

Transfer Matrices for the Flexibility Method

Matrix	*Order*	*Definition*
$\mathbf{A}_{ML}$	$m \times 1$	Member end-actions in the released structure due to the combined joint loads (m = number of members)
$\mathbf{A}_{MQ}$	$m \times q$	Member end-actions in the released structure due to unit values of the redundants (q = number of redundants)
$\mathbf{A}_{MJ}$	$m \times j$	Member end-actions in the released structure due to unit loads corresponding to the joint displacements (j = number of joint displacements to be found)
$\mathbf{A}_{RL}$	$r \times 1$	Reactions in the released structure due to the combined joint loads (r = number of reactions to be found)
$\mathbf{A}_{RQ}$	$r \times q$	Reactions in the released structure due to unit values of the redundants

NOTE: The sizes of the transfer matrices $\mathbf{A}_{ML}$, $\mathbf{A}_{MQ}$, and $\mathbf{A}_{MJ}$ given above are for a plane or space truss, both of which have one end-action per member. In the case of a beam, the number m must be replaced by $2m$; for a plane frame and a grid, the number becomes either $2m$ or $3m$, depending upon the number of end-actions that are considered; and, for a space frame, the number is $6m$.

The procedure to be followed in analyzing a structure by the flexibility method may now be summarized. The outline which follows is analogous to the one presented in Art. 2.7, except that it has been modified to provide for the use of transfer matrices. The steps in the analysis are as follows:

1. Replace the actual loads on the structure by combined joint loads (see Art. 3.2).
2. Select the redundants Q.
3. Obtain the vector $\mathbf{D}_Q$ if there are support displacements corresponding to one or more of the redundants; otherwise, $\mathbf{D}_Q$ is null.
4. Obtain the member flexibility matrices and construct the matrix $\mathbf{F}_M$.
5. Analyze the released structure by static equilibrium to obtain the transfer matrices $\mathbf{A}_{ML}$, $\mathbf{A}_{MQ}$, $\mathbf{A}_{MJ}$, $\mathbf{A}_{RL}$, and $\mathbf{A}_{RQ}$ (see Table 3-1).
6. Calculate the vector $\mathbf{D}_{QL}$ and the matrix $\mathbf{F}$ by means of Eqs. (3-25) and (3-26).
7. Solve for the vector $\mathbf{Q}$ of redundants from Eq. (3-23a).

8. Solve for the vector $\mathbf{A}_M$ of member end-actions from Eq. (3-23b).
9. Solve for the vector $\mathbf{A}_R$ of reactions from Eq. (3-23c).
10. Solve for the vector $\mathbf{D}_J$ of joint displacements from Eq. (3-29). (Alternatively, the vector $\mathbf{D}_J$ can be obtained by using Eqs. 3-27, 3-28, and 3-23d.)
11. Calculate member end-actions due to the actual loads on the structure by means of Eq. (3-24).

The steps outlined above can be divided into two principal phases. The first phase consists of "setting up" the calculations, and includes determining the transfer matrices and member flexibilities (steps 1 through 5). These steps normally must be performed by the structural analyst. The second phase (steps 6 through 11) involves matrix calculations based upon the equations given in this article. This phase proceeds in a systematic manner, and can be performed automatically.

Examples of the flexibility method illustrating the above steps are given in the next article. In each case it is assumed that the structure is subjected to a single system of loading. However, multiple load systems are easily handled by means of additional columns in the transfer matrices $\mathbf{A}_{ML}$ and $\mathbf{A}_{RL}$ (see Table 3-1). Of course, a corresponding number of additional columns also would have to be added to the matrices $\mathbf{Q}$, $\mathbf{A}_M$, $\mathbf{A}_R$, $\mathbf{D}_{JL}$, and $\mathbf{D}_J$.

3.9 Examples. The examples given in this article illustrate the analysis of several different types of structures by the flexibility method. The solutions are performed by following the procedure described in the preceding article (see steps 1 through 11), and only the effects of loads on the structures are considered. Other effects are described in the next article (Art. 3.10).

Example 1. The first example involves the analysis of the statically indeterminate truss shown in Fig. 3-19. The truss has six members, numbered as shown in the figure, and all members are assumed to have the same axial rigidity EA. There are two loads acting on the truss at joint A, and support restraints exist at joints B, C, and D. (The same truss was solved previously as Example 3 of Art. 2.3.)

In this problem all of the loads on the truss are in the form of joint loads; hence, there is no necessity to replace member loads by equivalent loads at the joints. However, if member loads are present the determination of equivalent joint loads is quite simple. They can be found from the formulas for pinned-end actions for truss members, which are given in Table B-5 of Appendix B. The equivalent loads are in the form of forces only, since there are no bending moments produced at the ends of a truss member.

The redundant actions Q_1 and Q_2 for the truss are selected as the horizontal reactive force at support B and the axial force in bar AD, respectively. The first of these redundants is assumed to be positive when in the positive x direction, and the second is positive when the member is in tension.

In addition to finding the redundants, it is assumed in this example that all member end-actions, reactions, and joint displacements are to be determined.

The member end-actions are axial forces at the k ends of all the members, as shown in Fig. 3-2 for a typical member. While the member end-actions in general must be taken at a designated k end for each member, it can be seen that, in the case of a truss with joint loads only, the end-actions are the same at both ends. Therefore, in this example it is not necessary to designate specifically the end of each member at which the end-action is determined; instead, either end can be used.

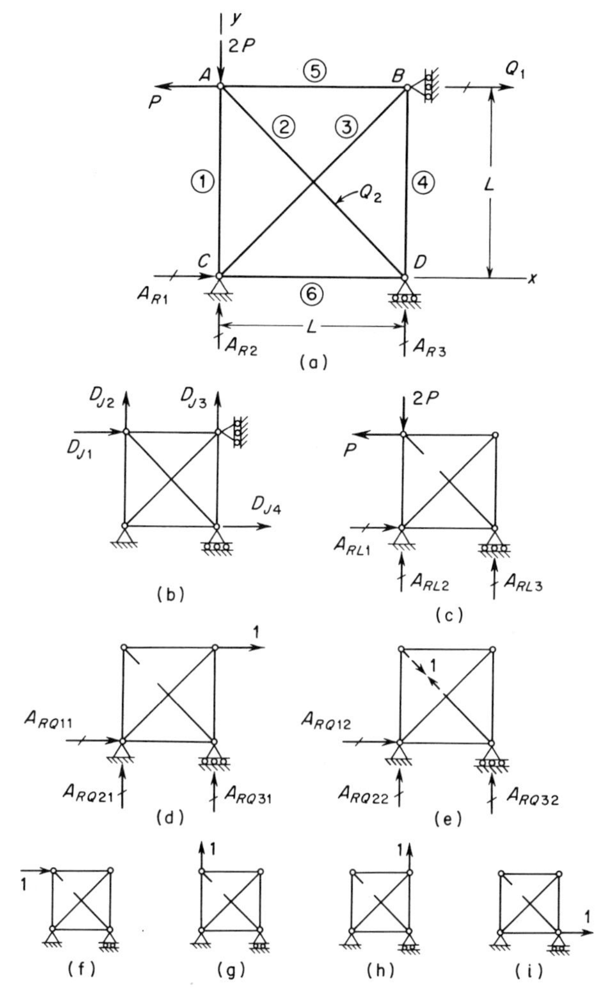

FIG. 3-19. Example 1: Plane truss.

The truss has four reactions to be determined; the reactions at joints C and D are denoted A_{R1}, A_{R2}, and A_{R3} (see Fig. 3-19a) while the remaining reaction is the redundant Q_1 itself. Also, the four unknown joint displacements (denoted $D_{J1}, \ldots, D_{J4}$) are shown in Fig. 3-19b. The reactions and joint displacements are assumed positive when in the positive directions of the x and y axes.

The next step in the solution is to determine the vector $\mathbf{D}_\mathrm{Q}$ of actual displacements corresponding to the redundants. However, the vector $\mathbf{D}_\mathrm{Q}$ is null in this example because it is assumed that there is no displacement at support B in the x direction.

The individual member flexibilities (see Eq. 3-1) are used in forming the matrix $\mathbf{F}_\mathrm{M}$, which is as follows (see Eq. 3-4):

$$\mathbf{F}_\mathrm{M} = \frac{L}{EA}\begin{bmatrix} 1 & 0 & 0 & 0 & 0 & 0 \\ 0 & 1.414 & 0 & 0 & 0 & 0 \\ 0 & 0 & 1.414 & 0 & 0 & 0 \\ 0 & 0 & 0 & 1 & 0 & 0 \\ 0 & 0 & 0 & 0 & 1 & 0 \\ 0 & 0 & 0 & 0 & 0 & 1 \end{bmatrix}$$

The determination of the transfer matrices listed in Table 3-1 is performed next. All of the transfer matrices are found by static equilibrium analyses of the released structure. Beginning with the effects of the loads, it can be seen that the released structure shown in Fig. 3-19c must be analyzed by statics in order to obtain the transfer matrices $\mathbf{A}_\mathrm{ML}$ and $\mathbf{A}_\mathrm{RL}$. These two matrices consist of six member end-actions and three reactions, respectively, as follows:

$$\mathbf{A}_\mathrm{ML} = P\{-2, 0, -1.414, 1, 1, 0\}$$

$$\mathbf{A}_\mathrm{RL} = P\{1, 3, -1\}$$

The effects of unit values of the two redundants acting on the released structure must be calculated in order to obtain the transfer matrices $\mathbf{A}_\mathrm{MQ}$ and $\mathbf{A}_\mathrm{RQ}$. A value of Q_1 equal to unity is shown acting on the released structure in Fig. 3-19d, and the resulting member end-actions and reactions constitute the first columns of the matrices $\mathbf{A}_\mathrm{MQ}$ and $\mathbf{A}_\mathrm{RQ}$, respectively. Similarly, the released structure acted upon by a unit value of Q_2 is shown in Fig. 3-19e, and the resulting end-actions and reactions constitute the second columns of $\mathbf{A}_\mathrm{MQ}$ and $\mathbf{A}_\mathrm{RQ}$; thus:

$$\mathbf{A}_\mathrm{MQ} = \begin{bmatrix} 0 & -.707 \\ 0 & 1 \\ 1.414 & 1 \\ -1 & -.707 \\ 0 & -.707 \\ 0 & -.707 \end{bmatrix} \qquad \mathbf{A}_\mathrm{RQ} = \begin{bmatrix} -1 & 0 \\ -1 & 0 \\ 1 & 0 \end{bmatrix}$$

The last transfer matrix to be calculated is $\mathbf{A}_\mathrm{MJ}$, the matrix of end-actions due to unit loads corresponding to the joint displacements D_J. A unit load corresponding to D_{J1} is shown in Fig. 3-19f, and the resulting end-actions appear in the first column of the matrix $\mathbf{A}_\mathrm{MJ}$, which is given below. The unit loads corresponding to D_{J2}, D_{J3}, and D_{J4} are shown in Figs. 3-19g, 3-19h, and 3-19i, and the resulting end-actions are given in the last three columns of $\mathbf{A}_\mathrm{MJ}$. Thus,

$$\mathbf{A}_\mathrm{MJ} = \begin{bmatrix} 0 & 1 & 0 & 0 \\ 0 & 0 & 0 & 0 \\ 1.414 & 0 & 0 & 0 \\ -1 & 0 & 1 & 0 \\ -1 & 0 & 0 & 0 \\ 0 & 0 & 0 & 1 \end{bmatrix}$$

Having found the matrix $\mathbf{F}_{\mathrm{M}}$ and the five transfer matrices, the remainder of the solution may be carried out using only routine matrix operations. Equations (3-25) and (3-26) give the matrices $\mathbf{D}_{\mathrm{QL}}$ and $\mathbf{F}$:

$$\mathbf{D}_{\mathrm{QL}} = \mathbf{A}'_{\mathrm{MQ}}\mathbf{F}_{\mathrm{M}}\mathbf{A}_{\mathrm{ML}} = \frac{PL}{EA}\begin{bmatrix} -3.828 \\ -2.000 \end{bmatrix}$$

$$\mathbf{F} = \mathbf{A}'_{\mathrm{MQ}}\mathbf{F}_{\mathrm{M}}\mathbf{A}_{\mathrm{MQ}} = \frac{L}{EA}\begin{bmatrix} 3.828 & 2.707 \\ 2.707 & 4.828 \end{bmatrix}$$

Then the redundants Q are obtained by solving Eq. (3-23a):

$$\mathbf{Q} = \mathbf{F}^{-1}(\mathbf{D}_{\mathrm{Q}} - \mathbf{D}_{\mathrm{QL}}) = \frac{EA}{L}\begin{bmatrix} 0.4328 & -0.2426 \\ -0.2426 & 0.3431 \end{bmatrix}\begin{bmatrix} 3.828 \\ 2.000 \end{bmatrix}\frac{PL}{EA} = P\begin{bmatrix} 1.172 \\ -0.243 \end{bmatrix}$$

These results for the redundants agree with those found earlier in Example 3 of Art. 2.3.

The next steps in the solution consist of finding the member end-actions and reactions from Eqs. (3-23b) and (3-23c), respectively. These equations give the following results:

$$\mathbf{A}_{\mathrm{M}} = \mathbf{A}_{\mathrm{ML}} + \mathbf{A}_{\mathrm{MQ}}\mathbf{Q} = P\{-1.828, -0.243, 0, 0, 1.172, 0.172\}$$

$$\mathbf{A}_{\mathrm{R}} = \mathbf{A}_{\mathrm{RL}} + \mathbf{A}_{\mathrm{RQ}}\mathbf{Q} = P\{-0.172, 1.828, 0.172\}$$

Finally, the vector of joint displacements D_J is calculated by means of Eq. (3-29):

$$\mathbf{D}_{\mathrm{J}} = \mathbf{A}'_{\mathrm{MJ}}\mathbf{F}_{\mathrm{M}}\mathbf{A}_{\mathrm{M}} = \frac{PL}{EA}\{-1.172, -1.828, 0, 0.172\}$$

Thus, the complete analysis of the truss, including the determination of all end-actions, reactions, and joint displacements, has been accomplished.

Example 2. The two-span continuous beam shown in Fig. 3-20a is subjected to a uniform load w (equal to $4P/L$) in span AB and two concentrated forces P in span BC. Both members have the same length L, but the flexural rigidity of member AB is twice that of member BC. The object of the analysis is to determine end-actions, reactions, and joint displacements for the beam.

Members AB and BC are denoted as members 1 and 2, respectively, and the k ends of the members are taken as ends B and C, respectively. Thus, the four end-actions used in the analysis are the shearing forces and bending moments at right-hand ends of the members. These end-actions will be numbered sequentially from 1 to 4, and their positive directions are in accordance with the axes shown in the figure (see Fig. 3-7 for the sign convention for end-actions).

The redundant actions are selected as the reactive moment at A and the bending moment at B (see Fig. 3-20b). Because there is a couple applied as a load at joint B when the actual loads are replaced by combined joint loads, ambiguity will be avoided if Q_2 is taken a small distance to one side of the joint itself. Therefore, in this example it is assumed that Q_2 is the bending moment to the right of joint B. The released structure which results from this selection of redundants is shown in Fig. 3-20b. Note that both redundants are assumed to be positive when they cause compression on the upper part of the beam.

The reactive forces in the y direction at the supports, as well as the two joint displacements (rotations at B and C), are identified in Fig. 3-20c. A fourth reactive action is the couple at support A, but since this reaction is chosen as one of the redundants it is not necessary to include it in the vector $\mathbf{A}_{\mathrm{R}}$.

The determination of the combined joint loads is shown in Fig. 3-20d, which shows the two fixed-end beams obtained by restraining the joints against displacements. The negatives of the fixed-end actions for these beams constitute the equivalent joint loads. Then, the equivalent loads are added to any actual

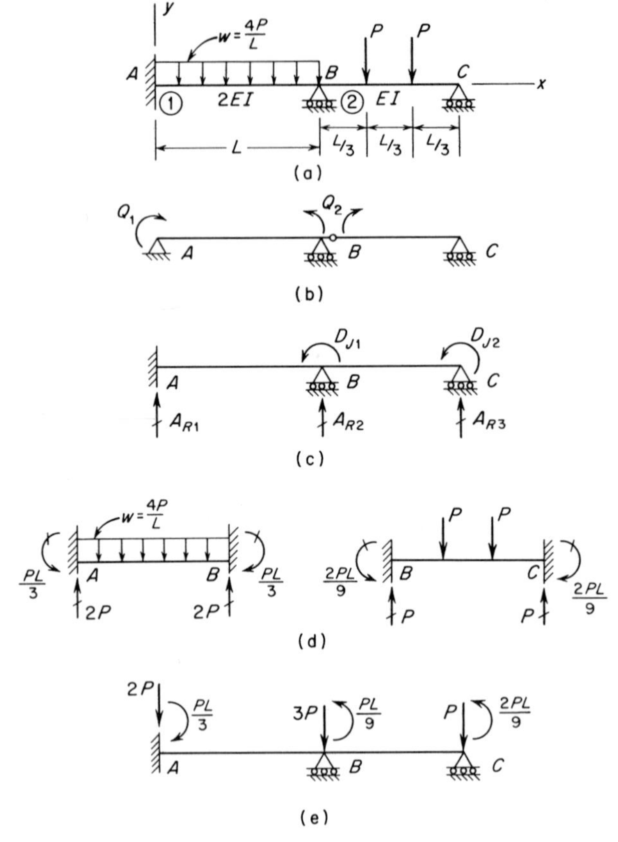

FIG. 3-20. Example 2: Continuous beam.

joint loads (none in this example) to obtain the combined joint loads, which are shown in Fig. 3-20e.

The flexibility matrices for the individual members can be found readily (see Eq. 3-13):

$$\mathbf{F}_{M2} = 2\mathbf{F}_{M1} = \frac{L}{12EI}\begin{bmatrix} 4L^2 & 6L \\ 6L & 12 \end{bmatrix}$$

and, therefore, the matrix $\mathbf{F}_M$ is

$$\mathbf{F}_M = \frac{L}{12EI}\begin{bmatrix} 2L^2 & 3L & 0 & 0 \\ 3L & 6 & 0 & 0 \\ 0 & 0 & 4L^2 & 6L \\ 0 & 0 & 6L & 12 \end{bmatrix}$$

as given by Eq. (3-16).

Consider now the calculation of the various transfer matrices for the released structure. The matrices $\mathbf{A}_{ML}$ and $\mathbf{A}_{RL}$, which consist of end-actions and reactions due to the combined joint loads, are found by static equilibrium analysis of the released structure shown in Fig. 3-21a. The vector $\mathbf{A}_{ML}$ is composed of the four end-actions for the members, as follows:

$$\mathbf{A}_{ML} = \frac{P}{9}\{2, L, -2, 2L\}$$

and the vector $\mathbf{A}_{RL}$ consists of the reactions:

$$\mathbf{A}_{RL} = \frac{P}{9}\{16, 31, 7\}$$

The transfer matrices $\mathbf{A}_{MQ}$ and $\mathbf{A}_{RQ}$ represent the end-actions and support reactions caused by unit values of the redundants. Therefore, the first column of each of these matrices is found from an equilibrium analysis of the released

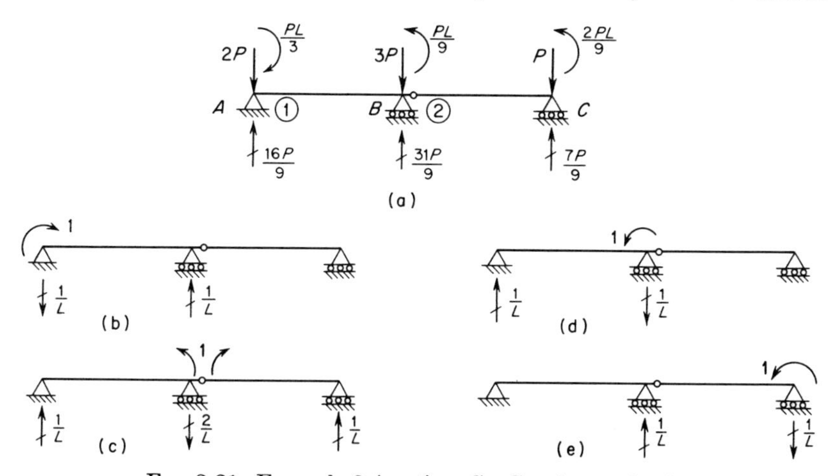

FIG. 3-21. Example 2 (continued): Continuous beam.

structure shown in Fig. 3-21b, while the second column is obtained in a similar manner from Fig. 3-21c. Thus, these matrices are

$$\mathbf{A}_{MQ} = \frac{1}{L}\begin{bmatrix} 1 & -1 \\ 0 & L \\ 0 & 1 \\ 0 & 0 \end{bmatrix} \qquad \mathbf{A}_{RQ} = \frac{1}{L}\begin{bmatrix} -1 & 1 \\ 1 & -2 \\ 0 & 1 \end{bmatrix}$$

The transfer matrix $\mathbf{A}_{MJ}$ consists of member end-actions caused by unit loads corresponding to the displacements D_{J1} and D_{J2}. Therefore, the first and second columns of $\mathbf{A}_{MJ}$ are found from analyses of the beams pictured in Figs. 3-21d and 3-21e, respectively. The matrix is

$$\mathbf{A}_{MJ} = \frac{1}{L}\begin{bmatrix} -1 & 0 \\ L & 0 \\ 0 & -1 \\ 0 & L \end{bmatrix}$$

After all of the transfer matrices and the flexibility matrix $\mathbf{F}_{\mathrm{M}}$ have been established, the remaining calculations proceed routinely. Thus, by following steps 6 through 10 of the outline in Art. 3.8, one obtains:

$$\mathbf{D}_{\mathrm{QL}} = \mathbf{A}'_{\mathrm{MQ}}\mathbf{F}_{\mathrm{M}}\mathbf{A}_{\mathrm{ML}} = \frac{PL^2}{108EI}\begin{bmatrix} 7 \\ 9 \end{bmatrix}$$

$$\mathbf{F} = \mathbf{A}'_{\mathrm{MQ}}\mathbf{F}_{\mathrm{M}}\mathbf{A}_{\mathrm{MQ}} = \frac{L}{12EI}\begin{bmatrix} 2 & 1 \\ 1 & 6 \end{bmatrix}$$

$$\mathbf{F}^{-1} = \frac{12EI}{11L}\begin{bmatrix} 6 & -1 \\ -1 & 2 \end{bmatrix}$$

$$\mathbf{Q} = \mathbf{F}^{-1}(\mathbf{D}_{\mathrm{Q}} - \mathbf{D}_{\mathrm{QL}}) = -\frac{PL}{9}\{3, 1\}$$

$$\mathbf{A}_{\mathrm{M}} = \mathbf{A}_{\mathrm{ML}} + \mathbf{A}_{\mathrm{MQ}}\mathbf{Q} = \frac{P}{9}\{0, 0, -3, 2L\}$$

$$\mathbf{A}_{\mathrm{R}} = \mathbf{A}_{\mathrm{RL}} + \mathbf{A}_{\mathrm{RQ}}\mathbf{Q} = \frac{P}{3}\{6, 10, 2\}$$

$$\mathbf{D}_{\mathrm{J}} = \mathbf{A}'_{\mathrm{MJ}}\mathbf{F}_{\mathrm{M}}\mathbf{A}_{\mathrm{M}} = \frac{PL^2}{18EI}\{0, 1\}$$

All of the above results pertain to the beam with combined joint loads (Fig. 3-20e). In addition, the vectors $\mathbf{A}_{\mathrm{R}}$ and $\mathbf{D}_{\mathrm{J}}$ also are valid for the beam with the actual loads (Fig. 3-20a). To obtain the member end-actions due to the actual loads, Eq. (3-24) is used. The calculations are as follows:

$$(\mathbf{A}_{\mathrm{M}})_{\mathrm{R}} = \frac{P}{9}\{18, -3L, 9, -2L\}$$

$$(\mathbf{A}_{\mathrm{M}})_{\mathrm{A}} = \mathbf{A}_{\mathrm{M}} + (\mathbf{A}_{\mathrm{M}})_{\mathrm{R}} = \frac{P}{9}\{18, -3L, 6, 0\}$$

The redundant Q_1, which is a reaction, is valid for the actual beam of Fig. 3-20a. However, this conclusion is not true for the second redundant, because Q_2 is an internal bending moment. If it is desired to obtain the value of Q_2 for the actual loads, it is necessary to make a calculation analogous to the one made for the end-actions; thus:

$$(Q_2)_A = Q_2 + (Q_2)_R = -\frac{PL}{9} - \frac{2PL}{9} = -\frac{PL}{3}$$

The minus sign in this result shows that there is tension on the upper side of the beam at joint B.

Example 3. The plane frame shown in Fig. 3-22a has span length L, height H equal to $L/2$, and constant EI for all members, and is subjected to two concentrated forces. Member numbers are indicated in the figure. Only flexural effects are to be considered in the analysis, and the reactive force in the x direction at support D is chosen as the redundant action.

The j and k ends of the three members are selected arbitrarily as follows: for member 1, the j and k ends are at A and B, respectively; for member 2, at joints B and C, respectively; and for member 3, at joints D and C, respectively. In accordance with this selection, the member end-actions are the shearing forces and bending moments shown in Fig. 3-22b.

The joint displacements for the frame are shown also in Fig. 3-22b. They consist of the rotation at joint A, the translation and rotation at B, the rotation at C, and the rotation at D. Because axial deformations are neglected in this example, there is only one independent joint translation (D_{J2}). The reactions for the frame also are identified in Fig. 3-22b; only three reactions need to be considered, inasmuch as the fourth reaction is the redundant Q.

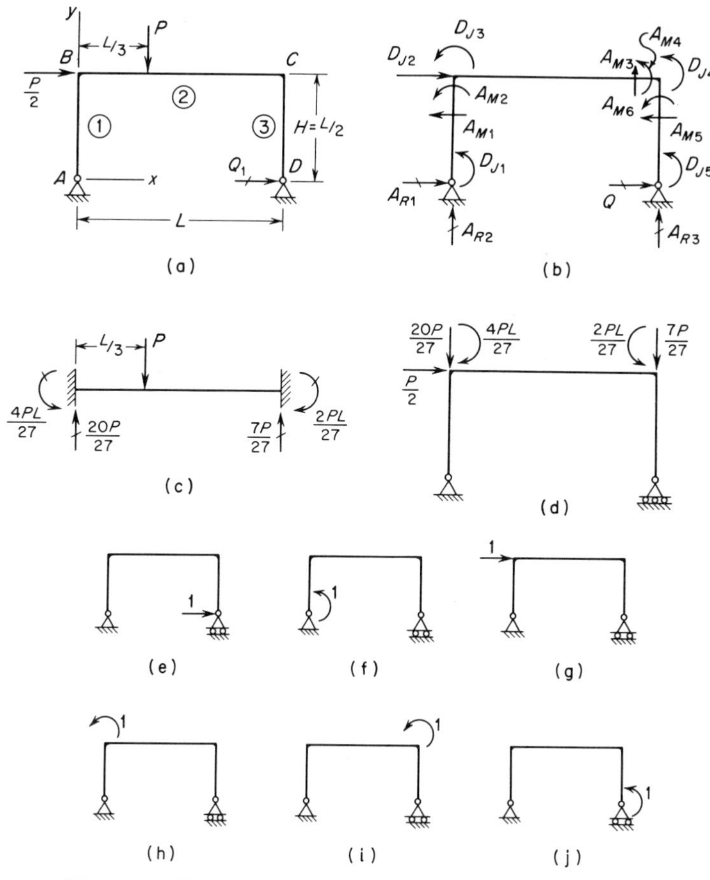

FIG. 3-22. Example 3: Plane frame (flexural effects only).

If not all of the reactions and joint displacements shown in Fig. 3-22b are desired, some of them can be omitted from the analysis. However, all of the member end-actions must be included, since they are required in the intermediate phases of the analysis.

The equivalent joint loads are found by restraining all joints of the structure, calculating the fixed-end actions caused by the member loads, and then reversing their directions. In this example the only member load is the force P acting on member BC; thus, when joints B and C are restrained, the fixed-end actions are those shown in Fig. 3-22c. The negatives of these fixed-end actions are added to the actual joint loads to give the combined loads shown in Fig 3-22d.

The member flexibility matrices are obtained by applying Eq. (3-13) to each member of the frame:

$$\mathbf{F}_{M1} = \mathbf{F}_{M3} = \begin{bmatrix} \dfrac{H^3}{3EI} & \dfrac{H^2}{2EI} \\ \dfrac{H^2}{2EI} & \dfrac{H}{EI} \end{bmatrix} \qquad \mathbf{F}_{M2} = \begin{bmatrix} \dfrac{L^3}{3EI} & \dfrac{L^2}{2EI} \\ \dfrac{L^2}{2EI} & \dfrac{L}{EI} \end{bmatrix}$$

The flexibility matrix $\mathbf{F}_M$ is formed by placing the above matrices on the principal diagonal (see Eq. 3-16). Inasmuch as $H = L/2$, this matrix becomes

$$\mathbf{F}_M = \frac{L}{24EI}\begin{bmatrix} L^2 & 3L & 0 & 0 & 0 & 0 \\ 3L & 12 & 0 & 0 & 0 & 0 \\ 0 & 0 & 8L^2 & 12L & 0 & 0 \\ 0 & 0 & 12L & 24 & 0 & 0 \\ 0 & 0 & 0 & 0 & L^2 & 3L \\ 0 & 0 & 0 & 0 & 3L & 12 \end{bmatrix}$$

The transfer matrices $\mathbf{A}_{ML}$ and $\mathbf{A}_{RL}$ can be obtained by statical analysis of the released structure subjected to the combined joint loads (see Fig. 3-22d). The first of these matrices consists of the six member end-actions, while the second consists of the three reactions. Thus, these vectors are as follows:

$$\mathbf{A}_{ML} = \frac{P}{108}\{-54, 27L, 35, 8L, 0, 0\}$$

$$\mathbf{A}_{RL} = \frac{P}{12}\{-6, 5, 7\}$$

The calculation of the transfer matrices $\mathbf{A}_{MQ}$ and $\mathbf{A}_{RQ}$ requires that the released structure be analyzed for a unit value of the redundant Q (see Fig. 3-22e). The member end-actions and reactions caused by this unit load also can be found by statics, thus giving the following vectors:

$$\mathbf{A}_{MQ} = \tfrac{1}{2}\{-2, L, 0, L, 2, -L\}$$

$$\mathbf{A}_{RQ} = \{-1, 0, 0\}$$

The transfer matrix $\mathbf{A}_{MJ}$ is composed of end-actions due to unit loads corresponding to the various joint displacements D_J. Since five displacements are to be obtained, there are five unit loads, as shown in Figs. 3-22f to 3-22j. Each of these loads results in a set of six member end-actions which constitutes the corresponding column of $\mathbf{A}_{MJ}$; thus:

$$\mathbf{A}_{MJ} = \frac{1}{2L}\begin{bmatrix} 0 & -2L & 0 & 0 & 0 \\ -2L & L^2 & 0 & 0 & 0 \\ -2 & L & -2 & -2 & -2 \\ 0 & 0 & 0 & 2L & 2L \\ 0 & 0 & 0 & 0 & 0 \\ 0 & 0 & 0 & 0 & -2L \end{bmatrix}$$

Next, the quantities $\mathbf{D}_{QL}$ and $\mathbf{F}$ are obtained from Eqs. (3-25) and (3-26):

$$\mathbf{D}_{QL} = \mathbf{A}'_{MQ}\mathbf{F}_M\mathbf{A}_{ML} = \frac{5PL^3}{36EI}$$

$$\mathbf{F} = \mathbf{A}'_{MQ}\mathbf{F}_M\mathbf{A}_{MQ} = \frac{L^3}{3EI}$$

and the redundant Q is found from Eq. (3-23a):

$$\mathbf{Q} = \mathbf{F}^{-1}(\mathbf{D}_Q - \mathbf{D}_{QL}) = -\frac{5P}{12}$$

The vectors of member end-actions $\mathbf{A}_M$, reactions $\mathbf{A}_R$, and joint displacements $\mathbf{D}_J$ are (see Eqs. 3-23b, 3-23c, and 3-29):

$$\mathbf{A}_M = \mathbf{A}_{ML} + \mathbf{A}_{MQ}\mathbf{Q} = \frac{P}{216}\{-18, 9L, 70, -29L, -90, 45L\}$$

$$\mathbf{A}_R = \mathbf{A}_{RL} + \mathbf{A}_{RQ}\mathbf{Q} = \frac{P}{12}\{-1, 5, 7\}$$

$$\mathbf{D}_J = \mathbf{A}'_{MJ}\mathbf{F}_M\mathbf{A}_M = \frac{PL^2}{2592EI}\{-133, 62L, -106, -34, -169\}$$

All of the preceding results are valid for the frame with combined joint loads (Fig. 3-22d); in addition, the vectors $\mathbf{Q}$, $\mathbf{A}_R$, and $\mathbf{D}_J$ also hold for the structure with the actual loads. However, the member end-actions $\mathbf{A}_M$ must be added to the end-actions $(\mathbf{A}_M)_R$ in the frame having restrained joints (see Fig. 3-22c) in order to obtain the end-actions $(\mathbf{A}_M)_A$ due to the actual loads (see Eq. 3-24). The end-actions for the restrained frame are

$$(\mathbf{A}_M)_R = \frac{P}{27}\{0, 0, 7, -2L, 0, 0\}$$

and when this vector is added to the vector $\mathbf{A}_M$ given previously, the end-actions for the actual frame (Fig. 3-22a) are found to be:

$$(\mathbf{A}_M)_A = \mathbf{A}_M + (\mathbf{A}_M)_R = \frac{P}{24}\{-2, L, 14, -5L, -10, 5L\}$$

Thus, all significant actions and displacements for the plane frame of Fig. 3-22a have been found, and therefore the solution can be considered as complete.

Example 4. The plane frame to be analyzed in this example is shown in Fig. 3-23a, and is the same structure discussed previously in Example 4 of Art. 2.3. The effects of both flexural and axial deformations are to be included in the analysis, and it is assumed that the rigidities of the members are EI and EA.

The redundant actions Q_1, Q_2, and Q_3 are obtained by releasing the structure at joint B, as shown in Fig. 3-23b. The member end-actions, reactions, and joint displacements are identified in Fig. 3-23c. Note that there are three end-actions for each member (axial force, shearing force, and bending moment, in that order), and that these are taken at end B of member AB and at end C of member BC. Thus, these ends have been selected as the k ends of the member (compare with Fig. 3-13).

The combined joint loads can be found without difficulty (see Fig. 3-23d) and are shown acting on the released structure in Fig. 3-23e. Because there are three sets of actions at joint B, namely, the end-actions for member 1, the combined joint loads, and the redundant actions, it is important to avoid ambiguity by deciding arbitrarily upon the relative positions of these actions at the joint. In this example, it is decided that the redundants are taken just below joint B.

The member flexibility matrix is obtained by referring to Eqs. (3-16) and (3-20):

$$\mathbf{F}_{\mathrm{M}} = \begin{bmatrix} \mathbf{F}_{\mathrm{M1}} & \mathbf{O} \\ \mathbf{O} & \mathbf{F}_{\mathrm{M2}} \end{bmatrix} = \begin{bmatrix} \frac{L}{EA} & 0 & 0 & 0 & 0 & 0 \\ 0 & \frac{L^3}{3EI} & \frac{L^2}{2EI} & 0 & 0 & 0 \\ 0 & \frac{L^2}{2EI} & \frac{L}{EI} & 0 & 0 & 0 \\ 0 & 0 & 0 & \frac{H}{EA} & 0 & 0 \\ 0 & 0 & 0 & 0 & \frac{H^3}{3EI} & \frac{H^2}{2EI} \\ 0 & 0 & 0 & 0 & \frac{H^2}{2EI} & \frac{H}{EI} \end{bmatrix}$$

The released structure with the combined joint loads acting on it is pictured in Fig. 3-23e. Because the redundants are taken just below joint B, as explained above, the loads at joint B are considered to act on member 1 rather than on member 2. The member end-actions and support reactions produced by these loads are used in forming the transfer matrices $\mathbf{A}_{\mathrm{ML}}$ and $\mathbf{A}_{\mathrm{RL}}$:

$$\mathbf{A}_{\mathrm{ML}} = \left\{0, -\frac{P}{2}, \frac{PL}{8}, 0, 0, 0\right\}$$

$$\mathbf{A}_{\mathrm{RL}} = \left\{0, P, \frac{PL}{2}, 0, 0, 0\right\}$$

Unit loads corresponding to the three redundants are shown in Figs. 3-23f, 3-23g, and 3-23h. Member end-actions and support reactions for each of these released structures are placed in the appropriate columns of the transfer matrices $\mathbf{A}_{\mathrm{MQ}}$ and $\mathbf{A}_{\mathrm{RQ}}$, as follows:

$$\mathbf{A}_{\mathrm{MQ}} = \begin{bmatrix} 1 & 0 & 0 \\ 0 & 1 & 0 \\ 0 & 0 & 1 \\ 0 & -1 & 0 \\ 1 & 0 & 0 \\ -H & 0 & 1 \end{bmatrix} \qquad \mathbf{A}_{\mathrm{RQ}} = \begin{bmatrix} -1 & 0 & 0 \\ 0 & -1 & 0 \\ 0 & -L & -1 \\ 1 & 0 & 0 \\ 0 & 1 & 0 \\ -H & 0 & 1 \end{bmatrix}$$

A unit load corresponding to D_{J1} is shown in Fig. 3-23i. This load can be considered to act upon member 1, as shown in the figure, or alternatively it could be placed upon member 2, since the horizontal displacement of both members at joint B must be the same. Similarly, unit loads corresponding to D_{J2} and D_{J3}

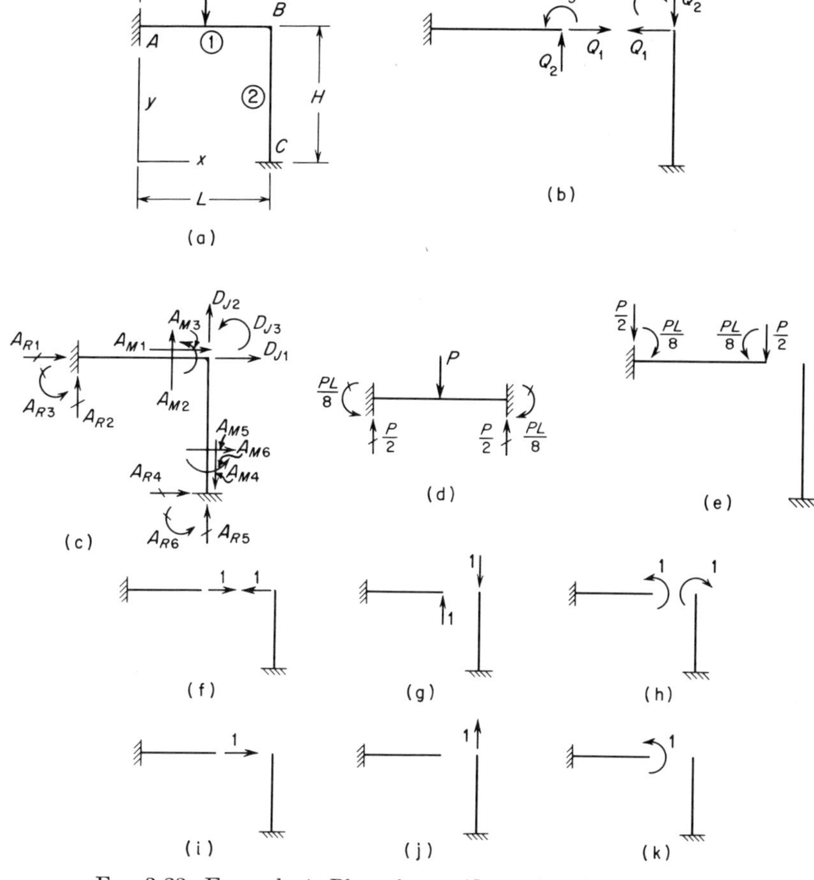

FIG. 3-23. Example 4: Plane frame (flexural and axial effects).

are shown in Figs. 3-23j and 3-23k. The matrix $\mathbf{A}_{MJ}$ constructed from these figures takes a very simple form:

$$\mathbf{A}_{MJ} = \begin{bmatrix} 1 & 0 & 0 \\ 0 & 0 & 0 \\ 0 & 0 & 1 \\ 0 & 1 & 0 \\ 0 & 0 & 0 \\ 0 & 0 & 0 \end{bmatrix}$$

As in previous examples, the matrices $\mathbf{D}_{QL}$ and $\mathbf{F}$ are found by substituting $\mathbf{F}_M$, $\mathbf{A}_{ML}$, and $\mathbf{A}_{MQ}$ into Eqs. (3-25) and (3-26). The results are:

$$\mathbf{D}_{QL} = \frac{PL^2}{48EI}\{0, -5L, -6\}$$

$$\mathbf{F} = \begin{bmatrix} \frac{L}{EA} + \frac{H^3}{3EI} & 0 & -\frac{H^2}{2EI} \\ 0 & \frac{L^3}{3EI} + \frac{H}{EA} & \frac{L^2}{2EI} \\ -\frac{H^2}{2EI} & \frac{L^2}{2EI} & \frac{L}{EI} + \frac{H}{EI} \end{bmatrix}$$

These matrices are the same as those obtained in the earlier solution for this frame (see Example 4, Art. 2.3).

The remainder of the solution can be performed more satisfactorily if numerical values are used; hence, the following data (the same as in the earlier solution) are assumed:

$$P = 10 \text{ kips} \qquad L = H = 12 \text{ ft} = 144 \text{ in.}$$

$$E = 30{,}000 \text{ ksi} \qquad I = 200 \text{ in.}^4 \qquad A = 10 \text{ in.}^2$$

Substitution of the above values into the matrices $\mathbf{D}_{QL}$ and $\mathbf{F}$ gives the matrices in numerical terms. These matrices, as well as $\mathbf{F}^{-1}$ and $\mathbf{Q}$, are given in the earlier solution and are not repeated here. The vectors $\mathbf{A}_M$, $\mathbf{A}_R$, and $\mathbf{D}_J$ are found as usual from Eqs. (3-23b), (3-23c), and (3-29); the results are

$$\mathbf{A}_M = \{-0.913, -0.97, 92.1, -4.03, -0.913, 43.6\}$$

$$\mathbf{A}_R = \{0.913, 5.97, 227.6, -0.913, 4.03, 43.6\}$$

$$\mathbf{D}_J = \{-0.000438, -0.00193, 0.00053\}$$

All numerical results in this solution are based upon units of kips, inches, and radians.

The member end-actions due to the actual loads (Fig. 3-23a) are found by adding the vector $(\mathbf{A}_M)_R$ of end-actions in the restrained structure to the vector $\mathbf{A}_M$ given above, in accordance with Eq. (3-24). The end-actions in the restrained structure (see Fig. 3-23d) are

$$(\mathbf{A}_M)_R = \left\{0, \frac{P}{2}, -\frac{PL}{8}, 0, 0, 0\right\} = \{0, 5, -180, 0, 0, 0\}$$

and, therefore,

$$(\mathbf{A}_M)_A = \mathbf{A}_M + (\mathbf{A}_M)_R = \{-0.913, 4.03, -87.9, -4.03, -0.913, 43.6\}$$

All of these results are in agreement with those found in the previous solution of the same frame.

The vector $\mathbf{Q}$ of redundants which was determined above for the structure with combined joint loads is valid also for the actual structure. This conclusion can be noted from the fact that there are no actions in the restrained structure corresponding to the redundants (see Fig. 3-23d).

The techniques illustrated by the preceding examples can easily be extended to other types of framed structures. However, the difficulty with the use of the flexibility method for hand computation lies in the tediousness of the calculations, especially if there are many members in the structure. This difficulty can be avoided by performing all of the matrix calculations (steps 6 through 11 in the outline given at the end of Art. 3.8) by a digital computer.

3.10 Effects of Temperature, Prestrain, and Support Displacement. The primary emphasis in the preceding articles of this chapter has been upon the effects of loads acting upon structures. However, the effects of temperature changes, prestrains in the members, and support displacements can also be taken into consideration. These effects were discussed previously for the flexibility method in Art. 2.4, where a method for incorporating them into the basic superposition equations was explained. It may be recalled from that discussion that support displacements are of two types: those that correspond to redundants and those that do not. The former are readily handled by including them in the vector $\mathbf{D}_{Q}$ of actual displacements corresponding to the redundants. The other support displacements require further consideration, and in the following discussion it is understood that any mention of support displacements refers to those of the latter type.

There are two general approaches that can be followed in dealing with temperature, prestrain, and support displacements. The first approach is based upon the use of equivalent joint loads and is the simpler and more direct method of the two. In this method the analysis begins by imposing the effect under consideration on the restrained structure. From the resulting fixed-end actions, a set of equivalent joint loads can be found. Thereafter, the structure can be analyzed as described in Art. 3.8 for a structure subjected to joint loads only. The final step in the analysis is to modify the member end-actions according to Eq. (3-24). The only difference from earlier solutions is that the end-actions $(A_M)_R$ in the restrained structure are due to the particular effect under consideration, rather than due to loads on the members.

The second approach makes use of the more general equations of the flexibility method (see Eqs. 2-13 and 2-17). The use of these equations requires the determination of various displacements in the released structure due to temperature, prestrain, and support displacement. Specifically, the displacement vectors $\mathbf{D}_{QT}$, $\mathbf{D}_{QP}$, $\mathbf{D}_{QR}$, $\mathbf{D}_{JT}$, $\mathbf{D}_{JP}$, and $\mathbf{D}_{JR}$ must be found (see Eqs. 2-11 and 2-18). Each of these displacement vectors can be formulated in matrix terms, analogous to the expression of $\mathbf{D}_{QL}$ and $\mathbf{D}_{JL}$ by the matrix multiplications indicated in Eqs. (3-25) and (3-27). However, since the required matrices are not of general interest, inasmuch as they are different for each type of effect, they will not be given here. Instead, the method of equivalent joint loads is suggested as a more convenient approach.

The flexibility method can be developed much further than is done in this chapter, and the interested reader should consult the references given at the back of the book.

3.11 Members with Symmetric and Nonsymmetric Cross-Sections. It has been assumed in all previous discussions that the cross-sectional shapes of the members met certain requirements of symmetry. For example, it is assumed that the plane of bending of a beam member (the x_M-y_M plane in Fig. 3-7) is a plane of symmetry. Thus, while both the y_M and z_M axes must be principal axes of the cross-sectional area, the y_M axis must also be an axis of symmetry. Under these conditions the beam will deflect only in the x_M-y_M plane, assuming of course that all forces act in that plane and that all couples have their vectors parallel to the z_M axis. The same conditions are required for a plane frame member.

In the case of a grid member subjected to torsion as well as bending, it has been assumed that both the y_M and z_M axes (see Fig. 3-14) are axes of symmetry. If this is the case, all deflections of a grid member will be in the x_M-y_M plane, since torsion of the member will produce rotation about the x_M axis, but no bending. Similarly, it has been assumed also that a space frame member (Fig. 3-17) has two axes of symmetry in the cross-section, so that torsion and bending occur independently. Several examples of sections having two axes of symmetry are given in Appendix C.

In the case of a truss member there is no requirement of symmetry in the cross-section. It is necessary, however, that the axial force in the member act through the centroid of the cross-sectional area; otherwise, there will be bending moments produced in the member.

Returning now to the case of the beam member, consider how its behavior is altered if the y_M axis (Fig. 3-7) is not an axis of symmetry. An example of such a member is a channel section (see Fig. 3-24a). In the figure, points C and O represent the centroid and shear center, respectively. If the loads on the member act through any point other than the shear center, the beam will twist as it bends. However, if the loads act through point O, the beam will bend in the x_M-y_M plane without twisting. Therefore, it can be concluded that a beam for which the

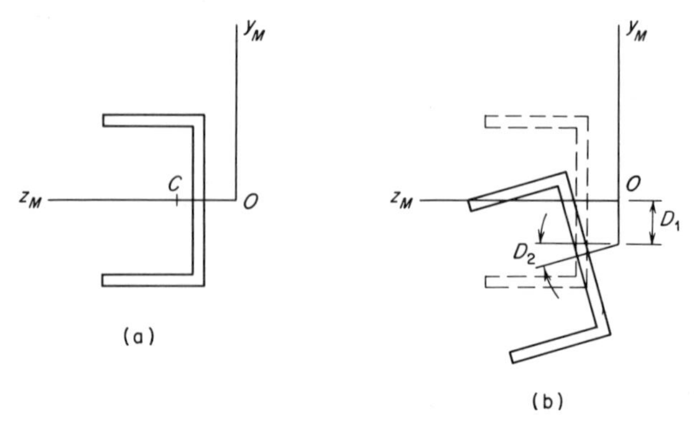

FIG. 3-24. Beam member with channel cross-section.

x_M-y_M plane is not a plane of symmetry can be analyzed in the same manner as one for which it is a plane of symmetry, provided that the x_M-y_M plane is taken through the shear center. Such a requirement means that the joints between members must be located at the intersections of the shear center axes, and that points of support must be on the axes of shear centers.

For a plane frame, a grid, and a space frame the same general comment can be made as in the case of a beam, namely, that the shear center axis must be taken as the axis of the member. The displacements that are determined in the analysis then will be the displacements of the shear center axis, and are not necessarily the same as those of the member itself. A case of this type is shown in Fig. 3-24b where it is assumed that the channel section has undergone a translation D_1 of the shear center in the negative y_M direction and a rotation D_2 about the x_M axis. It can be seen that the translation of any point in the member itself is composed of two parts: the translation of the shear center, plus an additional translation produced by the rotation about the shear center.*

When discussing torsion of a member, it has been assumed in this book that the effect of warping of the cross-sections is negligible. As a result, the torsional rigidity GJ has been used in all formulas and calculations involving torsion. In some instances a more exact theory of nonuniform torsion, which considers the effects of warping, should be used. In such cases the relationships between loads and displacements become more complicated than given in Appendix A, and they involve an additional property of the cross-section (namely, the warping constant). The interested reader should consult other references for a discussion of nonuniform torsion.**

Problems

In the following problems, assume that all displacements are positive when in the positive senses of the coordinate axes shown in the figures. Use the numbering systems for the members as shown in the figures.

3.3-1. Determine the displacements in the x and y directions at joint A of the plane truss shown in the figure. Assume that $H = 3L/4$ and that EA is constant for all members.

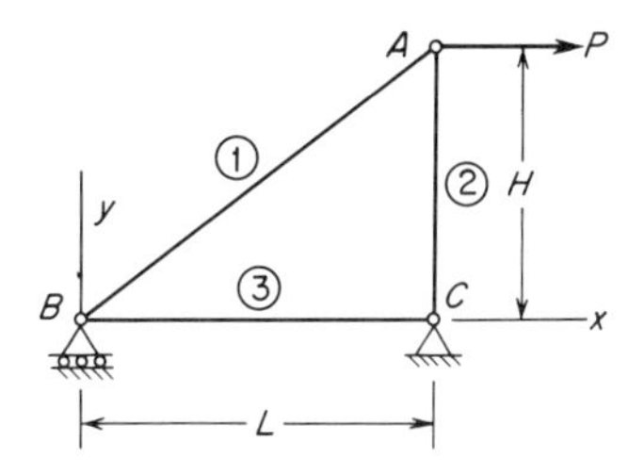

PROB. 3.3-1.

* For a discussion of bending of unsymmetrical beams, see S. P. Timoshenko, "Strength of Materials," 3rd ed., Part I, D. Van Nostrand Co., Inc., Princeton, N.J., 1955, pp. 235–244.

** See, for example, S. P. Timoshenko, *ibid.*, Part II, 1956, pp. 255–273.

3.3-2. The plane truss shown in the figure is subjected to two load systems: (1) load P_1 alone, and (2) load P_2 alone. The axial rigidity EA is the same for all members. Find the translations of joint D in the x direction and of joint B in the y direction for each load system.

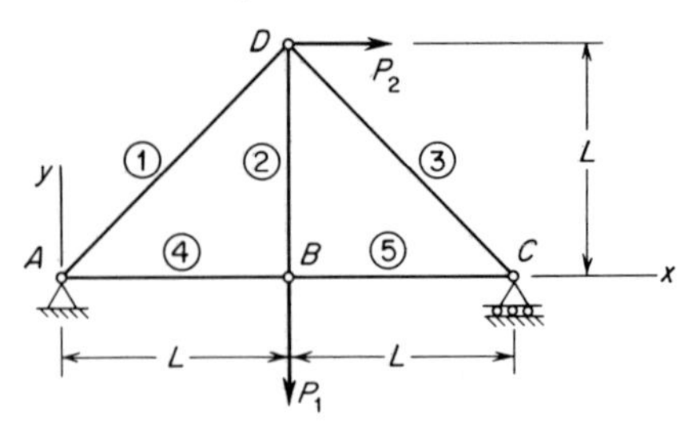

PROB. 3.3-2.

3.3-3. Find the translations of joint A in the x, y, and z directions for the space truss (see figure). The supports at B, C, and D lie in the x-y plane. Assume the following data: areas of members 1 and 2 are 1.0 in.² each, area of member 3 is 3.0 in.², $P = 6$ kips, and $E = 30{,}000$ ksi.

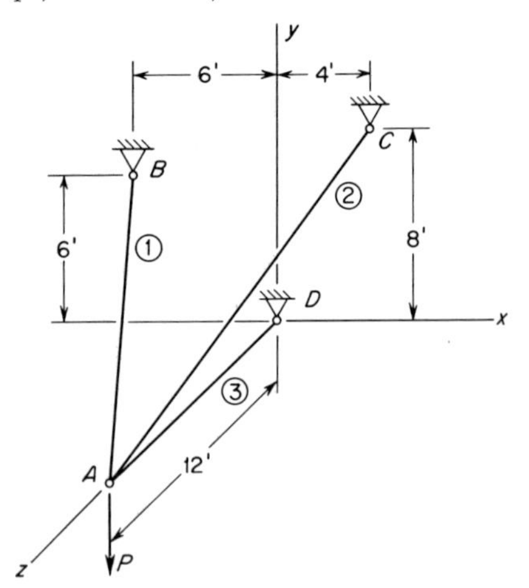

PROB. 3.3-3.

3.4-1. Determine the displacements of joint A for the beam with an overhang (see figure). The displacements are to be taken in the following order: translation in the y direction, and rotation in the z sense. Assume that there are two load systems (the uniform load w only, and the concentrated forces P only) and that the beam has constant EI.

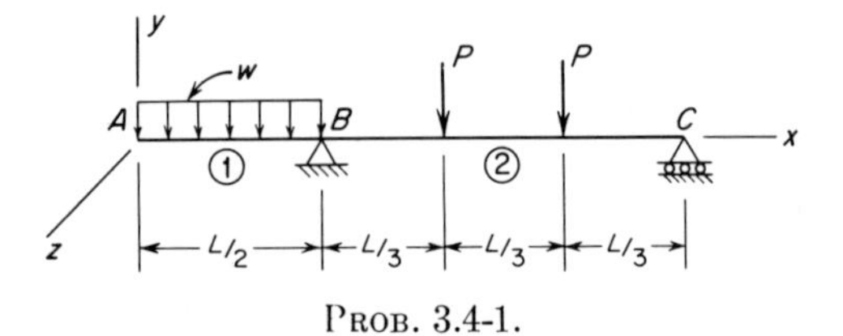

PROB. 3.4-1.

3.4-2. The cantilever beam ABC shown in the figure is reinforced along the region AB so that the moment of inertia of AB is $3I$ while that of BC is I. Find the displacements at points B and C (translation in the y direction and rotation in the z sense) due to the load P.

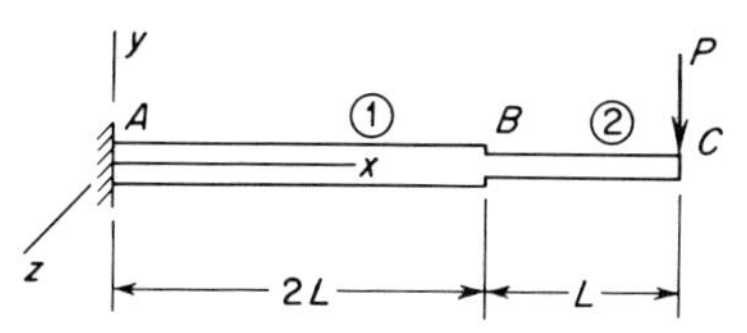

PROB. 3.4-2 and PROB. 3.4-3.

3.4-3. Solve the preceding problem including the effects of shearing deformations as well as flexural deformations. Assume that the shearing rigidity of both members of the beam is GA/f.

3.5-1. Determine the translation of joint B in the y direction and the translation of joint A in the x direction for the plane frame shown in the figure. Consider only the effects of flexural deformations, and assume EI is constant for all members.

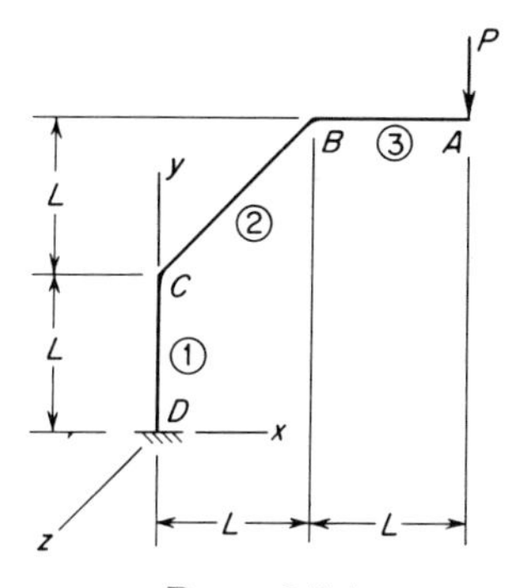

PROB. 3.5-1.

3.5-2. Find the translation in the y direction at point C of the plane frame (see figure). Consider both axial and flexural deformations, and assume that E, I, and A are the same for all members. Also, assume that $w = P/L$.

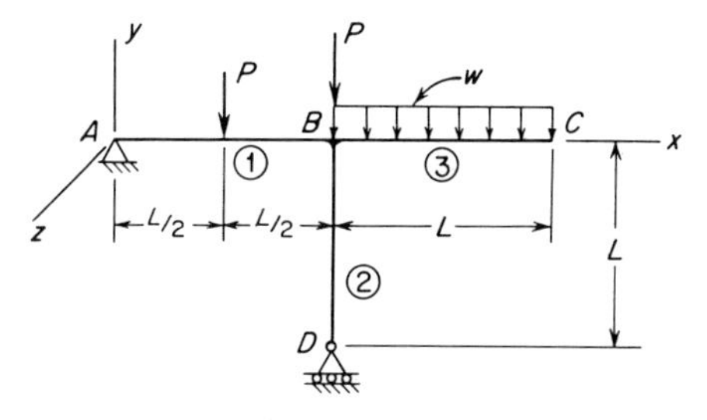

PROB. 3.5-2.

3.6-1. The grid shown in the figure lies in the x-z plane and is simply supported at A, B, and C. The loads on the grid consist of a concentrated force P

and a twisting couple M acting at the midpoint of member AB. (Such a set of loads is the statical equivalent of an eccentrically applied force P.) The joint at B is a rigid connection between the two members. Each member has flexural rigidity EI and torsional rigidity GJ. Find the rotations in the x and z senses at points A and C.

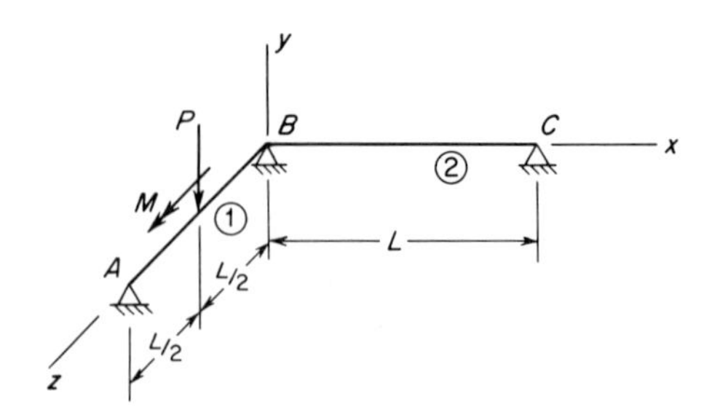

PROB. 3.6-1.

3.9-1. The plane truss shown in the figure supports a load P at joint A. The axial rigidity EA is the same for all members. Find the vector $\mathbf{A}_{\mathrm{M}}$ of axial forces in all members, and find the displacements at joint A (translations in the x and y directions).

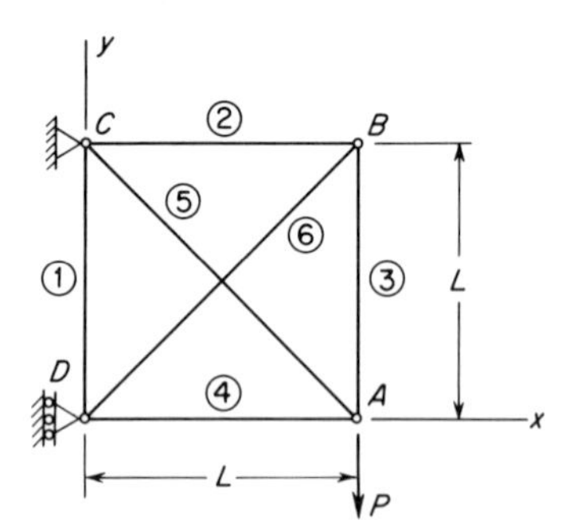

PROB. 3.9-1.

3.9-2. Determine the axial forces in all members of the truss (see figure). Assume $L = 20$ ft, E is constant for all members, the area of members 1, 2, and 3 is 4.0 in.^{2}, the area of all other members is 2.0 in.^{2}, and $P = 30$ kips.

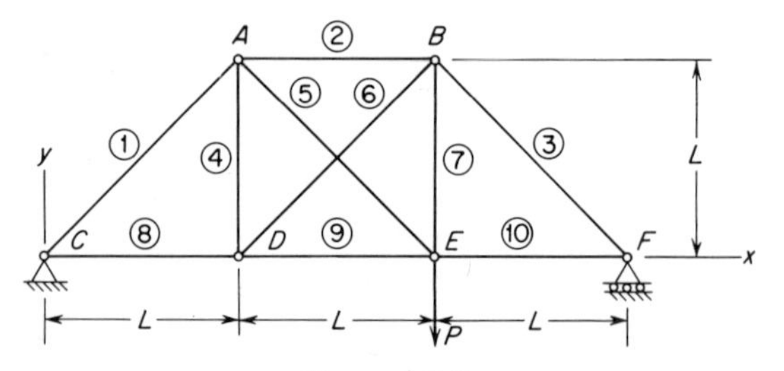

PROB. 3.9-2.

3.9-3. Find the axial forces in the bars and the translation of joint A in the plane truss shown in the figure. Assume L, E, and A are the same for all three bars.

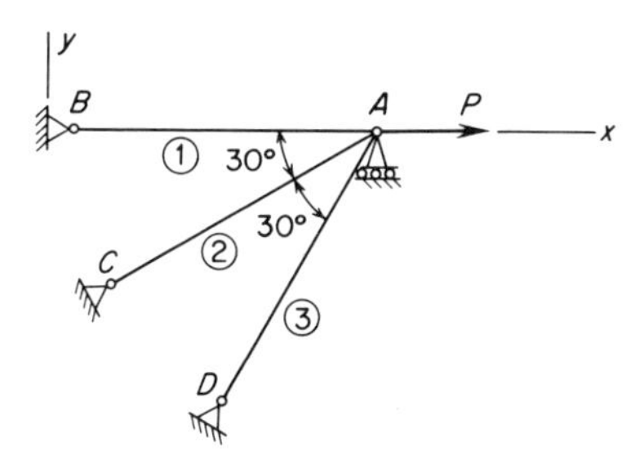

PROB. 3.9-3.

3.9-4. Determine the reactions (i.e., force in the y direction and couple in the z sense, at joints A and C) for the beam with fixed ends (see figure). Assume that $EI_1 = 2EI$ and $EI_2 = EI$. Consider AB and BC as members 1 and 2, respectively.

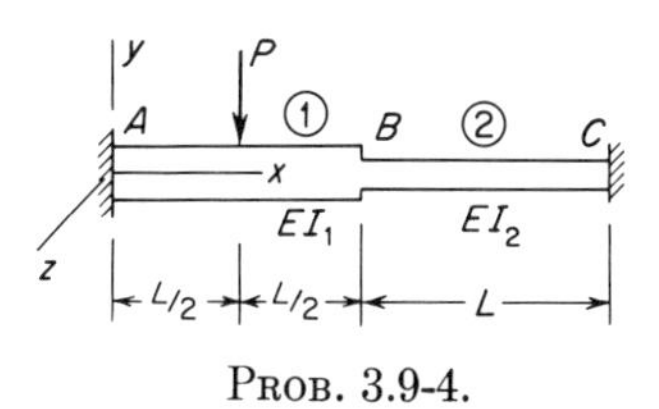

PROB. 3.9-4.

3.9-5. Calculate the rotation in the z sense at support A of the continuous beam shown in the figure. Assume that EI is constant for all spans, and that $P = wL$. Number the members from left to right along the beam.

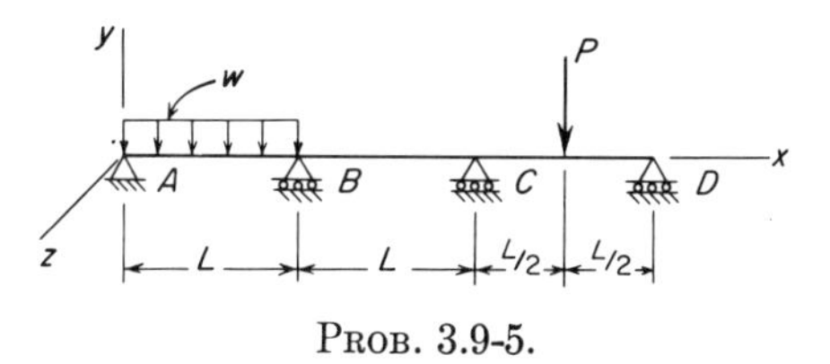

PROB. 3.9-5.

3.9-6. Find the shearing force and bending moment at end B of member AB in the plane frame shown in the figure. Note that the moments of inertia of members 1, 2, and 3 are $2I$, $8I$, and I, respectively. Also, calculate the translation in the x direction of joint C, if $E = 30{,}000$ ksi and $I = 200$ in.[4] Consider only the effects of flexural deformations.

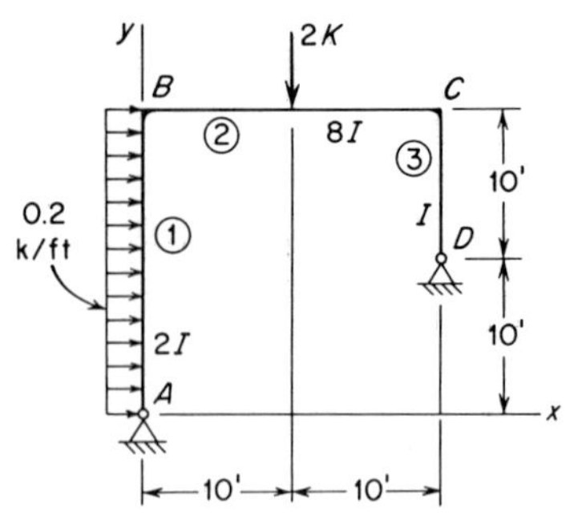

PROB. 3.9-6.

3.9-7. The plane frame shown in the figure is subjected to a uniform load w along span BC and a concentrated couple M at joint B. Determine the reactions at support A (forces in the x and y directions, and moment in the z sense) and the rotations at joints B and C. Select the reactions at support C as the redundants, and consider only the effects of flexural deformations. Assume that $M = 2wL^2$, $H = L$, and EI is constant.

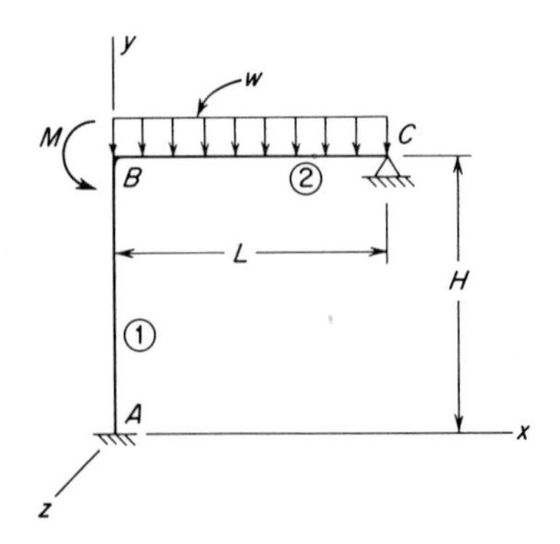

PROB. 3.9-7.

3.9-8. The plane frame shown in the figure is to be analyzed using the following redundants: Q_1 is the bending moment a small distance to the left of joint B (positive when the upper side of the member is in compression), and Q_2 is the reactive force at support C. Determine the reactions at supports A and D, and the rotation at joint B. Consider only flexural effects, and assume EI is constant for all members.

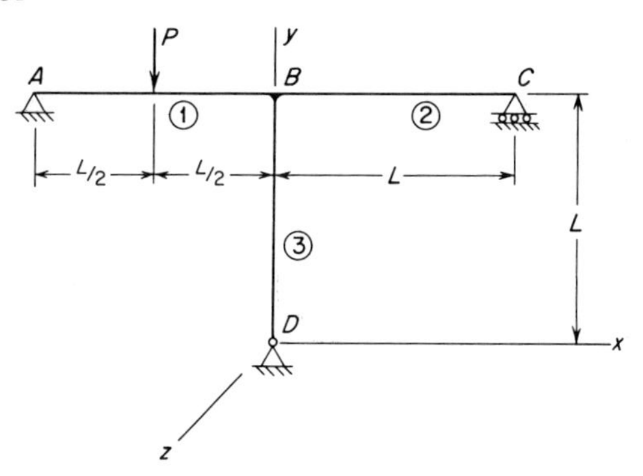

PROB. 3.9-8.

3.9-9. Determine the rotations at joint C in the x and z senses for the grid of Prob. 2.3-18. Assume that members AB, BC, and CD are members 1, 2, and 3, respectively. Each member has flexural rigidity EI, torsional rigidity GJ, and length L. The loads P act at the midpoints of members AB and BC.

Chapter 4

STIFFNESS METHOD

4.1 Introduction. The theory of the stiffness method of analysis was developed and applied in a manner suitable for hand calculations in Chapter 2. In the following articles the method is developed further and applied in a more formalized fashion. The objective is to cast the analysis into a form which may be readily programmed on a digital computer. In fact, the solutions described in this chapter for the six basic types of framed structures have their counterparts in the form of flow charts for computer programs in the next chapter. Thus, the detailed formalized approach of this chapter constitutes the intermediate step between hand calculations and computer programming.

In Art. 4.2 an outline is presented for the purpose of orienting the stiffness method of analysis toward computer programming. Since member stiffnesses play an essential role in the analyses of all types of framed structures, this topic is treated next in Art. 4.3. Thus, this information is available for use subsequently in the chapter. Then, in Arts. 4.4 through 4.7, certain aspects of the stiffness method of analysis are reorganized into a format which is suitable for programming purposes. The necessary reorganization is explained in conjunction with a familiar example problem which has been solved previously. The remaining articles of the chapter deal with applications to the various types of framed structures. For simplicity, only the effects of loads on the structure are considered in this chapter. The methods of treating temperature changes, prestrains, support displacements, and other effects are described in Chapter 6.

4.2 Outline of the Stiffness Method. The basic equations of the stiffness method of analysis were discussed previously in Chapter 2. When only the effects of loads on the structure are taken into account, the equations for the joint displacements, member end-actions, and reactions are:

$$\mathbf{A}_{\mathrm{D}} = \mathbf{A}_{\mathrm{DL}} + \mathbf{S}\mathbf{D} \qquad \text{(2-23) repeated}$$

$$\mathbf{A}_{\mathrm{M}} = \mathbf{A}_{\mathrm{ML}} + \mathbf{A}_{\mathrm{MD}}\mathbf{D} \qquad \text{(2-25) repeated}$$

$$\mathbf{A}_{\mathrm{R}} = \mathbf{A}_{\mathrm{RL}} + \mathbf{A}_{\mathrm{RD}}\mathbf{D} \qquad \text{(2-26) repeated}$$

In the first of these equations, the vector $\mathbf{A}_D$ consists of loads corresponding to the unknown displacements $\mathbf{D}$, the vector $\mathbf{A}_{DL}$ is composed of artificial restraint actions in the restrained structure corresponding to the displacements $\mathbf{D}$ and caused by all loads other than those in $\mathbf{A}_D$, and the stiffness matrix $\mathbf{S}$ corresponds to the displacements $\mathbf{D}$. In the second equation, the vector $\mathbf{A}_M$ consists of member end-actions in the actual structure, $\mathbf{A}_{ML}$ is a vector of member end-actions in the restrained structure due to loads, and $\mathbf{A}_{MD}$ is a matrix of end-actions due to unit values of the joint displacements. In the third equation, the vector $\mathbf{A}_R$ represents reactions at the supports of the actual structure, $\mathbf{A}_{RL}$ is the vector of corresponding quantities in the restrained structure subjected to loads, and $\mathbf{A}_{RD}$ is a matrix of support reactions due to unit values of the joint displacements $\mathbf{D}$.

When solving problems by hand, one may generate the matrices in the above equations in any convenient fashion with no loss of efficiency, as was done in Chapter 2. The equations are suitable for calculating certain selected member end-actions and reactions in relatively simple structures. If the structure to be analyzed is large and complicated, however, and if all member end-actions and reactions are to be determined, the above equations are not efficient. Moreover, large and complicated structures cannot be analyzed directly by hand, but the calculations must be carried out on a digital computer. In a computer program it becomes necessary to handle all information about the structure and the loads in a highly organized manner. Thus, a formalized approach must be developed which enables a computer to process large amounts of information by a routine procedure.

A computer program for the analysis of a structure by the stiffness method divides conveniently into several phases. These phases are not the same as outlined in Chapter 2 for hand calculations. The difference lies in the fact that when using a computer it is desirable to work with all of the data pertaining to the structure at the outset. These steps include the formation of the stiffness matrix, which is a property of the structure. Subsequently, the load data is manipulated, after which the final results of the analysis are computed. This sequence is particularly efficient if more than one load system is being considered, since the initial phases of the calculations need not be repeated. The phases to be considered in the subsequent discussions are the following:

(1) *Assembly of Structure Data.* Information pertaining to the structure itself must be assembled and recorded. This information includes the number of members, the number of joints, the number of degrees of freedom, and the elastic properties of the material. The locations of the joints of the structure are specified by means of geometric coordinates. In addition, the section properties of each member in the structure must be given. Finally, the conditions of restraint at the supports of the

structure must be identified. In computer programming, all such information is coded in some convenient way, as will be shown subsequently in this chapter and also in Chapter 5.

(2) *Generation and Inversion of Stiffness Matrix.* The stiffness matrix is an inherent property of the structure and is based upon the structure data only. In computer programming it is convenient to obtain the joint stiffness matrix by summing contributions from individual member stiffness matrices (a discussion of prismatic member stiffnesses is given in Art. 4.3). The essential change from the previous approach consists of generalizing the joint stiffness matrix from one that is related only to the degrees of freedom in the structure to one that is related to all possible joint displacements, including support displacements. This generalized stiffness matrix shall be called the *over-all joint stiffness matrix* and is described in Art. 4.4.

(3) *Assembly of Load Data.* All loads acting on the structure must be specified in a manner which is suitable for computer programming. Both joint loads and member loads must be given. The former may be handled directly, but the latter are handled indirectly by supplying as data the fixed-end actions caused by the loads on the members.

(4) *Generation of Vectors Associated with Loads.* The fixed-end actions due to loads on members may be converted to *equivalent joint loads,* as described previously in Art. 3.2. These equivalent joint loads may then be added to the actual joint loads to produce a problem in which the structure is imagined to be loaded at the joints only. This manipulation of load data is described in Art. 4.5.

(5) *Calculation of Results.* In the final phase of the analysis all of the joint displacements, reactions, and member end-actions are computed. In this phase there are also certain modifications of the previous approach. In particular, one performs the calculation of member end-actions member by member instead of considering the structure as a whole. Such calculations require the use of member stiffness matrices, a subject which is summarized in the next article. A simple calculation of this type is demonstrated for illustrative purposes in Art. 4.6.

It should be noted that there are many possible variations in organizing the stiffness method for computer programming. The phases of analysis listed above constitute an orderly approach which has certain essential features that are advantageous when dealing with large, complicated frameworks. Each of these phases will be discussed and illustrated by means of a familiar example in the articles which follow.

4.3 Prismatic Member Stiffnesses. A stiffness coefficient at any joint of a structure is composed of the sum of the stiffnesses of members which frame into that joint, as illustrated in the examples solved in Art. 2.9. Therefore, in generating the joint stiffness matrix for a structure it is convenient to sum the *member stiffnesses* in some systematic fashion. It is also convenient to make use of a member stiffness matrix when calculating the final actions A_M at the ends of a member after the joint displacements have been found.

Certain member stiffnesses have already been used in earlier problems. For example, the quantities $4EI/L$ and $2EI/L$ have been used repeatedly in beam analyses (see Fig. 2-15), and the term EA/L has appeared in truss analyses (see Example 3 of Art. 2.9). In general, member stiffnesses of all types are required when analyzing structures by the stiffness method. Therefore, all of the stiffnesses for a prismatic member are given in this article in order to have them available in subsequent discussions.

In order to begin the discussion, consider the prismatic member shown in Fig. 4-1. The member is assumed to be fully restrained at both ends, which are denoted as ends j and k. It is convenient to develop member stiffnesses in conjunction with a set of orthogonal axes which are oriented to the axis of the member. Such a set of *member-oriented axes* appears in Fig. 4-1. The x_M axis coincides with the centroidal axis of the

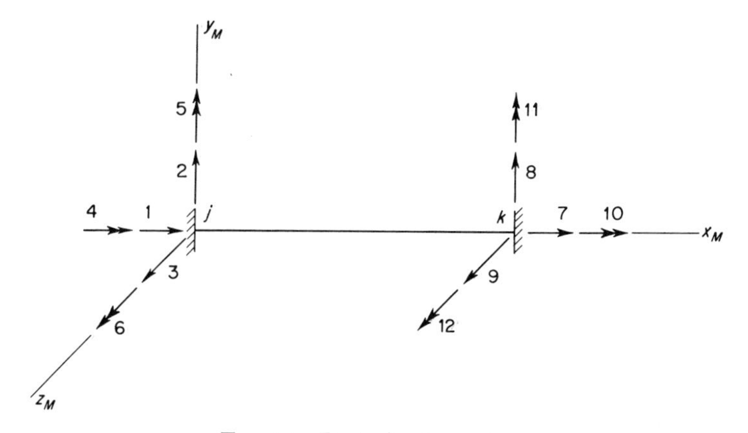

FIG. 4-1. Restrained member.

member and is positive in the sense from j to k. The y_M and z_M axes are principal axes for the member, that is, the x_M-y_M and x_M-z_M planes are principal planes of bending. It is assumed that the shear center and the centroid of the member coincide so that twisting and bending of the member are not coupled, but may occur independently of one another. This restriction is normally satisfied in framed structures; however, it is possible to make a more general analysis if necessary (see Art. 3.11).

It should be mentioned at this point that the analysis of trusses, plane frames, grids, and space structures requires the use of a set of reference axes oriented in some manner to the structure in its entirety. Such axes are known as *structure-oriented axes* and, in general, the member axes are skew with respect to the structure axes. However, it is always possible to obtain the member stiffnesses with respect to the member axes, as is done in this article, and then to transform these stiffnesses to the structure axes. The procedures for accomplishing this transformation are described later in the chapter for each type of structure.

The properties of the member shown in Fig. 4-1 will be defined in a systematic fashion* for the purpose of computer programming for complex structures. Let L denote the length of the member and A_X the area of the cross-section. (The symbol A without subscript is used later as an identifier for actions.) The principal moments of inertia of the cross-section of the member with respect to the y_M and z_M axes are denoted I_Y and I_Z, respectively. Also, let the torsion constant for the member be I_X, which is the same as the constant J that appears in Chapter 2 and the Appendices. Since the symbol J has use as an index for joints of the structure in computer programming, the use of the symbol I_X for the torsion constant is desirable. The torsion constant I_X is not to be interpreted as the polar moment of inertia of the cross-section except in the special case of a circular cylindrical member.

The member stiffnesses for the restrained member shown in Fig. 4-1 are the actions exerted on the member by the restraints when unit displacements (translations and rotations) are imposed at each end of the member. The values of these restraint actions may be obtained from Table B-4 in Appendix B. The unit displacements are considered to be induced one at a time while all other end displacements are retained at zero, and they are assumed to be positive in the x_M, y_M, and z_M directions. Thus, the positive senses of the three translations and the three rotations at each end of the member are indicated by arrows in Fig. 4-1. In the figure the single-headed arrows denote translations, whereas the double-headed arrows represent rotations. At joint j the translations are numbered 1, 2, and 3, and the rotations are numbered 4, 5, and 6. Similarly, at the k end of the member 7, 8, and 9 are translations, and 10, 11, and 12 are rotations. In all cases the displacements are taken in the order x_M, y_M, and z_M, respectively. In a space frame, which is the most general type of framed structure, the displacements at the ends of a member are numbered in exactly this order. However, in the members of other structures, such as plane frames, not all of the stiffnesses are needed. Therefore, some of the displacements are omitted from consideration, and the numbering is changed accordingly.

The member stiffnesses for the twelve possible types of end displacements (shown in Fig. 4-1) are summarized pictorially in Fig. 4-2. In each case the various restraint actions (or member stiffnesses) are shown

* This system of denoting member properties is used by A. S. Hall and R. W. Woodhead in *Frame Analysis*, John Wiley and Sons, Inc., New York, 1961, p. 7.

as vectors. An arrow with a single head represents a force vector, and an arrow with a double head represents a moment vector. All vectors are drawn in the positive senses, but in cases where the restraint actions are actually negative a minus sign precedes the expression for the stiffness coefficient.

In order to show how the member stiffnesses are determined, consider case (1) in Fig. 4-2. The restraint actions shown in the figure arise due to a unit displacement of the j end of the member in the positive x_M direction. All other displacements are zero. This displacement causes a pure compressive force EA_X/L in the member. At the j end of the member this compression force is equilibrated by a restraint action of EA_X/L in the positive x_M direction, and at the k end of the member the restraint action has the same value but is in the negative x_M direction. All other restraint actions are zero in this case.

Case (2) in Fig. 4-2 involves a unit displacement of the j end of the member in the positive y_M direction, while all other displacements are zero. This displacement causes both moment and shear in the member. At the j end, the restraint actions required to keep the member in equilibrium are a lateral force of $12EI_Z/L^3$ in the positive y_M direction and a couple $6EI_Z/L^2$ in the positive z_M sense (see Table B-4). At the k end of the member the restraint actions are the same except that the lateral force acts in the negative y_M direction.

All of the member stiffnesses shown in the figure are derived by determining the values of the restraint actions required to hold the distorted member in equilibrium. The reader should verify all of the expressions before proceeding further. These stiffnesses can be used for the purpose of formulating stiffness matrices for the members of different types of structures. In the most general case (a space frame), it is possible for the member to undergo any of the twelve displacements shown in Fig. 4-2. The stiffness matrix for such a member, denoted $\mathbf{S}_M$, is therefore of order 12×12, and each column in the matrix represents the actions caused by one of the unit displacements. The space frame member stiffness matrix is shown in Table 4-1; it is, of course, symmetrical. The rows and columns of the matrix are numbered down the side and across the top to assist the reader in identifying a particular element. In addition, the matrix is partitioned in order to delineate the portions which are associated with the two ends of the member.

The member stiffness matrices needed for other structures, such as continuous beams and plane frames, are of lesser order than the matrix shown in Table 4-1. This is because only certain of the end displacements shown in Figs. 4-1 and 4-2 are considered in the analyses of such structures. As an example of how such a member stiffness matrix is formed, the stiffness matrix for a member in a continuous beam will now be developed.

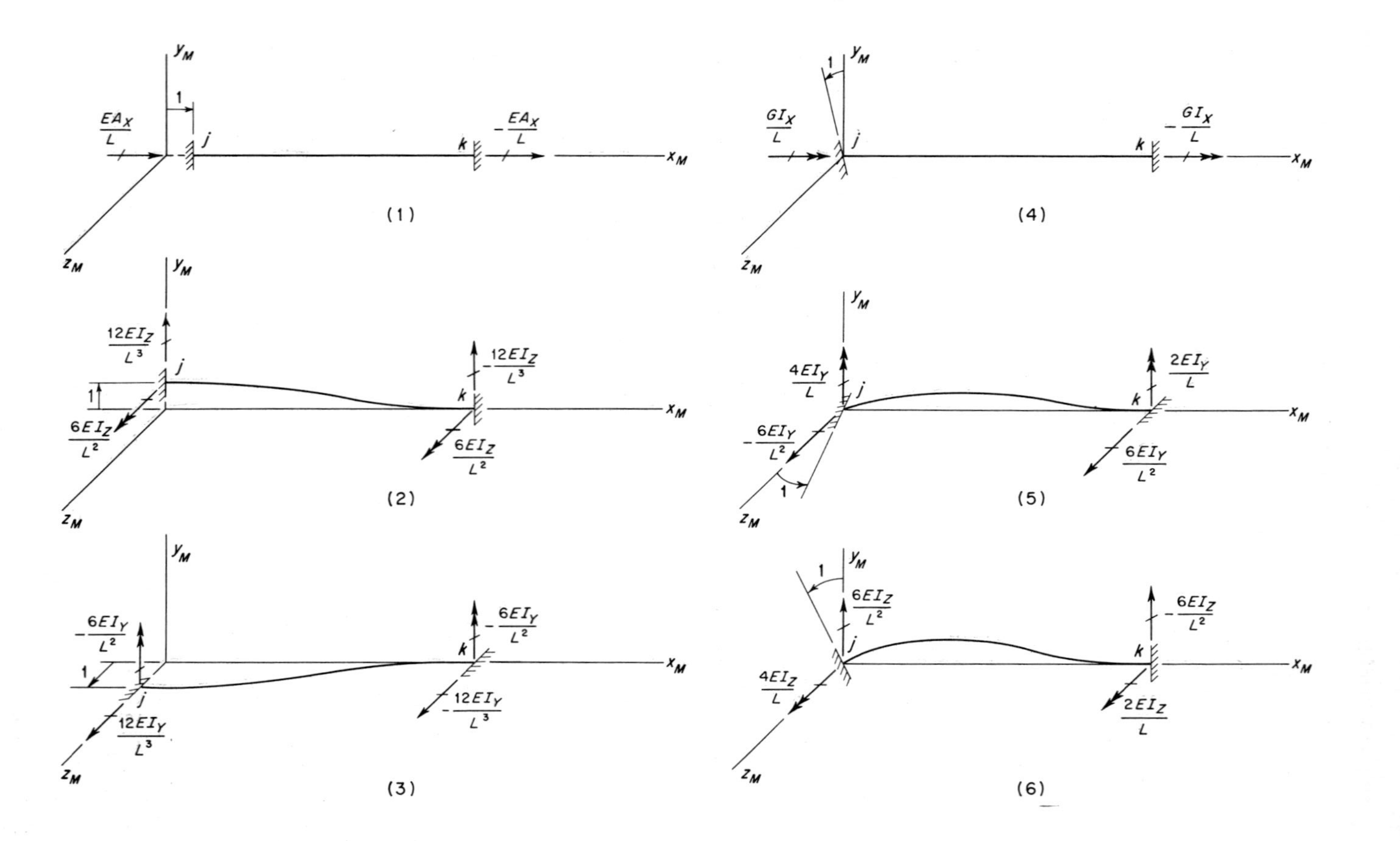
(1)
(2)
(3)
(4)
(5)
(6)
x_M
y_M
z_M
j
k
1
$\frac{EA_X}{L}$
$-\frac{EA_X}{L}$
$\frac{GI_X}{L}$
$-\frac{GI_X}{L}$
$\frac{12EI_Z}{L^3}$
$-\frac{12EI_Z}{L^3}$
$\frac{6EI_Z}{L^2}$
$\frac{4EI_Y}{L}$
$\frac{2EI_Y}{L}$
$-\frac{6EI_Y}{L^2}$
$\frac{6EI_Y}{L^2}$
$\frac{12EI_Y}{L^3}$
$-\frac{12EI_Y}{L^3}$
$\frac{6EI_Z}{L^2}$
$-\frac{6EI_Z}{L^2}$
$\frac{4EI_Z}{L}$
$\frac{2EI_Z}{L}$

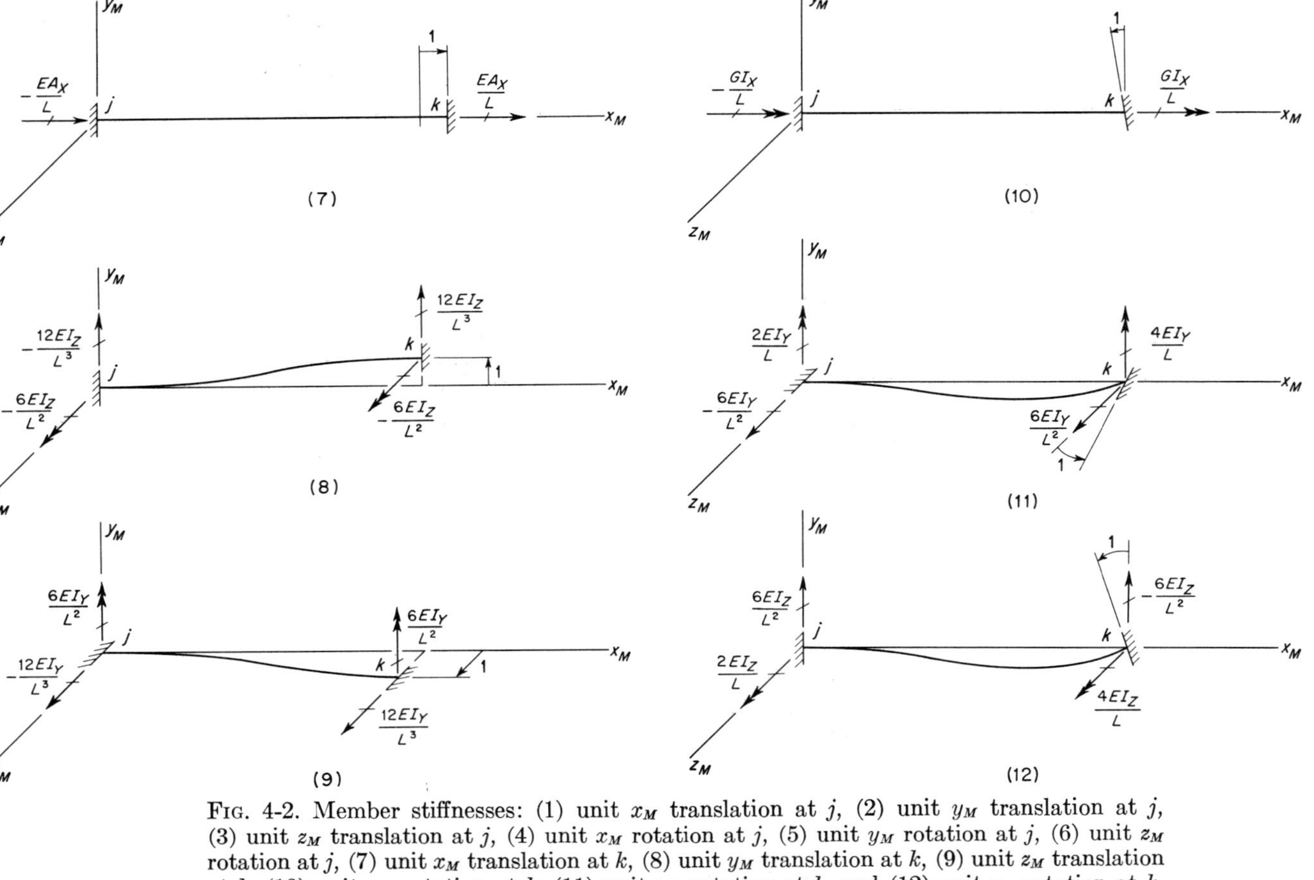

FIG. 4-2. Member stiffnesses: (1) unit x_M translation at j, (2) unit y_M translation at j, (3) unit z_M translation at j, (4) unit x_M rotation at j, (5) unit y_M rotation at j, (6) unit z_M rotation at j, (7) unit x_M translation at k, (8) unit y_M translation at k, (9) unit z_M translation at k, (10) unit x_M rotation at k, (11) unit y_M rotation at k, and (12) unit z_M rotation at k.

TABLE 4-1
Space Frame Member Stiffness Matrix

$$
\mathbf{S}_M =
\begin{array}{c}
\\ 1 \\ 2 \\ 3 \\ 4 \\ 5 \\ 6 \\ 7 \\ 8 \\ 9 \\ 10 \\ 11 \\ 12
\end{array}
\begin{array}{cccccc|cccccc}
1 & 2 & 3 & 4 & 5 & 6 & 7 & 8 & 9 & 10 & 11 & 12 \\
\frac{EA_X}{L} & 0 & 0 & 0 & 0 & 0 & -\frac{EA_X}{L} & 0 & 0 & 0 & 0 & 0 \\
0 & \frac{12EI_Z}{L^3} & 0 & 0 & 0 & \frac{6EI_Z}{L^2} & 0 & -\frac{12EI_Z}{L^3} & 0 & 0 & 0 & \frac{6EI_Z}{L^2} \\
0 & 0 & \frac{12EI_Y}{L^3} & 0 & -\frac{6EI_Y}{L^2} & 0 & 0 & 0 & -\frac{12EI_Y}{L^3} & 0 & -\frac{6EI_Y}{L^2} & 0 \\
0 & 0 & 0 & \frac{GI_X}{L} & 0 & 0 & 0 & 0 & 0 & -\frac{GI_X}{L} & 0 & 0 \\
0 & 0 & -\frac{6EI_Y}{L^2} & 0 & \frac{4EI_Y}{L} & 0 & 0 & 0 & \frac{6EI_Y}{L^2} & 0 & \frac{2EI_Y}{L} & 0 \\
0 & \frac{6EI_Z}{L^2} & 0 & 0 & 0 & \frac{4EI_Z}{L} & 0 & -\frac{6EI_Z}{L^2} & 0 & 0 & 0 & \frac{2EI_Z}{L} \\
\hline
-\frac{EA_X}{L} & 0 & 0 & 0 & 0 & 0 & \frac{EA_X}{L} & 0 & 0 & 0 & 0 & 0 \\
0 & -\frac{12EI_Z}{L^3} & 0 & 0 & 0 & -\frac{6EI_Z}{L^2} & 0 & \frac{12EI_Z}{L^3} & 0 & 0 & 0 & -\frac{6EI_Z}{L^2} \\
0 & 0 & -\frac{12EI_Y}{L^3} & 0 & \frac{6EI_Y}{L^2} & 0 & 0 & 0 & \frac{12EI_Y}{L^3} & 0 & \frac{6EI_Y}{L^2} & 0 \\
0 & 0 & 0 & -\frac{GI_X}{L} & 0 & 0 & 0 & 0 & 0 & \frac{GI_X}{L} & 0 & 0 \\
0 & 0 & -\frac{6EI_Y}{L^2} & 0 & \frac{2EI_Y}{L} & 0 & 0 & 0 & \frac{6EI_Y}{L^2} & 0 & \frac{4EI_Y}{L} & 0 \\
0 & \frac{6EI_Z}{L^2} & 0 & 0 & 0 & \frac{2EI_Z}{L} & 0 & -\frac{6EI_Z}{L^2} & 0 & 0 & 0 & \frac{4EI_Z}{L}
\end{array}
$$

Consider one member of a continuous beam between supports denoted as j and k, as shown in Fig. 4-3a. The x_M, y_M, and z_M axes are taken in the directions shown in the figure, so that the x_M-y_M plane is the plane of bending of the beam. In a member of a continuous beam there are four

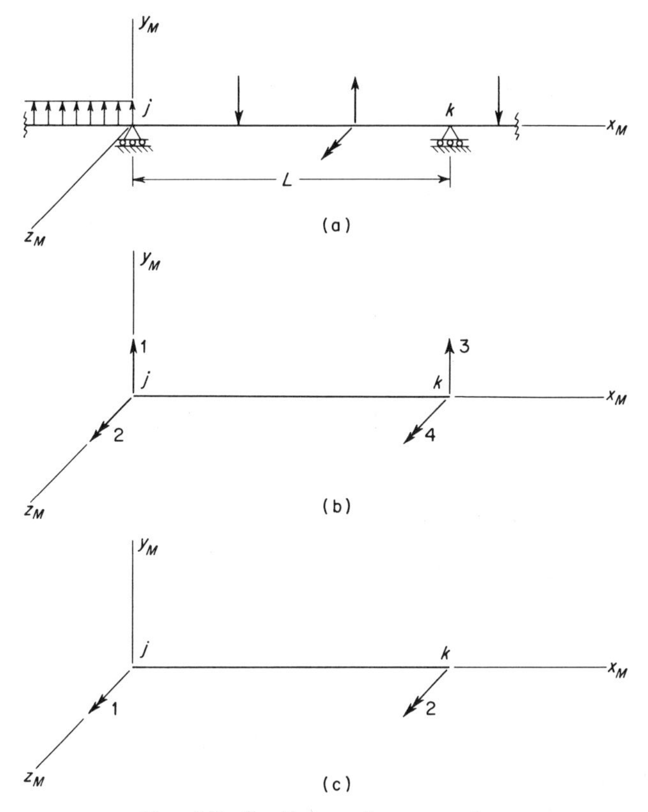

FIG. 4-3. Continuous beam member.

significant types of displacements that can occur at the ends of the member. These displacements are indicated in Fig. 4-3b by the vectors numbered 1 through 4. The corresponding member stiffness matrix is of order 4 × 4 and is shown in Table 4-2. The elements of this matrix are obtained from cases (2), (6), (8), and (12) of Fig. 4-2.

For continuous beam problems in which the supports do not allow translational displacements at joints, only the rotations shown in Fig. 4-3c are possible. In such a case the first and third rows and columns of $\mathbf{S}_M$ could be deleted, and the reduced member stiffness matrix would consist of the remaining elements as shown in Table 4-3.

The member stiffness matrix given in Table 4-2 will be used in the analysis of continuous beams. Member stiffness matrices for other types of framed structures will be discussed in later articles.

TABLE 4-2

Prismatic Beam Member Stiffness Matrix

$$\mathbf{S}_M = \begin{bmatrix} \frac{12EI_Z}{L^3} & \frac{6EI_Z}{L^2} & -\frac{12EI_Z}{L^3} & \frac{6EI_Z}{L^2} \\ \frac{6EI_Z}{L^2} & \frac{4EI_Z}{L} & -\frac{6EI_Z}{L^2} & \frac{2EI_Z}{L} \\ -\frac{12EI_Z}{L^3} & -\frac{6EI_Z}{L^2} & \frac{12EI_Z}{L^3} & -\frac{6EI_Z}{L^2} \\ \frac{6EI_Z}{L^2} & \frac{2EI_Z}{L} & -\frac{6EI_Z}{L^2} & \frac{4EI_Z}{L} \end{bmatrix}$$

TABLE 4-3

Reduced Member Stiffness Matrix

$$\mathbf{S}_M = \begin{bmatrix} \frac{4EI_Z}{L} & \frac{2EI_Z}{L} \\ \frac{2EI_Z}{L} & \frac{4EI_Z}{L} \end{bmatrix}$$

4.4 Over-all Joint Stiffness Matrix. The concept of an over-all joint stiffness matrix will be explained in conjunction with the two-span beam shown in Fig. 4-4a. This beam is the same one that was analyzed previously in Chapter 2, except that no particular load system is indicated. As in the earlier example, assume that the flexural rigidity EI_Z of the beam is constant throughout its length. In order to identify the various joint displacements, consider the unloaded beam to be completely restrained at all joints, as shown in Fig. 4-4b. In this figure a numbering system is indicated for the six possible joint displacements which could occur in the structure. The first two (the rotations at B and C) are actually free to occur, but the last four (the translation and rotation at A and the translations at B and C) are restrained by the supports.

One may think in terms of generating for this structure an *over-all joint stiffness matrix* $\mathbf{S}_J$, which contains terms for all of the possible joint displacements, including those restrained by supports. The construction of such a matrix for this beam involves the six unit displacements shown in Figs. 4-4c through 4-4h. These figures portray unit vertical translations (assumed positive upward) and unit rotations (assumed positive counterclockwise) in the sequence indicated by the numbering system (Fig. 4-4b). The resulting restraint actions shown in the figures become elements in the joint stiffness matrix (the sign convention for the restraint actions corresponds to that for displacements). The restraint actions in Fig. 4-4c, for example, provide the elements of the first

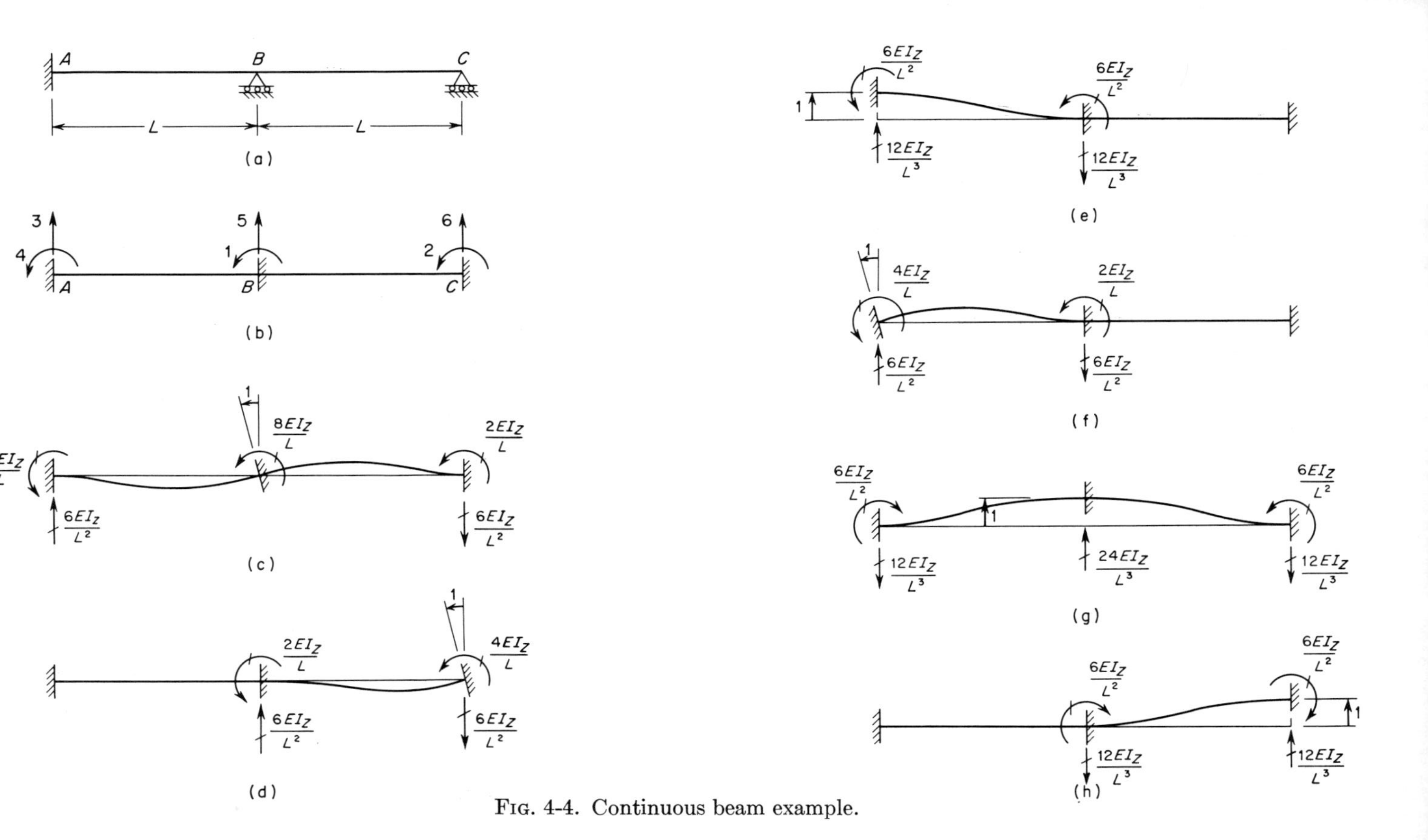

FIG. 4-4. Continuous beam example.

column in the joint stiffness matrix $\mathbf{S}_J$. The first element is $8EI/L$, the second element is $2EI/L$, the third element is $6EI/L^2$, and so on.* Similarly, the actions in Figs. 4-4d through 4-4h provide the elements of the second through the sixth columns of the joint stiffness matrix. Thus, the total array for the two-span beam is seen to be the matrix shown in Table 4-4. The matrix is of order 6×6 because there are six

TABLE 4-4

Joint Stiffness Matrix for Two-Span Beam

$$
\mathbf{S}_J = \left[\begin{array}{cc|cccc}
\frac{8EI}{L} & \frac{2EI}{L} & \frac{6EI}{L^2} & \frac{2EI}{L} & 0 & -\frac{6EI}{L^2} \\
\frac{2EI}{L} & \frac{4EI}{L} & 0 & 0 & \frac{6EI}{L^2} & -\frac{6EI}{L^2} \\
\hline
\frac{6EI}{L^2} & 0 & \frac{12EI}{L^3} & \frac{6EI}{L^2} & -\frac{12EI}{L^3} & 0 \\
\frac{2EI}{L} & 0 & \frac{6EI}{L^2} & \frac{4EI}{L} & -\frac{6EI}{L^2} & 0 \\
0 & \frac{6EI}{L^2} & -\frac{12EI}{L^3} & -\frac{6EI}{L^2} & \frac{24EI}{L^3} & -\frac{12EI}{L^3} \\
-\frac{6EI}{L^2} & -\frac{6EI}{L^2} & 0 & 0 & -\frac{12EI}{L^3} & \frac{12EI}{L^3}
\end{array}\right]
$$

Rows 1–2: Free to displace; rows 3–6: Restrained by supports. Columns 1–2: Free to displace; columns 3–6: Restrained by supports.

joint displacements to be considered. Note that the matrix is square and symmetrical. Expansion of its determinant would also show that it is singular because of the fact that certain rows and columns are linear combinations of others. Therefore, the boundary conditions in the problem must be recognized in the form of the actual supports before proceeding further.

A partitioning of the joint stiffness matrix $\mathbf{S}_J$ according to whether the displacements are free to displace or are restrained by supports is indicated in Table 4-4. With this partitioning, the matrix may be considered to consist of four parts, as follows:

$$
\mathbf{S}_J = \left[\begin{array}{c|c} \mathbf{S} & \mathbf{S}_{DR} \\ \hline \mathbf{S}_{RD} & \mathbf{S}_{RR} \end{array}\right] \tag{4-1}
$$

* For convenience, the subscript Z is omitted from the moment of inertia.

Each of the submatrices in Eq. (4-1) has a physical significance that will now be explained.

The stiffness matrix $\mathbf{S}$ in the upper left-hand portion of $\mathbf{S}_J$ is a square, symmetric stiffness matrix that corresponds to the unknown displacements in the structure, that is, to the degrees of freedom. It is the same stiffness matrix $\mathbf{S}$ obtained previously for this structure (see Art. 2.8):

$$\mathbf{S} = \frac{EI}{L}\begin{bmatrix} 8 & 2 \\ 2 & 4 \end{bmatrix}$$

The inverse of $\mathbf{S}$ may be obtained as usual:

$$\mathbf{S}^{-1} = \frac{L}{14EI}\begin{bmatrix} 2 & -1 \\ -1 & 4 \end{bmatrix}$$

and then substituted into the following equation (see Eq. 2-24).

$$\mathbf{D} = \mathbf{S}^{-1}(\mathbf{A}_D - \mathbf{A}_{DL}) \tag{4-2}$$

for the purpose of calculating the unknown displacements.

The matrix $\mathbf{S}_{RD}$ is a rectangular submatrix of $\mathbf{S}_J$ that contains actions corresponding to the support restraints, due to unit values of displacements corresponding to the degrees of freedom. In other words, this submatrix gives the reactions for the structure due to unit values of the unknown displacements $\mathbf{D}$. Thus, the matrix $\mathbf{S}_{RD}$ is the matrix that previously was denoted $\mathbf{A}_{RD}$ in Eq. (2-26). Therefore,

$$\mathbf{S}_{RD} = \mathbf{A}_{RD} \tag{4-3}$$

and Eq. (2-26) may be rewritten in terms of $\mathbf{S}_{RD}$ as follows:

$$\mathbf{A}_R = \mathbf{A}_{RL} + \mathbf{S}_{RD}\mathbf{D} \tag{4-4}$$

The submatrix $\mathbf{S}_{DR}$ represents actions corresponding to the degrees of freedom and caused by unit displacements corresponding to the support restraints. It can be seen that $\mathbf{S}_{DR}$ is the transpose of $\mathbf{S}_{RD}$:

$$\mathbf{S}_{DR} = \mathbf{S}'_{RD}$$

The matrix $\mathbf{S}_{RR}$ is a square, symmetric submatrix of $\mathbf{S}_J$ that contains actions corresponding to the support restraints due to unit displacements corresponding to the same set of restraints. The submatrices $\mathbf{S}_{DR}$ and $\mathbf{S}_{RR}$ may be used in analyzing structures having support displacements (see Art. 6.5).

In the two-span beam example the matrix $\mathbf{S}$ is only a small portion of the joint stiffness matrix $\mathbf{S}_J$ (see Table 4-4). This is a consequence of the fact that the structure is highly restrained. In large structures having many joints and relatively few supports, the matrix $\mathbf{S}$ constitutes the major portion of $\mathbf{S}_J$.

4.5 Loads. After finding the joint stiffness matrix, the next step in the analysis is to consider the loads on the structure, as mentioned previously in the outline of the method (see Art. 4.2). It is convenient initially to handle the loads at the joints and the loads on the members separately. The reason for doing so is that the joint loads and the member loads are treated in different ways. The joint loads are ready for immediate placement into a vector of actions to be used in the solution, but the loads on the members are taken into account by calculating the fixed-end actions that they produce. These fixed-end actions may then be transformed into equivalent joint loads and combined with the actual joint loads on the structure (see Art. 3.2 for a discussion of equivalent joint loads).

The loads applied at the joints may be listed in a vector $\mathbf{A}$, which contains the applied loads corresponding to all possible joint displacements, including those at support restraints. The elements in $\mathbf{A}$ are numbered in the same sequence as the joint displacements. For the two-span beam example, the numbering system is shown in Fig. 4-4b. If this beam is subjected to the loads shown in Fig. 4-5a (which shows the same loads as those in Fig. 2-14), the joint loads are the couple M at joint B and the force P_3 at joint C. Therefore, the vector $\mathbf{A}$ takes the form*

$$\mathbf{A} = \{M, 0, 0, 0, 0, P_3\}$$

or, substituting for M and P_3 the values shown in the figure:

$$\mathbf{A} = \{PL, 0, 0, 0, 0, P\}$$

The moment M and the force P_3 in the vector $\mathbf{A}$ are positive because M acts counterclockwise and P_3 acts upward. In general, the joint loads on a continuous beam may be a force and a couple at every joint.

The remaining loads on the structure (P_1 and P_2) act directly on the members and are shown acting on the two members of the restrained structure in Fig. 4-5b. Also shown in this figure are the fixed-end actions at the ends of the members. These fixed-end actions may be compiled into a rectangular matrix $\mathbf{A}_{ML}$ in which each row contains the end-actions for a given member. Previously, the matrix $\mathbf{A}_{ML}$ was handled as a vector (see Eq. 2-25). However, when all end-actions for all members are to be considered in the analysis, it becomes more convenient to handle $\mathbf{A}_{ML}$ as a rectangular matrix. Thus, the definition of the meaning of $\mathbf{A}_{ML}$ remains the same, but its format is changed. In a beam there are two significant types of fixed-end actions, namely, the shear forces and the moments. Therefore, each row of the matrix $\mathbf{A}_{ML}$ will contain for one member the shear force and moment at the left end and the shear force and moment at the right end, in that order. As shown in Fig. 4-5b, the fixed-end actions on member AB consist of the shear force $P_1/2$ and the couple $P_1L/8$ at the left-hand end, and the shear force $P_1/2$ and the

* As mentioned previously in Chapter 3, braces { } are used to denote a column vector that is written in a row.

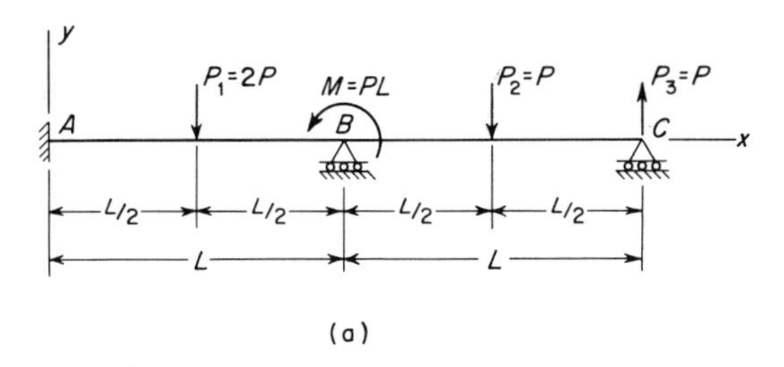

(a)

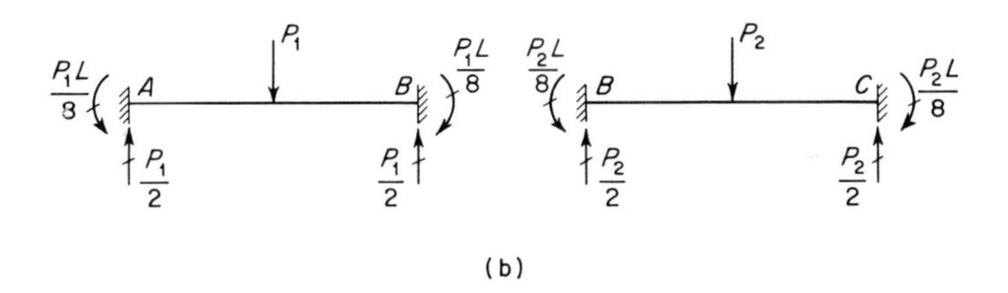

(b)

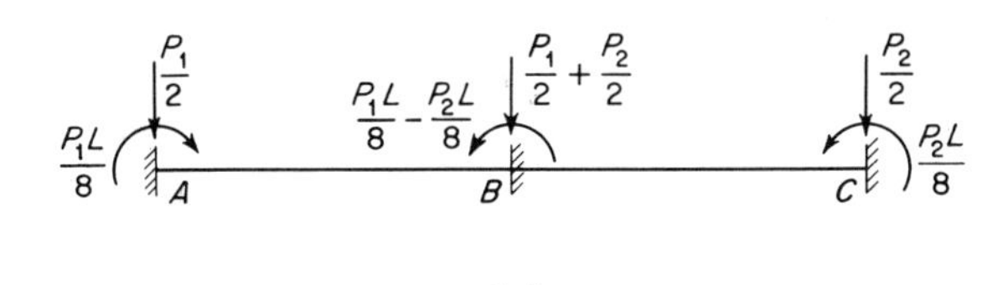

(c)

FIG. 4-5. Combined joint loads.

couple $-P_1L/8$ at the right-hand end. These actions constitute the first row of the matrix $\mathbf{A}_{\mathrm{ML}}$. The second row is constructed in a similar manner from the end-actions for member BC. Thus, the matrix is

$$\mathbf{A}_{\mathrm{ML}} = \begin{bmatrix} \dfrac{P_1}{2} & \dfrac{P_1L}{8} & \dfrac{P_1}{2} & -\dfrac{P_1L}{8} \\[2ex] \dfrac{P_2}{2} & \dfrac{P_2L}{8} & \dfrac{P_2}{2} & -\dfrac{P_2L}{8} \end{bmatrix}$$

In general, this matrix will have as many rows as there are members of the structure. When the values shown in Fig. 4-5a are substituted for the loads, the matrix $\mathbf{A}_{\mathrm{ML}}$ becomes

$$\mathbf{A}_{\mathrm{ML}} = \begin{bmatrix} P & \dfrac{PL}{4} & P & -\dfrac{PL}{4} \\[2ex] \dfrac{P}{2} & \dfrac{PL}{8} & \dfrac{P}{2} & -\dfrac{PL}{8} \end{bmatrix}$$

When the fixed-end actions $\mathbf{A}_{\mathrm{ML}}$ are reversed in direction, they constitute the equivalent joint loads, and for the two-span beam they are shown in Fig. 4-5c. The loads at joints A and C are simply the reverse of the actions in Fig. 4-5b, while at joint B the equivalent loads consist of the sums of the actions from members AB and BC. These equivalent joint loads may be put into a vector $\mathbf{A}_E$ which has the same form as the vector $\mathbf{A}$ of actual joint loads:

$$\mathbf{A}_E = \left\{ \frac{P_1L}{8} - \frac{P_2L}{8}, \ \frac{P_2L}{8}, \ -\frac{P_1}{2}, \ -\frac{P_1L}{8}, \ -\frac{P_1}{2} - \frac{P_2}{2}, \ -\frac{P_2}{2} \right\}$$

or, with the appropriate values substituted for the loads,

$$\mathbf{A}_E = \left\{ \frac{PL}{8}, \frac{PL}{8}, -P, -\frac{PL}{4}, -\frac{3P}{2}, -\frac{P}{2} \right\}$$

The equivalent joint loads are listed in this vector in the sequence denoted by the numbering system in Fig. 4-4b, and the signs are determined by the convention described previously.

Actual joint loads (vector $\mathbf{A}$) may be added to equivalent joint loads (vector $\mathbf{A}_E$) to produce the combined load vector $\mathbf{A}_C$:

$$\mathbf{A}_C = \mathbf{A} + \mathbf{A}_E \tag{4-5}$$

Substituting into Eq. (4-5) the values in vectors $\mathbf{A}$ and $\mathbf{A}_E$ obtained above yields:

$$\mathbf{A}_C = \left\{ \frac{9PL}{8}, \frac{PL}{8}, -P, -\frac{PL}{4}, -\frac{3P}{2}, \frac{P}{2} \right\}$$

The vector $\mathbf{A}_C$ may be seen to consist of two parts. The first part of the vector (the first two elements in the example) represents the sums of actual and equivalent joint loads corresponding to the known degrees of freedom in the problem. Thus, this part of $\mathbf{A}_C$ is actually the vector $\mathbf{A}_D$ (or $\mathbf{A}_D - \mathbf{A}_{DL}$ with $\mathbf{A}_{DL}$ a null vector) in Eqs. (2-23) and (4-2). The second part of the vector $\mathbf{A}_C$ (the last four elements in the example) consists of the sums of actual and equivalent joint loads corresponding to the support restraints on the structure. If the signs on the elements of this part of $\mathbf{A}_C$ are reversed, it becomes the vector $\mathbf{A}_{RL}$ in Eq. (4-4). In summary, the combined load vector $\mathbf{A}_C$ is seen to be composed as follows:

$$\mathbf{A}_C = \begin{bmatrix} \mathbf{A}_D \\ \hdashline -\mathbf{A}_{RL} \end{bmatrix} \tag{4-6}$$

In the example problem,

$$\mathbf{A}_D = \left\{ \frac{9PL}{8}, \frac{PL}{8} \right\}$$

and

$$\mathbf{A}_{RL} = \left\{ P, \frac{PL}{4}, \frac{3P}{2}, -\frac{P}{2} \right\}$$

The formation of the vectors $\mathbf{A}_D$ and $\mathbf{A}_{RL}$ in the above fashion sets the stage for the completion of the analysis of the structure. The fact that the effects of loads on the members have been put into the form of equivalent joint loads automatically implies that the vector $\mathbf{A}_{\mathrm{DL}}$ is null (all elements equal to zero). By this approach Eqs. (2-23) and (4-2) can be simplified as follows:

$$\mathbf{A}_{\mathrm{D}} = \mathbf{S}\mathbf{D} \tag{4-7}$$

$$\mathbf{D} = \mathbf{S}^{-1}\mathbf{A}_{\mathrm{D}} \tag{4-8}$$

4.6 Calculation of Results. In the final phase of the analysis, the matrices generated in the previous steps (see Arts. 4.4 and 4.5) are substituted into the appropriate equations for the purpose of calculating unknown joint displacements D, reactions A_R, and member end-actions A_M. The unknown displacements D are found by substituting the matrices $\mathbf{S}^{-1}$ and $\mathbf{A}_{\mathrm{D}}$ into Eq. (4-8). For the two-span beam (Fig. 4-5a), the solution becomes

$$\mathbf{D} = \mathbf{S}^{-1}\mathbf{A}_{\mathrm{D}} = \frac{L}{14EI}\begin{bmatrix} 2 & -1 \\ -1 & 4 \end{bmatrix}\begin{bmatrix} \frac{9}{8} \\ \frac{1}{8} \end{bmatrix} PL = \frac{PL^2}{112EI}\begin{bmatrix} 17 \\ -5 \end{bmatrix}$$

The reactions A_R are found by substituting the matrices $\mathbf{A}_{\mathrm{RL}}$, $\mathbf{S}_{\mathrm{RD}}$, and $\mathbf{D}$ into Eq. (4-4). This substitution for the example problem yields the following result:

$$\mathbf{A}_{\mathrm{R}} = \begin{bmatrix} P \\ \frac{PL}{4} \\ \frac{3P}{2} \\ \frac{-P}{2} \end{bmatrix} + \begin{bmatrix} \frac{6EI}{L^2} & 0 \\ \frac{2EI}{L} & 0 \\ 0 & \frac{6EI}{L^2} \\ -\frac{6EI}{L^2} & -\frac{6EI}{L^2} \end{bmatrix}\begin{bmatrix} 17 \\ -5 \end{bmatrix}\frac{PL^2}{112EI} = \frac{P}{56}\begin{bmatrix} 107 \\ 31L \\ 69 \\ -64 \end{bmatrix}$$

The results given above for the vectors $\mathbf{D}$ and $\mathbf{A}_{\mathrm{R}}$ are the same as obtained previously in Art. 2.8.

The member end-actions A_M could be obtained by using Eq. (2-25), as was done in Chapter 2. However, in the examples of Chapter 2 only a few selected end-actions were calculated. If all actions at the ends of all members are to be calculated, it is more convenient to work with one member at a time instead of with the whole structure. The final end-actions in a given member consist of the superposition of initial fixed-end actions and the additional effects caused by the displacements of the ends of the member. This superposition of actions is expressed for the i-th member in a structure by the following equation, which has a form similar to Eq. (2-25):

$$\{A_M\}_i = \{A_{ML}\}_i + [S_M]_i\{D_M\}_i \tag{4-9}$$

In this expression $\{A_M\}_i$ is the vector of final end-actions for the member. The first term on the right-hand side of the equation, which is $\{A_{ML}\}_i$, is a vector of fixed-end actions for the member. In other words, this vector contains the elements of the i-th row of the rectangular matrix $\mathbf{A}_{\mathrm{ML}}$ described in Art. 4.5. The second term on the right-hand side of Eq. (4-9) consists of the product of the member stiffness matrix $[S_M]_i$ and the vector $\{D_M\}_i$. The elements of the latter are the displacements of the ends of the member.*

Equation (4-9) will now be applied to the example problem. Let members AB and BC be designated as members 1 and 2, respectively. Then $\{A_M\}_1$ will consist of four elements, which are the shear and moment at the left end of member AB and the shear and moment at the right end, taken in that order. Similarly, $\{A_M\}_2$ is composed of the four end-actions for member BC.

In the example problem the vector $\{A_{ML}\}_1$, consisting of fixed-end actions for member AB, is drawn from the first row of the matrix $\mathbf{A}_{\mathrm{ML}}$; thus:

$$\{A_{ML}\}_1 = \left\{P, \frac{PL}{4}, P, -\frac{PL}{4}\right\}$$

Similarly, the vector $\{A_{ML}\}_2$ of fixed-end actions for member BC consists of the elements of the second row of $\mathbf{A}_{\mathrm{ML}}$:

$$\{A_{ML}\}_2 = \left\{\frac{P}{2}, \frac{PL}{8}, \frac{P}{2}, -\frac{PL}{8}\right\}$$

Consider now the displacements D_M at the ends of the members. The only end-displacement that member AB experiences is a rotation of its right-hand end. The translation and rotation at the left end and the translation at the right end are all zero because of the support restraints. Therefore, the vector of end-displacements for member AB is

$$\{D_M\}_1 = \left\{0, 0, 0, \frac{17PL^2}{112EI}\right\}$$

In this expression the subscript 1 denotes the fact that the vector refers to the first member in the structure. The order in which the end-displacements are listed in the vector follows the pattern given in Fig. 4-3b. Similarly, the end-displacements in member BC may be placed in a second vector as follows:

$$\{D_M\}_2 = \left\{0, \frac{17PL^2}{112EI}, 0, -\frac{5PL^2}{112EI}\right\}$$

Having the above vectors on hand, one may then apply Eq. (4-9) twice (once for each member) in order to evaluate the final end-actions in members AB and BC. In both applications the member stiffness

* Equation (4-9) is also analogous to Eq. (3-24) used previously in the flexibility method.

matrix $[S_M]_i$ is the same (because of equal spans and constant flexural rigidity), and is given in Table 4-2 (see Art. 4.3). Thus, the equations become

$$\{A_M\}_1 = \{A_{ML}\}_1 + [S_M]_1\{D_M\}_1$$

$$= \begin{bmatrix} P \\ \frac{PL}{4} \\ P \\ -\frac{PL}{4} \end{bmatrix} + [S_M]_1 \begin{bmatrix} 0 \\ 0 \\ 0 \\ 17 \end{bmatrix} \frac{PL^2}{112EI} = \frac{P}{56} \begin{bmatrix} 107 \\ 31L \\ 5 \\ 20L \end{bmatrix}$$

$$\{A_M\}_2 = \{A_{ML}\}_2 + [S_M]_2\{D_M\}_2$$

$$= \begin{bmatrix} \frac{P}{2} \\ \frac{PL}{8} \\ \frac{P}{2} \\ -\frac{PL}{8} \end{bmatrix} + [S_M]_2 \begin{bmatrix} 0 \\ 17 \\ 0 \\ -5 \end{bmatrix} \frac{PL^2}{112EI} = \frac{P}{56} \begin{bmatrix} 64 \\ 36L \\ -8 \\ 0 \end{bmatrix}$$

In this manner all of the member end-actions have been obtained.

The above calculations may appear at first to be unnecessarily extensive and cumbersome, and indeed they are for the small structure used as an example. Their virtue, however, lies in the fact that they represent an orderly procedure which can be readily programmed for a computer and, hence, may be applied repeatedly to any number of members.

4.7 Arbitrary Numbering Systems. In the preceding articles the two-span beam shown in Figs. 4-4 and 4-5 was discussed in connection with the particular numbering system for joint displacements indicated in Fig. 4-4b. This numbering system identifies first all of the displacements that are free to occur, and second it identifies the possible displacements corresponding to support restraints. In each of the two categories, the displacements are taken in convenient order with translations preceding rotations at each joint. However, it is generally preferable in a practical analysis to be able to number the displacements in an arbitrary manner. Then, all of the matrices required in the analysis may be generated in conformity with the arbitrary numbering system. Subsequently, these matrices may be rearranged in order to correspond to a numbering system of the type shown in Fig. 4-4b. In other words, the rearrangement of the matrices is equivalent to transforming from the arbitrary numbering system to the original numbering system. In doing so, the various matrices can be partitioned in a manner which segregates elements associated with the degrees of freedom from those associated with the support restraints.

Consider again the two-span beam in Fig. 4-4 and imagine it to be completely restrained at all joints as shown in Fig. 4-6. The structure may be considered to be disassociated from any particular set of support conditions for the purpose of generating the over-all joint stiffness matrix $\mathbf{S}_J$. If the displacements of the structure are numbered without regard to the actual degrees of freedom or the actual support conditions, a variety of numbering schemes could be used. A convenient numbering system is indicated in Fig. 4-6 (compare Fig. 4-6 with Fig. 4-4b). In

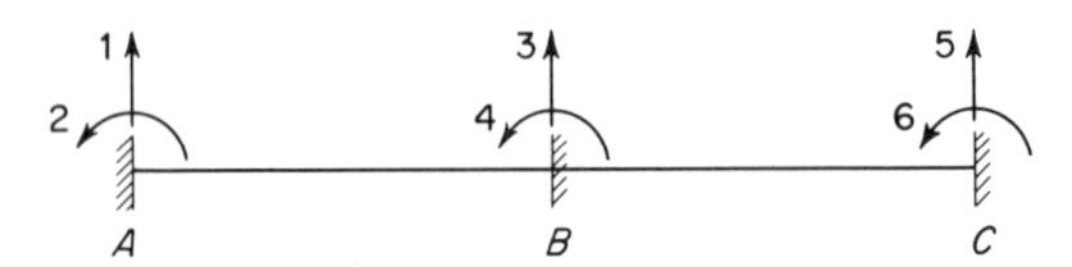

FIG. 4-6. Arbitrary numbering system.

this system the possible translation and rotation at each joint are numbered in sequence, and joints are taken in order from left to right. The construction of the joint stiffness matrix $\mathbf{S}_J$ involves the six unit displacements of the restrained structure shown previously in Fig. 4-4c through 4-4h, but now they must be taken in a different order. To comply with the arbitrary numbering system, each joint in turn from left to right is given a unit vertical translation and then a unit rotation. Thus, the elements of the six columns of $\mathbf{S}_J$ may be drawn from Figs. 4-4e, 4-4f, 4-4g, 4-4c, 4-4h, and 4-4d in that order. For instance, Fig. 4-4e provides the elements of the first column of the joint stiffness matrix. The first element is $12EI/L^3$, the second element is $6EI/L^2$, the third element is $-12EI/L^3$, the fourth element is $6EI/L^2$, and the last two elements are zero. The other five columns of $\mathbf{S}_J$ are obtained in a similar fashion. The resulting square, symmetrical 6×6 stiffness matrix is shown in Table 4-5. The reader should verify all of the terms in this matrix by reference to Fig. 4-4. The reason for generating the matrix $\mathbf{S}_J$ in the form shown in Table 4-5 is the fact that it can be done automatically by operating consecutively with every joint in the structure.

It is also important to note the pattern in which the member stiffnesses contribute to the joint stiffness matrix. This pattern is indicated in Table 4-5 by the two sections of $\mathbf{S}_J$ delineated by dashed lines and overlapping in the center portion. The 4×4 upper left-hand section receives contributions from the member stiffness matrix of member AB (see Table 4-2, Art. 4.3), and the 4×4 lower right-hand section receives contributions from the member stiffness matrix of member BC. Together, these two member stiffness matrices (which are both the same in this example) constitute the joint stiffness matrix for the two-span beam. In the overlapping portion of $\mathbf{S}_J$, elements consist of the sums of contributions from members AB and BC. For example, the element S_{J33} (equal to $24EI/L^3$) is the sum of S_{M33} (equal to $12EI/L^3$) for member AB plus

TABLE 4-5

Joint Stiffness Matrix for Beam of Fig. 4-6

Member AB

$$\mathbf{S}_J = \begin{bmatrix} \frac{12EI}{L^3} & \frac{6EI}{L^2} & -\frac{12EI}{L^3} & \frac{6EI}{L^2} & 0 & 0 \\ \frac{6EI}{L^2} & \frac{4EI}{L} & -\frac{6EI}{L^2} & \frac{2EI}{L} & 0 & 0 \\ -\frac{12EI}{L^3} & -\frac{6EI}{L^2} & \frac{24EI}{L^3} & 0 & -\frac{12EI}{L^3} & \frac{6EI}{L^2} \\ \frac{6EI}{L^2} & \frac{2EI}{L} & 0 & \frac{8EI}{L} & -\frac{6EI}{L^2} & \frac{2EI}{L} \\ 0 & 0 & -\frac{12EI}{L^3} & -\frac{6EI}{L^2} & \frac{12EI}{L^3} & -\frac{6EI}{L^2} \\ 0 & 0 & \frac{6EI}{L^2} & \frac{2EI}{L} & -\frac{6EI}{L^2} & \frac{4EI}{L} \end{bmatrix}$$

Member BC

S_{M11} (equal to $12EI/L^3$) for member BC. Similarly, S_{J34} (equal to 0) is the sum of S_{M34} (equal to $-6EI/L^2$) for member AB plus S_{M12} (equal to $6EI/L^2$) for member BC. Other elements of $\mathbf{S}_J$ outside of the overlapping portion but inside the dashed lines consist of single contributions from either member AB or member BC. Elements of $\mathbf{S}_J$ outside of the dashed lines are all zero due to the nature of the structure. Thus, the use of the arbitrary numbering system results in a joint stiffness matrix in which the contributions from member stiffnesses may be clearly seen. Indeed, one of the essential features of a computer program for analysis by the stiffness method involves the generation of $\mathbf{S}_J$ by combining the contributions from individual member stiffness matrices.

In order for the stiffness matrix $\mathbf{S}_J$ in Table 4-5 to be useful, the actual degrees of freedom and support restraints in a given structure must be recognized, and the matrix must be put into the form given in Table 4-4 (see Art. 4.4). In other words, the joint stiffness matrix must be rearranged by interchanging rows and columns in such a manner that the stiffnesses corresponding to the actual degrees of freedom are listed first, and those corresponding to support restraints are listed second. In the present problem the actual degrees of freedom are the rotations at joints B and C, which correspond to the fourth and sixth rows and columns of the matrix $\mathbf{S}_J$. If the fourth and sixth rows are switched to the first and second rows while all others are moved downward without changing their order, the matrix takes the form shown in Table 4-6.

TABLE 4-6

Joint Stiffness Matrix with Rows Rearranged

$$\begin{bmatrix} \frac{6EI}{L^2} & \frac{2EI}{L} & 0 & \frac{8EI}{L} & -\frac{6EI}{L^2} & \frac{2EI}{L} \\ 0 & 0 & \frac{6EI}{L^2} & \frac{2EI}{L} & -\frac{6EI}{L^2} & \frac{4EI}{L} \\ \frac{12EI}{L^3} & \frac{6EI}{L^2} & -\frac{12EI}{L^3} & \frac{6EI}{L^2} & 0 & 0 \\ \frac{6EI}{L^2} & \frac{4EI}{L} & -\frac{6EI}{L^2} & \frac{2EI}{L} & 0 & 0 \\ -\frac{12EI}{L^3} & -\frac{6EI}{L^2} & \frac{24EI}{L^3} & 0 & -\frac{12EI}{L^3} & \frac{6EI}{L^2} \\ 0 & 0 & -\frac{12EI}{L^3} & -\frac{6EI}{L^2} & \frac{12EI}{L^3} & -\frac{6EI}{L^2} \end{bmatrix}$$

Next, the fourth and sixth columns of the stiffness matrix are moved to the first two columns, and all other columns are moved to the right without changing their order. This rearrangement produces the same symmetrical joint stiffness matrix that was generated previously and is shown in Table 4-4. This matrix will hereafter be referred to as the *rearranged joint stiffness matrix*. As before, it may be partitioned in the manner denoted by Eq. (4-1).

The load vectors $\mathbf{A}$, $\mathbf{A}_E$, and $\mathbf{A}_C$ may also be constructed in conjunction with an arbitrary numbering system and then rearranged and partitioned in accordance with the actual degrees of freedom and the support restraints. Since the last of these vectors is the sum of the first two (see Eq. 4-5), it is possible to rearrange in either of two ways. The first two vectors ($\mathbf{A}$ and $\mathbf{A}_E$) may be generated in conformity with the arbitrary numbering system, rearranged, and then added. As a result, the vector $\mathbf{A}_C$ is automatically in rearranged form. Alternatively, the vectors $\mathbf{A}$ and $\mathbf{A}_E$ may be generated with the arbitrary numbering system and then added to give $\mathbf{A}_C$. Then the vector $\mathbf{A}_C$ must be rearranged as a final step.

To illustrate the latter of these two procedures, imagine that the load vectors $\mathbf{A}$ and $\mathbf{A}_E$ are obtained for the beam of Fig. 4-5a by using the arbitrary numbering system shown in Fig. 4-6. From the data in Art. 4-5 it can be seen that these vectors now become

$$\mathbf{A} = \{0, 0, 0, PL, P, 0\}$$

$$\mathbf{A}_E = \left\{-P, \ -\frac{PL}{4}, \ -\frac{3P}{2}, \frac{PL}{8}, \ -\frac{P}{2}, \frac{PL}{8}\right\}$$

The sum of the vectors (see Eq. 4-5) is

$$\mathbf{A}_C = \left\{-P, \ -\frac{PL}{4}, \ -\frac{3P}{2}, \frac{9PL}{8}, \frac{P}{2}, \frac{PL}{8}\right\}$$

which is also in accord with the arbitrary numbering system. The vector $\mathbf{A}_C$ then may be rearranged to conform to the numbering system of Fig. 4-4b by moving the fourth and sixth elements to the first and second positions and moving all others toward the end without changing their order. This rearrangement produces $\mathbf{A}_C$ in the same form in which it was constructed originally in Art. 4.5:

$$\mathbf{A}_C = \left\{\frac{9PL}{8}, \frac{PL}{8}, \ -P, \ -\frac{PL}{4}, \ -\frac{3P}{2}, \frac{P}{2}\right\}$$

This vector may be partitioned as shown previously in Eq. (4-6).

After the matrices are rearranged and partitioned as described above, the remainder of the analysis is the same as that described in Art. 4.6.

4.8 Analysis of Continuous Beams. The continuous beams to be discussed in this article are assumed to consist of prismatic members rigidly connected to each other and supported at various points along their lengths. The joints of a continuous beam will be selected normally at points of support and at any free (or overhanging) ends. However, in those cases where there are changes in the cross-section between points of support, so that the member consists of two or more prismatic segments, it is always possible to analyze the beam by considering a joint to exist at the change in section (see Prob. 2.9-14). This technique of solution will be illustrated in the example of the next article. Another method of analyzing a beam with changes in section, and which does not require the introduction of additional joints to the structure, is described later in Art. 6.8.

A continuous beam having m members and $m + 1$ joints is shown in Fig. 4-7a. The x-y plane is the plane of bending of the beam. The members are numbered above the beam in the figure, and the joints are numbered below the beam. In each case the numbering is from left to right along the beam. The central part of Fig. 4-7a shows the i-th member framing at the left end into a joint designated as joint j, and framing at the right end into a joint designated as joint k. Note that the member and joint numbering systems are such that the joint number j

must be equal numerically to the member number i, and the joint number k must be equal to $i + 1$. These particular relationships between the member numbers and the joint numbers are, of course, valid only for a continuous beam. Also, it can be seen that in this case the total

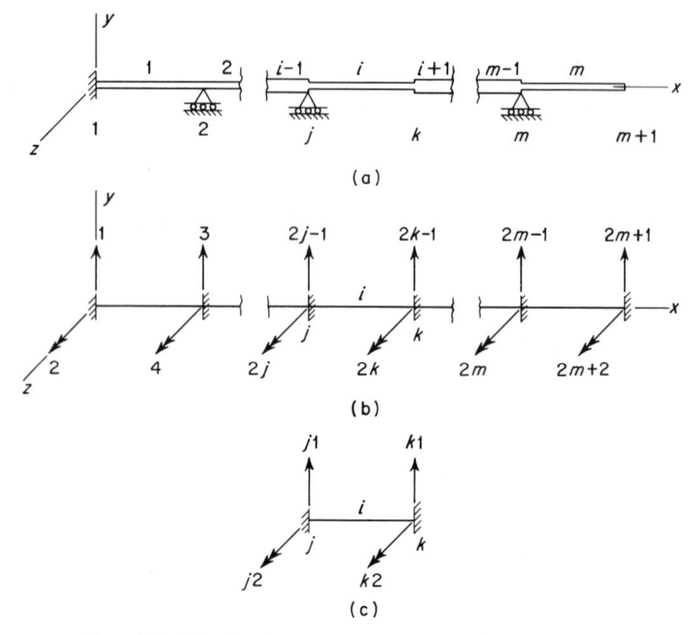

FIG. 4-7. Numbering system for a continuous beam.

number of joints, denoted henceforth as n_j, is always one more than the number of members; thus, $n_j = m + 1$.

Support restraints of two types may exist at any joint in a continuous beam. These are a restraint against translation in the y direction and a restraint against rotation about the z axis (see Fig. 4-7a for directions of axes). For example, joint 1 of the beam in the figure is assumed to be fixed, and therefore is restrained against both translation and rotation; joints 2, j, and m are restrained against translation only; and joints k and $m + 1$ are not restrained at all.

In a continuous beam the displacements are due primarily to flexural deformations, and only such deformations will be considered in this article. The effects of shearing deformations can be included in the analysis if necessary, as described later in Art. 6.11. In either case, however, the omission of axial deformations means that a maximum of two possible displacements may occur at any joint. These are the translation in the y direction and the rotation about the z axis. A numbering system for all of the possible joint displacements is indicated in Fig. 4-7b. Starting at the left-hand end of the structure and proceeding to the right, the translation and rotation at each joint are numbered in sequence. Since the translation at a particular joint is numbered before the rotation, it follows that in all cases the number representing the translation is equal to twice the joint number minus one, while the number for the rotation is twice the joint number. For instance, at joint j the translation and rotation are numbered $2j - 1$ and $2j$, respectively. Similar comments apply to the other joints, such as joints k, m, and $m + 1$. It is evident that the total number of possible joint displacements is twice the number of joints, or $2n_j$. Furthermore, if the total number of support restraints against translation and rotation is denoted n_r, the number of actual displacements, or degrees of freedom, is

$$n = 2n_j - n_r = 2m + 2 - n_r \tag{4-10}$$

in which n is the number of degrees of freedom.

The analysis of a continuous beam consists of setting up the necessary stiffness and load matrices and applying the equations of the stiffness method to the particular problem at hand. For this purpose, the most important matrix to be generated is the over-all joint stiffness matrix $\mathbf{S_J}$. The joint stiffness matrix consists of contributions from the beam stiffnesses $\mathbf{S_M}$ given previously in Art. 4.3 (see Table 4-2). For example, the i-th member in the continuous beam of Fig. 4-7 contributes to the stiffnesses of the joints j and k at the left and right ends of the member, respectively. Therefore, it is necessary to relate the end-displacements of member i to the displacements of joints j and k by means of an appropriate indexing system.

In order to relate the end-displacements of a particular member to the displacements of the joints, consider the typical member i as shown in Fig. 4-7c. The displacements of the ends of this member will be denoted $j1$ and $j2$ at the left end, and $k1$ and $k2$ at the right end. In both cases the translation is numbered before the rotation. These are the same member end-displacements that were numbered 1, 2, 3, and 4 in Fig. 4-3b. However, the use of a new notation, such as $j1$, $j2$, $k1$, and $k2$, is desired in order to have a symbol that can be used in computer programming to represent the number for the end-displacement. The four end-displacements of member i are related to the corresponding joint displacements by the following expressions (compare Figs. 4-7b and c):

$$\begin{aligned} j1 &= 2j - 1 \qquad & j2 &= 2j \\ k1 &= 2k - 1 \qquad & k2 &= 2k \end{aligned} \tag{4-11}$$

However, since in the continuous beam of Fig. 4-7 the joint numbers j and k are equal numerically to i and $i + 1$, respectively, the end-displacements are also given by

$$\begin{aligned} j1 &= 2i - 1 \qquad & j2 &= 2i \\ k1 &= 2i + 1 \qquad & k2 &= 2i + 2 \end{aligned} \tag{4-12}$$

Thus, the above equations serve to index the possible joint displacements at the left and right ends of any member i in terms of either the joint numbers (Eqs. 4-11) or the member numbers (Eqs. 4-12). Such an indexing system is necessary for the purpose of constructing the joint stiffness matrix from the member stiffness matrices. The system also proves to be useful when calculating end-actions in the members due to displacements of the joints, as shown later.

As mentioned previously, the over-all joint stiffness matrix $\mathbf{S}_J$ is made up of contributions from individual member stiffnesses. Hence, it is convenient to assess such contributions for one typical member i in the beam, and then to repeat the process for all members from 1 through m. A typical member i from a continuous beam is shown again in Fig. 4-8, with adjacent members $i - 1$ and $i + 1$ also indicated. In part (a) of the figure the beam is shown with a unit displacement corresponding to $j1$; that is, a y translation at the left end of the member. The four actions developed at joints j and k at the two ends of member i are stiffness coefficients S_J and are elements of the over-all joint stiffness matrix. Each such stiffness will have two subscripts as it appears in the joint stiffness matrix. The first subscript is the number (or index) that denotes the location of the action itself, and the second is the index for the unit displacement causing the action. Thus, the stiffness at joint j in the y direction has subscripts $j1$ and $j1$, meaning that the action corresponds to a displacement of type $j1$ and is caused by a unit displacement of type $j1$. This stiffness is denoted, therefore, by the symbol $(S_J)_{j1,j1}$, as shown in Fig. 4-8a. Of course, the actual value of the index $j1$ is obtained from Eqs. (4-12). Similarly, each of the remaining three stiffnesses in Fig. 4-8a has a first subscript that identifies the type of displacement to which the stiffness corresponds. However, each of them has the same second subscript, denoting the unit displacement of type $j1$.

The joint stiffnesses due to the remaining three possible joint displacements at the ends of member i are shown in Figs. 4-8b, 4-8c, and 4-8d. In each figure the four joint stiffnesses are shown with the appropriate subscripts that indicate (1) the type of action and (2) the unit displacement. The reader should verify the notation for each of the stiffnesses shown in Fig. 4-8.

The next step is to express the joint stiffness coefficients shown in Fig. 4-8 in terms of the various member stiffnesses that contribute to the joint stiffnesses. This step requires that the member stiffnesses be obtained from Table 4-2, which gives the stiffnesses for a continuous beam member (see Fig. 4-3b for the indexing system to Table 4-2). For example, the contribution to the joint stiffness $(S_J)_{j1,j1}$ from member $i - 1$ (see Fig. 4-8a) is the stiffness S_{M33} for that member. Similarly, the contribution to $(S_J)_{j1,j1}$ from member i is the stiffness S_{M11} for member i. These two contributions will be denoted $(S_{M33})_{i-1}$ and $(S_{M11})_i$, respec-

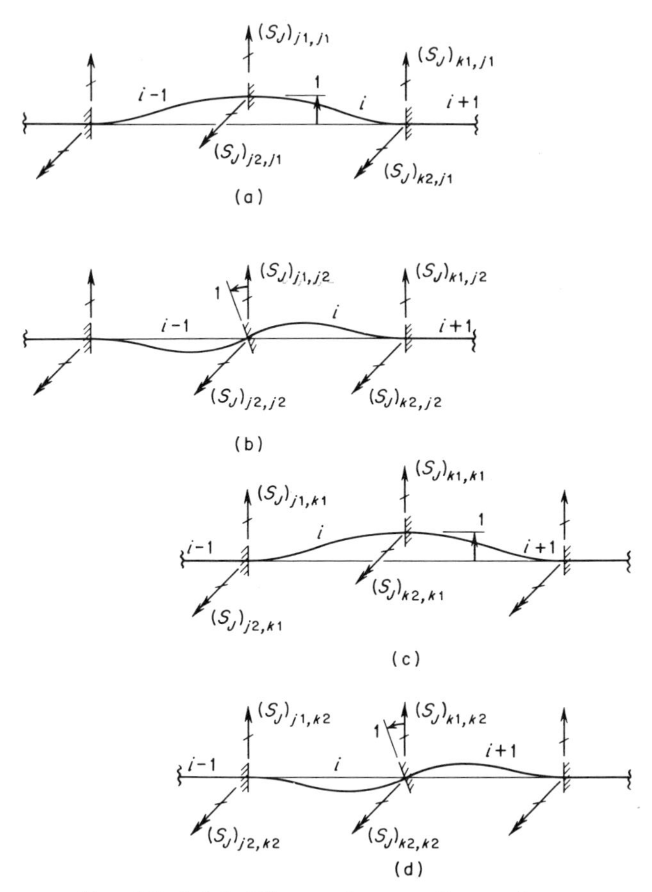

FIG. 4-8. Joint stiffnesses for a continuous beam.

tively. In general, the contribution of one member to a particular joint stiffness will be denoted by appending the member subscript to the member stiffness itself. The latter quantity is obtained from the appropriate member stiffness matrix, which is Table 4-2 for a continuous beam having two possible displacements at each joint. From this discussion it can be seen that the joint stiffness coefficients shown in Fig. 4-8a are given in terms of the member stiffnesses by the following expressions:

$$\begin{aligned}
(S_J)_{j1,j1} &= (S_{M33})_{i-1} + (S_{M11})_i \\
(S_J)_{j2,j1} &= (S_{M43})_{i-1} + (S_{M21})_i \\
(S_J)_{k1,j1} &= (S_{M31})_i \\
(S_J)_{k2,j1} &= (S_{M41})_i
\end{aligned} \tag{4-13}$$

which represent the transfer of elements of the first column of the member stiffness matrix to the appropriate locations in $\mathbf{S_J}$. The first two joint stiffnesses consist of the sum of contributions from members $i-1$ and i. The last two stiffnesses involve contributions from member i only. This pattern of multiple terms when the stiffness is at the near end of the member, and single terms when the stiffness is at the far end of the member, is typical of all types of framed structures. Of course, there are occasional exceptions, such as at the end of a continuous beam, where there is only a single contribution even though the stiffness is at the near end.

Expressions that are analogous to Eqs. (4-13) are easily obtained for a unit rotation about the z axis at joint j. This rotation is a unit displacement of type $j2$ for member i, as shown in Fig. 4-8b. The expressions for the joint stiffnesses in terms of the member stiffnesses are (from the second column of the member stiffness matrix):

$$
\begin{aligned}
(S_J)_{j1,j2} &= (S_{M34})_{i-1} + (S_{M12})_i \\
(S_J)_{j2,j2} &= (S_{M44})_{i-1} + (S_{M22})_i \\
(S_J)_{k1,j2} &= (S_{M32})_i \\
(S_J)_{k2,j2} &= (S_{M42})_i
\end{aligned}
\qquad (4\text{-}14)
$$

Similarly, for a unit y displacement at joint k (see Fig. 4-8c) the stiffnesses are (from the third column):

$$
\begin{aligned}
(S_J)_{j1,k1} &= (S_{M13})_i \\
(S_J)_{j2,k1} &= (S_{M23})_i \\
(S_J)_{k1,k1} &= (S_{M33})_i + (S_{M11})_{i+1} \\
(S_J)_{k2,k1} &= (S_{M43})_i + (S_{M21})_{i+1}
\end{aligned}
\qquad (4\text{-}15)
$$

Finally, the expressions for a unit z rotation at joint k (see Fig. 4-8d) are (from the fourth column):

$$
\begin{aligned}
(S_J)_{j1,k2} &= (S_{M14})_i \\
(S_J)_{j2,k2} &= (S_{M24})_i \\
(S_J)_{k1,k2} &= (S_{M34})_i + (S_{M12})_{i+1} \\
(S_J)_{k2,k2} &= (S_{M44})_i + (S_{M22})_{i+1}
\end{aligned}
\qquad (4\text{-}16)
$$

Equations (4-13) through (4-16) show that the sixteen elements of the 4×4 member stiffness matrix $[S_M]_i$ for member i contribute to sixteen of the joint stiffnesses in a very regular pattern. This pattern may be observed schematically in Fig. 4-9b, which indicates the formation of the joint stiffness matrix for a six-span continuous beam shown with restrained joints in Fig. 4-9a. For this structure the number of joints is seven, the number of possible joint displacements is fourteen, and therefore the joint stiffness matrix is of order 14×14. The indexing scheme is shown down the left-hand edge and across the upper edge of the matrix in Fig. 4-9. The contributions of individual members are indicated by the hatched blocks, each of which is of the order 4 × 4. The blocks are numbered in the upper right-hand corner to identify the member which is being considered. The overlapping blocks, which are of order 2 × 2 in this example, denote elements of $\mathbf{S}_J$ which receive contributions from two adjacent members. All elements outside the shaded blocks are zero.

The stiffness matrix illustrated in Fig. 4-9b is the over-all matrix for all possible joint displacements. In order to analyze a given beam, how-

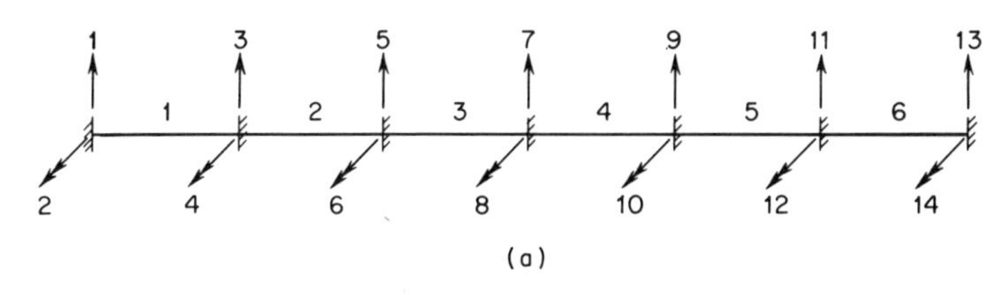

(a)

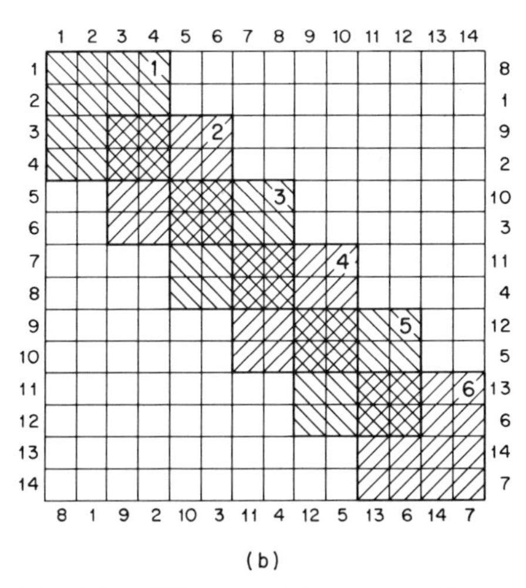

(b)

FIG. 4-9. Joint stiffness matrix for a continuous beam.

ever, the matrix must be rearranged in the partitioned form given by Eq. (4-1). Suppose, for example, that the actual beam has simple supports at all joints, as shown in Fig. 4-10a. The rearranged and partitioned stiffness matrix for this case is indicated in Fig. 4-10b. To obtain this rearranged matrix, rows and columns of the original matrix have been switched in proper sequence in order to place the stiffnesses pertaining to the actual degrees of freedom in the first seven rows and columns. At the same time, the stiffnesses pertaining to the support restraints have been placed in the last seven rows and columns. As an aid in the rearranging process, the new row and column designations are listed in Fig. 4-9b down the right side and across the bottom of the array. The rearrangement of the original stiffness matrix in this manner is equivalent to numbering the degrees of freedom and support restraints as shown in Fig. 4-10a. However, the general approach in this chapter, as explained earlier, is to generate the stiffness matrix using a numbering system like that shown in Fig. 4-9a, and then to rearrange rows and columns as was done in the example of Figs. 4-9 and 4-10.

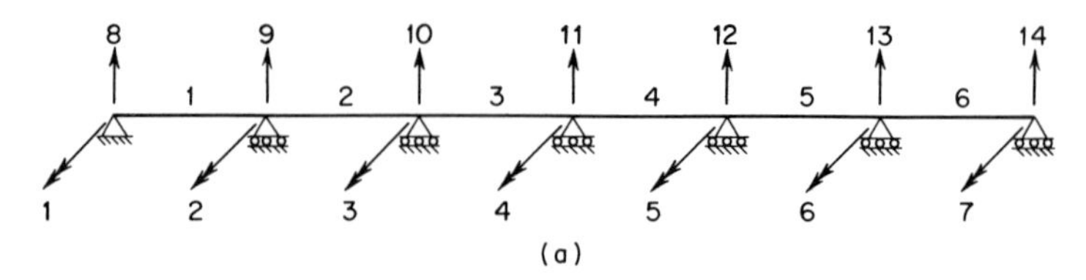

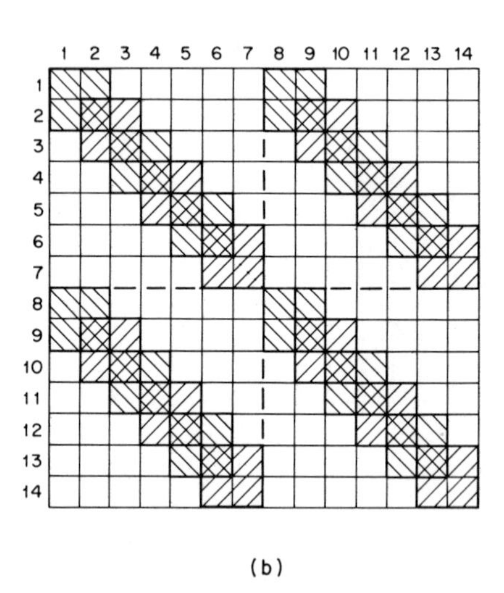

FIG. 4-10. Rearranged joint stiffness matrix for a continuous beam.

In summary, the procedure to be followed in generating the joint stiffness matrix $\mathbf{S}_J$ consists of taking the members in sequence and evaluating their contributions one at a time. For the typical member i, the procedure is as follows. First, the possible displacements at ends j and k of the member are related to the member number by calculating the indexes $j1$, $j2$, $k1$, and $k2$ (see Eqs. 4-12). Then the member stiffness matrix $[S_M]_i$ is generated, and the elements of this matrix are transferred to $\mathbf{S}_J$ as indicated by the terms with subscripts i in Eqs. (4-13) through (4-16). After all members have been processed in this manner, the matrix $\mathbf{S}_J$ is complete. It may then be rearranged and partitioned in order to isolate the stiffness matrix $\mathbf{S}$ (see Eq. 4-1). The inverse matrix $\mathbf{S}^{-1}$ is then determined in preparation for calculating the final results.

After obtaining the stiffness matrix, the next step is to obtain the load vectors. Consider first the vector $\mathbf{A}$ of actual joint loads. This vector contains $2n_j$ elements, each of which corresponds to one of the possible joint displacements shown in Fig. 4-7b. Thus, at each joint there are two possible applied loads, namely, a force in the y direction and a couple in the z sense. These loads are shown in Fig. 4-11 and are denoted A_{2k-1} and A_{2k}, respectively. The subscripts used in identifying these actions are the same as the numbering system for the possible joint displacements (see Fig. 4-7b). Thus, the vector $\mathbf{A}$ takes the form

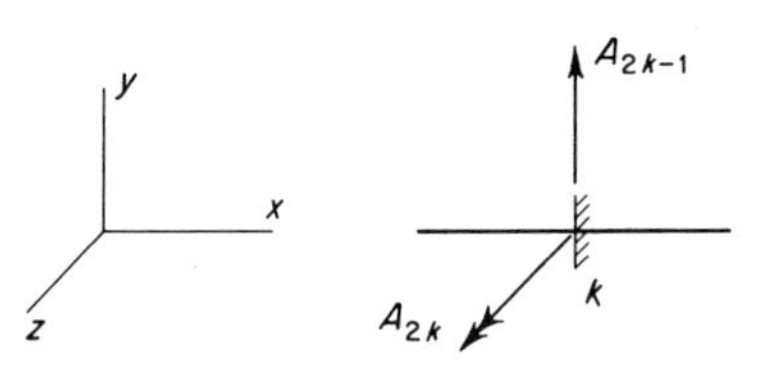

FIG. 4-11. Joint loads for a continuous beam.

$$\mathbf{A} = \{A_1, A_2, \ldots, A_{2k-1}, A_{2k}, \ldots, A_{2m+1}, A_{2m+2}\} \tag{4-17}$$

The elements in this vector are known immediately from the given loads on the beam.

Consider next the formation of the matrix $\mathbf{A}_{\mathrm{ML}}$ of fixed-end actions due to loads and the construction of the vector $\mathbf{A}_{\mathrm{E}}$ of equivalent joint loads. Figure 4-12b shows again the i-th member of a continuous beam, but

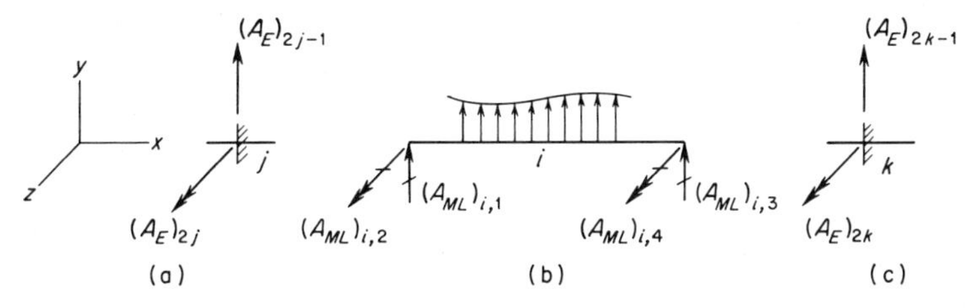

FIG. 4-12. Loads on a continuous beam member.

with lateral loads applied along the length. The actions at the ends of member i when its ends are fixed are denoted as follows:

$(A_{ML})_{i,1}$ = force in the y direction at the left end

$(A_{ML})_{i,2}$ = couple in the z sense at the left end

$(A_{ML})_{i,3}$ = force in the y direction at the right end

$(A_{ML})_{i,4}$ = couple in the z sense at the right end

In general, the first subscript for a fixed-end action denotes the member, and the second denotes the action itself. The latter is identified by using the numbering scheme shown in Fig. 4-12b, which is the same as the one used previously for member stiffnesses (see Fig. 4-3b). When this system of notation is used, the matrix $\mathbf{A}_{\mathrm{ML}}$ of fixed-end actions will be of order $m \times 4$, as follows:

$$\mathbf{A}_{\mathrm{ML}} = \begin{bmatrix} (A_{ML})_{1,1} & (A_{ML})_{1,2} & (A_{ML})_{1,3} & (A_{ML})_{1,4} \\ \cdots & \cdots & \cdots & \cdots \\ (A_{ML})_{i,1} & (A_{ML})_{i,2} & (A_{ML})_{i,3} & (A_{ML})_{i,4} \\ \cdots & \cdots & \cdots & \cdots \\ (A_{ML})_{m,1} & (A_{ML})_{m,2} & (A_{ML})_{m,3} & (A_{ML})_{m,4} \end{bmatrix} \tag{4-18}$$

The elements in this matrix are fixed-end actions that can be found from the formulas given in Appendix B.

The vector $\mathbf{A}_{\mathrm{E}}$ of equivalent joint loads may be constructed from the elements of the matrix $\mathbf{A}_{\mathrm{ML}}$. The process for doing this can be visualized by referring again to Fig. 4-12. The member i shown in part (b) of the figure contributes to the equivalent loads at joints j and k, which are at the ends of the member. The equivalent loads at these joints are shown in Figs. 4-12a and 4-12c, and are denoted by the same subscripts used previously for the possible joint displacements (see Fig. 4-7b). Thus the vector $\mathbf{A}_{\mathrm{E}}$ has the general form

$$\mathbf{A}_{\mathrm{E}} = \{(A_E)_1, (A_E)_2, \ldots, (A_E)_{2j-1}, (A_E)_{2j}, (A_{\mathrm{E}})_{2k-1}, (A_E)_{2k}, \ldots, (A_E)_{2m+2}\} \tag{4-19}$$

The individual elements of the vector $\mathbf{A}_{\mathrm{E}}$ consist of contributions from the two adjoining members. Consider first the force $(A_E)_{2j-1}$ in the y direction at joint j (see Fig. 4-12a). This action, which also may be denoted $(A_E)_{2i-1}$, is composed of the negative of the end-action $(A_{ML})_{i,1}$ from member i (see Fig. 4-12b) and a similar contribution from the adjacent member $i - 1$ on the left. The latter contribution is the negative of the force $(A_{ML})_{i-1,3}$ at the right-hand end of member $i - 1$. By analogous reasoning the expressions for the other three equivalent loads can be obtained, thus giving the following results:

$$\begin{aligned} (A_E)_{2j-1} &= (A_{\mathrm{E}})_{2i-1} = -(A_{ML})_{i-1,3} - (A_{ML})_{i,1} \\ (A_E)_{2j} &= (A_E)_{2i} = -(A_{ML})_{i-1,4} - (A_{ML})_{i,2} \\ (A_E)_{2k-1} &= (A_E)_{2i+1} = - (A_{ML})_{i,3} - (A_{ML})_{i+1,1} \\ (A_E)_{2k} &= (A_E)_{2i+2} = - (A_{ML})_{i,4} - (A_{ML})_{i+1,2} \end{aligned} \tag{4-20}$$

The method for obtaining the vector $\mathbf{A}_{\mathrm{E}}$ consists of taking the members in sequence and evaluating their contributions one at a time. Thus, successive rows of the matrix $\mathbf{A}_{\mathrm{ML}}$ are considered, and the elements of each row are transferred to the appropriate elements in the vector $\mathbf{A}_{\mathrm{E}}$ as indicated by the terms with subscripts i in Eqs. (4-20). After all members of the beam have been considered in this fashion, the vector $\mathbf{A}_{\mathrm{E}}$ is complete. It may then be added to the vector $\mathbf{A}$ according to Eq. (4-5) to form the vector $\mathbf{A}_{\mathrm{C}}$ of combined joint loads. Lastly, the vector $\mathbf{A}_{\mathrm{C}}$ can be rearranged so that the first part becomes $\mathbf{A}_{\mathrm{D}}$ and the second part becomes $-\mathbf{A}_{\mathrm{RL}}$, as explained in Art. 4.5 (see Eq. 4-6).

After the required matrices and vectors have been formulated by the methods described above, the solution for the joint displacements $\mathbf{D}$ and support reactions $\mathbf{A}_R$ consists simply of substituting into Eqs. (4-8) and (4-4), respectively, and performing the indicated matrix multiplications. After finding the joint displacements $\mathbf{D}$, which correspond only to the degrees of freedom of the structure, it is possible to form an overall joint displacement vector $\mathbf{D}_J$ containing elements that correspond to all possible joint displacements (see Fig. 4-7b). Thus, the vector $\mathbf{D}_J$ will be of order $2n_j \times 1$. It will contain values given by the vector $\mathbf{D}$ (corresponding to degrees of freedom), and the remaining elements (corresponding to support restraints) will be zero. In general, the form of the vector $\mathbf{D}_J$ is

$$\mathbf{D}_J = \{(D_J)_1, (D_J)_2, \ldots, (D_J)_{2j-1}, (D_J)_{2j}, (D_J)_{2k-1}, (D_J)_{2k}, \ldots, (D_J)_{2m+2}\} \quad (4\text{-}21)$$

Elements from this vector are used in calculating member end-actions, as described in the following.

The evaluation of member end-actions requires a repetitive application of Eq. (4-9), which is repeated below:

$$\{A_M\}_i = \{A_{ML}\}_i + [S_M]_i\{D_M\}_i \quad (4\text{-}9) \text{ repeated}$$

This equation must be applied once for each member of the structure. When written in detailed form, the equation is the following:

$$\begin{bmatrix} (A_M)_{i,1} \\ (A_M)_{i,2} \\ (A_M)_{i,3} \\ (A_M)_{i,4} \end{bmatrix} = \begin{bmatrix} (A_{ML})_{i,1} \\ (A_{ML})_{i,2} \\ (A_{ML})_{i,3} \\ (A_{ML})_{i,4} \end{bmatrix} + \begin{bmatrix} S_{M11} & S_{M12} & S_{M13} & S_{M14} \\ S_{M21} & S_{M22} & S_{M23} & S_{M24} \\ S_{M31} & S_{M32} & S_{M33} & S_{M34} \\ S_{M41} & S_{M42} & S_{M43} & S_{M44} \end{bmatrix} \begin{bmatrix} D_{M1} \\ D_{M2} \\ D_{M3} \\ D_{M4} \end{bmatrix} \quad (4\text{-}22)$$

The vector $\{\mathbf{A}_{ML}\}_i$ is obtained from the i-th row of the matrix $\mathbf{A}_{ML}$ given previously, and the stiffness matrix $[\mathbf{S}_M]_i$ is obtained from Table 4-2. The vector $\{D_M\}_i$ represents the end-displacements for member i. These displacements are obtained from the vector $\mathbf{D}_J$ by taking from that vector the four consecutive displacements that are associated with member i. In general, the four displacements shown in Eq. (4-22), that is, D_{M1}, D_{M2}, D_{M3}, and D_{M4}, are equal to the displacements $(D_J)_{j1}$, $(D_J)_{j2}$, $(D_J)_{k1}$, and $(D_J)_{k2}$, respectively, from the vector $\mathbf{D}_J$. Thus, the four end-displacements for any member can be extracted from the vector $\mathbf{D}_J$ without difficulty.

Equation (4-22) can be further expanded by substituting into it the elements of $[S_M]_i$ and $\{D_M\}_i$ and performing the matrix multiplication. When this is done, the following four equations are obtained:

$$(A_M)_{i,1} = (A_{ML})_{i,1} + \frac{12EI_{Zi}}{L_i^3}[(D_J)_{j1} - (D_J)_{k1}] + \frac{6EI_{Zi}}{L_i^2}[(D_J)_{j2} + (D_J)_{k2}]$$

$$(A_M)_{i,2} = (A_{ML})_{i,2} + \frac{6EI_{Zi}}{L_i^2}[(D_J)_{j1} - (D_J)_{k1}] + \frac{4EI_{Zi}}{L_i}[(D_J)_{j2} + \tfrac{1}{2}(D_J)_{k2}] \quad (4\text{-}23)$$

$$(A_M)_{i,3} = (A_{ML})_{i,3} - \frac{12EI_{Zi}}{L_i^3}[(D_J)_{j1} - (D_J)_{k1}] - \frac{6EI_{Zi}}{L_i^2}[(D_J)_{j2} + (D_J)_{k2}]$$

$$(A_M)_{i,4} = (A_{ML})_{i,4} + \frac{6EI_{Zi}}{L_i^2}[(D_J)_{j1} - (D_J)_{k1}] + \frac{4EI_{Zi}}{L_i}[\tfrac{1}{2}(D_J)_{j2} + (D_J)_{k2}]$$

Equations (4-22) and (4-23) are equivalent, of course, and either may be used for the purpose of computing the member end-actions.

The analysis of continuous beams using the highly organized method described above is demonstrated by an example in the next article. In the solution, the steps described in this article are followed as much as possible. Naturally, such procedures are very cumbersome for a hand solution, but they are used here deliberately in order to illustrate the way in which the solution is carried out by means of a computer program. A computer program for the analysis of continuous beams is presented in Art. 5.4.

4.9 Example. The continuous beam shown in Fig. 4-13a is to be analyzed by the formalized approach of the preceding article. The beam is restrained

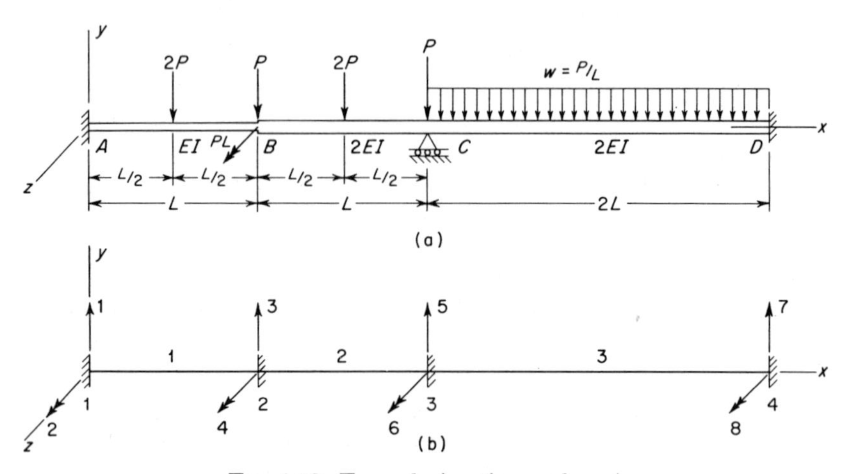

FIG. 4-13. Example (continuous beam).

against translation at support C and against both translation and rotation at points A and D. At point B the flexural rigidity of the beam changes from EI to $2EI$. Therefore, point B is taken to be a joint in the structure.

The member and joint numbering systems are shown in Fig. 4-13b, and it is seen that the number of members m is three, the number of joints n_j is four, and the number of support restraints n_r is five. Therefore, the number of degrees of freedom n is equal to three (see Eq. 4-10).

The member properties and the indexes computed from Eq. (4-12) are given in Table 4-7. The moment of inertia and length of each member can be considered as given data (or input) to the analysis. The remaining quantities in the table are computed from the member numbers. Joint restraints for the beam are indicated in Table 4-8. In this table the index numbers of all possible displacements are listed for each joint of the beam. This is followed by a restraint list, in which the numeral 1 is used to indicate the existence of a restraint, while a zero indicates no restraint (or a degree of freedom). The last column of the table shows the numbers for the displacements when they are rearranged so that the

TABLE 4-7

Member Information for Beam of Fig. 4-13

Member Number	Joint Numbers at Ends of Member		Indexes for End-Displacements of Member				I_Z	Length
i	j	k	j1	j2	k1	k2		
1	1	2	1	2	3	4	I	L
2	2	3	3	4	5	6	2I	L
3	3	4	5	6	7	8	2I	2L

TABLE 4-8

Joint Information for Beam of Fig. 4-13

Joint Number	Indexes for Possible Displacements	Restraint List	Indexes for Rearrangement
1	1	1	4
	2	1	5
2	3	0	1
	4	0	2
3	5	1	6
	6	0	3
4	7	1	7
	8	1	8

TABLE 4-9

Member Stiffness Matrices

$$[S_M]_1 = \frac{EI}{L^3}\begin{bmatrix} 12 & 6L & -12 & 6L \\ 6L & 4L^2 & -6L & 2L^2 \\ -12 & -6L & 12 & -6L \\ 6L & 2L^2 & -6L & 4L^2 \end{bmatrix}$$

$$[S_M]_2 = \frac{EI}{L^3}\begin{bmatrix} 24 & 12L & -24 & 12L \\ 12L & 8L^2 & -12L & 4L^2 \\ -24 & -12L & 24 & -12L \\ 12L & 4L^2 & -12L & 8L^2 \end{bmatrix}$$

$$[S_M]_3 = \frac{EI}{L^3}\begin{bmatrix} 3 & 3L & -3 & 3L \\ 3L & 4L^2 & -3L & 2L^2 \\ -3 & -3L & 3 & -3L \\ 3L & 2L^2 & -3L & 4L^2 \end{bmatrix}$$

TABLE 4-10

Joint Stiffness Matrix for Beam of Fig. 4-13

(All elements in the matrix are to be multiplied by the factor EI/L^3)

Member 1 (rows/columns 1–4), Member 2 (rows/columns 3–6), Member 3 (rows/columns 5–8)

	1	2	3	4	5	6	7	8	
1	12	6L	–12	6L	0	0	0	0	4
2	6L	4L²	–6L	2L²	0	0	0	0	5
3	–12	–6L	36	6L	–24	12L	0	0	1
4	6L	2L²	6L	12L²	–12L	4L²	0	0	2
5	0	0	–24	–12L	27	–9L	–3	3L	6
6	0	0	12L	4L²	–9L	12L²	–3L	2L²	3
7	0	0	0	0	–3	–3L	3	–3L	7
8	0	0	0	0	3L	2L²	–3L	4L²	8
	4	5	1	2	6	3	7	8	

degrees of freedom precede the actual restraints. The two columns containing joint numbers and the restraint list can be considered as input, while the other two columns can be derived from the data in the input columns.

In preparation for generating the over-all joint stiffness matrix $\mathbf{S}_J$, the member stiffness matrices for all three members of the structure are given in Table 4-9. These are formed from Table 4-2 given in Art. 4.3. The elements of the first member stiffness matrix $[S_M]_1$ for member 1 are transferred to $\mathbf{S}_J$ according to Eqs. (4-13) through (4-16). The contributions to $\mathbf{S}_J$ from member 1 appear in Table 4-10 within the upper left-hand portion enclosed by dashed lines. In a similar manner, elements from $[S_M]_2$ and $[S_M]_3$ are transferred to $\mathbf{S}_J$ as shown by the two remaining portions of Table 4-10 that are delineated by dashed lines. In the regions where the portions enclosed by dashed lines overlap, the elements shown in the table are the sums of two terms, one from each of two member stiffness matrices. Note that all elements in the table are to be multiplied by the factor EI/L^3. The numbering system for the stiffness matrix appears on the left side and across the top of the matrix and is in accordance with the numbering system shown in Fig. 4-13b.

Next, the matrix $\mathbf{S}_J$ is rearranged according to the rearranged list of displacements given in Table 4-8. This list is indicated by the numbers on the right side and across the bottom of the matrix in Table 4-10. The third, fourth, and sixth rows and columns are to be shifted to the first three positions, and all other rows and columns are to be shifted downward and to the right without changing their order. The resulting rearranged joint stiffness matrix is given in Table 4-11.

TABLE 4-11

Rearranged Joint Stiffness Matrix for Beam of Fig. 4-13
(All elements in the matrix are to be multiplied by the factor EI/L^3)

	1	2	3	4	5	6	7	8
1	36	6L	12L	−12	−6L	−24	0	0
2	6L	$12L^2$	$4L^2$	6L	$2L^2$	−12L	0	0
3	12L	$4L^2$	$12L^2$	0	0	−9L	−3L	$2L^2$
4	−12	6L	0	12	6L	0	0	0
5	−6L	$2L^2$	0	6L	$4L^2$	0	0	0
6	−24	−12L	−9L	0	0	27	−3	3L
7	0	0	−3L	0	0	−3	3	−3L
8	0	0	$2L^2$	0	0	3L	−3L	$4L^2$

Then this matrix is partitioned, as shown by the dashed lines, in the manner given by Eq. (4-1).

The 3 × 3 stiffness matrix **S** is drawn from Table 4-11 as follows:

$$\mathbf{S} = \frac{EI}{L^3}\begin{bmatrix} 36 & 6L & 12L \\ 6L & 12L^2 & 4L^2 \\ 12L & 4L^2 & 12L^2 \end{bmatrix}$$

and subsequently inverted to obtain $\mathbf{S}^{-1}$:

$$\mathbf{S}^{-1} = \frac{L}{756EI}\begin{bmatrix} 32L^2 & -6L & -30L \\ -6L & 72 & -18 \\ -30L & -18 & 99 \end{bmatrix}$$

Thus, all of the calculations involving the properties of the structure have been completed, and information concerning the loads on the structure may now be processed.

Inspection of Fig. 4-13a shows that loads are applied at joints B and C only (joints 2 and 3 in Fig. 4-13b). These joint loads are listed in Table 4-12 in a

TABLE 4-12

Actions Applied at Joints

Joint	Force in y Direction	Couple in z Sense
1	0	0
2	–P	PL
3	–P	0
4	0	0

manner that is suitable for input to a computer program. Next, these actions are placed in the vector **A** as follows (see Eq. 4-17):

$$\mathbf{A} = \{0, 0, -P, PL, -P, 0, 0, 0\}$$

Fixed-end actions are tabulated in Table 4-13 in the same arrangement in which they appear in the matrix $\mathbf{A}_{ML}$ (see Eq. 4-18). The elements of $\mathbf{A}_{ML}$ are

TABLE 4-13

Fixed-End Actions Due to Loads

Member	$(A_{ML})_{i,1}$	$(A_{ML})_{i,2}$	$(A_{ML})_{i,3}$	$(A_{ML})_{i,4}$
1	P	PL/4	P	–PL/4
2	P	PL/4	P	–PL/4
3	P	PL/3	P	–PL/3

transferred to the vector of equivalent joint loads $\mathbf{A}_E$ as indicated by Eqs. (4-20). First, the fixed-end actions for member 1 are transferred to the first four elements of $\mathbf{A}_E$; then the actions for member 2 are transferred to elements three to six of $\mathbf{A}_E$; and lastly, the actions for member 3 are transferred to the last four elements of $\mathbf{A}_E$. When the vector has been generated in this manner, the result is:

$$\mathbf{A}_E = \{-P, -PL/4, -2P, 0, -2P, -PL/12, -P, PL/3\}$$

The vectors $\mathbf{A}$ and $\mathbf{A}_E$ are now combined using Eq. (4-5) to obtain the vector $\mathbf{A}_C$:

$$\mathbf{A}_C = \{-P, -PL/4, -3P, PL, -3P, -PL/12, -P, PL/3\}$$

This vector must be rearranged by placing the third, fourth, and sixth elements into the first three positions and moving all other elements toward the end without changing their order. This rearrangement results in the following vector:

$$\mathbf{A}_C = \{-3P, PL, -PL/12, -P, -PL/4, -3P, -P, PL/3\}$$

in which the first three elements are the vector $\mathbf{A}_D$:

$$\mathbf{A}_D = \{-3P, PL, -PL/12\}$$

and the last five elements are the negatives of the elements of $\mathbf{A}_{RL}$:

$$\mathbf{A}_{RL} = \{P, PL/4, 3P, P, -PL/3\}$$

Having all of the required matrices on hand, one may complete the solution by first calculating the joint displacements D using Eq. (4-8):

$$\mathbf{D} = \mathbf{S}^{-1}\mathbf{A}_D = \frac{L}{756EI}\begin{bmatrix} 32L^2 & -6L & -30L \\ -6L & 72 & -18 \\ -30L & -18 & 99 \end{bmatrix}\begin{bmatrix} -36 \\ 12L \\ -L \end{bmatrix}\frac{P}{12}$$

$$= \frac{PL^2}{3024EI}\begin{bmatrix} -398L \\ 366 \\ 255 \end{bmatrix}$$

The vector $\mathbf{D}$ may now be used in obtaining the vector $\mathbf{D}_J$ of all possible joint displacements by referring to the indexes for rearrangement shown in Table 4-8. These indexes show that $D_{J3} = D_1$, $D_{J4} = D_2$, and $D_{J6} = D_3$. All other elements of $\mathbf{D}_J$ are zero because they correspond to support restraints. Thus, the vector $\mathbf{D}_J$ becomes the following:

$$\mathbf{D}_J = \frac{PL^2}{3024EI}\{0, 0, -398L, 366, 0, 255, 0, 0\}$$

Next, the reactions A_R may be determined from Eq. (4-4). For this purpose the matrix $\mathbf{S}_{RD}$ is obtained from the lower left-hand portion of Table 4-11. The calculation for $\mathbf{A}_R$ is as follows:

$$\mathbf{A}_R = \mathbf{A}_{RL} + \mathbf{S}_{RD}\mathbf{D}$$

$$= \frac{P}{12}\begin{bmatrix} 12 \\ 3L \\ 36 \\ 12 \\ -4L \end{bmatrix} + \frac{EI}{L^3}\begin{bmatrix} -12 & 6L & 0 \\ -6L & 2L^2 & 0 \\ -24 & -12L & -9L \\ 0 & 0 & -3L \\ 0 & 0 & 2L^2 \end{bmatrix}\begin{bmatrix} -398L \\ 366 \\ 255 \end{bmatrix}\frac{PL^2}{3024EI}$$

$$= \frac{P}{1008}\begin{bmatrix} 3332 \\ 1292L \\ 3979 \\ 753 \\ -166L \end{bmatrix}$$

Finally, the member end-actions A_M are obtained by repeated applications of Eq. (4-9) (see also Eq. 4-22). Since there are three members in this example, three calculations of this type are required. For member 1 the calculations are as follows:

$$\{A_M\}_1 = \{A_{ML}\}_1 + [S_M]_1\{D_M\}_1$$

$$= \frac{P}{4}\begin{bmatrix} 4 \\ L \\ 4 \\ -L \end{bmatrix} + [S_M]_1 \begin{bmatrix} 0 \\ 0 \\ -398L \\ 366 \end{bmatrix} \frac{PL^2}{3024EI} = \frac{P}{252}\begin{bmatrix} 833 \\ 323L \\ -329 \\ 258L \end{bmatrix}$$

In the above equation the vector $\{D_M\}_1$ consists of the first four elements of $\mathbf{D}_J$. Similarly, the equations for $\{A_M\}_2$ and $\{A_M\}_3$ are

$$\{A_M\}_2 = \{A_{ML}\}_2 + [S_M]_2\{D_M\}_2$$

$$= \frac{P}{4}\begin{bmatrix} 4 \\ L \\ 4 \\ -L \end{bmatrix} + [S_M]_2 \begin{bmatrix} -398L \\ 366 \\ 0 \\ 255 \end{bmatrix} \frac{PL^2}{3024EI} = \frac{P}{252}\begin{bmatrix} 77 \\ -6L \\ 427 \\ -169L \end{bmatrix}$$

$$\{A_M\}_3 = \{A_{ML}\}_3 + [S_M]_3\{D_M\}_3$$

$$= \frac{P}{3}\begin{bmatrix} 3 \\ L \\ 3 \\ -L \end{bmatrix} + [S_M]_3 \begin{bmatrix} 0 \\ 255 \\ 0 \\ 0 \end{bmatrix} \frac{PL^2}{3024EI} = \frac{P}{1008}\begin{bmatrix} 1263 \\ 676L \\ 753 \\ -166L \end{bmatrix}$$

in which the vector $\{D_M\}_2$ is made up of the third through the sixth elements of $\mathbf{D}_J$, and the vector $\{D_M\}_3$ consists of the last four elements of $\mathbf{D}_J$. Thus, all of the joint displacements, support reactions, and member end-actions are determined, and the problem is considered to be completed.

4.10 Plane Truss Member Stiffnesses. The determination of the member stiffness matrix for a typical truss member is a preliminary to the analysis of plane trusses. A typical member i in a plane truss is shown in Fig. 4-14a. The joints at the ends of this member are denoted as joints j and k. The plane truss itself is assumed to lie in the x-y plane, where x and y are reference axes for the structure. The joint translations are the unknown displacements in the analysis, and all of these translations may be expressed conveniently by their components in the x and y directions. For the typical member i, the positive directions of the four displacement components at its ends (with respect to the structure-oriented axes) are indicated in Fig. 4-14b.

It is convenient when dealing with inclined members in a framed structure to make use of direction cosines. The direction cosines for the member shown in Fig. 4-14b are the cosines of the angles γ_1 and γ_2 between the axis of the member and the x and y axes, respectively. These angles will always be taken at the j end of the member. The corresponding direction cosines are denoted as C_X and C_Y and are given as follows:

$$C_X = \cos\gamma_1 \qquad C_Y = \cos\gamma_2$$

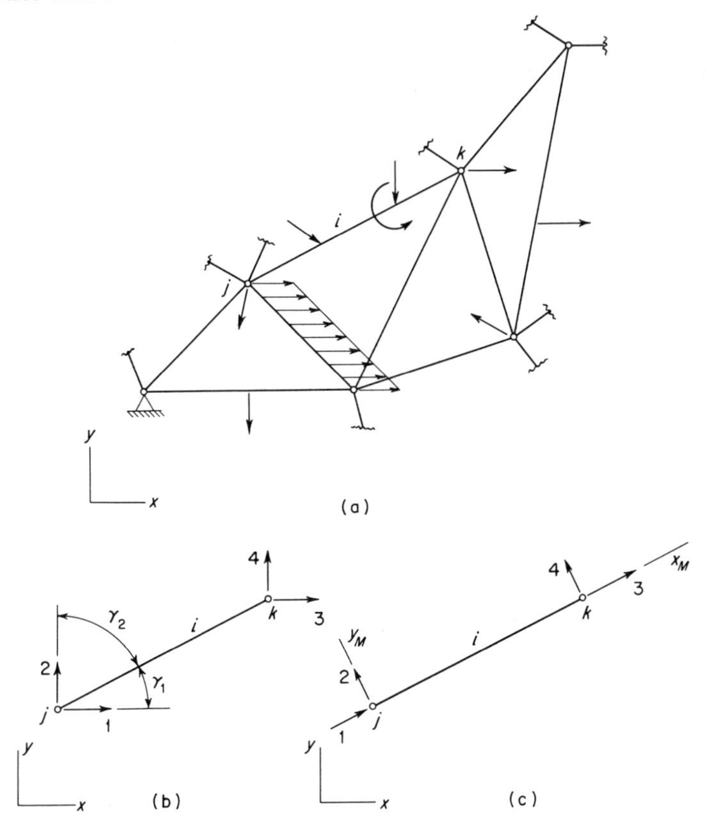

FIG. 4-14. Numbering system for a plane truss member.

The direction cosines for the member can also be expressed in terms of the coordinates of the joints j and k. Denoting the x and y coordinates of joints j and k as (x_j, y_j) and (x_k, y_k), respectively, the direction cosines become

$$C_X = \frac{x_k - x_j}{L} \qquad C_Y = \frac{y_k - y_j}{L} \tag{4-24}$$

in which L is the length of the member. The length L may be computed from the coordinates of the joints at the ends of the member as follows:

$$L = \sqrt{(x_k - x_j)^2 + (y_k - y_j)^2} \tag{4-25}$$

Both direction cosines are positive when the member is oriented as shown in Fig. 4-14b, that is, when the x and y coordinates of joint k are greater than those for joint j. If the angle γ_1 is greater than 90°, the formulas given above are still valid, and one or both of the direction cosines will be negative.

In the analysis of a plane truss, as in the case of any other type of framed structure, it is convenient to generate the joint stiffness matrix

$\mathbf{S}_J$ by assessing the contributions from member stiffnesses. In the case of a continuous beam this operation is straightforward (see Art. 4.8) because the member-oriented axes may be taken parallel to, or coincident with, the structure-oriented axes. Thus, the beam member stiffness matrix for member-oriented axes, as derived in Art. 4.3, can be used directly when obtaining joint stiffnesses for the structure axes. For any other type of framed structure, however, the member axes will not necessarily be parallel to the structure axes. For example, the x-y axes in Fig. 4-14b are structure-oriented axes, whereas the member axes x_M and y_M are assumed to be along the axis of the member and perpendicular to the member, as shown in Fig. 4-14c. The stiffness matrix $\mathbf{S}_M$ for the member-oriented axes shown in Fig. 4-14c can be obtained readily by referring to cases (1) and (7) of Fig. 4-2. Using the numbering system shown in Fig. 4-14c for the end-displacements with respect to the x_M-y_M axes, it becomes evident that the stiffness matrix $\mathbf{S}_M$ for those axes has the form shown in Table 4-14. In the table, the axial rigidity of the bar is denoted EA_X.

TABLE 4-14

Plane Truss Member Stiffness Matrix for Member Axes (Fig. 4-14c)

$$\mathbf{S}_M = \frac{EA_X}{L}\begin{bmatrix} 1 & 0 & -1 & 0 \\ 0 & 0 & 0 & 0 \\ -1 & 0 & 1 & 0 \\ 0 & 0 & 0 & 0 \end{bmatrix}$$

However, since the matrix $\mathbf{S}_J$ is based upon axes oriented to the structure, it also becomes necessary to obtain member stiffnesses for the structure axes. The member stiffnesses for the structure axes may be found in either of two ways. The first method consists of a direct formulation of the stiffnesses. In this approach, unit displacements in the directions of the structure axes (see Fig. 4-14b) are induced at the ends of the member, and the corresponding restraint actions in the same directions are calculated. These actions become the elements of the member stiffness matrix for the structure axes. The second method for finding the member stiffness matrix consists of first obtaining the stiffness matrix for member-oriented axes (see Table 4-14) and then transforming this matrix to the structure-oriented axes by a process of rotation of axes. Using an appropriate transformation matrix, the rotation of axes may be executed by matrix multiplications, as will be described in a later article.

The direct approach to obtaining the member stiffness matrix is demonstrated in this article for a plane truss member. The method also could be used for the other types of framed structures. However, the direct formulation technique becomes quite involved for the more complex types of structures, especially the space frame. On the other hand, the rotation of axes method is a formalized approach which is no more difficult in theory for a complicated structure than for a simple one. In order to explain this method, the subject of rotation of axes is introduced in Art. 4.13, and then the method is used in Art. 4.14 to derive the stiffness matrix for a plane truss member with respect to the structure axes. This approach makes it possible to compare the derivation of the stiffness matrix by both methods. Subsequently, the member stiffnesses for all other types of structures are derived by using the rotation of axes method.

The member stiffness matrix for a plane truss member will now be developed by the direct method. For this purpose it is necessary to consider unit displacements in the x and y directions at both ends of the member. The first such displacement is shown in Fig. 4-15a, and consists

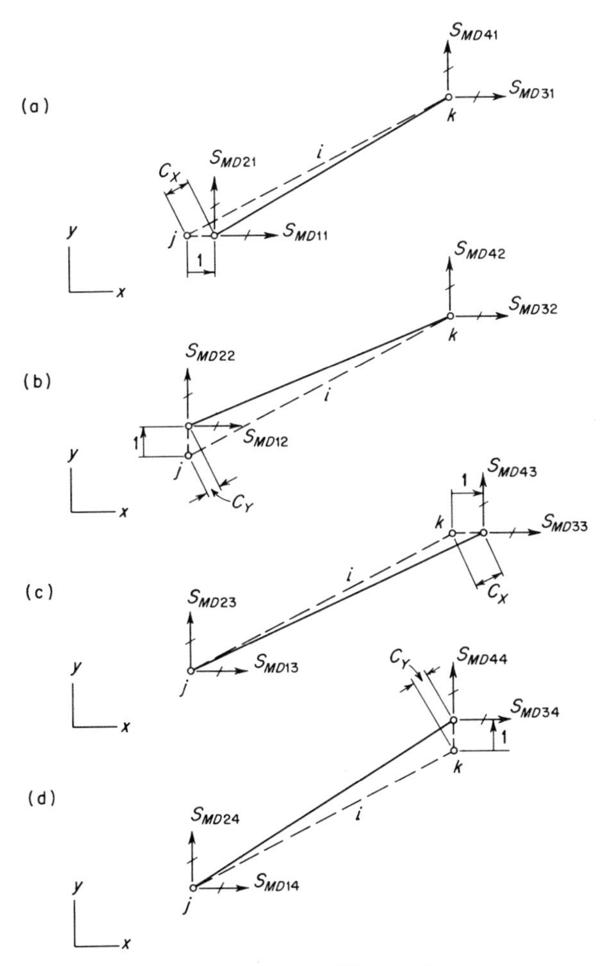

Fig. 4-15. Plane truss member stiffnesses for structure axes.

of a unit translation in the x direction at the j end of the member. As a result of this displacement, an axial force is induced in the member. This force can be calculated from the axial shortening of the member, which is numerically equal to the x direction cosine C_X for the member (see Fig. 4-15a). The axial compressive force in the member due to this change of length is equal to

$$\frac{EA_X}{L} C_X$$

The restraint actions at the ends of the member in the x and y directions, which are equal to the components of the axial force, are the desired member stiffnesses for the structure axes. Such stiffnesses are identified by the symbol S_{MD} (see Fig. 4-15a) in order to distinguish them from the stiffnesses S_M for the member axes (see Table 4-14). The numbering system for the stiffnesses S_{MD} is shown in Fig. 4-14b. The restraint action at end j in the x direction, denoted as S_{MD11}, must be equal to the x component of the force in the member. Therefore, this stiffness is equal to the axial force times the x direction cosine, as follows:

$$S_{MD11} = \frac{EA_X}{L} C_X^2$$

Also, the restraint action at j in the y direction is equal to the y component of the force in the member:

$$S_{MD21} = \frac{EA_X}{L} C_X C_Y$$

The restraint actions at the k end of the member in Fig. 4-15a are readily found by static equilibrium, as follows:

$$S_{MD31} = -S_{MD11} = -\frac{EA_X}{L} C_X^2$$

$$S_{MD41} = -S_{MD21} = -\frac{EA_X}{L} C_X C_Y$$

The expressions given above for the four stiffnesses shown in Fig. 4-15a constitute the elements of the first column of the matrix $\mathbf{S_{MD}}$. The second, third, and fourth columns of $\mathbf{S_{MD}}$ may be obtained in a similar manner from Figs. 4-15b, 4-15c, and 4-15d, respectively, to give the 4×4 stiffness matrix shown in Table 4-15. The reader should verify for himself the remaining elements in this matrix.

TABLE 4-15

Plane Truss Member Stiffness Matrix for Structure Axes (Fig. 4-14b)

$$\mathbf{S_{MD}} = \frac{EA_X}{L} \begin{bmatrix} C_X^2 & C_X C_Y & -C_X^2 & -C_X C_Y \\ C_X C_Y & C_Y^2 & -C_X C_Y & -C_Y^2 \\ -C_X^2 & -C_X C_Y & C_X^2 & C_X C_Y \\ -C_X C_Y & -C_Y^2 & C_X C_Y & C_Y^2 \end{bmatrix}$$

4.11 Analysis of Plane Trusses. As an initial step in the analysis of a plane truss, all of the joints and members must be numbered. The joints of the structure are numbered consecutively from 1 through n_j, where n_j is the total number of joints. In addition, the members are numbered from 1 through m, where m is the total number of members. The order in which the joints and members are numbered is immaterial. However, after the numbering is completed, it is necessary to record the two joint numbers that are associated with each member. This association of joint numbers with member numbers is necessary in order to ascertain which elements of the joint stiffness matrix $\mathbf{S}_J$ and the equivalent load vector $\mathbf{A}_E$ receive contributions from each member.

To illustrate the arbitrary numbering of the joints and members, a plane truss is shown in Fig. 4-16a. In the figure the joint numbers appear

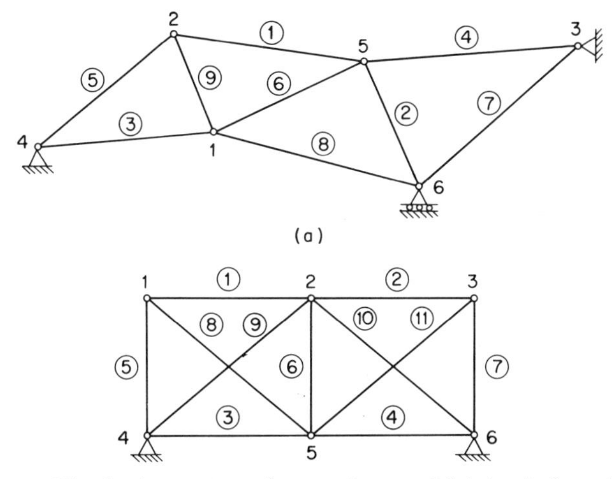

FIG. 4-16. Numbering systems for members and joints of plane trusses.

adjacent to the joints, while the member numbers are enclosed in circles adjacent to the members. The numbering system for this structure is shown in Table 4-16. It is necessary for purposes of analysis to identify a j end and a k end for each member, as shown in the table, although the selection itself is arbitrary.

TABLE 4-16

Listing of Members and Joints for Truss of Fig. 4-16a

Member	1	2	3	4	5	6	7	8	9
Joint j	2	6	4	3	4	1	6	1	2
Joint k	5	5	1	5	2	5	3	6	1

While the methods of analysis in this chapter do not require any particular order of numbering joints and members, it is only natural to number them in a systematic pattern whenever feasible. For example, in some structures a natural order may be from left to right or from top to bottom, as shown by the plane truss in Fig. 4-16b. The geometry of this truss suggests that a systematic pattern for numbering the joints is to proceed from left to right across the top and then from left to right across the bottom. A systematic pattern for the members consists of numbering consecutively the horizontal members, the vertical members, and then the diagonal members, as shown in the figure. The numbering system for this structure is summarized in Table 4-17. Note again that

TABLE 4-17

Listing of Members and Joints for Truss of Fig. 4-16b

Member	1	2	3	4	5	6	7	8	9	10	11
Joint j	1	2	4	5	1	2	3	1	4	2	5
Joint k	2	3	5	6	4	5	6	5	2	6	3

it is necessary to identify one end of each member as being the j end and one end as the k end.

Many possible schemes for numbering the joints and members of a truss can be organized, depending mainly on personal preference. If it is desired to improve the efficiency of the solution, the numbering system may be selected to minimize the amount of rearrangement of the over-all joint stiffness matrix $\mathbf{S}_J$. This procedure might be advantageous for hand calculations, but the advantage disappears when the analysis is programmed for a computer.

After numbering the members and joints, the next step in the analysis is to identify all possible joint displacements and the degrees of freedom. The number of the former will be twice the number of joints, or $2n_j$, since each joint may undergo a translation in both the x and y directions. The number of degrees of freedom n is given by the expression

$$n = 2n_j - n_r \tag{4-26}$$

in which n_r denotes the number of support restraints.

The possible joint displacements will be numbered in the same order as the joints, taking the x translation before the y translation at each joint. Thus, the x translation at joint 1 becomes displacement number 1, the y translation at joint 1 becomes displacement number 2, the x translation at joint 2 becomes displacement number 3, and so forth, until at the last joint the x and y translations are numbered $2n_j - 1$

and $2n_j$, respectively. In general, at joint j of the truss the x and y translations have the indexes $2j - 1$ and $2j$, respectively.

In order to construct the stiffness matrix $\mathbf{S}_J$ from the member stiffnesses, it is useful to relate the indexes for the possible joint displacements to the end displacements for a particular member, as was done in the analysis of continuous beams. For this purpose consider the typical member i, shown in Fig. 4-17, which frames into joints j and k at the ends. The x and y axes in the figure are assumed to be structure-oriented axes. The end-displacements of this member may be identified by the indexes $j1$, $j2$, $k1$, and $k2$, as shown in the figure. These indexes for the member are related to the corresponding indexes for the possible joint displacements by the following relations:

$$
\begin{aligned}
j1 &= 2j - 1 \qquad & j2 &= 2j \\
k1 &= 2k - 1 \qquad & k2 &= 2k
\end{aligned}
\tag{4-27}
$$

These relations follow directly from the numbering system for the possible joint displacements, which is described above.

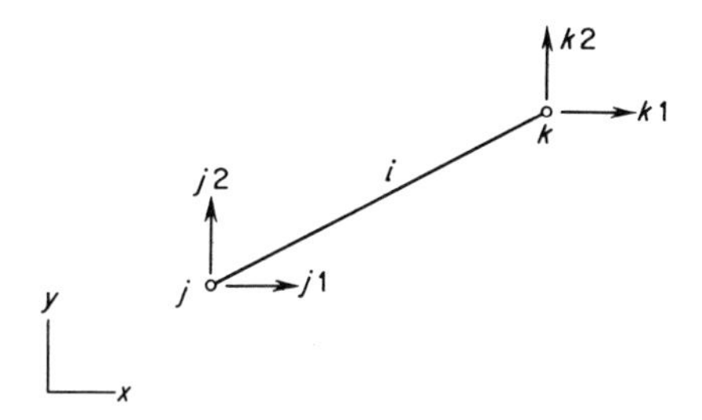

FIG. 4-17. End-displacements for plane truss member.

The joint stiffness matrix consists of contributions from the individual members. Therefore, the member stiffness matrix $\mathbf{S}_{MD}$ (see Art. 4.10, Table 4-15) for each member must be determined and its elements placed in the proper position in the matrix $\mathbf{S}_J$. In order to see how this is accomplished, consider again a typical member i (Fig. 4-18a). This member contributes to the stiffnesses of joints j and k at its ends. If a unit displacement in the x direction is induced in the restrained structure, as shown in the figure, there will be restraint actions in the x and y directions at both joints. At joint j the action in the x direction is the joint stiffness $(S_J)_{j1,j1}$, and the action in the y direction is the stiffness $(S_J)_{j2,j1}$. Similarly, the actions at joint k are denoted $(S_J)_{k1,j1}$ and $(S_J)_{k2,j1}$. The two actions at joint j consist of contributions from the member i, plus contributions from all other members that frame into joint j. The latter contributions will be denoted for simplicity by the indefinite symbol $\sum S_{MD}$. The contributions from member i, however, will be written in precise form since these are the terms that will be used later in forming the stiffness matrix. In the case of the stiffness $(S_J)_{j1,j1}$ the contribution from member i is the member stiffness S_{MD11}, and in the case of the stiffness $(S_J)_{j2,j1}$ the contribution is S_{MD21}. For the stiffnesses at joint k, only contributions from member i are involved. These contributions are S_{MD31} and S_{MD41} for the stiffnesses in the x and y directions,

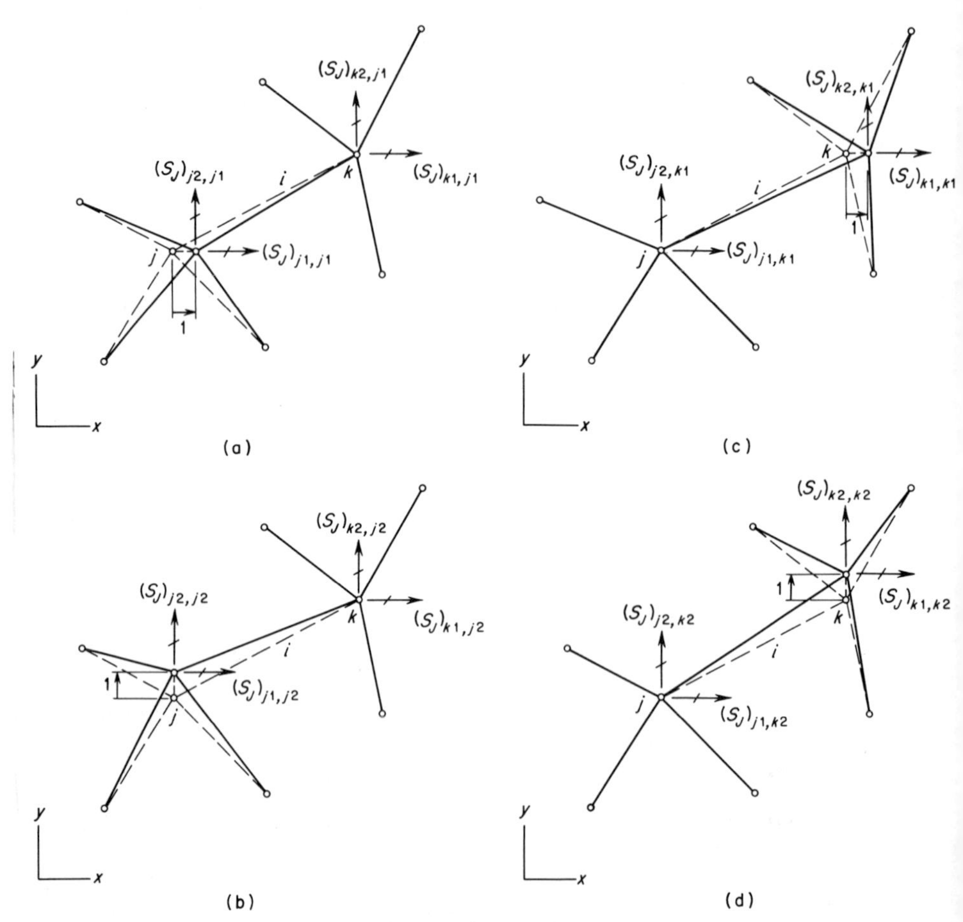

FIG. 4-18. Joint stiffnesses for a plane truss.

respectively, at joint k. Thus, the expressions for the joint stiffnesses shown in Fig. 4-18a are:

$$
\begin{aligned}
(S_J)_{j1,j1} &= \sum S_{MD} + (S_{MD11})_i \\
(S_J)_{j2,j1} &= \sum S_{MD} + (S_{MD21})_i \\
(S_J)_{k1,j1} &= (S_{MD31})_i \\
(S_J)_{k2,j1} &= (S_{MD41})_i
\end{aligned}
\tag{4-28}
$$

Each of the above stiffnesses is due to a unit displacement at joint j, and each receives a contribution from the first column of the matrix $\mathbf{S}_{\mathrm{MD}}$ for member i.

Expressions similar to Eqs. (4-28) may be written for a unit displacement of the restrained structure in the y direction at joint j, as shown in Fig. 4-18b. These expressions involve the second column of $\mathbf{S}_{\mathrm{MD}}$ and are as follows:

$$\begin{aligned}
(S_J)_{j1,j2} &= \sum S_{MD} + (S_{MD12})_i \\
(S_J)_{j2,j2} &= \sum S_{MD} + (S_{MD22})_i \\
(S_J)_{k1,j2} &= \qquad\qquad (S_{MD32})_i \\
(S_J)_{k2,j2} &= \qquad\qquad (S_{MD42})_i
\end{aligned} \tag{4-29}$$

Also, for a unit x displacement at k (Fig. 4-18c) the joint stiffnesses which receive contributions from the third column of $\mathbf{S}_{\mathrm{MD}}$ are:

$$\begin{aligned}
(S_J)_{j1,k1} &= \qquad\qquad (S_{MD13})_i \\
(S_J)_{j2,k1} &= \qquad\qquad (S_{MD23})_i \\
(S_J)_{k1,k1} &= \sum S_{MD} + (S_{MD33})_i \\
(S_J)_{k2,k1} &= \sum S_{MD} + (S_{MD43})_i
\end{aligned} \tag{4-30}$$

Finally, the stiffnesses for a unit y displacement at k (see Fig. 4-18d) receive contributions from the fourth column of $\mathbf{S}_{\mathrm{MD}}$, as follows:

$$\begin{aligned}
(S_J)_{j1,k2} &= \qquad\qquad (S_{MD14})_i \\
(S_J)_{j2,k2} &= \qquad\qquad (S_{MD24})_i \\
(S_J)_{k1,k2} &= \sum S_{MD} + (S_{MD34})_i \\
(S_J)_{k2,k2} &= \sum S_{MD} + (S_{MD44})_i
\end{aligned} \tag{4-31}$$

The pattern of contributions of member stiffnesses to the joint stiffness matrix, as expressed by the above equations, is determined by the geometry of the truss and the system for numbering joints and members. The pattern for the structure shown in Fig. 4-19 will be used as an example. Joints and members for the truss are numbered as shown in the figure, and the possible joint displacements (shown by arrows) are numbered according to Eqs. (4-27). For this structure there are six members and four joints, and the over-all joint stiffness matrix is of order 8×8. The positions of the member contributions to the joint stiffness matrix are indicated by crosses in Figs. 4-20a through 4-20f, and the final stiffness matrix is a composite of these contributions. As an example, the stiffness S_{J11} is made up of contributions from members 1, 2, and 3, as shown in Fig. 4-20a, 4-20b, and 4-20c. The stiffness is, therefore, as follows:

$$S_{J11} = (S_{MD11})_1 + (S_{MD11})_2 + (S_{MD11})_3$$

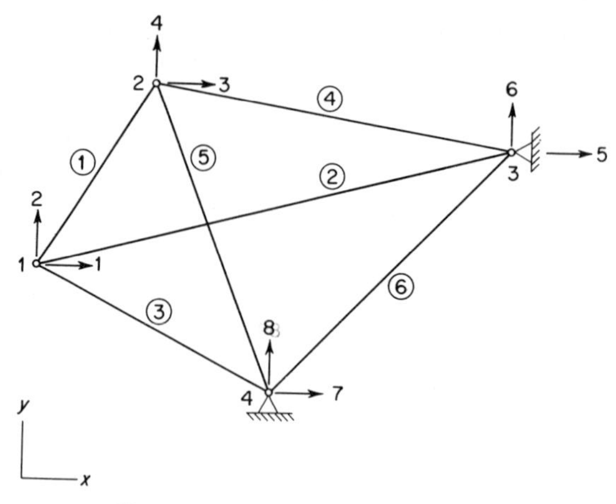

FIG. 4-19. Example (plane truss).

As another illustration, consider the stiffness S_{J34}, which is composed of member stiffnesses from members 1, 4, and 5. This stiffness is

$$S_{J34} = (S_{MD34})_1 + (S_{MD12})_4 + (S_{MD12})_5$$

as can be seen by referring to Figs. 4-20a, 4-20d, and 4-20e.

In the particular example of Fig. 4-19 the stiffness matrix $\mathbf{S}_J$ does not have to be rearranged because of the sequence in which the joints are numbered. The first four rows and columns of $\mathbf{S}_J$ are associated with the degrees of freedom, and the last four rows and columns pertain to the

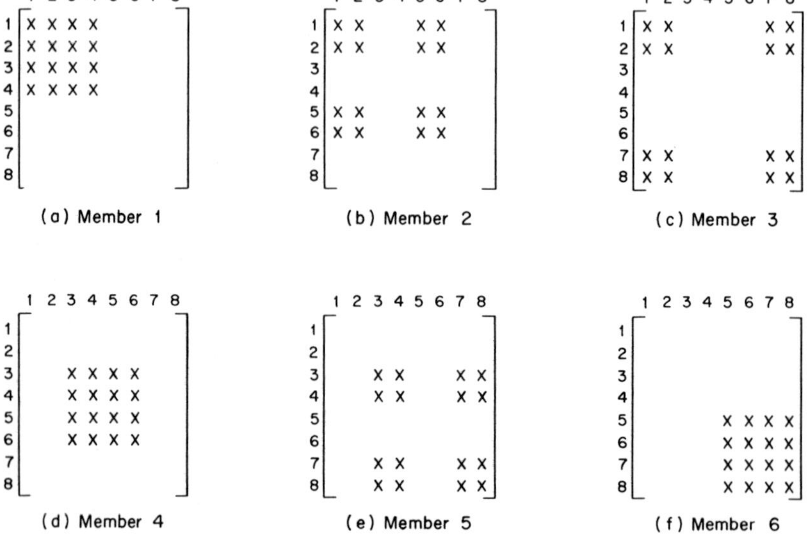

FIG. 4-20. Contributions of member stiffnesses to joint stiffnesses for plane truss example.

support restraints. However, this is a special case, and even a small change in the structure, such as placing a roller support at joint 3, would make a rearrangement necessary.

The vectors associated with the loads on a truss will be discussed next. The first one to be considered is the vector **A** of actions (or loads) applied at the joints. At a typical joint k, two orthogonal force components may exist, as shown in Fig. 4-21. The action A_{2k-1} is the force in the positive x direction, and the action A_{2k} is the force in the positive y direction. The vector **A**, therefore, will have the following form:

$$\mathbf{A} = \{A_1, A_2, \ldots, A_{2k-1}, A_{2k}, \ldots, A_{2n_j}\} \tag{4-32}$$

in which there are $2n_j$ elements corresponding to the $2n_j$ possible joint displacements.

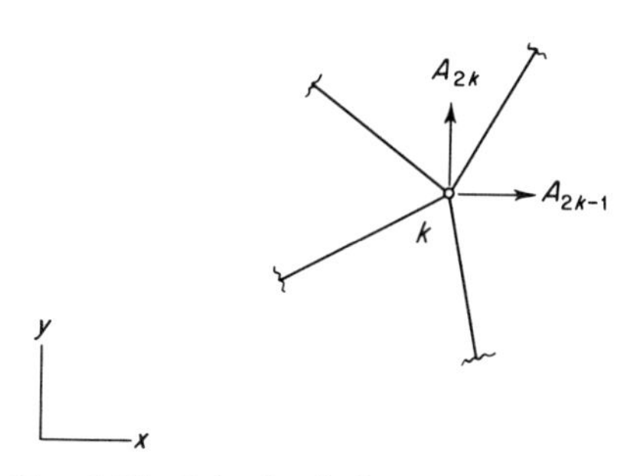

FIG. 4-21. Joint loads for a plane truss

Consider next the matrix of actions $\mathbf{A}_{ML}$ due to loads acting on the members when the joints of the truss are restrained against translation. Loads acting on member i are illustrated in Fig. 4-22b, which also shows a set of member-oriented axes x_M and y_M. The actions A_{ML} for member i are defined with respect to the x_M and y_M axes, as follows:

$(A_{ML})_{i,1}$ = force in the x_M direction at the j end

$(A_{ML})_{i,2}$ = force in the y_M direction at the j end

$(A_{ML})_{i,3}$ = force in the x_M direction at the k end

$(A_{ML})_{i,4}$ = force in the y_M direction at the k end

These end-actions may be obtained for any particular loading conditions by referring to the tables in Appendix B. The matrix $\mathbf{A}_{ML}$ is of order $m \times 4$ and is of the same form as that given by Eq. (4-18) for continuous beams.

The vector $\mathbf{A}_E$ of equivalent joint loads may be constructed from the elements of the matrix $\mathbf{A}_{ML}$. This vector is of the same form as that given by Eq. (4-32), except that A_E replaces A. To illustrate the calculation of the equivalent joint loads, consider the action $(A_E)_{2j-1}$ shown in Fig. 4-22a. This action is made up of contributions from member i, plus contributions from all other members meeting at the joint. The latter contributions will be denoted by the indefinite symbol $\sum A_{ML}$. The contributions from member i, however, can be readily expressed in terms of the end-actions $(A_{ML})_i$. All that is necessary is to take the components in the x direction of the end-actions at joint j, and to reverse their signs. Thus, the contribution of member i to the equivalent load $(A_E)_{2j-1}$ is

$$-(A_{ML})_{i,1}C_{Xi} + (A_{ML})_{i,2}C_{Yi}$$

in which C_{Xi} and C_{Yi} are the direction cosines for member i. By proceed-

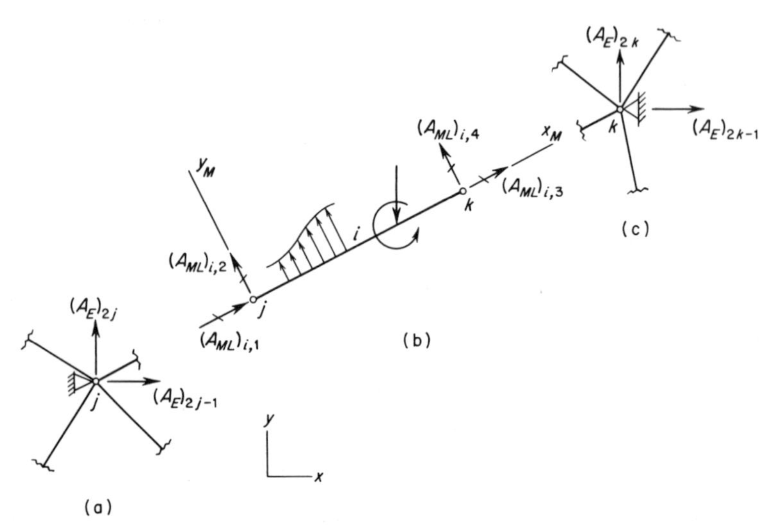

FIG. 4-22. Loads on a plane truss member.

ing in this manner, the equivalent loads at joints j and k (see Figs. 4-22a and 4-22c) can be found:

$$\begin{aligned}
(A_E)_{2j-1} &= \sum A_{ML} - (A_{ML})_{i,1}C_{Xi} + (A_{ML})_{i,2}C_{Yi} \\
(A_E)_{2j} &= \sum A_{ML} - (A_{ML})_{i,1}C_{Yi} - (A_{ML})_{i,2}C_{Xi} \\
(A_E)_{2k-1} &= \sum A_{ML} - (A_{ML})_{i,3}C_{Xi} + (A_{ML})_{i,4}C_{Yi} \\
(A_E)_{2k} &= \sum A_{ML} - (A_{ML})_{i,3}C_{Yi} - (A_{ML})_{i,4}C_{Xi}
\end{aligned} \tag{4-33}$$

These equations give the contributions of member i to the equivalent joint loads. By applying the equations consecutively to all members of the truss, the equivalent joint loads can be obtained for all joints.

The contributions of member i to the vector $\mathbf{A}_E$ may also be ascertained by the method of rotation of axes. This procedure is demonstrated for plane truss structures in Art. 4.14.

The vectors $\mathbf{A}$ and $\mathbf{A}_E$ may be added together (see Eq. 4-5) to form the vector $\mathbf{A}_C$. The latter vector is then rearranged if necessary in order to isolate the vector $\mathbf{A}_D$ (see Eq. 4-6). Then the vector $\mathbf{A}_D$ and the inverse matrix $\mathbf{S}^{-1}$ are used in calculating the joint displacements $\mathbf{D}$, as given by Eq. (4-8). Next, the vector of joint displacements $\mathbf{D}$ can be expanded into the vector $\mathbf{D}_J$ of all possible joint displacements. The vector $\mathbf{D}_J$ has $2n_j$ elements, and those elements which correspond to the support restraints are zero. As the next step, the solution for the support reactions is obtained in the usual manner by means of Eq. (4-4).

The final task of solving for the member end-actions may be carried out by means of Eq. (4-9), which is repeated below:

$$\{A_M\}_i = \{A_{ML}\}_i + [S_M]_i \{D_M\}_i \tag{4-9}$$

repeated

This equation must be applied once for each member of the structure. Note that when written in expanded form, the equation has the same form as Eq. (4-22). The vector $\{A_{ML}\}_i$ is obtained from the i-th row of the matrix $\mathbf{A}_{\mathrm{ML}}$ described above. The matrix $[S_M]_i$ is the stiffness matrix for the i-th member with respect to member axes (see Table 4-14).

The vector $\{D_M\}_i$ in Eq. (4-9) consists of the end-displacements for member i in the directions of the member axes. These displacements can be calculated from the displacements D_J, which are in the directions of the structure axes. For instance, the first element D_{M1} in the vector $\{D_M\}_i$ represents the displacement of joint j in the x_M direction (see Fig. 4-22b). This displacement is given by the following expression:

$$D_{M1} = (D_J)_{j1}C_{Xi} + (D_J)_{j2}C_{Yi}$$

in which $(D_J)_{j1}$ and $(D_J)_{j2}$ are the displacements of joint j in the x and y directions, respectively (see Fig. 4-17). Similarly, the displacement of joint j in the y_M direction can be expressed in terms of $(D_J)_{j1}$ and $(D_J)_{j2}$ by the following:

$$D_{M2} = -(D_J)_{j1}C_{Yi} + (D_J)_{j2}C_{Xi}$$

Similarly, the end-displacements at joint k may be obtained as follows:

$$D_{M3} = (D_J)_{k1}C_{Xi} + (D_J)_{k2}C_{Yi}$$

$$D_{M4} = -(D_J)_{k1}C_{Yi} + (D_J)_{k2}C_{Xi}$$

The above expressions for the elements of $\{D_M\}_i$ can now be substituted into Eq. (4-9). When similar substitutions for the elements of $[S_M]_i$ are made, the equation can be expanded into the following four separate equations:

$$\begin{aligned}
(A_M)_{i,1} &= (A_{ML})_{i,1} + \frac{EA_{Xi}}{L_i}\{[(D_J)_{j1} - (D_J)_{k1}]C_{Xi} \\
&\qquad + [(D_J)_{j2} - (D_J)_{k2}]C_{Yi}\} \\
(A_M)_{i,2} &= (A_{ML})_{i,2} \\
(A_M)_{i,3} &= (A_{ML})_{i,3} - \frac{EA_{Xi}}{L_i}\{[(D_J)_{j1} - (D_J)_{k1}]C_{Xi} \\
&\qquad + [(D_J)_{j2} - (D_J)_{k2}]C_{Yi}\} \\
(A_M)_{i,4} &= (A_{ML})_{i,4}
\end{aligned} \tag{4-34}$$

These equations may now be applied repeatedly for all members of the truss. For each member the displacements D_J which appear in the equations must be extracted in the appropriate manner from the vector $\mathbf{D}_J$ of all joint displacements.

Another method for obtaining the vector $\{D_M\}_i$ which appears in Eq. (4-9) is to use rotation of axes. This method is described later in Art. 4.14.

An example analysis of a plane truss structure appears in the next article, and a computer program for the analysis of plane trusses is presented in Art. 5.5.

4.12 Example. The plane truss shown in Fig. 4-23a is to be analyzed using the methods described in the previous article. The truss is restrained at points C and D by hinge supports which prevent translations in both the x and y directions. The loads on the truss consist of both joint loads and member loads.

An arbitrary system for numbering the members and joints is given in Fig. 4-23b, which shows the restrained structure. Member numbers are indicated in circles adjacent to the members, and joint numbers are indicated by numbers adjacent to the joints. The numbering system for joint displacements is repre-

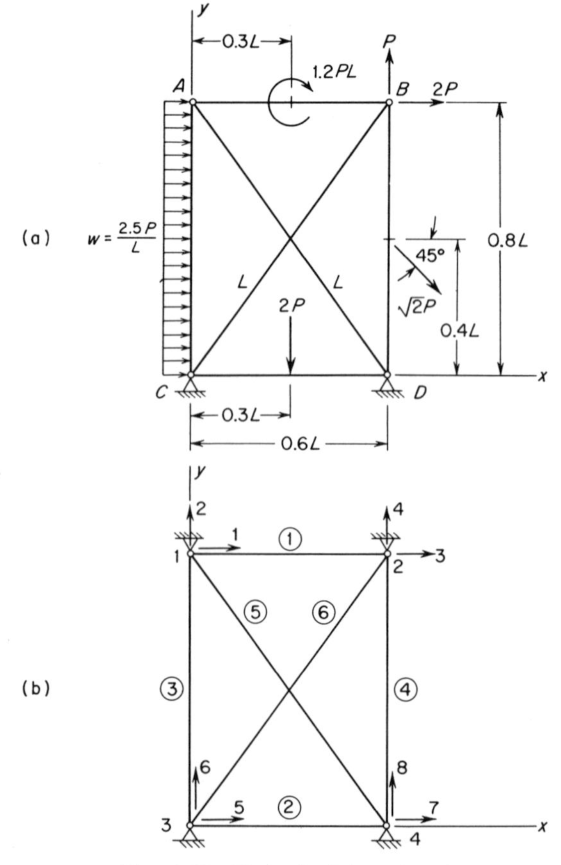

FIG. 4-23. Example (plane truss).

sented by numbers adjacent to the arrows which denote the positive directions of the possible displacements. Of course, the displacement numbering system derives from the joint numbering system according to Eqs. (4-27). Note that the system shown in Fig. 4-23b for numbering the joints (and hence the displacements) obviates the necessity of rearranging the matrices in the analysis. However, this will not be the case in general.

With the origin of coordinates selected at joint 3, as shown in Fig. 4-23, the x and y coordinates of that point are both zero. Coordinates for all of the joints in the structure are given in Table 4-18 in terms of the length L of one of the diagonal members. Table 4-18 also gives the restraint list for the structure. As in the continuous beam example, the integer 1 in the restraint list indicates that a restraint exists, and the presence of a zero indicates that a restraint does not exist.

TABLE 4-18

Joint Information for Truss of Fig. 4-23

Joint	Coordinates		Restraint List	
	x	y	x	y
1	0	0.8L	0	0
2	0.6L	0.8L	0	0
3	0	0	1	1
4	0.6L	0	1	1

Table 4-19 contains the member information for the truss of Fig. 4-23. The

TABLE 4-19

Member Information for Truss of Fig. 4-23

Member	Joint j	Joint k	Area	Length	Direction Cosines	
					C_X	C_Y
1	1	2	$0.6A_X$	0.6L	1.0	0
2	3	4	$0.6A_X$	0.6L	1.0	0
3	3	1	$0.8A_X$	0.8L	0	1.0
4	4	2	$0.8A_X$	0.8L	0	1.0
5	1	4	A_X	L	0.6	−0.8
6	3	2	A_X	L	0.6	0.8

member numbers, joint numbers, and cross-sectional areas are essential data, but the lengths of members and their direction cosines can be computed from the coordinates of the joints at the ends of the members (see Eqs. 4-24 and 4-25). Note that the arbitrary choice of which end of a member is to be denoted j or k determines the signs of the direction cosines.

TABLE 4-20

Member Stiffness Matrices for Structure Axes

$$[S_{MD}]_1 = \frac{EA_X}{L}\begin{array}{c}\left[\begin{array}{cccc} 1 & 0 & -1 & 0 \\ 0 & 0 & 0 & 0 \\ -1 & 0 & 1 & 0 \\ 0 & 0 & 0 & 0 \end{array}\right] \\ \begin{array}{cccc} 1 & 2 & 3 & 4 \end{array}\end{array}\begin{array}{c} 1 \\ 2 \\ 3 \\ 4 \\ \end{array} \qquad [S_{MD}]_2 = \frac{EA_X}{L}\begin{array}{c}\left[\begin{array}{cccc} 1 & 0 & -1 & 0 \\ 0 & 0 & 0 & 0 \\ -1 & 0 & 1 & 0 \\ 0 & 0 & 0 & 0 \end{array}\right] \\ \begin{array}{cccc} 5 & 6 & 7 & 8 \end{array}\end{array}\begin{array}{c} 5 \\ 6 \\ 7 \\ 8 \\ \end{array}$$

$$[S_{MD}]_3 = \frac{EA_X}{L}\begin{array}{c}\left[\begin{array}{cccc} 0 & 0 & 0 & 0 \\ 0 & 1 & 0 & -1 \\ 0 & 0 & 0 & 0 \\ 0 & -1 & 0 & 1 \end{array}\right] \\ \begin{array}{cccc} 5 & 6 & 1 & 2 \end{array}\end{array}\begin{array}{c} 5 \\ 6 \\ 1 \\ 2 \\ \end{array} \qquad [S_{MD}]_4 = \frac{EA_X}{L}\begin{array}{c}\left[\begin{array}{cccc} 0 & 0 & 0 & 0 \\ 0 & 1 & 0 & -1 \\ 0 & 0 & 0 & 0 \\ 0 & -1 & 0 & 1 \end{array}\right] \\ \begin{array}{cccc} 7 & 8 & 3 & 4 \end{array}\end{array}\begin{array}{c} 7 \\ 8 \\ 3 \\ 4 \\ \end{array}$$

$$[S_{MD}]_5 = \frac{EA_X}{L}\begin{array}{c}\left[\begin{array}{cccc} 0.36 & -0.48 & -0.36 & 0.48 \\ -0.48 & 0.64 & 0.48 & -0.64 \\ -0.36 & 0.48 & 0.36 & -0.48 \\ 0.48 & -0.64 & -0.48 & 0.64 \end{array}\right] \\ \begin{array}{cccc} 1 & 2 & 7 & 8 \end{array}\end{array}\begin{array}{c} 1 \\ 2 \\ 7 \\ 8 \\ \end{array}$$

$$[S_{MD}]_6 = \frac{EA_X}{L}\begin{array}{c}\left[\begin{array}{cccc} 0.36 & 0.48 & -0.36 & -0.48 \\ 0.48 & 0.64 & -0.48 & -0.64 \\ -0.36 & -0.48 & 0.36 & 0.48 \\ -0.48 & -0.64 & 0.48 & 0.64 \end{array}\right] \\ \begin{array}{cccc} 5 & 6 & 3 & 4 \end{array}\end{array}\begin{array}{c} 5 \\ 6 \\ 3 \\ 4 \\ \end{array}$$

As a preliminary step for generating the over-all joint stiffness matrix $\mathbf{S}_J$, the individual member stiffness matrices $\mathbf{S}_{MD}$ are summarized in Table 4-20. These matrices are formed from Table 4-15 given in Art. 4.10. The elements of each of these matrices may be transferred to the appropriate locations in $\mathbf{S}_J$ by calculating the indexes $j1$, $j2$, $k1$, and $k2$ given by Eqs. (4-27) and then making the transfers according to Eqs. (4-28) through (4-31). As an aid in this process, the numerical values of the indexes $j1$, $j2$, $k1$, and $k2$ are given in Table 4-20 down the right-hand side and across the bottom of each matrix $\mathbf{S}_{MD}$. As an example of the transferring process, consider the element $(S_{MD13})_2$, which is the element in the first row and third column of the matrix $\mathbf{S}_{MD}$ for member 2. This element is encircled in Table 4-20, and its value is $-EA_X/L$. The row and column indexes of $\mathbf{S}_J$ for the element are also encircled in Table 4-20. They indicate that the element is to be transferred to the position in the fifth row and the seventh column of the matrix $\mathbf{S}_J$. This element is shown encircled in Table 4-21, which

TABLE 4-21

Joint Stiffness Matrix for Truss of Fig. 4-23

$$
\mathbf{S}_J = \frac{EA_X}{L}\left[\begin{array}{cccc|cccc}
1.36 & -0.48 & -1.00 & 0 & 0 & 0 & -0.36 & 0.48 \\
-0.48 & 1.64 & 0 & 0 & 0 & -1.00 & 0.48 & -0.64 \\
-1.00 & 0 & 1.36 & 0.48 & -0.36 & -0.48 & 0 & 0 \\
0 & 0 & 0.48 & 1.64 & -0.48 & -0.64 & 0 & -1.00 \\
\hline
0 & 0 & -0.36 & -0.48 & 1.36 & 0.48 & \textcircled{-1.00} & 0 \\
0 & -1.00 & -0.48 & -0.64 & 0.48 & 1.64 & 0 & 0 \\
-0.36 & 0.48 & 0 & 0 & -1.00 & 0 & 1.36 & -0.48 \\
0.48 & -0.64 & 0 & -1.00 & 0 & 0 & -0.48 & 1.64
\end{array}\right]
$$

shows the appearance of the matrix $\mathbf{S}_J$ after the transferring process is completed. The matrix is partitioned in accordance with Eq. (4-1), and the 4×4 stiffness matrix $\mathbf{S}$ appears automatically in the upper left-hand portion with no necessity for rearrangement in this example. The inverse of the matrix $\mathbf{S}$ is shown in Table 4-22 for use in subsequent calculations.

TABLE 4-22

Inverse of Stiffness Matrix

$$
\mathbf{S}^{-1} = \frac{L}{EA_X}\begin{bmatrix}
2.503 & 0.733 & 2.053 & -0.601 \\
0.733 & 0.824 & 0.601 & -0.176 \\
2.053 & 0.601 & 2.503 & -0.733 \\
-0.601 & -0.176 & -0.733 & 0.824
\end{bmatrix}
$$

The loads applied to the structure appear in Fig. 4-23a. Those applied at joints are shown in Table 4-23, and the actions at the ends of the members in

TABLE 4-23

Actions Applied at Joints

Joint	Force in x Direction	Force in y Direction
1	0	0
2	2P	P
3	0	0
4	0	0

the restrained structure caused by loads are summarized in Table 4-24. The joint loads are placed in the vector **A** (see Eq. 4-32):

$$\mathbf{A} = P\{0, 0, 2, 1, 0, 0, 0, 0\}$$

and the actions in Table 4-24 form the matrix $\mathbf{A}_{\mathrm{ML}}$. Next, the elements of $\mathbf{A}_{\mathrm{ML}}$

TABLE 4-24

Actions at Ends of Restrained Members Due to Loads

Member	$(A_{ML})_{i,1}$	$(A_{ML})_{i,2}$	$(A_{ML})_{i,3}$	$(A_{ML})_{i,4}$
1	0	−2P	0	2P
2	0	P	0	P
3	0	P	0	P
4	0.5P	0.5P	0.5P	0.5P
5	0	0	0	0
6	0	0	0	0

are transferred to the vector of equivalent joint loads $\mathbf{A}_{\mathrm{E}}$ as indicated by Eqs. (4-33). The resulting vector is as follows:

$$\mathbf{A}_{\mathrm{E}} = P\{1.0, 2.0, 0.5, -2.5, 1.0, -1.0, 0.5, -1.5\}$$

Then the vectors **A** and $\mathbf{A}_{\mathrm{E}}$ are added together (see Eq. 4-5) to form the combined load vector $\mathbf{A}_{\mathrm{C}}$:

$$\mathbf{A}_{\mathrm{C}} = P\{1.0, 2.0, 2.5, -1.5, 1.0, -1.0, 0.5, -1.5\}$$

As mentioned earlier, there is no need to rearrange this vector. The first four elements constitute the vector $\mathbf{A}_{\mathrm{D}}$:

$$\mathbf{A}_{\mathrm{D}} = P\{1.0, 2.0, 2.5, -1.5\}$$

and the last four elements are the negatives of the elements of $\mathbf{A}_{RL}$:

$$\mathbf{A}_{RL} = P\{-1.0, 1.0, -0.5, 1.5\}$$

The solution may be completed by first calculating the joint displacements D, using Eq. (4-8), with the following result:

$$\mathbf{D} = \mathbf{S}^{-1}\mathbf{A}_D = \frac{PL}{EA_X}\{10.001, 4.147, 10.611, -4.020\}$$

The vector $\mathbf{D}_J$ for this structure contains the vector $\mathbf{D}$ in the first part and zeros in the latter part:

$$\mathbf{D}_J = \frac{PL}{EA_X}\{10.001, 4.147, 10.611, -4.020, 0, 0, 0, 0\}$$

In the next step the reactions are computed using Eq. (4-4) with the matrix $\mathbf{S}_{RD}$ obtained from the lower left-hand portion of Table 4-21:

$$\mathbf{A}_R = \mathbf{A}_{RL} + \mathbf{S}_{RD}\mathbf{D} = P\{-2.89, -5.67, -2.11, 7.67\}$$

As the last step in the analysis, the member end-actions A_M are obtained by applying Eq. (4-9) or Eqs. (4-34) to each member of the structure. The results

TABLE 4-25

Final Member End-Actions

Member	$(A_M)_{i,1}$	$(A_M)_{i,2}$	$(A_M)_{i,3}$	$(A_M)_{i,4}$
1	−0.610P	−2.000P	0.610P	2.000P
2	0.0	1.000P	0.0	1.000P
3	−4.147P	1.000P	4.147P	1.000P
4	4.520P	0.500P	−3.520P	0.500P
5	2.683P	0.0	−2.683P	0.0
6	−3.150P	0.0	3.150P	0.0

of these calculations are summarized in Table 4-25. This step completes the analysis of the plane truss structure for the given loads.

4.13 Rotation of Axes in Two Dimensions. As mentioned previously in Art. 4.10, the direct method of formulating member stiffnesses is satisfactory for continuous beams and trusses. On the other hand, a method employing rotation of axes is well suited to more complex structures. In this article rotation of axes for two-dimensional vectors is formulated on a geometric basis, and in the next article the method is applied in the analysis of plane trusses. Then in Art. 4.15 the subject of rotation of axes in three dimensions is discussed, and in subsequent articles other types of structures are treated using this technique.

In order to begin the discussion, consider an action A acting in the x-y plane (see Fig. 4-24). Two sets of orthogonal axes with origin at 0 are shown in the figure. The x_S, y_S axes will later be taken parallel to a set of structure reference axes, and the x_M, y_M set will be taken as a pair of member-oriented axes. The x_M, y_M axes are rotated from the x_S, y_S axes by the angle γ. Let the direction cosines of the x_M axis with respect to the axes x_S and y_S be λ_{11} and λ_{12}, respectively. It is evident from Fig.

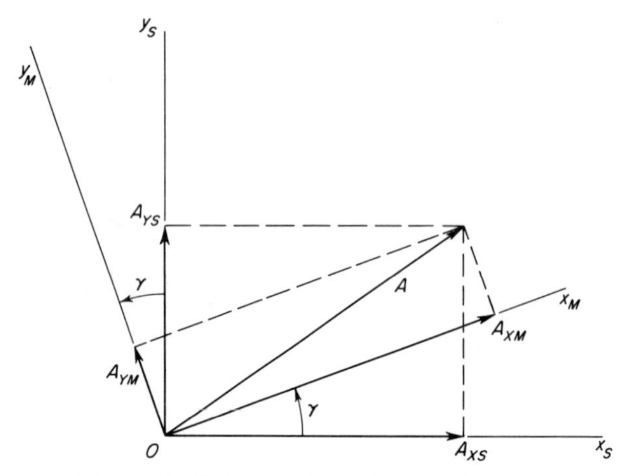

FIG. 4-24. Rotation of axes in two dimensions.

4-24 that these direction cosines may be expressed in terms of the angle γ as follows:

$$\lambda_{11} = \cos\gamma \qquad \lambda_{12} = \cos(90° - \gamma) = \sin\gamma \tag{a}$$

Also, let the direction cosines of the y_M axis with respect to the axes x_S and y_S be λ_{21} and λ_{22}, respectively. These direction cosines may also be expressed in terms of the angle γ:

$$\lambda_{21} = \cos(90° + \gamma) = -\sin\gamma \qquad \lambda_{22} = \cos\gamma \tag{b}$$

For any one of the direction cosines above, the first subscript refers to the x_M, y_M axes, and the second subscript refers to the x_S, y_S axes. Moreover, the number 1 denotes the x direction (either x_M or x_S), and the number 2 denotes the y direction (either y_M or y_S). For example, λ_{12} is the direction cosine of the x_M axis with respect to the y_S axis.

The action A may be resolved into two orthogonal components A_{XS} and A_{YS} in the x_S and y_S directions, respectively, as shown in Fig. 4-24. Alternatively, A may also be resolved into two orthogonal components A_{XM} and A_{YM} in the directions x_M and y_M, respectively, as shown also in the figure. The latter set of components may be expressed in terms of the former set by inspection of the geometry in Fig. 4-24. It may be observed that A_{XM} is equal to the sum of the projections of A_{XS} and

A_{YS} on the x_M axis. In addition, A_{YM} is equal to the sum of the projections of A_{XS} and A_{YS} on the y_M axis. Therefore, the expressions for A_{XM} and A_{YM} are:

$$\begin{aligned} A_{XM} &= \lambda_{11}A_{XS} + \lambda_{12}A_{YS} \\ A_{YM} &= \lambda_{21}A_{XS} + \lambda_{22}A_{YS} \end{aligned} \tag{c}$$

In matrix form, these equations become:

$$\begin{bmatrix} A_{XM} \\ A_{YM} \end{bmatrix} = \begin{bmatrix} \lambda_{11} & \lambda_{12} \\ \lambda_{21} & \lambda_{22} \end{bmatrix} \begin{bmatrix} A_{XS} \\ A_{YS} \end{bmatrix} \tag{4-35}$$

Substitution of expressions (a) and (b) into Eq. (4-35) yields the following alternative form:

$$\begin{bmatrix} A_{XM} \\ A_{YM} \end{bmatrix} = \begin{bmatrix} \cos\gamma & \sin\gamma \\ -\sin\gamma & \cos\gamma \end{bmatrix} \begin{bmatrix} A_{XS} \\ A_{YS} \end{bmatrix} \tag{4-36}$$

Equations (4-35) and (4-36) may also be stated in concise form by the expression

$$\mathbf{A}_{\mathrm{M}} = \mathbf{R}\mathbf{A}_{\mathrm{S}} \tag{4-37}$$

In this equation, $\mathbf{A}_{\mathrm{M}}$ is a vector consisting of the components of the action A parallel to the x_M, y_M axes, $\mathbf{A}_{\mathrm{S}}$ is a vector containing the components of the action A parallel to the x_S, y_S axes, and $\mathbf{R}$ is a matrix of direction cosines which will be referred to as the *rotation matrix*. As shown by Eqs. (4-35) and (4-36), the rotation matrix in a two-dimensional problem is as follows:

$$\mathbf{R} = \begin{bmatrix} \lambda_{11} & \lambda_{12} \\ \lambda_{21} & \lambda_{22} \end{bmatrix} = \begin{bmatrix} \cos\gamma & \sin\gamma \\ -\sin\gamma & \cos\gamma \end{bmatrix} \tag{4-38}$$

It is also possible to express the x_S, y_S set of components of the action A in terms of the x_M, y_M set of components. This transformation may be accomplished by observing that A_{XS} is equal to the sum of the projections of A_{XM} and A_{YM} on the x_S axis, and that A_{YS} is equal to the sum of the projections of A_{XM} and A_{YM} on the y_S axis. Thus,

$$\begin{aligned} A_{XS} &= \lambda_{11}A_{XM} + \lambda_{21}A_{YM} \\ A_{YS} &= \lambda_{12}A_{XM} + \lambda_{22}A_{YM} \end{aligned} \tag{d}$$

When expressed in matrix form, these equations become:

$$\begin{bmatrix} A_{XS} \\ A_{YS} \end{bmatrix} = \begin{bmatrix} \lambda_{11} & \lambda_{21} \\ \lambda_{12} & \lambda_{22} \end{bmatrix} \begin{bmatrix} A_{XM} \\ A_{YM} \end{bmatrix} \tag{4-39}$$

or

$$\begin{bmatrix} A_{XS} \\ A_{YS} \end{bmatrix} = \begin{bmatrix} \cos\gamma & -\sin\gamma \\ \sin\gamma & \cos\gamma \end{bmatrix} \begin{bmatrix} A_{XM} \\ A_{YM} \end{bmatrix} \tag{4-40}$$

Equations (4-39) and (4-40) may be represented succinctly by the matrix equation

$$\mathbf{A}_{\mathrm{S}} = \mathbf{R}'\mathbf{A}_{\mathrm{M}} \tag{4-41}$$

in which $\mathbf{R}'$ is the transpose of the rotation matrix $\mathbf{R}$.

Finally, from Eqs. (4-37) and (4-41) it is apparent that the transpose of $\mathbf{R}$ is equal to its inverse:

$$\mathbf{R}' = \mathbf{R}^{-1} \tag{4-42}$$

Therefore, the rotation matrix $\mathbf{R}$ is an orthogonal matrix.*

Since small displacements as well as actions may be treated as vectors, the relationships formulated above for the action A may be applied equally well to a displacement D. Thus, equations similar to Eqs. (4-37) and (4-41) can be written for displacements, as follows:

$$\mathbf{D}_{\mathrm{M}} = \mathbf{R}\mathbf{D}_{\mathrm{S}} \tag{4-43}$$

$$\mathbf{D}_{\mathrm{S}} = \mathbf{R}'\mathbf{D}_{\mathrm{M}} \tag{4-44}$$

In these equations the vector $\mathbf{D}_{\mathrm{M}}$ consists of the components of the displacement D parallel to the member axes, and the vector $\mathbf{D}_{\mathrm{S}}$ contains the components of the displacement D parallel to the structure axes.

The concepts of rotation of axes discussed above are applied to plane truss structures in the next article.

4.14 Application to Plane Truss Members. A typical plane truss member i, framing into joints j and k at the ends, is shown in Fig. 4-25. For convenience, the member is oriented in such a manner that its

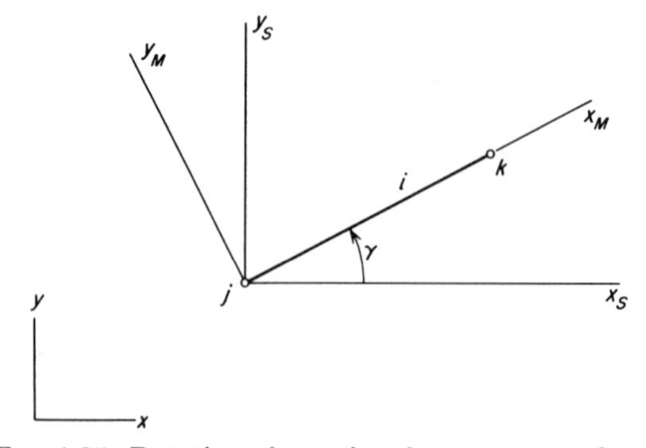

FIG. 4-25. Rotation of axes for plane truss member.

direction cosines are positive. The member axes x_M, y_M are rotated through the angle γ from the axes x_S, y_S, which are parallel to the reference axes x, y for the structure as a whole. For the purpose of rotational transformations it is immaterial whether one refers to the x_S, y_S set of axes or the parallel x, y set. For efficiency in notation, the subscript S will usually be omitted when reference to the structure axes is intended.

* For further discussion of orthogonal matrices, see any text on matrix algebra; for example, *Matrix Algebra for Engineers*, by the authors, D. Van Nostrand Co., Inc., Princeton, N. J., 1965.

As a first step in the use of rotation of axes in a plane truss analysis, the rotation matrix $\mathbf{R}$ will be expressed in terms of the direction cosines of the member i. Inspection of Fig. 4-25 shows that the direction cosines of the member in terms of the angle γ are:

$$C_X = \cos \gamma \qquad C_Y = \sin \gamma$$

Then the rotation matrix (see Eq. 4-38) becomes:

$$\mathbf{R} = \begin{bmatrix} \cos \gamma & \sin \gamma \\ -\sin \gamma & \cos \gamma \end{bmatrix} = \begin{bmatrix} C_X & C_Y \\ -C_Y & C_X \end{bmatrix} \tag{4-45}$$

Next, the action-displacement relationships in the x_M and y_M directions at the ends of member i may be expressed by the following:

$$\{A_M\}_i = [S_M]_i\{D_M\}_i \tag{4-46}$$

Equation (4-46) is the same as Eq. (4-9) when the vector $\{A_{ML}\}_i$, which appears in the latter equation, is null. Thus, Eq. (4-46) gives the actions at the ends of the member due to the displacements of the ends. The member stiffness matrix $[S_M]_i$ (see Eq. 4-46) was developed in Art. 4.10 and presented in Table 4-14. The objective now is to transform this matrix for member axes into the matrix $[S_{MD}]_i$ for structures axes (see Table 4-15).

As a preliminary step in transforming to structure axes, Eq. (4-46) may be written in partitioned form (omitting the subscript i), as follows:

$$\begin{bmatrix} A_{M1} \\ A_{M2} \\ \hline A_{M3} \\ A_{M4} \end{bmatrix} = \left[\begin{array}{cc|cc} S_{M11} & S_{M12} & S_{M13} & S_{M14} \\ S_{M21} & S_{M22} & S_{M23} & S_{M24} \\ \hline S_{M31} & S_{M32} & S_{M33} & S_{M34} \\ S_{M41} & S_{M42} & S_{M43} & S_{M44} \end{array}\right] \begin{bmatrix} D_{M1} \\ D_{M2} \\ \hline D_{M3} \\ D_{M4} \end{bmatrix} \tag{a}$$

The subscripts 1, 2, 3, and 4 used in this equation refer to the member-oriented directions shown in Fig. 4-14c. Equation (a) may also be written in the following manner:

$$\begin{bmatrix} \mathbf{A}_{Mj} \\ \mathbf{A}_{Mk} \end{bmatrix} = \begin{bmatrix} \mathbf{S}_{Mjj} & \mathbf{S}_{Mjk} \\ \mathbf{S}_{Mkj} & \mathbf{S}_{Mkk} \end{bmatrix} \begin{bmatrix} \mathbf{D}_{Mj} \\ \mathbf{D}_{Mk} \end{bmatrix} \tag{b}$$

In this equation the subscripts j and k attached to the submatrices refer to the j and k ends of the member. The terms $\mathbf{A}_{Mj}$, $\mathbf{A}_{Mk}$, $\mathbf{D}_{Mj}$, and $\mathbf{D}_{Mk}$ in Eq. (b) represent two-dimensional vectors (either actions or displacements) at the ends of the member in the directions of member axes (see Fig. 4-14c). Therefore, they can be expressed with respect to the structure axes (see Fig. 4-14b) by using the appropriate rotation formulas from the preceding article. These formulas are Eqs. (4-37) and (4-43). When these relations are substituted into Eq. (b), it becomes

$$\begin{bmatrix} \mathbf{RA}_j \\ \mathbf{RA}_k \end{bmatrix} = \begin{bmatrix} \mathbf{S}_{Mjj} & \mathbf{S}_{Mjk} \\ \mathbf{S}_{Mkj} & \mathbf{S}_{Mkk} \end{bmatrix} \begin{bmatrix} \mathbf{RD}_j \\ \mathbf{RD}_k \end{bmatrix} \tag{c}$$

The submatrices $\mathbf{A}_j$, $\mathbf{A}_k$, $\mathbf{D}_j$, and $\mathbf{D}_k$ represent the two-dimensional action and displacement vectors at the ends of the member with respect to the structure axes.

An equivalent form of Eq. (c) is the following:

$$\begin{bmatrix} \mathbf{R} & \mathbf{O} \\ \mathbf{O} & \mathbf{R} \end{bmatrix} \begin{bmatrix} \mathbf{A}_j \\ \mathbf{A}_k \end{bmatrix} = \begin{bmatrix} \mathbf{S}_{Mjj} & \mathbf{S}_{Mjk} \\ \mathbf{S}_{Mkj} & \mathbf{S}_{Mkk} \end{bmatrix} \begin{bmatrix} \mathbf{R} & \mathbf{O} \\ \mathbf{O} & \mathbf{R} \end{bmatrix} \begin{bmatrix} \mathbf{D}_j \\ \mathbf{D}_k \end{bmatrix} \tag{d}$$

To simplify the writing of this equation, let $\mathbf{R}_T$ be the *rotation transformation matrix* for actions and displacements at both ends of the member:

$$\mathbf{R}_T = \begin{bmatrix} \mathbf{R} & \mathbf{O} \\ \mathbf{O} & \mathbf{R} \end{bmatrix} \tag{4-47}$$

Equation (d) then may be rewritten as follows:

$$\mathbf{R}_T\mathbf{A} = \mathbf{S}_M\mathbf{R}_T\mathbf{D} \tag{e}$$

The vectors $\mathbf{A}$ and $\mathbf{D}$ in Eq. (e) consist of the actions and displacements at the ends of the member in the directions of structure axes (see Fig. 4-14b).

Premultiplying both sides of Eq. (e) by the inverse of $\mathbf{R}_T$ gives:

$$\mathbf{A} = \mathbf{R}_T^{-1}\mathbf{S}_M\mathbf{R}_T\mathbf{D} \tag{f}$$

Since the submatrix $\mathbf{R}$ is orthogonal, the matrix $\mathbf{R}_T$ is also orthogonal. This fact can be seen by multiplying $\mathbf{R}_T$ by its transpose, as follows:

$$\mathbf{R}_T'\mathbf{R}_T = \begin{bmatrix} \mathbf{R}' & \mathbf{O} \\ \mathbf{O} & \mathbf{R}' \end{bmatrix} \begin{bmatrix} \mathbf{R} & \mathbf{O} \\ \mathbf{O} & \mathbf{R} \end{bmatrix} = \begin{bmatrix} \mathbf{R}'\mathbf{R} & \mathbf{O} \\ \mathbf{O} & \mathbf{R}'\mathbf{R} \end{bmatrix}$$

$$= \begin{bmatrix} \mathbf{R}^{-1}\mathbf{R} & \mathbf{O} \\ \mathbf{O} & \mathbf{R}^{-1}\mathbf{R} \end{bmatrix} = \begin{bmatrix} \mathbf{I} & \mathbf{O} \\ \mathbf{O} & \mathbf{I} \end{bmatrix} = \mathbf{I}$$

Hence, the transpose of $\mathbf{R}_T$ is also the inverse of $\mathbf{R}_T$:

$$\mathbf{R}_T^{-1} = \mathbf{R}_T' \tag{4-48}$$

and substitution of Eq. (4-48) into Eq. (f) yields:

$$\mathbf{A} = \mathbf{R}_T'\mathbf{S}_M\mathbf{R}_T\mathbf{D} \tag{4-49}$$

Since the action-displacement equation that relates the actions $\mathbf{A}$ and displacements $\mathbf{D}$ is

$$\mathbf{A} = \mathbf{S}_{MD}\mathbf{D} \tag{4-50}$$

in which $\mathbf{S}_{MD}$ is the member stiffness matrix for structure axes, it is readily seen by comparing Eqs. (4-49) and (4-50) that

$$\mathbf{S}_{MD} = \mathbf{R}_T'\mathbf{S}_M\mathbf{R}_T \tag{4-51}$$

Evaluation of the matrix $\mathbf{S}_{MD}$ from this equation is performed as follows:

$\mathbf{S}_{MD} =$

$$
\begin{bmatrix} C_X & -C_Y & 0 & 0 \\ C_Y & C_X & 0 & 0 \\ 0 & 0 & C_X & -C_Y \\ 0 & 0 & C_Y & C_X \end{bmatrix}
\begin{bmatrix} \frac{EA_X}{L} & 0 & -\frac{EA_X}{L} & 0 \\ 0 & 0 & 0 & 0 \\ -\frac{EA_X}{L} & 0 & \frac{EA_X}{L} & 0 \\ 0 & 0 & 0 & 0 \end{bmatrix}
\begin{bmatrix} C_X & C_Y & 0 & 0 \\ -C_Y & C_X & 0 & 0 \\ 0 & 0 & C_X & C_Y \\ 0 & 0 & -C_Y & C_X \end{bmatrix}
$$

When this matrix multiplication is executed, the result is the matrix $\mathbf{S}_{\mathrm{MD}}$ (see Table 4-15) which was previously obtained by the method of direct formulation.

In addition to the transformation of the member stiffness matrix from member axes to structure axes, the rotation of axes concept can also be used for other purposes in the stiffness method of analysis. One important application arises in the construction of the equivalent load vector $\mathbf{A}_{\mathrm{E}}$ from elements of the matrix $\mathbf{A}_{\mathrm{ML}}$. Contributions to the former array (the elements of which are aligned with structure axes) from the latter array (which has elements aligned with member axes) may be obtained by the following transformation:

$$
[R_T']_i\{A_{ML}\}_i = \begin{bmatrix} C_{Xi} & -C_{Yi} & 0 & 0 \\ C_{Yi} & C_{Xi} & 0 & 0 \\ 0 & 0 & C_{Xi} & -C_{Yi} \\ 0 & 0 & C_{Yi} & C_{Xi} \end{bmatrix} \begin{bmatrix} (A_{ML})_{i,1} \\ (A_{ML})_{i,2} \\ (A_{ML})_{i,3} \\ (A_{ML})_{i,4} \end{bmatrix}
= \begin{bmatrix} (A_{ML})_{i,1}C_{Xi} - (A_{ML})_{i,2}C_{Yi} \\ (A_{ML})_{i,1}C_{Yi} + (A_{ML})_{i,2}C_{Xi} \\ (A_{ML})_{i,3}C_{Xi} - (A_{ML})_{i,4}C_{Yi} \\ (A_{ML})_{i,3}C_{Yi} + (A_{ML})_{i,4}C_{Xi} \end{bmatrix}
$$

in which the vector $\{A_{ML}\}_i$ consists of the elements in the i-th row of the matrix $\mathbf{A}_{\mathrm{ML}}$. These expressions, with the signs reversed, represent the incremental portions of $\mathbf{A}_{\mathrm{E}}$ given previously in Eqs. (4-33).

Another significant application of the rotation of axes concept appears in the computation of final member end-actions. This computation consists of the superposition of the initial actions in member i and the effects of joint displacements. This superposition procedure is expressed by Eq. (4-9), which is repeated below:

$$
\{A_M\}_i = \{A_{ML}\}_i + [S_M]_i\{D_M\}_i \tag{4-9 repeated}
$$

The vector $\{D_M\}_i$ in this equation must be ascertained from the vector of joint displacements $\mathbf{D}_{\mathrm{J}}$. The latter displacements are in the directions of structure axes, but the former displacements are in the directions of member axes. Therefore, the vector $\{D_M\}_i$ may be obtained by the following transformation:

$$
\{D_M\}_i = [R_T]_i\{D_J\}_i \tag{4-52}
$$

in which $\{D_J\}_i$ is the vector of joint displacements for the ends of member i. Substitution of Eq. (4-52) into Eq. (4-9) produces the following expression:

$$\{A_M\}_i = \{A_{ML}\}_i + [S_M]_i[R_T]_i\{D_J\}_i \tag{4-53}$$

Equation (4-53) may also be written in expanded form:

$$\begin{bmatrix} (A_M)_{i,1} \\ (A_M)_{i,2} \\ (A_M)_{i,3} \\ (A_M)_{i,4} \end{bmatrix} = \begin{bmatrix} (A_{ML})_{i,1} \\ (A_{ML})_{i,2} \\ (A_{ML})_{i,3} \\ (A_{ML})_{i,4} \end{bmatrix}$$

$$+ \frac{EA_{Xi}}{L_i}\begin{bmatrix} 1 & 0 & -1 & 0 \\ 0 & 0 & 0 & 0 \\ -1 & 0 & 1 & 0 \\ 0 & 0 & 0 & 0 \end{bmatrix}\begin{bmatrix} C_{Xi} & C_{Yi} & 0 & 0 \\ -C_{Yi} & C_{Xi} & 0 & 0 \\ 0 & 0 & C_{Xi} & C_{Yi} \\ 0 & 0 & -C_{Yi} & C_{Xi} \end{bmatrix}\begin{bmatrix} (D_J)_{j1} \\ (D_J)_{j2} \\ (D_J)_{k1} \\ (D_J)_{k2} \end{bmatrix} \tag{4-54}$$

In Eq. (4-54) the subscripts $j1$, $j2$, $k1$, and $k2$ carry the definitions given previously by Eqs. (4-27). When the matrix multiplications indicated in Eq. (4-54) are performed, the resulting four equations are the same as Eqs. (4-34) obtained in Art. 4.11 by the method of direct formulation.

In summary, the concept of rotation of axes has several useful applications in the stiffness method of analysis. The member stiffness matrix for member axes $\mathbf{S}_\mathrm{M}$ may be transformed into the member stiffness matrix for structure axes $\mathbf{S}_\mathrm{MD}$ by means of Eq. (4-51). In addition, the contributions to the equivalent load vector $\mathbf{A}_\mathrm{E}$ from a given member may be evaluated conveniently by rotation of axes as discussed above. Finally, the member end-actions can be obtained by the rotation of axes formulation given by Eq. (4-53).

It will be seen later that the matrix equations given above for rotation of axes in plane trusses can be generalized and applied to more complicated types of structures.

4.15 Rotation of Axes in Three Dimensions. Consider the action A shown in three dimensions in Fig. 4-26. The two sets of orthogonal axes x_S, y_S, z_S and x_M, y_M, z_M are analogous to the two sets of axes in the two-dimensional case in Fig. 4-24. Let the direction cosines of the x_M axis with respect to the x_S, y_S, z_S axes be λ_{11}, λ_{12}, and λ_{13}. These direction cosines are the cosines of the angles between the axis x_M and the three axes x_S, y_S, and z_S, respectively. Also, let the direction cosines for the y_M axis be λ_{21}, λ_{22}, and λ_{23}, and those for the z_M axis be λ_{31}, λ_{32}, and λ_{33}.

The action A may be represented by a set of three orthogonal components A_{XS}, A_{YS}, and A_{ZS} in the x_S, y_S, and z_S directions, respectively, as shown in Fig. 4-26. Alternatively, this action may be represented by a second set of components A_{XM}, A_{YM}, and A_{ZM} in the x_M, y_M, and z_M

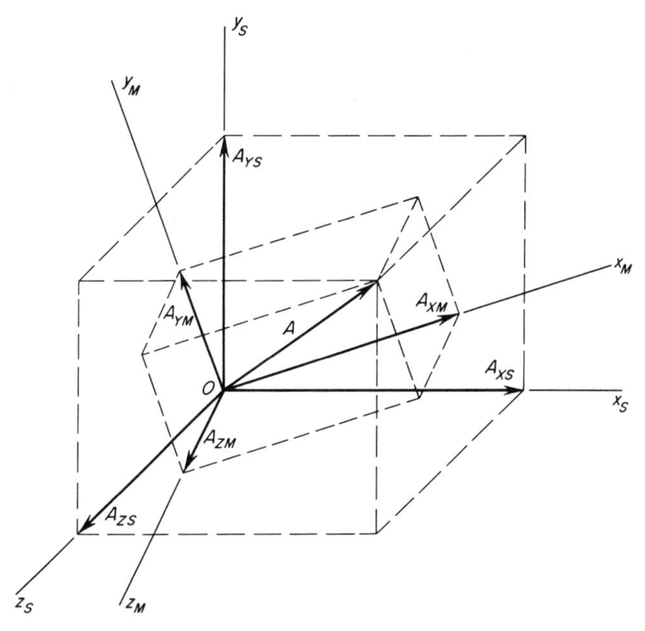

FIG. 4-26. Rotation of axes in three dimensions.

directions; these are also indicated in the figure. The latter components may be related to the former as in the two-dimensional case. For example, A_{XM} is equal to the sum of the projections of A_{XS}, A_{YS}, and A_{ZS} on the x_M axis. The components A_{YM} and A_{ZM} can be expressed in a similar manner, and the following relationship results:

$$\mathbf{A}_{\mathrm{M}} = \begin{bmatrix} A_{XM} \\ A_{YM} \\ A_{ZM} \end{bmatrix} = \begin{bmatrix} \lambda_{11} & \lambda_{12} & \lambda_{13} \\ \lambda_{21} & \lambda_{22} & \lambda_{23} \\ \lambda_{31} & \lambda_{32} & \lambda_{33} \end{bmatrix} \begin{bmatrix} A_{XS} \\ A_{YS} \\ A_{ZS} \end{bmatrix} \tag{4-55}$$

If a three-dimensional rotation matrix $\mathbf{R}$ is defined as follows,

$$\mathbf{R} = \begin{bmatrix} \lambda_{11} & \lambda_{12} & \lambda_{13} \\ \lambda_{21} & \lambda_{22} & \lambda_{23} \\ \lambda_{31} & \lambda_{32} & \lambda_{33} \end{bmatrix} \tag{4-56}$$

then Eq. (4-55) can be written in the form

$$\mathbf{A}_{\mathrm{M}} = \mathbf{R}\mathbf{A}_{\mathrm{S}} \tag{4-57}$$

which is the same form as Eq. (4-37) for the two-dimensional case.

It is also possible to express the x_S, y_S, z_S set of components of the action A in terms of the x_M, y_M, z_M set of components. For example, A_{XS} is equal to the sum of the projections of A_{XM}, A_{YM}, and A_{ZM} on the x_S axis. Expressing the components in this manner leads to the following relationship:

$$\mathbf{A}_S = \begin{bmatrix} A_{XS} \\ A_{YS} \\ A_{ZS} \end{bmatrix} = \begin{bmatrix} \lambda_{11} & \lambda_{21} & \lambda_{31} \\ \lambda_{12} & \lambda_{22} & \lambda_{32} \\ \lambda_{13} & \lambda_{23} & \lambda_{33} \end{bmatrix} \begin{bmatrix} A_{XM} \\ A_{YM} \\ A_{ZM} \end{bmatrix} \tag{4-58}$$

This equation may be written concisely as:

$$\mathbf{A}_S = \mathbf{R}'\mathbf{A}_M \tag{4-59}$$

which is the same form as Eq. (4-41) in the two-dimensional case.

It is apparent from Eqs. (4-57) and (4-59) that the transpose of the 3×3 matrix $\mathbf{R}$ is equal to its inverse. Therefore, this matrix is orthogonal, as in the two-dimensional case.

Relationships corresponding to the above equations, which are derived for actions, hold also for the transformation of displacements in three dimensions. Thus, the components $\mathbf{D}_M$ of a displacement D in the x_M, y_M, z_M directions may be expressed in terms of the components $\mathbf{D}_S$ in the x_S, y_S, z_S directions as follows:

$$\mathbf{D}_M = \mathbf{R}\mathbf{D}_S \tag{4-60}$$

Also, the vector $\mathbf{D}_S$ may be expressed in terms of the vector $\mathbf{D}_M$ by the relationship:

$$\mathbf{D}_S = \mathbf{R}'\mathbf{D}_M \tag{4-61}$$

Equations (4-60) and (4-61) are of the same form as Eqs. (4-43) and (4-44) for the two-dimensional case. Thus, the relationships for rotation of axes in three dimensions are completely analogous to those for two dimensions, except that the matrix $\mathbf{R}$ is of order 3×3 instead of 2×2, and the vectors are of order 3×1 instead of 2×1.

4.16 Plane Frame Member Stiffnesses. In preparation for the analysis of plane frames, the member stiffness matrix for a typical plane frame member is developed in this article. The matrix is first formulated with respect to member axes and then transformed to structure axes by the method of rotation of axes.

Figure 4-27a shows a typical member i in a plane frame. The joints at the ends of the member are denoted j and k as in previous structures. The orthogonal set of axes x, y, and z shown in Fig. 4-27 are reference axes for the structure. The plane frame lies in the x-y plane, which is assumed to be a principal plane of bending for all of the members. The members of the frame are assumed to be rigidly connected, and the significant displacements of the joints consist of translations in the x-y plane and rotations in the z sense.

The possible displacements of the ends of a typical member i are indicated in Fig. 4-27b for member-oriented axes x_M, y_M, and z_M. The member axes are rotated from the structure axes about the z_M axis through the angle γ. The six end-displacements, shown in their positive senses, consist of translations in the x_M and y_M directions and a rotation in the z_M (or z) sense at the ends j and k, respectively. If unit displacements of these types are induced at each end of the member one at a time, the resulting restraint actions will constitute the elements of the member stiffness matrix $\mathbf{S}_{\mathrm{M}}$ for member axes. These restraint actions

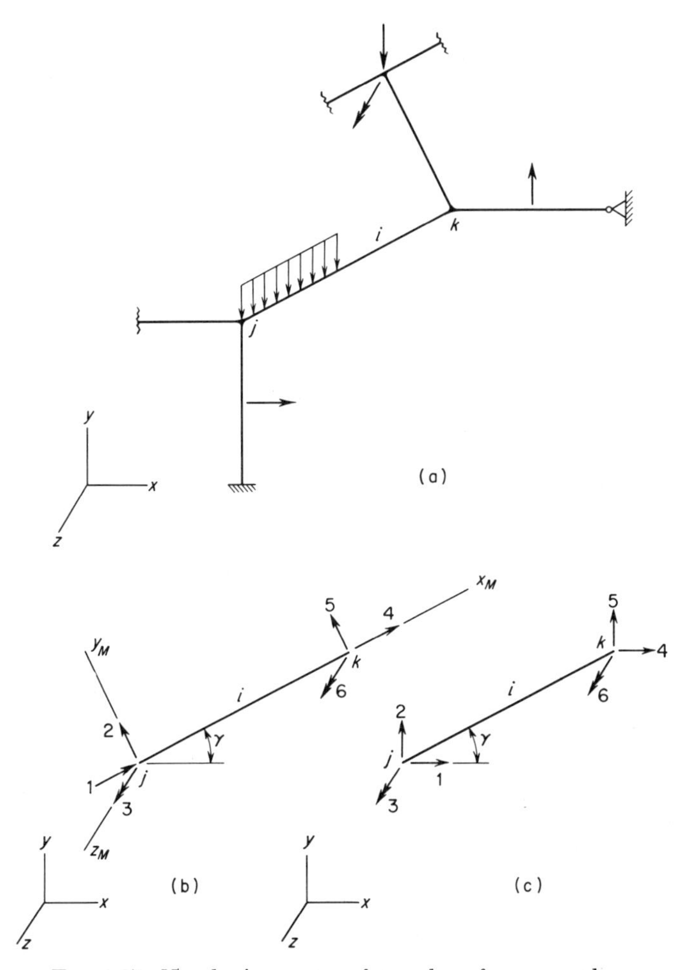

FIG. 4-27. Numbering system for a plane frame member.

may be drawn from cases (1), (2), (6), (7), (8), and (12) of Fig. 4-2 in Art. 4.3. The resulting 6 × 6 member stiffness matrix for member axes is given in Table 4-26.

Consider next the task of transforming the member stiffness matrix $\mathbf{S}_{\mathrm{M}}$ to the stiffness matrix for structure axes $\mathbf{S}_{\mathrm{MD}}$. Figure 4-27c indicates the six possible displacements at the ends of member i in the directions of the structure axes. In order to transform the member stiffness matrix from member axes to structure axes, the rotation transformation matrix

TABLE 4-26

Plane Frame Member Stiffness Matrix for Member Axes (Fig. 4-27b)

$$\mathbf{S_M} = \begin{bmatrix} \frac{EA_X}{L} & 0 & 0 & -\frac{EA_X}{L} & 0 & 0 \\ 0 & \frac{12EI_Z}{L^3} & \frac{6EI_Z}{L^2} & 0 & -\frac{12EI_Z}{L^3} & \frac{6EI_Z}{L^2} \\ 0 & \frac{6EI_Z}{L^2} & \frac{4EI_Z}{L} & 0 & -\frac{6EI_Z}{L^2} & \frac{2EI_Z}{L} \\ -\frac{EA_X}{L} & 0 & 0 & \frac{EA_X}{L} & 0 & 0 \\ 0 & -\frac{12EI_Z}{L^3} & -\frac{6EI_Z}{L^2} & 0 & \frac{12EI_Z}{L^3} & -\frac{6EI_Z}{L^2} \\ 0 & \frac{6EI_Z}{L^2} & \frac{2EI_Z}{L} & 0 & -\frac{6EI_Z}{L^2} & \frac{4EI_Z}{L} \end{bmatrix}$$

$\mathbf{R_T}$ for a plane frame member is required. As a first step, the 3×3 rotation matrix $\mathbf{R}$ will be expressed in terms of the direction cosines of the member i shown in Fig. 4-27b (or 4-27c). This may be accomplished by expressing the direction cosines of the member axes in terms of the angle γ and then substituting the direction cosines for the member, as follows:

$$\mathbf{R} = \begin{bmatrix} \lambda_{11} & \lambda_{12} & \lambda_{13} \\ \lambda_{21} & \lambda_{22} & \lambda_{23} \\ \lambda_{31} & \lambda_{32} & \lambda_{33} \end{bmatrix} = \begin{bmatrix} \cos\gamma & \sin\gamma & 0 \\ -\sin\gamma & \cos\gamma & 0 \\ 0 & 0 & 1 \end{bmatrix} = \begin{bmatrix} C_X & C_Y & 0 \\ -C_Y & C_X & 0 \\ 0 & 0 & 1 \end{bmatrix} \tag{4-62}$$

The rotation transformation matrix $\mathbf{R_T}$ for a plane frame member can be shown to take the same form as Eq. (4-47):

$$\mathbf{R_T} = \begin{bmatrix} \mathbf{R} & \mathbf{O} \\ \mathbf{O} & \mathbf{R} \end{bmatrix} \tag{4-63}$$

In Eq. (4-63) the matrix $\mathbf{R}$ is the 3×3 rotation matrix given by Eq. (4-62).

Having the rotation transformation matrix on hand, one may then calculate the member stiffness matrix for structure axes using the type of operation shown previously by Eq. (4-51):

$$\mathbf{S_{MD}} = \mathbf{R'_T S_M R_T} \tag{4-64}$$

in which the matrix $\mathbf{R_T}$ is given by Eq. (4-63). The member stiffness matrix which results from this transformation is presented in Table 4-27. This matrix will be used in the analysis of plane frames in the next article.

TABLE 4-27

Plane Frame Member Stiffness Matrix for Structure Axes (Fig. 4-27c)

$$
\mathbf{S}_{MD} = \begin{bmatrix}
\frac{EA_X}{L} C_X^2 + \frac{12EI_Z}{L^3} C_Y^2 & \left(\frac{EA_X}{L} - \frac{12EI_Z}{L^3}\right) C_X C_Y & -\frac{6EI_Z}{L^2} C_Y & -\left(\frac{EA_X}{L} C_X^2 + \frac{12EI_Z}{L^3} C_Y^2\right) & -\left(\frac{EA_X}{L} - \frac{12EI_Z}{L^3}\right) C_X C_Y & -\frac{6EI_Z}{L^2} C_Y \\
\left(\frac{EA_X}{L} - \frac{12EI_Z}{L^3}\right) C_X C_Y & \frac{EA_X}{L} C_Y^2 + \frac{12EI_Z}{L^3} C_X^2 & \frac{6EI_Z}{L^2} C_X & -\left(\frac{EA_X}{L} - \frac{12EI_Z}{L^3}\right) C_X C_Y & -\left(\frac{EA_X}{L} C_Y^2 + \frac{12EI_Z}{L^3} C_X^2\right) & \frac{6EI_Z}{L^2} C_X \\
-\frac{6EI_Z}{L^2} C_Y & \frac{6EI_Z}{L^2} C_X & \frac{4EI_Z}{L} & \frac{6EI_Z}{L^2} C_Y & -\frac{6EI_Z}{L^2} C_X & \frac{2EI_Z}{L} \\
-\left(\frac{EA_X}{L} C_X^2 + \frac{12EI_Z}{L^3} C_Y^2\right) & -\left(\frac{EA_X}{L} - \frac{12EI_Z}{L^3}\right) C_X C_Y & \frac{6EI_Z}{L^2} C_Y & \frac{EA_X}{L} C_X^2 + \frac{12EI_Z}{L^3} C_Y^2 & \left(\frac{EA_X}{L} - \frac{12EI_Z}{L^3}\right) C_X C_Y & \frac{6EI_Z}{L^2} C_Y \\
-\left(\frac{EA_X}{L} - \frac{12EI_Z}{L^3}\right) C_X C_Y & -\left(\frac{EA_X}{L} C_Y^2 + \frac{12EI_Z}{L^3} C_X^2\right) & -\frac{6EI_Z}{L^2} C_X & \left(\frac{EA_X}{L} - \frac{12EI_Z}{L^3}\right) C_X C_Y & \frac{EA_X}{L} C_Y^2 + \frac{12EI_Z}{L^3} C_X^2 & -\frac{6EI_Z}{L^2} C_X \\
-\frac{6EI_Z}{L^2} C_Y & \frac{6EI_Z}{L^2} C_X & \frac{2EI_Z}{L} & \frac{6EI_Z}{L^2} C_Y & -\frac{6EI_Z}{L^2} C_X & \frac{4EI_Z}{L}
\end{bmatrix}
$$

4.17 Analysis of Plane Frames. In this article a procedure is presented for analyzing plane frames of the type illustrated in Fig. 4-27a. Actions applied to such a frame are assumed to be forces in the plane of the structure (the x-y plane) or moment vectors normal to the plane.

As a preliminary step in the analysis, the members and joints of the structure must be numbered. The numbering techniques described in Art. 4.11 for plane trusses may also be applied to plane frames. The joints are numbered consecutively 1 through n_j, and the members are numbered consecutively 1 through m. The sequence of numbering is arbitrary, but each member and each joint must have a number.

Since both axial and flexural deformations will be taken into account in the analyses of plane frames, the possibility exists for three independent displacements at each joint. These displacements are taken to be the translations of the joint in the x and y directions and the rotation in the z sense. Thus, the possible displacements at a joint j may be designated as follows:

$$3j - 2 = \text{index for translation in the } x \text{ direction}$$

$$3j - 1 = \text{index for translation in the } y \text{ direction}$$

$$3j \quad = \text{index for rotation in the } z \text{ sense}$$

In addition, the number of degrees of freedom n in a plane frame is calculated from the number of joints n_j and the number of restraints n_r by the following expression:

$$n = 3n_j - n_r \tag{4-65}$$

A particular member i in a plane frame will have joint numbers j and k at its ends, as shown in Fig. 4-28. The possible displacements of the

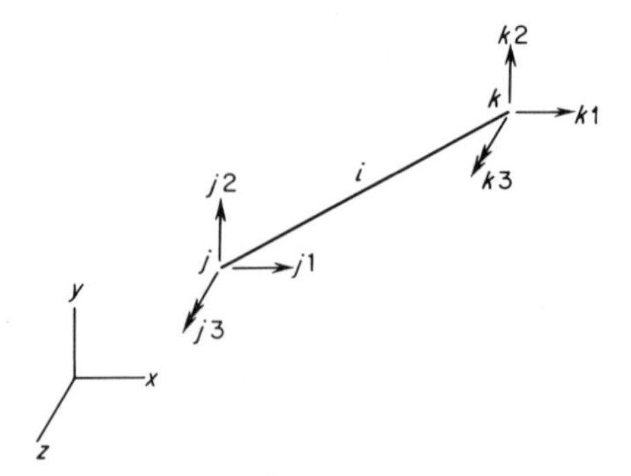

FIG. 4-28. End-displacements for plane frame member.

joints associated with this member are also indicated in Fig. 4-28. These displacements may be indexed by the following expressions:

$$j1 = 3j - 2 \qquad j2 = 3j - 1 \qquad j3 = 3j$$
$$k1 = 3k - 2 \qquad k2 = 3k - 1 \qquad k3 = 3k \tag{4-66}$$

The indexes in Eqs. (4-66) prove to be convenient for the purpose of assessing the contributions of member stiffnesses to the joint stiffness matrix, as in the case of beams and trusses. They are also useful for computing member end-actions due to joint displacements.

In order to construct the joint stiffness matrix in an orderly fashion, the following procedure is recommended. First, the 6×6 stiffness matrix $\mathbf{S}_{\mathrm{MD}}$ for structure axes is generated for the i-th member in the frame (see Art. 4.16, Table 4-27). Member i contributes to the stiffnesses of joints j and k at the ends of the member. Therefore, appropriate elements from the matrix $\mathbf{S}_{\mathrm{MD}}$ for this member may be transferred to the over-all joint stiffness matrix $\mathbf{S}_{\mathrm{J}}$ through an organized handling of subscripts. The first column in the matrix $\mathbf{S}_{\mathrm{MD}}$ consists of restraint actions at j and k due to a unit translation at the j end of member i in the x direction (index $j1$). This column is transferred to the matrix $\mathbf{S}_{\mathrm{J}}$ as follows:

$$\begin{aligned}
(S_J)_{j1,j1} &= \sum S_{MD} + (S_{MD11})_i \\
(S_J)_{j2,j1} &= \sum S_{MD} + (S_{MD21})_i \\
(S_J)_{j3,j1} &= \sum S_{MD} + (S_{MD31})_i \\
(S_J)_{k1,j1} &= (S_{MD41})_i \\
(S_J)_{k2,j1} &= (S_{MD51})_i \\
(S_J)_{k3,j1} &= (S_{MD61})_i
\end{aligned} \tag{4-67}$$

In these equations the first three stiffness coefficients consist of the sums of contributions from all members which frame into joint j, including member i. The last three stiffnesses involve contributions from member i only.

Expressions similar to Eqs. (4-67) may also be written for a unit translation of joint j in the y direction (index $j2$):

$$\begin{aligned}
(S_J)_{j1,j2} &= \sum S_{MD} + (S_{MD12})_i \\
\cdots \quad & \cdots \quad \cdots \\
(S_J)_{k3,j2} &= (S_{MD62})_i
\end{aligned} \tag{4-68}$$

Thus, the elements of the second column of $\mathbf{S}_{\mathrm{MD}}$ for member i are transferred as contributions to the matrix $\mathbf{S}_{\mathrm{J}}$.

Similarly, for a unit rotation of joint j in the z sense (index $j3$), the expressions for transferring the third column of $\mathbf{S}_{\mathrm{MD}}$ are:

$$\begin{aligned}
(S_J)_{j1,j3} &= \sum S_{MD} + (S_{MD13})_i \\
\cdots \quad & \cdots \quad \cdots \\
(S_J)_{k3,j3} &= (S_{MD63})_i
\end{aligned} \tag{4-69}$$

Expressions for transferring the fourth column of $\mathbf{S}_{MD}$ to the matrix $\mathbf{S}_J$ are similar to the above equations, except that the first three stiffnesses consist of contributions from member i only and the last three are sums of contributions from all members framing into joint k. Thus, for a unit translation of joint k in the x direction (index $k1$), the expressions are as follows:

$$\begin{aligned} (S_J)_{j1,k1} &= \qquad (S_{MD14})_i \\ \cdots \quad &\cdots \quad \cdots \\ (S_J)_{k3,k1} &= \sum S_{MD} + (S_{MD64})_i \end{aligned} \tag{4-70}$$

Similarly, for a unit translation of joint k in the y direction (index $k2$), the expressions for transferring the fifth column of $\mathbf{S}_{MD}$ are:

$$\begin{aligned} (S_J)_{j1,k2} &= \qquad (S_{MD15})_i \\ \cdots \quad &\cdots \quad \cdots \\ (S_J)_{k3,k2} &= \sum S_{MD} + (S_{MD65})_i \end{aligned} \tag{4-71}$$

Finally, for a unit rotation of joint k in the z sense (index $k3$), the following equations apply:

$$\begin{aligned} (S_J)_{j1,k3} &= \qquad (S_{MD16})_i \\ \cdots \quad &\cdots \quad \cdots \\ (S_J)_{k3,k3} &= \sum S_{MD} + (S_{MD66})_i \end{aligned} \tag{4-72}$$

In the process of transferring elements from the member stiffness matrix $\mathbf{S}_{MD}$ to the over-all joint stiffness matrix $\mathbf{S}_J$, as described above, there is no attempt to take advantage of the symmetry which exists in the matrices. Recognition of this symmetry allows short cuts in the transfer procedure which can readily be made.

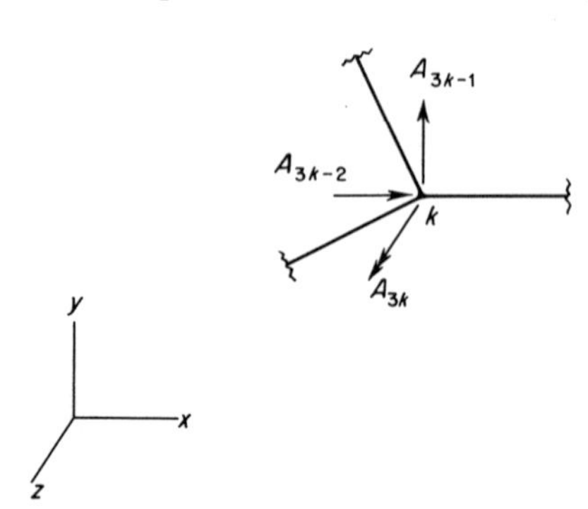

FIG. 4-29. Joint loads for a plane frame.

Construction of the complete matrix $\mathbf{S}_J$ consists of generating and transferring the matrix $\mathbf{S}_{MD}$ for all members (1 through m) of the structure. After the matrix $\mathbf{S}_J$ is generated, it must be rearranged if necessary into the form given by Eq. (4-1).

In the next phase of the analysis, vectors associated with loads on the frame are formed. External actions applied at joints constitute the vector $\mathbf{A}$. Figure 4-29 shows the actions at a typical joint k in a plane frame. The action A_{3k-2} is the x component of the force applied at k, A_{3k-1} is the y component of the applied force, and A_{3k} represents a couple in the z sense applied at the joint. Thus, the vector $\mathbf{A}$ will take the following form:

$$\mathbf{A} = \{A_1, A_2, A_3, \ldots, A_{3k-2}, A_{3k-1}, A_{3k}, \ldots, A_{3n_j-2}, A_{3n_j-1}, A_{3n_j}\} \qquad (4\text{-}73)$$

Figure 4-30b delineates the actions at the ends of a restrained plane

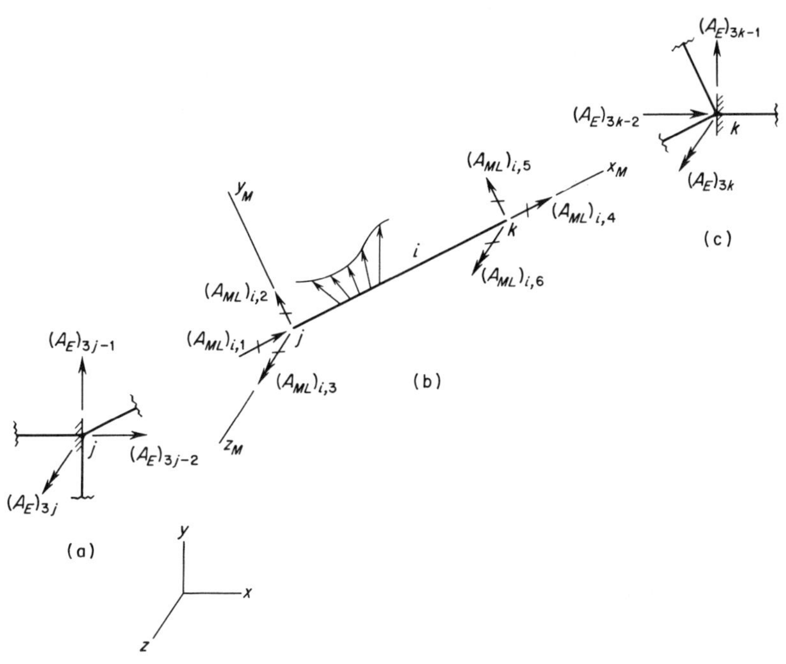

FIG. 4-30. Loads on a plane frame member.

frame member due to loads. End-actions for member i, with respect to member-oriented axes, are defined as follows:

$(A_{ML})_{i,1}$ = force in the x_M direction at the j end
$(A_{ML})_{i,2}$ = force in the y_M direction at the j end
$(A_{ML})_{i,3}$ = couple in the z_M sense at the j end
$(A_{ML})_{i,4}$ = force in the x_M direction at the k end
$(A_{ML})_{i,5}$ = force in the y_M direction at the k end
$(A_{ML})_{i,6}$ = couple in the z_M sense at the k end

These end-actions may be obtained from Appendix B for particular loading conditions. The matrix $\mathbf{A}_{\mathrm{ML}}$ is an array of order $m \times 6$, in which each row consists of the elements listed above for a given member; thus:

$$\mathbf{A}_{\mathrm{ML}} = \begin{bmatrix} (A_{ML})_{1,1} & (A_{ML})_{1,2} & (A_{ML})_{1,3} & (A_{ML})_{1,4} & (A_{ML})_{1,5} & (A_{ML})_{1,6} \\ \cdots & \cdots & \cdots & \cdots & \cdots & \cdots \\ (A_{ML})_{i,1} & (A_{ML})_{i,2} & (A_{ML})_{i,3} & (A_{ML})_{i,4} & (A_{ML})_{i,5} & (A_{ML})_{i,6} \\ \cdots & \cdots & \cdots & \cdots & \cdots & \cdots \\ (A_{ML})_{m,1} & (A_{ML})_{m,2} & (A_{ML})_{m,3} & (A_{ML})_{m,4} & (A_{ML})_{m,5} & (A_{ML})_{m,6} \end{bmatrix} \qquad (4\text{-}74)$$

The construction of the equivalent load vector $\mathbf{A}_E$ may be executed by the method of rotation of axes. This method was demonstrated previously for plane trusses in Art. 4.14. Figures 4-30a and c show the equivalent loads at j and k which receive contributions from member i. The negatives of these contributions may be evaluated as follows:

$$[R_T']_i\{A_{ML}\}_i = \begin{bmatrix} C_{Xi} & -C_{Yi} & 0 & 0 & 0 & 0 \\ C_{Yi} & C_{Xi} & 0 & 0 & 0 & 0 \\ 0 & 0 & 1 & 0 & 0 & 0 \\ 0 & 0 & 0 & C_{Xi} & -C_{Yi} & 0 \\ 0 & 0 & 0 & C_{Yi} & C_{Xi} & 0 \\ 0 & 0 & 0 & 0 & 0 & 1 \end{bmatrix} \begin{bmatrix} (A_{ML})_{i,1} \\ (A_{ML})_{i,2} \\ (A_{ML})_{i,3} \\ (A_{ML})_{i,4} \\ (A_{ML})_{i,5} \\ (A_{ML})_{i,6} \end{bmatrix}$$

The expressions which result from this multiplication, with their signs reversed, represent the incremental portions of $\mathbf{A}_E$ contributed by the i-th member. Thus,

$$\begin{aligned} (A_E)_{3j-2} &= \sum A_{ML} - (A_{ML})_{i,1}C_{Xi} + (A_{ML})_{i,2}C_{Yi} \\ (A_E)_{3j-1} &= \sum A_{ML} - (A_{ML})_{i,1}C_{Yi} - (A_{ML})_{i,2}C_{Xi} \\ (A_E)_{3j} &= \sum A_{ML} - (A_{ML})_{i,3} \\ (A_E)_{3k-2} &= \sum A_{ML} - (A_{ML})_{i,4}C_{Xi} + (A_{ML})_{i,5}C_{Yi} \\ (A_E)_{3k-1} &= \sum A_{ML} - (A_{ML})_{i,4}C_{Yi} - (A_{ML})_{i,5}C_{Xi} \\ (A_E)_{3k} &= \sum A_{ML} - (A_{ML})_{i,6} \end{aligned} \tag{4-75}$$

Addition of the vectors $\mathbf{A}$ and $\mathbf{A}_E$ produces the combined load vector $\mathbf{A}_C$ as given by Eq. (4-5). The vector $\mathbf{A}_C$ may then be rearranged if necessary into the form of Eq. (4-6).

After the generation of the required matrices is accomplished, substitution into Eqs. (4-8) and (4-4) yields the solution for joint displacements D (which is expanded into the vector $\mathbf{D}_J$) and support reactions A_R. Member end-actions in the plane frame may then be calculated by the same method as that indicated by Eq. (4-53) in Art. 4.14 for plane trusses. The equation which corresponds to Eq. (4-53) is the following:

$$\{A_M\}_i = \{A_{ML}\}_i + [S_M]_i[R_T]_i\{D_J\}_i \tag{4-76}$$

in which the 6×6 matrix $\mathbf{R}_T$ is given by Eq. (4-63). Substitution of $[S_M]_i$ from Table (4-26) and $[R_T]_i$ for a plane frame member into this expression yields the following:

$$\begin{aligned} (A_M)_{i,1} &= (A_{ML})_{i,1} + \frac{EA_{Xi}}{L_i}\{[(D_J)_{j1} - (D_J)_{k1}]C_{Xi} \\ &\quad + [(D_J)_{j2} - (D_J)_{k2}]C_{Yi}\} \\ (A_M)_{i,2} &= (A_{ML})_{i,2} + \frac{12EI_{Zi}}{L_i^3}\{-[(D_J)_{j1} - (D_J)_{k1}]C_{Yi} + \end{aligned}$$

$$+ [(D_J)_{j2} - (D_J)_{k2}]C_{Xi}\} + \frac{6EI_{Zi}}{L_i^2}[(D_J)_{j3} + (D_J)_{k3}]$$

$$(A_M)_{i,3} = (A_{ML})_{i,3} + \frac{6EI_{Zi}}{L_i^2}\{-[(D_J)_{j1} - (D_J)_{k1}]C_{Yi}$$

$$+ [(D_J)_{j2} - (D_J)_{k2}]C_{Xi}\} + \frac{4EI_{Zi}}{L_i}[(D_J)_{j3} + \tfrac{1}{2}(D_J)_{k3}]$$

$$(4\text{-}77)$$

$$(A_M)_{i,4} = (A_{ML})_{i,4} + \frac{EA_{Xi}}{L_i}\{-[(D_J)_{j1} - (D_J)_{k1}]C_{Xi}$$

$$- [(D_J)_{j2} - (D_J)_{k2}]C_{Yi}\}$$

$$(A_M)_{i,5} = (A_{ML})_{i,5} + \frac{12EI_{Zi}}{L_i^3}\{[(D_J)_{j1} - (D_J)_{k1}]C_{Yi}$$

$$- [(D_J)_{j2} - (D_J)_{k2}]C_{Xi}\} - \frac{6EI_{Zi}}{L_i^2}[(D_J)_{j3} + (D_J)_{k3}]$$

$$(A_M)_{i,6} = (A_{ML})_{i,6} + \frac{6EI_{Zi}}{L_i^2}\{-[(D_J)_{j1} - (D_J)_{k1}]C_{Yi}$$

$$+ [(D_J)_{j2} - (D_J)_{k2}]C_{Xi}\} + \frac{4EI_{Zi}}{L_i}[\tfrac{1}{2}(D_J)_{j3} + (D_J)_{k3}]$$

In Eqs. (4-77) the subscripts $j1$, $j2$, $j3$, $k1$, $k2$, and $k3$ carry the meanings defined previously in Eqs. (4-66).

The next article contains an example analysis of a plane frame by the method described above, and a computer program for plane frames appears in Art. 5.6.

4.18 Example. Figure 4-31a shows a plane frame having two members, three joints, six restraints, and three degrees of freedom. This frame is to be analyzed by the methods given in the previous article. For this purpose, assume that the cross-sectional area A_X and the moment of inertia I_Z are constant throughout the structure. Assume also that the parameters in the problem have the following numerical values:

$E = 10{,}000$ ksi $\quad L = 100$ in. $\quad I_Z = 1000$ in.^{4} $\quad P = 10$ kips $\quad A_X = 10$ in.^{2}

Units of kips, inches, and radians are used throughout the analysis.

A numbering system for members, joints, and displacements is given in Fig. 4-31b, which shows the restrained structure. The sequence for numbering the joints is selected in such a manner that the matrices in the analysis do not require rearrangement.

Joint information is summarized in Table 4-28, which contains the joint numbers, joint coordinates, and the conditions of restraint. The member information for the frame is presented in Table 4-29. Substitution of the direction cosines for the members into Eq. (4-62) results in the rotation matrices $\mathbf{R}_1$ and $\mathbf{R}_2$ shown in Table 4-30.

In preparation for generating the over-all joint stiffness matrix $\mathbf{S}_J$, the member stiffness matrices are calculated. This may be done for each member by first generating the member stiffness matrix $\mathbf{S}_M$ for member axes (see Table 4-26, Art. 4.16) and then calculating the matrix $\mathbf{S}_{MD}$ for structure axes by the rotation transformation of Eq. (4-64). For this purpose the matrix $\mathbf{R}_T$ for each member

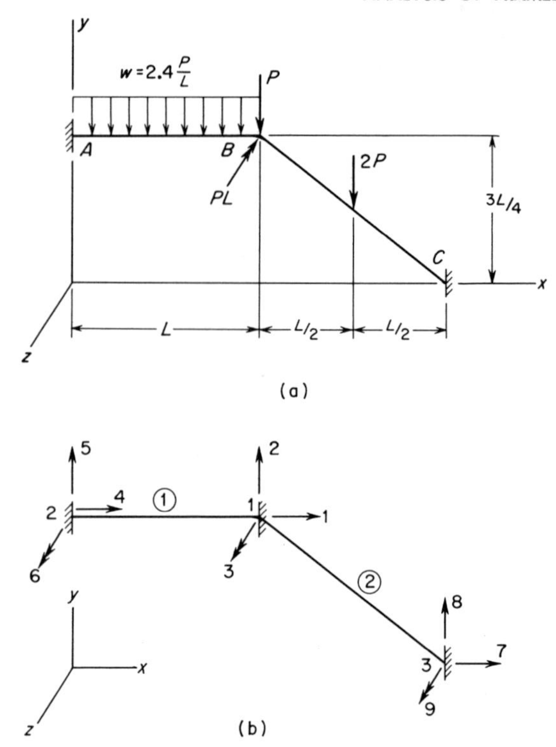

FIG. 4-31. Example (plane frame).

is composed as shown by Eq. (4-63). Alternatively, the matrix $\mathbf{S}_{MD}$ may be calculated directly for each member using Table 4-27. The resulting matrices are given in Table 4-31. The indexes $j1$ through $k3$ (computed by Eqs. 4-66) are also indicated in Table 4-31 down the right-hand side and across the bottom of each matrix $\mathbf{S}_{MD}$. These indexes may be used as a guide for the purpose of transferring elements to the matrix $\mathbf{S}_J$. After the transferring process is accomplished, the over-all joint stiffness matrix which results is shown in Table 4-32. This matrix is partitioned in the usual manner, thereby isolating the 3×3 stiffness matrix $\mathbf{S}$. The inverse of the matrix $\mathbf{S}$ is given in Table 4-33.

TABLE 4-28

Joint Information for Frame of Fig. 4-31

Joint	Coordinates (in.)		Restraint List		
	x	y	x	y	z
1	100	75	0	0	0
2	0	75	1	1	1
3	200	0	1	1	1

TABLE 4-29

Member Information for Frame of Fig. 4-31

Member	Joint j	Joint k	Area (in.^{2})	Moment of Inertia (in.^{4})	Length (in.)	Direction Cosines	
						C_X	C_Y
1	2	1	10	1000	100	1.0	0
2	1	3	10	1000	125	0.8	−0.6

TABLE 4-30

Rotation Matrices for Members of Frame of Fig. 4-31

$$\mathbf{R}_1 = \begin{bmatrix} 1.0 & 0 & 0 \\ 0 & 1.0 & 0 \\ 0 & 0 & 1.0 \end{bmatrix} \qquad \mathbf{R}_2 = \begin{bmatrix} 0.8 & -0.6 & 0 \\ 0.6 & 0.8 & 0 \\ 0 & 0 & 1.0 \end{bmatrix}$$

TABLE 4-31

Member Stiffness Matrices for Structure Axes

$$[S_{MD}]_1 = \begin{bmatrix} 1000.0 & 0 & 0 & -1000.0 & 0 & 0 \\ 0 & 120.0 & 6000.0 & 0 & -120.0 & 6000.0 \\ 0 & 6000.0 & 400000.0 & 0 & -6000.0 & 200000.0 \\ -1000.0 & 0 & 0 & 1000.0 & 0 & 0 \\ 0 & -120.0 & -6000.0 & 0 & 120.0 & -6000.0 \\ 0 & 6000.0 & 200000.0 & 0 & -6000.0 & 400000.0 \end{bmatrix} \begin{matrix} 4 \\ 5 \\ 6 \\ 1 \\ 2 \\ 3 \end{matrix}$$
$$\begin{matrix} 4 & 5 & 6 & 1 & 2 & 3 \end{matrix}$$

$$[S_{MD}]_2 = \begin{bmatrix} 534.1 & -354.5 & 2304.0 & -534.1 & 354.5 & 2304.0 \\ -354.5 & 327.3 & 3072.0 & 354.5 & -327.3 & 3072.0 \\ 2304.0 & 3072.0 & 320000.0 & -2304.0 & -3072.0 & 160000.0 \\ -534.1 & 354.5 & -2304.0 & 534.1 & -354.5 & -2304.0 \\ 354.5 & -327.3 & -3072.0 & -354.5 & 327.3 & -3072.0 \\ 2304.0 & 3072.0 & 160000.0 & -2304.0 & -3072.0 & 320000.0 \end{bmatrix} \begin{matrix} 1 \\ 2 \\ 3 \\ 7 \\ 8 \\ 9 \end{matrix}$$
$$\begin{matrix} 1 & 2 & 3 & 7 & 8 & 9 \end{matrix}$$

Next, the load information is processed, beginning with the joint loads shown in Table 4-34. The actions in this table are placed in the vector $\mathbf{A}$ as indicated by Eq. (4-73):

$$\mathbf{A} = \{0, -10, -1000, 0, 0, 0, 0, 0, 0\}$$

In addition, the actions in Table 4-35 are placed in the matrix $\mathbf{A}_{ML}$ and then transferred to the vector of equivalent joint loads $\mathbf{A}_E$ in accordance with Eqs. (4-75). The following vector results:

$$\mathbf{A}_E = \{0, -22, -50, 0, -12, -200, 0, -10, 250\}$$

Adding the vectors $\mathbf{A}$ and $\mathbf{A}_E$ produces the combined load vector $\mathbf{A}_C$ as follows:

$$\mathbf{A}_C = \{0, -32, -1050, 0, -12, -200, 0, -10, 250\}$$

TABLE 4-32

Joint Stiffness Matrix for Frame of Fig. 4-31

$$
\mathbf{S}_J = \left[\begin{array}{ccc|cccccc}
1534.1 & -354.5 & 2304.0 & -1000.0 & 0 & 0 & -534.1 & 354.5 & 2304.0 \\
-354.5 & 447.3 & -2928.0 & 0 & -120.0 & -6000.0 & 354.5 & -327.3 & 3072.0 \\
2304.0 & -2928.0 & 720000.0 & 0 & 6000.0 & 200000.0 & -2304.0 & -3072.0 & 160000.0 \\
\hline
-1000.0 & 0 & 0 & 1000.0 & 0 & 0 & 0 & 0 & 0 \\
0 & -120.0 & 6000.0 & 0 & 120.0 & 6000.0 & 0 & 0 & 0 \\
0 & -6000.0 & 200000.0 & 0 & 6000.0 & 400000.0 & 0 & 0 & 0 \\
-534.1 & 354.5 & -2304.0 & 0 & 0 & 0 & 534.1 & -354.5 & -2304.0 \\
354.5 & -327.3 & -3072.0 & 0 & 0 & 0 & -354.5 & 327.3 & -3072.0 \\
2304.0 & 3072.0 & 160000.0 & 0 & 0 & 0 & -2304.0 & 3072.0 & 320000.0
\end{array}\right]
$$

TABLE 4-33

Inverse of Stiffness Matrix

$$S^{-1} = \begin{bmatrix} 798.0 & 632.5 & 0.01877 \\ 632.5 & 2798.0 & 9.355 \\ 0.01877 & 9.355 & 1.426 \end{bmatrix} \times 10^{-6}$$

TABLE 4-34

Actions Applied at Joints

Joint	Force in x Direction (kips)	Force in y Direction (kips)	Couple in z Sense (kip-in.)
1	0	−10	−1000
2	0	0	0
3	0	0	0

TABLE 4-35

Actions at Ends of Restrained Members Due to Loads

Member	$(A_{ML})_{i,1}$ (kips)	$(A_{ML})_{i,2}$ (kips)	$(A_{ML})_{i,3}$ (kip-in.)	$(A_{ML})_{i,4}$ (kips)	$(A_{ML})_{i,5}$ (kips)	$(A_{ML})_{i,6}$ (kip-in.)
1	0	12	200	0	12	−200
2	−6	8	250	−6	8	−250

The first three elements of this vector constitute the vector $\mathbf{A}_D$:

$$\mathbf{A}_D = \{0, -32, -1050\}$$

and the last six elements are the negatives of the elements of the vector $\mathbf{A}_{RL}$:

$$\mathbf{A}_{RL} = \{0, 12, 200, 0, 10, -250\}$$

Having all of the required matrices on hand, one may complete the solution by calculating the joint displacements D by Eq. (4-8) with the following result:

$$\mathbf{D} = \mathbf{S}^{-1}\mathbf{A}_D = \{-0.02026, -0.09936, -0.001797\}$$

The first two elements in the vector $\mathbf{D}$ are the translations (inches) in the x and y directions at joint 1, and the last element is the rotation (radians) of the joint in the z sense.

The vector $\mathbf{D}_J$ for this structure consists of the vector $\mathbf{D}$ in the first part and zeros in the latter part:

$$\mathbf{D}_J = \{-0.02026, -0.09936, -0.001797, 0, 0, 0, 0, 0, 0\}$$

In the next step the support reactions are computed, using Eq. (4-4) with the matrix $\mathbf{S}_{RD}$ obtained from the lower left-hand portion of Table 4-32:

$$\mathbf{A}_R = \mathbf{A}_{RL} + \mathbf{S}_{RD}\mathbf{D} = \{20.26, 13.14, 436.6, -20.26, 40.86, -889.5\}$$

The elements in the vector $\mathbf{A}_R$ consist of the forces (kips) in the x and y directions and the couple (kip-inches) in the z sense at points 2 and 3, respectively, in the frame.

As the final step in the analysis, the member end-actions $\mathbf{A}_M$ are calculated, using either Eq. (4-76) or Eqs. (4-77). The results of these calculations are given in Table 4-36, which completes the analysis of the plane frame structure.

TABLE 4-36

Final Member End-Actions

Member	$(A_M)_{i,1}$ (kips)	$(A_M)_{i,2}$ (kips)	$(A_M)_{i,3}$ (kip-in.)	$(A_M)_{i,4}$ (kips)	$(A_M)_{i,5}$ (kips)	$(A_M)_{i,6}$ (kip-in.)
1	20.26	13.14	436.6	−20.26	10.86	−322.9
2	28.72	−4.53	−677.1	−40.73	20.53	−889.5

4.19 Grid Member Stiffnesses. A grid structure resembles a plane frame in several respects. All of the members and joints lie in the same plane, and the members are assumed to be rigidly connected at the joints (see Fig. 4-32a). Flexural effects tend to predominate in the analysis of both types of structures, with the effects of torsion often being secondary in the grid analysis and axial effects often being secondary in the plane frame analysis. The most important difference between a plane frame and a grid is that the former is assumed to be loaded in its own plane, whereas the loads on the latter are normal to its own plane. In addition, couples applied to a plane frame are normal to the plane of the structure, whereas couples applied to grids are assumed to lie in the plane of the structure. Both structures could be called plane frames, and the difference between them could be denoted by stating the nature of the loading system. Furthermore, if the applied loads were to have general orientations in space, the analysis of the structure could be divided into two parts. In the first part, the frame would be analyzed for the components of loads in the plane of the structure, and the second part would consist of analyzing for the components of loads normal to the plane. Superposition of these two analyses would then produce the total solution of the problem. Such a structure might be considered as a special case of a space frame in which all of the members and joints lie in a common plane.

In analyzing a grid structure, the coordinate axes will be taken as shown in Fig. 4-32a. The structure lies in the x-y plane, and all applied forces act parallel to the z axis. Loads in the form of couples have their moment vectors in the x-y plane. The figure shows a typical member i framing into joints j and k. The significant displacements of the joints are rotations in the x and y senses and translations in the z direction. The six possible displacements at the ends of member i in the directions of structure axes are shown in Fig. 4-32c. The numbering system for these displacements is in the order mentioned above. The reason for numbering the rotations before the translation at each joint is to maintain a parallelism with the analysis of a plane frame, as can be seen by comparing the computer programs for plane frames and grids (see Arts. 5.6 and 5.7).

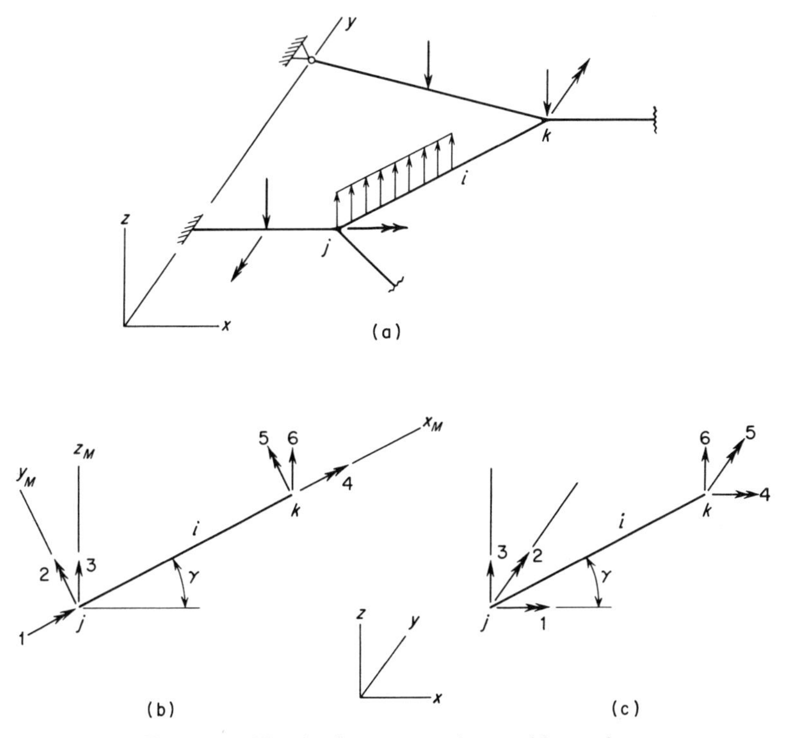

FIG. 4-32. Numbering system for a grid member.

Figure 4-32b depicts the member i in conjunction with a set of member-oriented axes x_M, y_M, and z_M. These axes are rotated from the structure axes about the z_M axis through the angle γ. The x_M-z_M plane for each member in the grid is assumed to be a plane of symmetry (and hence a principal plane of bending). The possible displacements of the ends of member i in the directions of member axes are also indicated in Fig. 4-32b. The six end-displacements, shown in their positive senses, consist of rotations in the x_M and y_M senses and a translation in the z_M (or z) direction at the ends j and k, respectively. Unit displacements of these six types may be induced at the ends of the member one at a time for the purpose of developing the member stiffness matrix $\mathbf{S}_M$ for member axes. This matrix is of order 6×6, and it may be obtained from cases (3), (4), (5), (9), (10), and (11) of Fig. 4-2 in Art. 4.3. The matrix which results is shown in Table 4-37.

TABLE 4-37

Grid Member Stiffness Matrix for Member Axes (Fig. 4-32b)

$$\mathbf{S}_M = \begin{bmatrix} \frac{GI_X}{L} & 0 & 0 & -\frac{GI_X}{L} & 0 & 0 \\ 0 & \frac{4EI_Y}{L} & -\frac{6EI_Y}{L^2} & 0 & \frac{2EI_Y}{L} & \frac{6EI_Y}{L^2} \\ 0 & -\frac{6EI_Y}{L^2} & \frac{12EI_Y}{L^3} & 0 & -\frac{6EI_Y}{L^2} & -\frac{12EI_Y}{L^3} \\ -\frac{GI_X}{L} & 0 & 0 & \frac{GI_X}{L} & 0 & 0 \\ 0 & \frac{2EI_Y}{L} & -\frac{6EI_Y}{L^2} & 0 & \frac{4EI_Y}{L} & \frac{6EI_Y}{L^2} \\ 0 & \frac{6EI_Y}{L^2} & -\frac{12EI_Y}{L^3} & 0 & \frac{6EI_Y}{L^2} & \frac{12EI_Y}{L^3} \end{bmatrix}$$

Transformation of the member stiffness matrix from member axes (Fig. 4-32b) to structure axes (Fig. 4-32c) follows the same pattern as that for a plane frame (see Art. 4.16). The rotation transformation matrix, which involves a rotation about the z_M axis, is exactly the same for both a plane frame member and a grid member because of the choice of numbering system mentioned above. This agreement between the two cases can be seen physically by comparing the orientation of the grid member in Fig. 4-32 with that of the plane frame member in Fig. 4-27. Thus, Eqs. (4-62) through (4-64) in Art. 4.16 may be applied to a grid member as well as a plane frame member. Substitution of the matrix $\mathbf{S}_M$ from Table 4-37 into Eq. (4-64) results in the member stiffness matrix $\mathbf{S}_{MD}$ for structure axes. This matrix is given in Table 4-38 and will be used subsequently in the analysis of grid structures.

TABLE 4-38

Grid Member Stiffness Matrix for Structure Axes (Fig. 4-32c)

$$
\mathbf{S}_{MD} = \begin{bmatrix}
\frac{GI_X}{L}C_X^2 + \frac{4EI_Y}{L}C_Y^2 & \left(\frac{GI_X}{L} - \frac{4EI_Y}{L}\right)C_XC_Y & \frac{6EI_Y}{L^2}C_Y & -\frac{GI_X}{L}C_X^2 + \frac{2EI_Y}{L}C_Y^2 & -\left(\frac{GI_X}{L} + \frac{2EI_Y}{L}\right)C_XC_Y & -\frac{6EI_Y}{L^2}C_Y \\
\left(\frac{GI_X}{L} - \frac{4EI_Y}{L}\right)C_XC_Y & \frac{GI_X}{L}C_Y^2 + \frac{4EI_Y}{L}C_X^2 & -\frac{6EI_Y}{L^2}C_X & -\left(\frac{GI_X}{L} + \frac{2EI_Y}{L}\right)C_XC_Y & -\frac{GI_X}{L}C_Y^2 + \frac{2EI_Y}{L}C_X^2 & \frac{6EI_Y}{L^2}C_X \\
\frac{6EI_Y}{L^2}C_Y & -\frac{6EI_Y}{L^2}C_X & \frac{12EI_Y}{L^3} & \frac{6EI_Y}{L^2}C_Y & -\frac{6EI_Y}{L^2}C_X & -\frac{12EI_Y}{L^3} \\
-\frac{GI_X}{L}C_X^2 + \frac{2EI_Y}{L}C_Y^2 & -\left(\frac{GI_X}{L} + \frac{2EI_Y}{L}\right)C_XC_Y & \frac{6EI_Y}{L^2}C_Y & \frac{GI_X}{L}C_X^2 + \frac{4EI_Y}{L}C_Y^2 & \left(\frac{GI_X}{L} - \frac{4EI_Y}{L}\right)C_XC_Y & -\frac{6EI_Y}{L^2}C_Y \\
-\left(\frac{GI_X}{L} + \frac{2EI_Y}{L}\right)C_XC_Y & -\frac{GI_X}{L}C_Y^2 + \frac{2EI_Y}{L}C_X^2 & -\frac{6EI_Y}{L^2}C_X & \left(\frac{GI_X}{L} - \frac{4EI_Y}{L}\right)C_XC_Y & \frac{GI_X}{L}C_Y^2 + \frac{4EI_Y}{L}C_X^2 & \frac{6EI_Y}{L^2}C_X \\
-\frac{6EI_Y}{L^2}C_Y & \frac{6EI_Y}{L^2}C_X & -\frac{12EI_Y}{L^3} & -\frac{6EI_Y}{L^2}C_Y & \frac{6EI_Y}{L^2}C_X & \frac{12EI_Y}{L^3}
\end{bmatrix}
$$

4.20 Analysis of Grids. The first step in the analysis of a grid is to number the joints and members. This is done in the same manner as for a plane truss or plane frame. As mentioned in the preceding article, there are three possible displacements at each joint. These are the joint rotations in the x and y senses and the joint translation in the z direction. Thus, the possible displacements at a joint j may be denoted by the following:

$$3j - 2 = \text{index for rotation in the } x \text{ sense}$$

$$3j - 1 = \text{index for rotation in the } y \text{ sense}$$

$$3j \qquad = \text{index for translation in the } z \text{ direction}$$

Note that these indexes are the same as those for the plane frame (see Art. 4.17) but that the meanings are different in the grid structure. Because there are three possible displacements at each joint in both a grid and a plane frame, the number of degrees of freedom in a grid may be determined using Eq. (4-65) in Art. 4.17. Moreover, the indexes

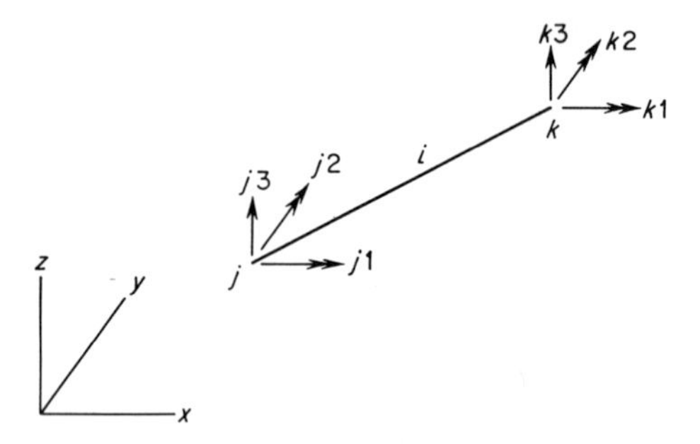

FIG. 4-33. End-displacements for grid member.

for the possible displacements of the two joints associated with member i may be calculated according to Eqs. (4-66) for either type of structure. These indexes are indicated for a grid member in Fig. 4-33. Since several of the steps in the analysis of a grid are symbolically the same as in the analysis of a plane frame, it is sufficient in much of the following discussion merely to refer to the plane frame analysis described in Art. 4.17.

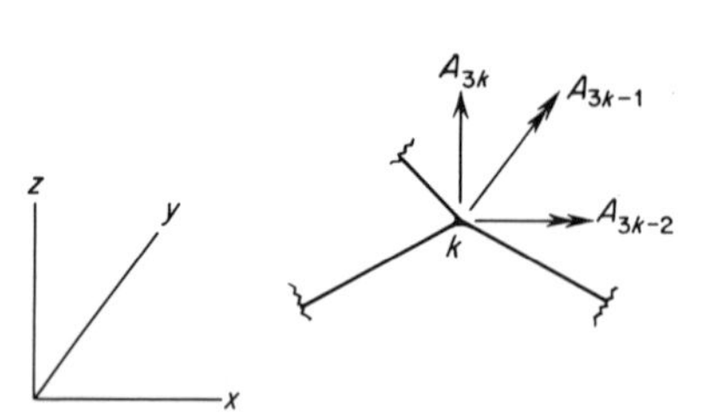

FIG. 4-34. Joint loads for a grid.

The analysis of the restrained grid for joint stiffnesses follows the same pattern as in the analysis of a plane frame. The 6×6 member stiffness matrix $\mathbf{S}_{MD}$ for each member is generated (see Table 4-38), and the elements of this array are transferred systematically to the joint stiffness matrix $\mathbf{S}_J$.

Equations (4-67) through (4-72) serve this purpose for a grid as well as for a plane frame.

The analysis of the restrained structure subjected to loads is also analogous to the plane frame, except that the types of actions involved are not the same. Consider first the vector of actions **A** applied at joints. Figure 4-34 shows the actions which may be imposed at a typical joint k in a grid. The action A_{3k-2} is the x component of a moment vector

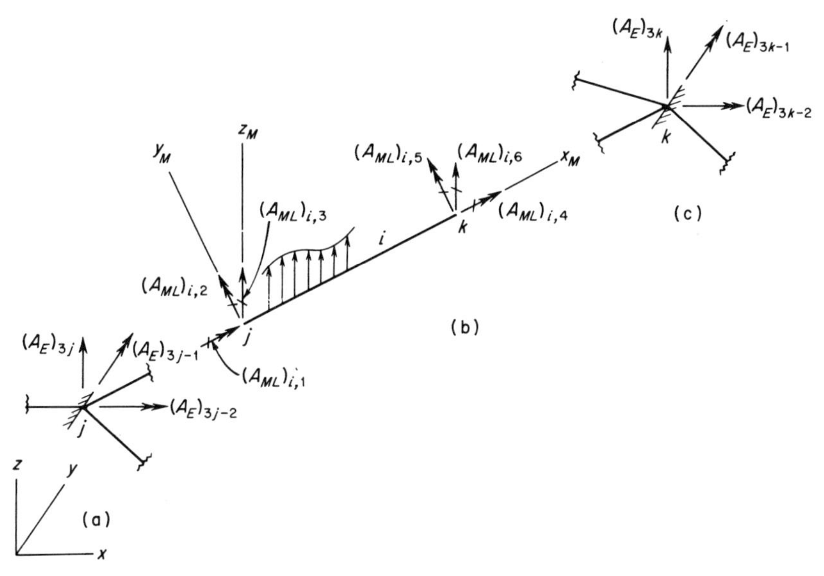

FIG. 4-35. Loads on a grid member.

applied at k, A_{3k-1} is the y component of the moment vector, and A_{3k} represents a force in the z direction applied at the joint. Thus, the vector **A** may be formed in the sequence denoted by Eq. (4-73).

Actions A_{ML} at the ends of a restrained grid member (due to loads) appear in Fig. 4-35b. The end-actions for the i-th member, with respect to member axes, are defined as follows:

$(A_{ML})_{i,1}$ = couple in the x_M sense at the j end

$(A_{ML})_{i,2}$ = couple in the y_M sense at the j end

$(A_{ML})_{i,3}$ = force in the z_M direction at the j end

$(A_{ML})_{i,4}$ = couple in the x_M sense at the k end

$(A_{ML})_{i,5}$ = couple in the y_M sense at the k end

$(A_{ML})_{i,6}$ = force in the z_M direction at the k end

The matrix $\mathbf{A}_{ML}$ may be formulated as a rectangular array of order $m \times 6$ of the type given by Eq. (4-74).

Construction of the equivalent load vector $\mathbf{A}_E$ from the matrix $\mathbf{A}_{ML}$ is executed as described in Art. 4.17 for plane frames. Figures 4-35a and 4-35c show the equivalent loads at joints j and k which receive contributions from member i. Equations (4-75) may be used for the purpose of computing these actions.

The calculation of displacements, reactions, and end-actions for a grid structure follows the same steps given previously for a plane frame (see Art. 4.17). Joint displacements D are computed, using Eq. (4-8), and then expanded into the vector $\mathbf{D}_J$. Support reactions are calculated, using Eq. (4-4), and member end-actions are determined, using Eq. (4-76). In the latter equation the matrix $[\mathbf{R}_T]_i$ is given by Eq. (4-63), and the matrix $[\mathbf{S}_M]_i$ is obtained from Table 4-37. Substitution of these matrices into Eq. (4-76) produces the following expressions for end-actions:

$$
\begin{aligned}
(A_M)_{i,1} &= (A_{ML})_{i,1} + \frac{GI_{Xi}}{L_i}\{[(D_J)_{j1} - (D_J)_{k1}]C_{Xi} \\
&\quad + [(D_J)_{j2} - (D_J)_{k2}]C_{Yi}\} \\
(A_M)_{i,2} &= (A_{ML})_{i,2} + \frac{4EI_{Yi}}{L_i}\{-[(D_J)_{j1} + \tfrac{1}{2}(D_J)_{k1}]\,C_{Yi} \\
&\quad + [(D_J)_{j2} + \tfrac{1}{2}(D_J)_{k2}]\,C_{Xi}\} - \frac{6EI_{Yi}}{L_i^2}[(D_J)_{j3} - (D_J)_{k3}] \\
(A_M)_{i,3} &= (A_{ML})_{i,3} + \frac{6EI_{Yi}}{L_i^2}\{[(D_J)_{j1} + (D_J)_{k1}]C_{Yi} \\
&\quad - [(D_J)_{j2} + (D_J)_{k2}]C_{Xi}\} + \frac{12EI_{Yi}}{L_i^3}[(D_J)_{j3} - (D_J)_{k3}] \qquad (4\text{-}78) \\
(A_M)_{i,4} &= (A_{ML})_{i,4} + \frac{GI_{Xi}}{L_i}\{-[(D_J)_{j1} - (D_J)_{k1}]C_{Xi} \\
&\quad - [(D_J)_{j2} - (D_J)_{k2}]C_{Yi}\} \\
(A_M)_{i,5} &= (A_{ML})_{i,5} + \frac{4EI_{Yi}}{L_i}\{-[\tfrac{1}{2}(D_J)_{j1} + (D_J)_{k1}]\,C_{Yi} \\
&\quad + [\tfrac{1}{2}(D_J)_{j2} + (D_J)_{k2}]\,C_{Xi}\} - \frac{6EI_{Yi}}{L_i^2}[(D_J)_{j3} - (D_J)_{k3}] \\
(A_M)_{i,6} &= (A_{ML})_{i,6} + \frac{6EI_{Yi}}{L_i^2}\{-[(D_J)_{j1} + (D_J)_{k1}]C_{Yi} \\
&\quad + [(D_J)_{j2} + (D_J)_{k2}]C_{Xi}\} - \frac{12EI_{Yi}}{L_i^3}[(D_J)_{j3} - (D_J)_{k3}]
\end{aligned}
$$

A computer program for the analysis of grids by the method described above is presented in Art. 5.7. Numerical examples of the analysis of grid structures also appear in that article.

4.21 Space Truss Member Stiffnesses. Figure 4-36 shows a portion of a space truss structure in conjunction with a set of structure axes x, y, and z. A typical member i, framing into joints j and k, is indicated in the figure. All joints in the space truss are assumed to be

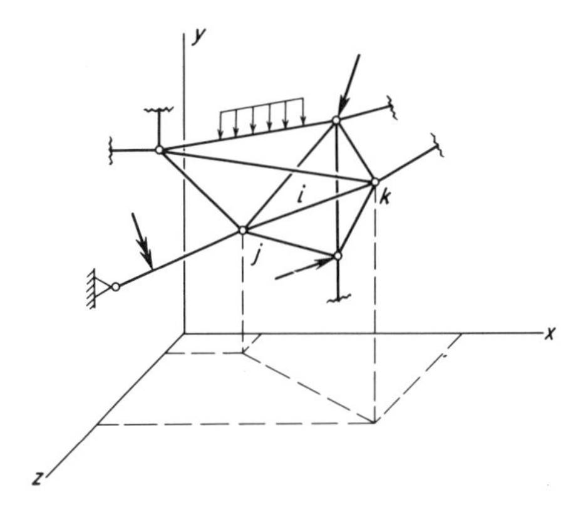

Fig. 4-36. Space truss.

universal hinges. Because of this idealization, rotations of the ends of the members are considered to be immaterial to the analysis. The significant joint displacements are translations, and these translations may be expressed conveniently by their components in the x, y, and z directions.

The possible displacements at the ends of a typical member i are indicated in Fig. 4-37a for member-oriented axes and in Fig. 4-37b for

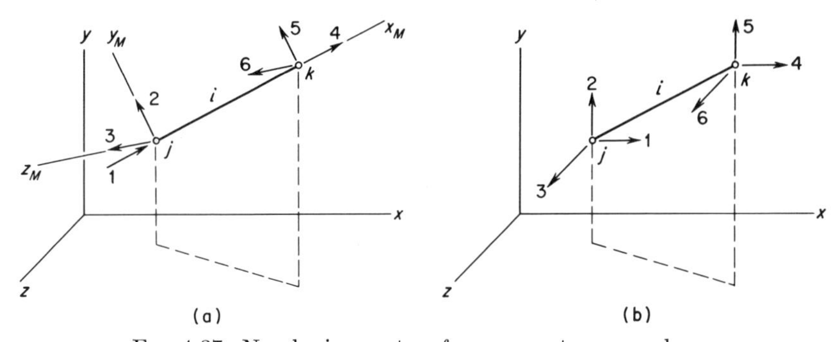

Fig. 4-37. Numbering system for a space truss member.

the structure axes. The member axes in Fig. 4-37a are arranged in such a manner that the x_M axis coincides with the axis of the member and has its positive sense from j to k. The y_M and z_M axes lie in a plane that is perpendicular to the axis of the member and passes through end j. However, the precise orientation of these axes in the plane is immaterial at this stage of the discussion, inasmuch as the position of these axes has no effect on the stiffness matrix for a truss member. In the next article, a particular choice for the directions of these axes will be made in order to facilitate the handling of load data.

The six end-displacements shown in Fig. 4-37a consist of translations in the x_M, y_M, and z_M directions at the ends j and k, respectively. The member stiffness matrix with respect to the member axes may be readily deduced from cases (1) and (7) of Fig. 4-2 in Art. 4.3, and the resulting 6×6 matrix is given in Table 4-39. It can be seen that the only nonzero

TABLE 4-39

Space Truss Member Stiffness Matrix for Member Axes (Fig. 4-37a)

$$\mathbf{S_M} = \frac{EA_X}{L}\begin{bmatrix} 1 & 0 & 0 & -1 & 0 & 0 \\ 0 & 0 & 0 & 0 & 0 & 0 \\ 0 & 0 & 0 & 0 & 0 & 0 \\ -1 & 0 & 0 & 1 & 0 & 0 \\ 0 & 0 & 0 & 0 & 0 & 0 \\ 0 & 0 & 0 & 0 & 0 & 0 \end{bmatrix}$$

elements in this matrix are associated with displacements 1 and 4, which are in the direction of the x_M axis (Fig. 4-37a). Thus, as mentioned above, the stiffness matrix for member axes is independent of the directions selected for the y_M and z_M axes.

In order to transform the member stiffness matrix from member axes to structure axes, the rotation transformation matrix $\mathbf{R_T}$ for a space truss member is required. This matrix takes the same form as that given by Eq. (4-63) for plane frames, and the transformation of $\mathbf{S_M}$ to $\mathbf{S_{MD}}$ is shown in Eq. (4-64). The 3×3 rotation matrix $\mathbf{R}$ required for $\mathbf{R_T}$ is explained in the following discussion.

The general form of the rotation matrix $\mathbf{R}$ is given by Eq. (4-56). The three elements λ_{11}, λ_{12}, and λ_{13} in the first row of $\mathbf{R}$ are the direction cosines for the x_M axis with respect to the structure axes. Therefore, these three elements are the same as the direction cosines for the member itself ($\lambda_{11} = C_X$, $\lambda_{12} = C_Y$, $\lambda_{13} = C_Z$) and can be found from the coordinates of the ends of the member:

$$C_X = \frac{x_k - x_j}{L} \qquad C_Y = \frac{y_k - y_j}{L} \qquad C_Z = \frac{z_k - z_j}{L} \tag{4-79}$$

The length L of the member may also be computed from the coordinates of its end points:

$$L = \sqrt{(x_k - x_j)^2 + (y_k - y_j)^2 + (z_k - z_j)^2} \tag{4-80}$$

The elements in the last two rows of $\mathbf{R}$ (the direction cosines for the y_M and z_M axes, respectively) can be left in indefinite form so that the rotation matrix becomes:

$$\mathbf{R} = \begin{bmatrix} C_X & C_Y & C_Z \\ \lambda_{21} & \lambda_{22} & \lambda_{23} \\ \lambda_{31} & \lambda_{32} & \lambda_{33} \end{bmatrix} \tag{4-81}$$

If this rotation matrix is substituted into Eq. (4-63) for $\mathbf{R}_\mathrm{T}$ and then both $\mathbf{R}_\mathrm{T}$ and $\mathbf{S}_\mathrm{M}$ (see Table 4-39) are substituted into Eq. (4-64), the result will be the 6×6 member stiffness matrix $\mathbf{S}_\mathrm{MD}$ for structure axes. This matrix, which contains terms involving only the direction cosines for the member itself, is given in Table 4-40.

TABLE 4-40

Space Truss Member Stiffness Matrix for Structure Axes (Fig. 4-37b)

$$\mathbf{S}_\mathrm{MD} = \frac{EA_X}{L} \begin{bmatrix} C_X^2 & C_YC_X & C_ZC_X & -C_X^2 & -C_YC_X & -C_ZC_X \\ C_XC_Y & C_Y^2 & C_ZC_Y & -C_XC_Y & -C_Y^2 & -C_ZC_Y \\ C_XC_Z & C_YC_Z & C_Z^2 & -C_XC_Z & -C_YC_Z & -C_Z^2 \\ -C_X^2 & -C_YC_X & -C_ZC_X & C_X^2 & C_YC_X & C_ZC_X \\ -C_XC_Y & -C_Y^2 & -C_ZC_Y & C_XC_Y & C_Y^2 & C_ZC_Y \\ -C_XC_Z & -C_YC_Z & -C_Z^2 & C_XC_Z & C_YC_Z & C_Z^2 \end{bmatrix}$$

4.22 Selection of Space Truss Member Axes. For the purpose of handling loads acting directly on the members of a space truss, it is necessary to select a specific orientation for all three member axes x_M, y_M, and z_M. Since the x_M axis has already been selected as the axis of the member (see Fig. 4-37a), it remains only to establish directions for y_M and z_M. For this purpose, the typical member i is shown again in Fig. 4-38. The structure axes x_S, y_S, and z_S are taken parallel to the x, y, and z axes (shown in Fig. 4-37) and through end j of the member. While many choices are possible for the directions of the y_M and z_M axes, a convenient one is to take the z_M axis as being horizontal (that is, lying in the x_S-z_S plane), as shown in the figure. It follows that the y_M axis is located in a vertical plane passing through the x_M and y_S axes.

When the member axes are specified in the manner just described, there is no ambiguity about their orientations except in the case of a vertical member. When the member is vertical it follows automatically that z_M will be in a horizontal plane, but its position in that plane is not defined uniquely. To overcome this difficulty, the additional specification will be made that the z_M axis is always taken to be along the z_S axis if the member is vertical. The two possibilities for this occurrence are shown in Figs. 4-39a and 4-39b.

In order to perform axis transformations, the rotation matrix $\mathbf{R}$ (see Eq. 4-81) is required. The direction cosines for the y_M and z_M axes (the last two rows of $\mathbf{R}$) can be found directly from Fig. 4-38 by geometrical considerations. However, an alternate approach involves successive rotations of axes. In using the latter method, the transformation from the structure axes (see Fig. 4-38) to the member axes may be considered to take place in two steps. The first of these is a rotation through an angle β about the y_S axis. This rotation places the x axis in the position denoted as x_β, which is the intersection of the x_S-z_S plane and the

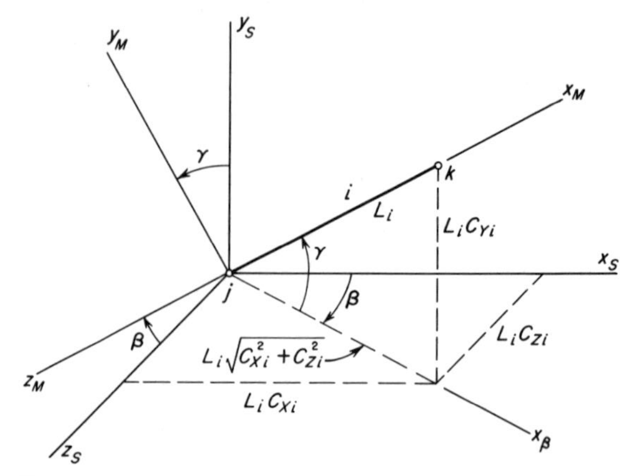

FIG. 4-38. Rotation of axes for a space truss member.

x_M-y_S plane. Also, this rotation places the z_M axis in its final position at the angle β with the z_S axis. The second step in the transformation consists of a rotation through an angle γ about the z_M axis. This rotation places the x_M and y_M axes in their final positions, as shown in the figure.

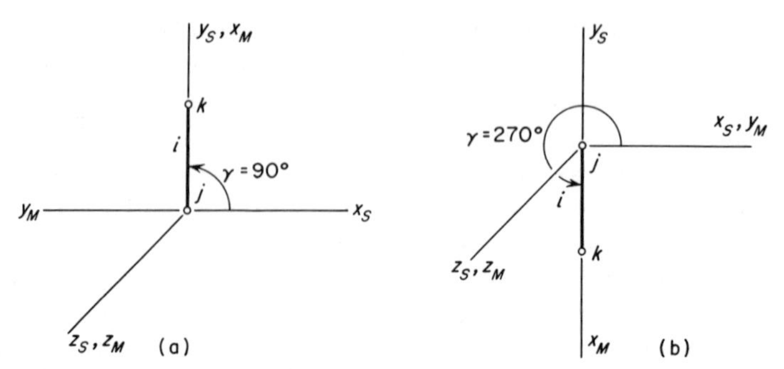

FIG. 4-39. Rotation of axes for a vertical space truss member.

The rotation matrix $\mathbf{R}$ which is used in transforming from the structure to the member axes can be developed following the two steps described above. Consider first the rotation about the y_S axis through the angle β. The 3×3 rotation matrix $\mathbf{R}_\beta$ for this transformation consists of the direction cosines of the β-axes (that is, the axes x_β, y_S, z_M) with respect to the structure axes (x_S, y_S, z_S):

$$\mathbf{R}_\beta = \begin{bmatrix} \cos\beta & 0 & \sin\beta \\ 0 & 1 & 0 \\ -\sin\beta & 0 & \cos\beta \end{bmatrix} \tag{a}$$

The functions $\cos\beta$ and $\sin\beta$ may be expressed in terms of the direction cosines of the member i by referring to the geometry of Fig. 4-38. Thus, these quantities become

$$\cos\beta = \frac{C_X}{\sqrt{C_X^2 + C_Z^2}} \qquad \sin\beta = \frac{C_Z}{\sqrt{C_X^2 + C_Z^2}} \tag{b}$$

in which the subscripts i are omitted from the direction cosines. Substitution of these expressions into Eq. (a) gives the matrix $\mathbf{R}_\beta$ in terms of the direction cosines:

$$\mathbf{R}_\beta = \begin{bmatrix} \frac{C_X}{\sqrt{C_X^2 + C_Z^2}} & 0 & \frac{C_Z}{\sqrt{C_X^2 + C_Z^2}} \\ 0 & 1 & 0 \\ \frac{-C_Z}{\sqrt{C_X^2 + C_Z^2}} & 0 & \frac{C_X}{\sqrt{C_X^2 + C_Z^2}} \end{bmatrix} \tag{4-82}$$

The physical significance of the matrix $\mathbf{R}_\beta$ is that it can be used to relate two alternative orthogonal sets of components of a vector (action or displacement) in the directions of the β-axes and the structure axes. For example, consider an action A that can be represented by either the vector $\mathbf{A}_\beta$ (consisting of components in the directions of the β axes) or the vector $\mathbf{A}_\mathrm{S}$ (consisting of components in the directions of the structure axes). The transformation of the latter vector into the former is performed by the method of Art. 4.15, as follows:

$$\mathbf{A}_\beta = \mathbf{R}_\beta \mathbf{A}_\mathrm{S} \tag{c}$$

where $\mathbf{R}_\beta$ is given by Eq. (4-82).

The second rotation about the axis z_M through the angle γ may be handled in a similar manner. A 3×3 matrix $\mathbf{R}_\gamma$ relates two orthogonal sets of components of the same vector for the two sets of axes. In the case of an action A, the transformation is

$$\mathbf{A}_\mathrm{M} = \mathbf{R}_\gamma \mathbf{A}_\beta \tag{d}$$

In this equation the vector $\mathbf{A}_\mathrm{M}$ consists of the components of the action A in the directions of the member axes, and the matrix $\mathbf{R}_\gamma$ contains the direction cosines of the member axes with respect to the β-axes. Writing the matrix $\mathbf{R}_\gamma$ in expanded form gives:

$$\mathbf{R}_\gamma = \begin{bmatrix} \cos\gamma & \sin\gamma & 0 \\ -\sin\gamma & \cos\gamma & 0 \\ 0 & 0 & 1 \end{bmatrix} \tag{e}$$

The functions $\cos \gamma$ and $\sin \gamma$ may be expressed in terms of the direction cosines of the member as follows (see Fig. 4-38):

$$\cos \gamma = \sqrt{C_X^2 + C_Z^2} \qquad \sin \gamma = C_Y \tag{f}$$

Substitution of these expressions into Eq. (e) yields the matrix $\mathbf{R}_\gamma$ in the following form:

$$\mathbf{R}_\gamma = \begin{bmatrix} \sqrt{C_X^2 + C_Z^2} & C_Y & 0 \\ -C_Y & \sqrt{C_X^2 + C_Z^2} & 0 \\ 0 & 0 & 1 \end{bmatrix} \tag{4-83}$$

Now that the two separate rotations have been expressed in matrix form, the single transformation matrix $\mathbf{R}$ from the structure axes to the member axes can also be obtained. Again considering the case of an action A, it can be seen that the vector $\mathbf{A}_\mathrm{M}$ may be expressed in terms of the vector $\mathbf{A}_\mathrm{S}$ by substituting $\mathbf{A}_\beta$ (see Eq. c) into Eq. (d):

$$\mathbf{A}_\mathrm{M} = \mathbf{R}_\gamma \mathbf{R}_\beta \mathbf{A}_\mathrm{S} \tag{g}$$

Comparing Eq. (g) with Eq. (4-57) in Art. 4.15 shows that the desired rotation matrix $\mathbf{R}$ consists of the product of $\mathbf{R}_\gamma$ and $\mathbf{R}_\beta$. Thus,

$$\mathbf{R} = \mathbf{R}_\gamma \mathbf{R}_\beta \tag{4-84}$$

If Eqs. (4-82) and (4-83) are substituted into Eq. (4-84) and multiplied, the following matrix results:

$$\mathbf{R} = \begin{bmatrix} C_X & C_Y & C_Z \\ \dfrac{-C_X C_Y}{\sqrt{C_X^2 + C_Z^2}} & \sqrt{C_X^2 + C_Z^2} & \dfrac{-C_Y C_Z}{\sqrt{C_X^2 + C_Z^2}} \\ \dfrac{-C_Z}{\sqrt{C_X^2 + C_Z^2}} & 0 & \dfrac{C_X}{\sqrt{C_X^2 + C_Z^2}} \end{bmatrix} \tag{4-85}$$

This is the rotation matrix $\mathbf{R}$ for a space truss member. It may be used whenever actions or displacements are to be transformed between the member axes and the structure axes.

The preceding rotation matrix $\mathbf{R}$ is valid for all positions of the member i except when it is vertical (see Fig. 4-39). In either of the cases shown in Fig. 4-39 the direction cosines of the member axes with respect to the structure axes can be determined by inspection. Thus, the rotation matrix is seen to be

$$\mathbf{R}_\mathrm{vert} = \begin{bmatrix} 0 & C_Y & 0 \\ -C_Y & 0 & 0 \\ 0 & 0 & 1 \end{bmatrix} \tag{4-86}$$

This expression for $\mathbf{R}$ is valid for both cases shown in Fig. 4-39. All that is necessary is to substitute for the direction cosine C_Y its appropriate value, which is 1 for the member in Fig. 4-39a and -1 for the member in Fig. 4-39b.

Substitution of the appropriate rotation matrix $\mathbf{R}$ (either Eq. 4-85 or 4-86) into Eq. (4-63) produces the desired rotation transformation matrix $\mathbf{R}_T$ for the set of member-oriented axes specified in this article.

4.23 Analysis of Space Trusses. The analysis of space truss structures is presented in this article in a manner analogous to the previous discussions for other types of structures. Loads on a space truss structure usually consist of concentrated forces applied at the joints, but in some instances actions of a more general nature may be applied to the individual members, as indicated in Fig. 4-36. As with previous structures, members are numbered 1 through m, and joints are numbered 1 through n_j.

Only axial deformations are taken into account in the analysis of a space truss, but there exists the possibility of three independent displacements at each joint. The possible displacements at a joint j in the structure are signified by the following indexes:

$$3j - 2 = \text{index for translation in the } x \text{ direction}$$

$$3j - 1 = \text{index for translation in the } y \text{ direction}$$

$$3j \quad = \text{index for translation in the } z \text{ direction}$$

The number of degrees of freedom in a space truss may be calculated using Eq. (4-65) in Art. 4.17. In addition, the indexes for the possible displacements of the ends of member i may be determined using

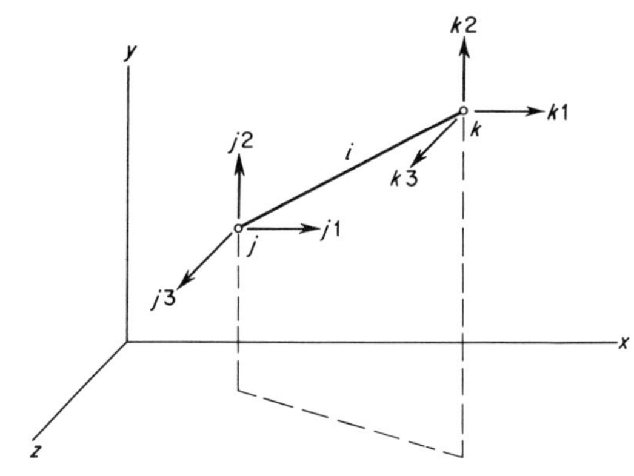

FIG. 4-40. End-displacements for space truss member.

Eqs. (4-66). These indexes are indicated in Fig. 4-40 for a space truss member.

The analysis of space trusses is symbolically similar to the analyses of plane frames (Art. 4.17) and grids (Art. 4.20). The reason for the parallelism is that all three of these types of structures have three possible displacements per joint. The similarity begins in the first phase of the analysis, in which the 6×6 member stiffness matrix $\mathbf{S}_{MD}$ (see Art. 4.21 for space truss member stiffnesses) is generated for each member in sequence. The elements of each of these member arrays are transferred systematically to the joint stiffness matrix $\mathbf{S}_J$ in the manner represented by Eqs. (4-67) through (4-72).

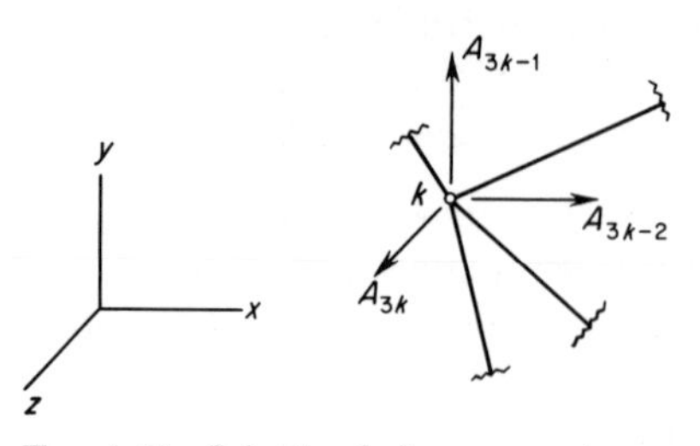

FIG. 4-41. Joint loads for a space truss.

In the next phase of the analysis of a space truss, loads applied to the structure are processed. Figure 4-41 shows the actions applied at a typical joint k in a space truss. The actions A_{3k-2}, A_{3k-1}, and A_{3k} are the x, y, and z components of the concentrated force applied at the joint. These actions are placed in the vector $\mathbf{A}$, which takes the form given by Eq. (4-73).

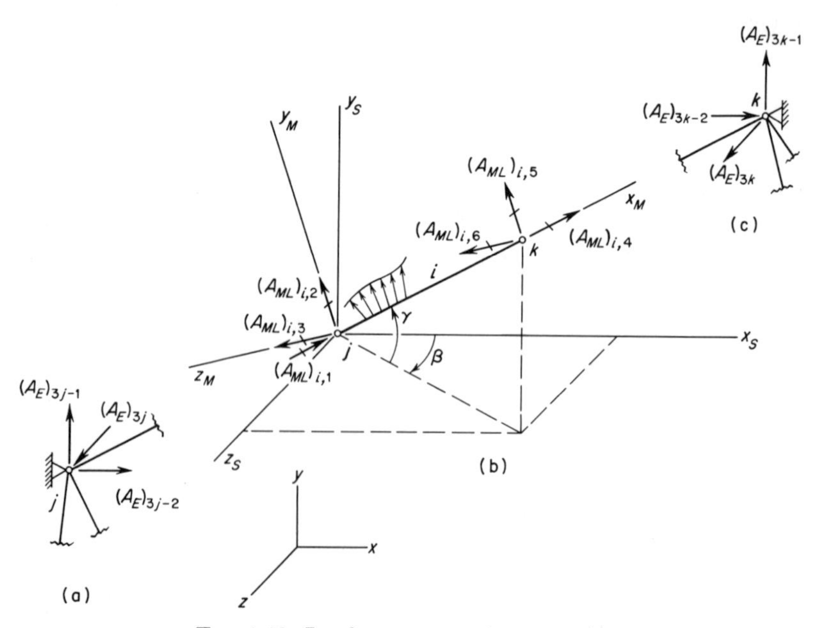

FIG. 4-42. Loads on a space truss member.

Consider next the matrix of actions $\mathbf{A}_{ML}$ at the ends of restrained members due to loads. In a truss structure these actions are determined with the ends of members restrained against translation but not against rotation (see Table B-5 of Appendix B). Figure 4-42b shows such a member i and the six end-actions caused by loading applied along its length. These end-actions are computed as the following:

$$(A_{ML})_{i,1} = \text{force in the } x_M \text{ direction at the } j \text{ end}$$
$$(A_{ML})_{i,2} = \text{force in the } y_M \text{ direction at the } j \text{ end}$$
$$(A_{ML})_{i,3} = \text{force in the } z_M \text{ direction at the } j \text{ end}$$
$$(A_{ML})_{i,4} = \text{force in the } x_M \text{ direction at the } k \text{ end}$$
$$(A_{ML})_{i,5} = \text{force in the } y_M \text{ direction at the } k \text{ end}$$
$$(A_{ML})_{i,6} = \text{force in the } z_M \text{ direction at the } k \text{ end}$$

The matrix $\mathbf{A}_{\mathrm{ML}}$ is conveniently formulated as a rectangular array of order $m \times 6$, as represented by Eq. (4-74).

Transfer of elements from the matrix $\mathbf{A}_{\mathrm{ML}}$ to the equivalent load vector $\mathbf{A}_{\mathrm{E}}$ may be accomplished by the method of rotation of axes. Figures 4-42a and 4-42c show the restraint actions at joints j and k which receive contributions from member i. The negatives of these contributions are determined by the calculations indicated in Table 4-41.

TABLE 4-41

Negatives of Contributions to $\mathbf{A}_{\mathrm{E}}$ for Skew Member

$$[R_T']_i\{A_{ML}\}_i = \begin{bmatrix} C_X & \frac{-C_XC_Y}{\sqrt{C_X^2 + C_Z^2}} & \frac{-C_Z}{\sqrt{C_X^2 + C_Z^2}} & 0 & 0 & 0 \\ C_Y & \sqrt{C_X^2 + C_Z^2} & 0 & 0 & 0 & 0 \\ C_Z & \frac{-C_YC_Z}{\sqrt{C_X^2 + C_Z^2}} & \frac{C_X}{\sqrt{C_X^2 + C_Z^2}} & 0 & 0 & 0 \\ 0 & 0 & 0 & C_X & \frac{-C_XC_Y}{\sqrt{C_X^2 + C_Z^2}} & \frac{-C_Z}{\sqrt{C_X^2 + C_Z^2}} \\ 0 & 0 & 0 & C_Y & \sqrt{C_X^2 + C_Z^2} & 0 \\ 0 & 0 & 0 & C_Z & \frac{-C_YC_Z}{\sqrt{C_X^2 + C_Z^2}} & \frac{C_X}{\sqrt{C_X^2 + C_Z^2}} \end{bmatrix}_i \begin{bmatrix} (A_{ML})_{i,1} \\ (A_{ML})_{i,2} \\ (A_{ML})_{i,3} \\ (A_{ML})_{i,4} \\ (A_{ML})_{i,5} \\ (A_{ML})_{i,6} \end{bmatrix}$$

Note that the transpose of the rotation transformation matrix $[R_T']_i$ in Table 4-41 incorporates the transpose of the rotation matrix $\mathbf{R}$ obtained from Eq. (4-85) for a skew member. If a particular member happens to be vertical, the transpose of the rotation matrix of Eq. (4-86) must be used instead. In this case the calculation takes the form given in Table 4-42.

In the final phase of the analysis, joint displacements D (translations in the x, y, and z directions) are computed by Eq. (4-8) and expanded into the vector $\mathbf{D}_{\mathrm{J}}$. Next, Eq. (4-4) provides the solution for support reactions (forces in the x, y, and z directions at supports). For the purpose of computing final end-actions, substitute the member stiffness

TABLE 4-42

Negatives of Contributions to A_E for Vertical Member

$$([R'_T]_i)_{\text{vert}}\{A_{ML}\}_i = \begin{bmatrix} 0 & -C_Y & 0 & 0 & 0 & 0 \\ C_Y & 0 & 0 & 0 & 0 & 0 \\ 0 & 0 & 1 & 0 & 0 & 0 \\ 0 & 0 & 0 & 0 & -C_Y & 0 \\ 0 & 0 & 0 & C_Y & 0 & 0 \\ 0 & 0 & 0 & 0 & 0 & 1 \end{bmatrix}_i \begin{bmatrix} (A_{ML})_{i,1} \\ (A_{ML})_{i,2} \\ (A_{ML})_{i,3} \\ (A_{ML})_{i,4} \\ (A_{ML})_{i,5} \\ (A_{ML})_{i,6} \end{bmatrix}$$

matrix $[S_M]_i$ from Table 4-39 and the rotation transformation matrix $[R_T]_i$ for a space truss into Eq. (4-76). The equations which result from the matrix multiplications are the following:

$$\begin{aligned} (A_M)_{i,1} &= (A_{ML})_{i,1} + \frac{EA_{Xi}}{L_i}\{[(D_J)_{j1} - (D_J)_{k1}]C_{Xi} \\ &\quad + [(D_J)_{j2} - (D_J)_{k2}]C_{Yi} + [(D_J)_{j3} - (D_J)_{k3}]C_{Zi}\} \\ (A_M)_{i,2} &= (A_{ML})_{i,2}; \quad (A_M)_{i,3} = (A_{ML})_{i,3} \\ (A_M)_{i,4} &= (A_{ML})_{i,4} - \frac{EA_{Xi}}{L_i}\{[(D_J)_{j1} - (D_J)_{k1}]C_{Xi} \\ &\quad + [(D_J)_{j2} - (D_J)_{k2}]C_{Yi} + [(D_J)_{j3} - (D_J)_{k3}]C_{Zi}\} \\ (A_M)_{i,5} &= (A_{ML})_{i,5}; \quad (A_M)_{i,6} = (A_{ML})_{i,6} \end{aligned} \tag{4-87}$$

This set of equations is valid for a member having any orientation, including a vertical member. In the latter case, the direction cosines C_{Xi} and C_{Zi} appearing in the equations will have the value zero.

A computer program for the analysis of space trusses is given in Art. 5.8. Numerical examples may also be found in that article.

4.24 Space Frame Member Stiffnesses. The space frame member stiffness matrix $\mathbf{S_M}$ for member axes was formulated previously in Art. 4.3. The objective in this article is to develop the member stiffness matrix $\mathbf{S_{MD}}$ for structure axes. The latter matrix is obtained from the former by the method of rotation of axes. The rotation matrix required for this transformation may take one of several forms (of varying complexity), depending upon the orientation of the member in space.

Figure 4-43 shows a portion of a space frame structure and a set of reference axes x, y, and z. A typical member i having positive direction cosines is indicated in the figure with joints at its ends denoted by j and k. The members of the frame are assumed to be rigidly connected at the joints, and each joint that is not restrained is assumed to translate and rotate in a completely general manner in space. Thus, all possible types of joint displacements must be considered and, for convenience, they are taken to be the translations and rotations in the directions of the x, y, and z axes (six possible displacements per joint).

The twelve possible displacements of the two ends of a member were discussed previously in Art. 4.3 (see Fig. 4-1) for member-oriented axes and are indicated again in Fig. 4-44a for the typical member i. In this

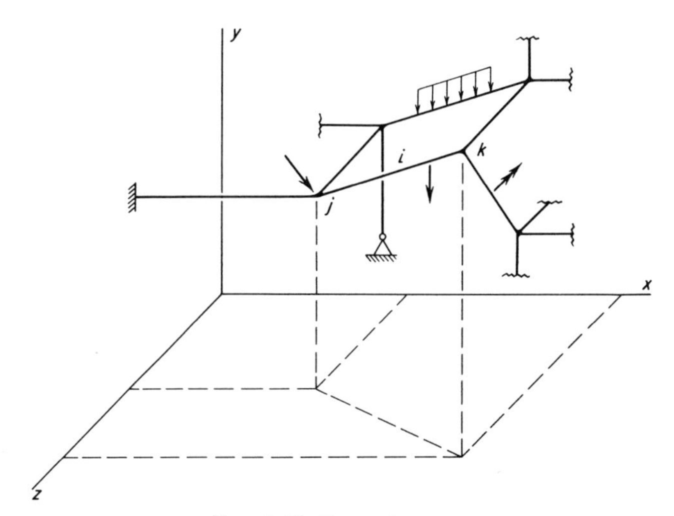

FIG. 4-43. Space frame.

figure, the x_M axis is taken along the axis of the member, in the same manner as for a space truss (compare with Fig. 4-38). The y_M and z_M axes are selected as the principal axes of the cross-section at end j of the member. The 12×12 stiffness matrix $\mathbf{S}_{\mathrm{M}}$ for the member axes was given earlier in Table 4-1. This matrix must be transformed, by means of a rotation transformation matrix, into the 12×12 matrix $\mathbf{S}_{\mathrm{MD}}$. The latter matrix corresponds to the twelve types of displacements indicated in Fig. 4-44b in the directions of structure axes.

The form of the rotation matrix $\mathbf{R}$ depends upon the particular orientation of the member axes. In many instances a space frame member will be oriented so that the principal axes of the cross-section lie in horizontal and vertical planes (for example, an I beam with its web in a vertical plane). Under these conditions the y_M and z_M axes can be selected exactly the same as for a space truss member (see Fig. 4-38), and the rotation matrix $\mathbf{R}$ given in Eq. (4-85) can also be used for the space frame member.

There are other instances in which a space frame member has two axes of symmetry in the cross-section and the same moment of inertia about each axis (for example, a circular or square member, either tubular or solid). In such cases the y_M and z_M axes can again be selected as described in the preceding paragraph. This choice can be made because all axes in the cross-section are principal axes, and any pair of axes can be selected for y_M and z_M.

In general, however, a space frame member may have its principal axes y_M and z_M in skew directions, such as indicated in Fig. 4-44a. There

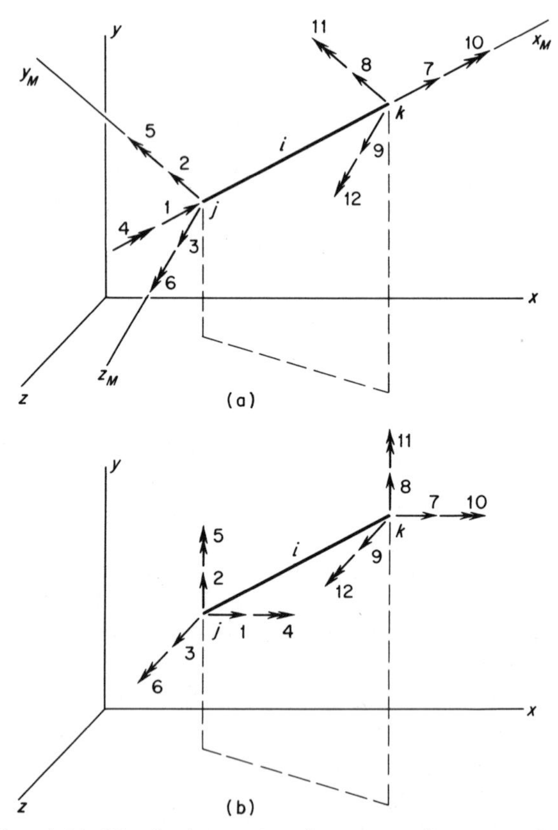

FIG. 4-44. Numbering system for a space frame member.

are various ways in which the position of these axes can be specified, and two methods will be described. The first method involves specifying the orientation of the principal axes by means of an angle of rotation about the x_M axis. In order to visualize clearly how such an angle is measured, consider the three successive rotations from the structure axes to the member axes shown in Fig. 4-45. The first two rotations through the angles β and γ (about the y_S and z_β axes, respectively) are exactly the same as shown in Fig. 4-38 for the special case described above. The third transformation consists of a rotation through the angle α about the x_M axis, causing the y_M and z_M axes to coincide with the principal

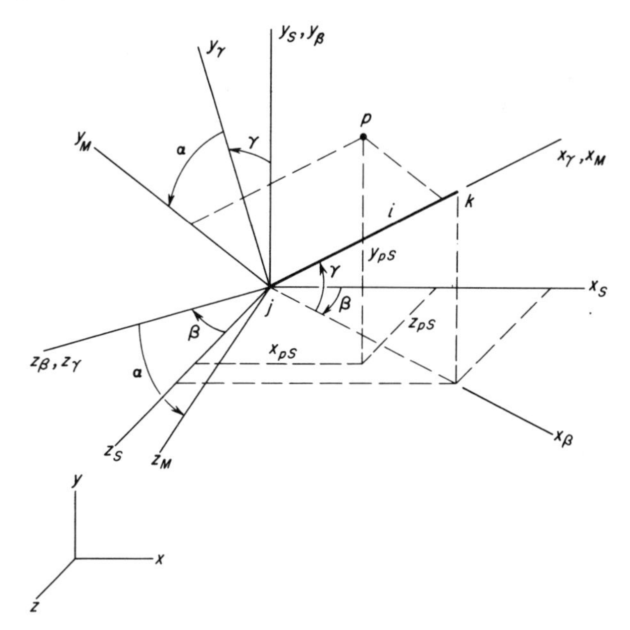

FIG. 4-45. Rotation of axes for a space frame member.

axes of the cross-section. This last rotation is also indicated in Fig. 4-46, which shows a cross-sectional view of the member looking in the negative x_M sense. The x_M-y_γ plane is a vertical plane through the axis of the member, and the angle α is measured (in the positive sense) from that plane to one of the principal axes of the cross-section.

The rotation of axes through the angle α requires the introduction of a rotation matrix $\mathbf{R}_\alpha$ in which the elements are the direction cosines of the final axes (x_M, y_M, z_M) with respect to the γ axes:

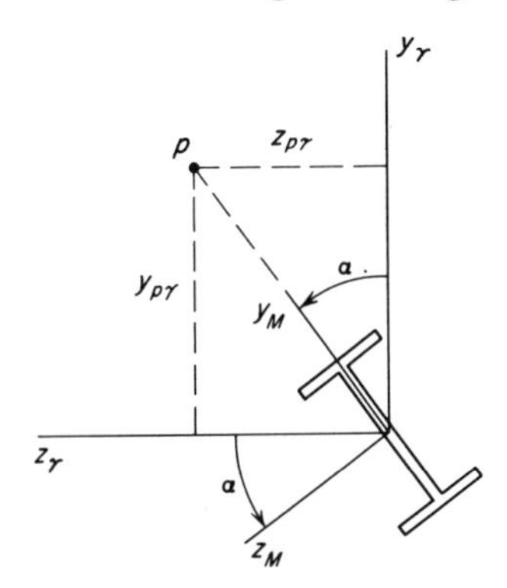

FIG. 4-46. Rotation of a space frame member about x_M axis.

$$\mathbf{R}_\alpha = \begin{bmatrix} 1 & 0 & 0 \\ 0 & \cos\alpha & \sin\alpha \\ 0 & -\sin\alpha & \cos\alpha \end{bmatrix} \tag{4-88}$$

Premultiplication of $\mathbf{R}_\gamma\mathbf{R}_\beta$ (see Eq. 4-84 and accompanying discussion) by $\mathbf{R}_\alpha$ yields the rotation matrix $\mathbf{R}$ for the three successive rotations shown in Fig. 4-45:

$$\mathbf{R} = \mathbf{R}_\alpha \mathbf{R}_\gamma \mathbf{R}_\beta \tag{4-89}$$

When the three rotation matrices $\mathbf{R}_\alpha$, $\mathbf{R}_\gamma$, and $\mathbf{R}_\beta$ (see Eqs. 4-88, 4-83, and 4-82) for a skew member are substituted into Eq. (4-89) and multiplied as indicated, the rotation matrix $\mathbf{R}$ becomes the following:

$$\mathbf{R} = \begin{bmatrix} C_X & C_Y & C_Z \\ \dfrac{-C_X C_Y \cos\alpha - C_Z \sin\alpha}{\sqrt{C_X^2 + C_Z^2}} & \sqrt{C_X^2 + C_Z^2}\cos\alpha & \dfrac{-C_Y C_Z \cos\alpha + C_X \sin\alpha}{\sqrt{C_X^2 + C_Z^2}} \\ \dfrac{C_X C_Y \sin\alpha - C_Z \cos\alpha}{\sqrt{C_X^2 + C_Z^2}} & -\sqrt{C_X^2 + C_Z^2}\sin\alpha & \dfrac{C_Y C_Z \sin\alpha + C_X \cos\alpha}{\sqrt{C_X^2 + C_Z^2}} \end{bmatrix} \tag{4-90}$$

This rotation matrix is expressed in terms of the direction cosines of the member (which are readily computed from the coordinates of the joints) and the angle α, which must be given as part of the description of the structure itself. Note that if α is equal to zero, the matrix $\mathbf{R}$ reduces to the form given previously for a space truss member (Eq. 4-85). The special case of a vertical member will be discussed later.

The rotation transformation matrix $\mathbf{R}_\mathrm{T}$ for a space frame member can be shown to take the following form:

$$\mathbf{R}_\mathrm{T} = \begin{bmatrix} \mathbf{R} & \mathbf{O} & \mathbf{O} & \mathbf{O} \\ \mathbf{O} & \mathbf{R} & \mathbf{O} & \mathbf{O} \\ \mathbf{O} & \mathbf{O} & \mathbf{R} & \mathbf{O} \\ \mathbf{O} & \mathbf{O} & \mathbf{O} & \mathbf{R} \end{bmatrix} \tag{4-91}$$

which is analogous to the matrix given in Eq. (4-63). Finally, the member stiffness matrix $\mathbf{S}_\mathrm{MD}$ for structure axes may be computed by the usual matrix multiplications:

$$\mathbf{S}_\mathrm{MD} = \mathbf{R}_\mathrm{T}' \mathbf{S}_\mathrm{M} \mathbf{R}_\mathrm{T} \tag{4-92}$$

The resulting matrix is quite complicated when expressed in literal form, and for that reason it is not expanded in the text. The fact that such a lengthy set of relationships can be represented so concisely by Eq. (4-92) is one of the principal advantages of matrix methods in the analysis of structures.

In some structures the orientation of a particular member may be such that the angle α which specifies the location of the principal axes for that member may not be readily available. In that event, a different technique for describing their location can be used. A suitable method is to give the coordinates of a point that lies in one of the principal planes of the member but is not on the axis of the member itself. This point and the x_M axis will define without ambiguity a plane in space, and that plane can be taken as the x_M-y_M plane. All that is necessary is to obtain expressions for the angle of rotation α, which appears in the rotation matrix **R** (Eq. 4-90), in terms of the coordinates of the given point and the coordinates of the ends of the member itself. Once this is accomplished, it becomes possible for the structural analyst to describe the position of the principal axes of a member either by giving the angle α directly or by giving the coordinates of a suitable point.

An arbitrary point p in the x_M-y_M plane is shown in Fig. 4-45. The x, y, and z coordinates of this point (denoted as x_p, y_p, and z_p) are assumed to be given. Since the structure axes x_S, y_S, and z_S have their origin at end j of the member, the coordinates of point p with respect to the structure axes (denoted as x_{pS}, y_{pS}, and z_{pS}) are

$$x_{pS} = x_p - x_j \qquad y_{pS} = y_p - y_j \qquad z_{pS} = z_p - z_j \tag{4-93}$$

in which x_j, y_j, and z_j are the coordinates of end j of the member.

The point p is also shown in Fig. 4-46 in conjunction with the rotation angle α. The coordinates of p with respect to the axes x_γ, y_γ, and z_γ can be obtained in terms of those for the axes x_S, y_S, and z_S by a rotation of axes through the angles β and γ. Let the coordinates of point p with respect to the γ-axes be $x_{p\gamma}$, $y_{p\gamma}$, and $z_{p\gamma}$ (the latter two coordinates are shown positive in Fig. 4-46). Then the desired transformation of coordinates becomes the following:

$$\begin{bmatrix} x_{p\gamma} \\ y_{p\gamma} \\ z_{p\gamma} \end{bmatrix} = \mathbf{R}_\gamma \mathbf{R}_\beta \begin{bmatrix} x_{pS} \\ y_{pS} \\ z_{pS} \end{bmatrix} \tag{a}$$

Substitution of $\mathbf{R}_\gamma \mathbf{R}_\beta$ from Eq. (4-85) into Eq. (a) yields the following expressions:

$$\begin{aligned} x_{p\gamma} &= C_X x_{pS} + C_Y y_{pS} + C_Z z_{pS} \\ y_{p\gamma} &= -\frac{C_X C_Y}{\sqrt{C_X^2 + C_Z^2}} x_{pS} + \sqrt{C_X^2 + C_Z^2}\, y_{pS} - \frac{C_Y C_Z}{\sqrt{C_X^2 + C_Z^2}} z_{pS} \\ z_{p\gamma} &= -\frac{C_Z}{\sqrt{C_X^2 + C_Z^2}} x_{pS} + \frac{C_X}{\sqrt{C_X^2 + C_Z^2}} z_{pS} \end{aligned} \tag{4-94}$$

These equations give the coordinates of point p with respect to the γ axes.

From the geometry of Fig. 4-46 the following expressions for the sine and cosine of the angle α are obtained:

$$\sin \alpha = \frac{z_{p\gamma}}{\sqrt{y_{p\gamma}^2 + z_{p\gamma}^2}} \qquad \cos \alpha = \frac{y_{p\gamma}}{\sqrt{y_{p\gamma}^2 + z_{p\gamma}^2}} \tag{4-95}$$

By using Eqs. (4-93), (4-94), and (4-95) the quantities $\sin \alpha$ and $\cos \alpha$ can be computed from the coordinates of point p and then substituted into the rotation matrix **R** (Eq. 4-90). This substitution produces the rotation matrix **R** in a form which involves only the direction cosines of the member itself and the coordinates of point p.

The preceding discussion dealt with a member that was not vertical. As shown in Art. 4.22 for a space truss member, a vertical member is always a special case that must be treated separately. The rotation matrix **R** given previously for a vertical space truss member (see Eq. 4-86) can also be used for certain types of vertical space frame members. The only requirement is that one of the principal axes of the cross-section of the member be in the direction of the z_S axis. In some cases the member will be oriented in such a manner that this requirement is met (for example, an I beam with its web in the x_S-y_S plane). The requirement is also satisfied by a member having a cross-section that is either circular or square (solid or tubular). In these cases all axes of the cross-section are principal axes, and hence one of them can arbitrarily be selected in the z_S direction.

In a more general case of a vertical member, the principal axes of the cross-section will be rotated about the x_M axis so that they form an angle α with the directions of the structure axes. The orientation of such a vertical member can best be visualized by considering the successive rotations of axes shown in Fig. 4-47. There is no rotation through the angle β (about the y_S axis) in this case. Instead, the first rotation is through the angle γ, which may be either 90° or 270° (see Figs. 4-47a and 4-47b, respectively). The second rotation is through the angle α and about the x_M axis. The rotation matrix for either of the cases shown in the figure can be obtained by inspection. It consists of the direction cosines of the x_M, y_M, and z_M axes with respect to the structure axes:

$$\mathbf{R}_{\text{vert}} = \begin{bmatrix} 0 & C_Y & 0 \\ -C_Y \cos\alpha & 0 & \sin\alpha \\ C_Y \sin\alpha & 0 & \cos\alpha \end{bmatrix} \tag{4-96}$$

This expression is valid for both orientations shown in Fig. 4-47 pro-

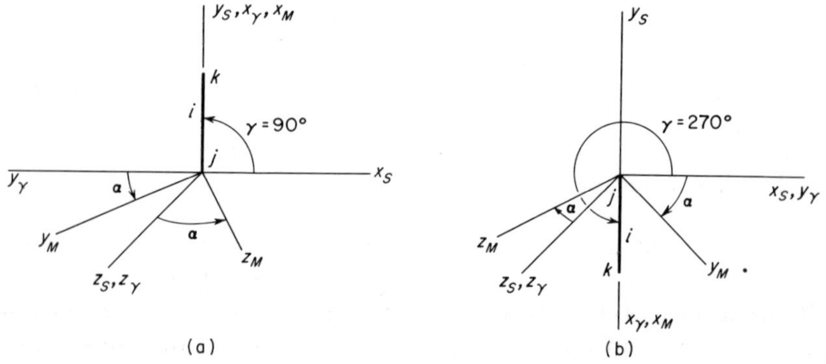

FIG. 4-47. Rotation of axes for a vertical space frame member.

vided the appropriate value of C_Y is substituted (+1 in the first case and -1 in the second case). It should be noted also that if α is equal to zero, the rotation matrix given in Eq. (4-96) reduces to the rotation matrix for a vertical space truss member (Eq. 4-86).

In those cases where it is desired to begin with the coordinates of a point p that is known to lie in a principal plane, it is possible to calculate $\sin \alpha$ and $\cos \alpha$ for use in Eq. (4-96) directly from the coordinates of the point. Figure 4-48a shows a vertical member with the lower end designated as joint j and the upper end designated as joint k. Also shown in the figure is the point p. When this point is located so that its coordinates with respect to the S-axes are positive, the angle α will be between 90° and 180°. The sine and cosine of this angle are as follows:

$$\sin \alpha = \frac{z_{pS}}{\sqrt{x_{pS}^2 + z_{pS}^2}} \qquad \cos \alpha = \frac{-x_{pS}}{\sqrt{x_{pS}^2 + z_{pS}^2}} \tag{b}$$

On the other hand, Fig. 4-48b represents a vertical member with the lower end designated as joint k and the upper end designated as joint j.

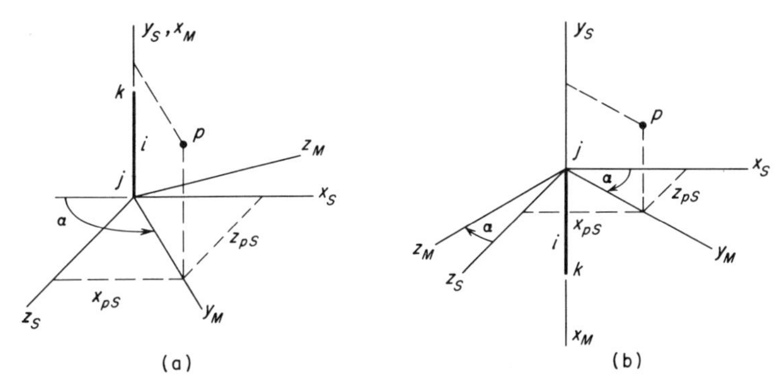

FIG. 4-48. Use of point p for a vertical space frame member.

In this case the angle α is between 0° and 90° when p has positive coordinates, and $\sin \alpha$ and $\cos \alpha$ are given by the following expressions:

$$\sin \alpha = \frac{z_{pS}}{\sqrt{x_{pS}^2 + z_{pS}^2}} \qquad \cos \alpha = \frac{x_{pS}}{\sqrt{x_{pS}^2 + z_{pS}^2}} \tag{c}$$

Expressions (b) and (c) may be combined into one set by introducing the direction cosine C_Y for the member, as follows:

$$\sin \alpha = \frac{z_{pS}}{\sqrt{x_{pS}^2 + z_{pS}^2}} \qquad \cos \alpha = \frac{-x_{pS}}{\sqrt{x_{pS}^2 + z_{pS}^2}} C_Y \tag{4-97}$$

For the cases shown in Fig. 4-48a and 4-48b the direction cosine C_Y has the values +1 and -1, respectively. Equations (4-97) can be used to calculate $\sin \alpha$ and $\cos \alpha$ from the coordinates of point p, which must be located in a principal plane of the member. These functions can then be substituted into the rotation matrix **R** (Eq. 4-96).

In summary, the member stiffness matrix $\mathbf{S}_{\mathrm{M}}$ for a space frame member is first obtained for the member axes using Table 4-1 in Art. 4.3. Next, a rotation matrix **R** is constructed in a form which depends upon the case under consideration, using either the angle α or the coordinates of a point p to identify a principal plane. The rotation matrix **R** either has the form given in Eq. (4-90) or Eq. (4-96), depending upon whether the member is inclined or vertical. In both cases, the angle α may be zero, which means that **R** reduces to one of the forms given previously for a space truss member (see Eqs. 4-85 and 4-86). In all cases, the rotation transformation matrix $\mathbf{R}_{\mathrm{T}}$ is formed according to Eq. (4-91), and the member stiffness matrix for structure axes is computed by Eq. (4-92).

4.25 Analysis of Space Frames. The space frame shown in Fig. 4-43 contains rigidly connected members oriented in a general manner in space. The loads on the structure may be of any type and orientation. The numbering system to be adopted for members and joints is the same as for structures discussed previously.

Axial, flexural, and torsional deformations are all considered in the analysis of space frames. The unknown displacements at the joints consist of six types, namely, the x, y, and z components of the joint translations and the x, y, and z components of the joint rotations. The six possible displacements at a particular joint j are denoted as follows:

$6j - 5$ = index for translation in the x direction

$6j - 4$ = index for translation in the y direction

$6j - 3$ = index for translation in the z direction

$6j - 2$ = index for rotation in the x sense

$6j - 1$ = index for rotation in the y sense

$6j$ = index for rotation in the z sense

Also, the number of degrees of freedom n in a space frame may be determined from the number of joints n_j and the number of restraints n_r by the following expression:

$$n = 6n_j - n_r \tag{4-98}$$

A member i in a space frame will have joint numbers j and k at its ends, as shown in Fig. 4-49. The twelve possible displacements of the joints associated with this member are also indicated in Fig. 4-49. These displacements are indexed as follows:

$$
\begin{array}{lll}
j1 = 6j - 5 & j2 = 6j - 4 & j3 = 6j - 3 \\
j4 = 6j - 2 & j5 = 6j - 1 & j6 = 6j \\
k1 = 6k - 5 & k2 = 6k - 4 & k3 = 6k - 3 \\
k4 = 6k - 2 & k5 = 6k - 1 & k6 = 6k
\end{array} \tag{4-99}
$$

Construction of the joint stiffness matrix follows the same general pattern as with plane frames (see Art. 4.17), except that the process is

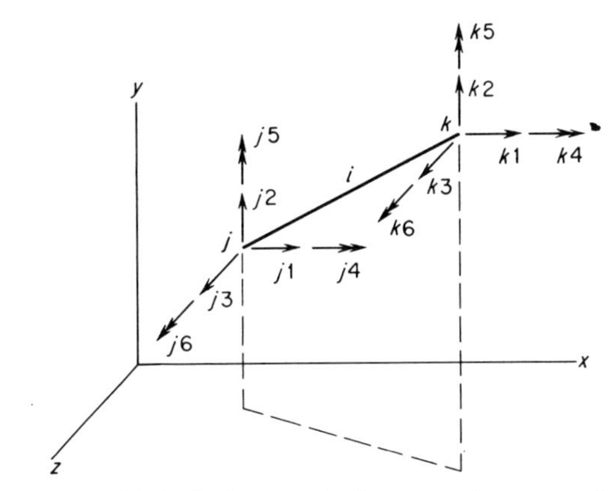

FIG. 4-49. End-displacements for space frame member.

complicated by the fact that more terms are involved. The 12 × 12 stiffness matrix $\mathbf{S_{MD}}$ is generated for each member in the frame (see Art. 4.24), and its contributions to the stiffnesses of joints j and k are assessed. For example, the first column in the matrix $\mathbf{S_{MD}}$ contributes to the joint stiffness matrix $\mathbf{S_J}$ as follows:

$$
\begin{array}{ll}
(S_J)_{j1,j1} = \sum S_{MD} + (S_{MD1,1})_i & (S_J)_{k1,j1} = (S_{MD7,1})_i \\
(S_J)_{j2,j1} = \sum S_{MD} + (S_{MD2,1})_i & (S_J)_{k2,j1} = (S_{MD8,1})_i \\
(S_J)_{j3,j1} = \sum S_{MD} + (S_{MD3,1})_i & (S_J)_{k3,j1} = (S_{MD9,1})_i \\
(S_J)_{j4,j1} = \sum S_{MD} + (S_{MD4,1})_i & (S_J)_{k4,j1} = (S_{MD10,1})_i \\
(S_J)_{j5,j1} = \sum S_{MD} + (S_{MD5,1})_i & (S_J)_{k5,j1} = (S_{MD11,1})_i \\
(S_J)_{j6,j1} = \sum S_{MD} + (S_{MD6,1})_i & (S_J)_{k6,j1} = (S_{MD12,1})_i
\end{array} \tag{4-100}
$$

Eleven other sets of expressions similar to Eqs. (4-100) may be written to make a total of twelve sets of equations. Each set involves transferring elements from a given column in the matrix $\mathbf{S_{MD}}$ to the appropriate locations in the matrix $\mathbf{S_J}$.

Actions applied at a typical joint k are shown in Fig. 4-50. In the figure, the actions A_{6k-5}, A_{6k-4}, and A_{6k-3} are the x, y, and z components of the concentrated force applied at the joint. In addition, the actions

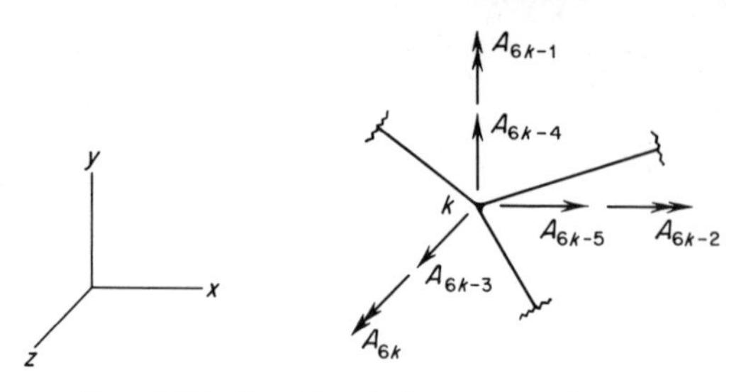

FIG. 4-50. Joint loads for a space frame.

A_{6k-2}, A_{6k-1}, and A_{6k} are the x, y, and z components of the moment vector acting at the joint. These actions are placed in the vector of applied actions $\mathbf{A}$, which assumes the following form:

$$\mathbf{A} = \{A_1, A_2, A_3, A_4, A_5, A_6, \ldots$$
$$\ldots, A_{6k-5}, A_{6k-4}, A_{6k-3}, A_{6k-2}, A_{6k-1}, A_{6k}, \ldots$$
$$\ldots, A_{6n_j-5}, A_{6n_j-4}, A_{6n_j-3}, A_{6n_j-2}, A_{6n_j-1}, A_{6n_j}\} \quad (4\text{-}101)$$

Figure 4-51b shows the actions at the ends of a restrained space frame member i due to loads on the member itself. The end-actions $\mathbf{A}_{\mathrm{ML}}$ for member-oriented axes are defined as follows:

$$(A_{ML})_{i,1} = \text{force in the } x_M \text{ direction at the } j \text{ end}$$
$$(A_{ML})_{i,2} = \text{force in the } y_M \text{ direction at the } j \text{ end}$$
$$(A_{ML})_{i,3} = \text{force in the } z_M \text{ direction at the } j \text{ end}$$
$$(A_{ML})_{i,4} = \text{couple in the } x_M \text{ sense at the } j \text{ end}$$
$$(A_{ML})_{i,5} = \text{couple in the } y_M \text{ sense at the } j \text{ end}$$
$$(A_{ML})_{i,6} = \text{couple in the } z_M \text{ sense at the } j \text{ end}$$
$$(A_{ML})_{i,7} = \text{force in the } x_M \text{ direction at the } k \text{ end}$$
$$(A_{ML})_{i,8} = \text{force in the } y_M \text{ direction at the } k \text{ end}$$
$$(A_{ML})_{i,9} = \text{force in the } z_M \text{ direction at the } k \text{ end}$$
$$(A_{ML})_{i,10} = \text{couple in the } x_M \text{ sense at the } k \text{ end}$$
$$(A_{ML})_{i,11} = \text{couple in the } y_M \text{ sense at the } k \text{ end}$$
$$(A_{ML})_{i,12} = \text{couple in the } z_M \text{ sense at the } k \text{ end}$$

The matrix $\mathbf{A}_{\mathrm{ML}}$ may be treated as a rectangular array of order $m \times 12$, in which each row is made up of the twelve elements enumerated above for a given member. Thus,

$$\mathbf{A}_{\mathrm{ML}} = \begin{bmatrix} (A_{ML})_{1,1} & (A_{ML})_{1,2} & \cdots & (A_{ML})_{1,11} & (A_{ML})_{1,12} \\ \cdots & \cdots & \cdots & \cdots & \cdots \\ (A_{ML})_{i,1} & (A_{ML})_{i,2} & \cdots & (A_{ML})_{i,11} & (A_{ML})_{i,12} \\ \cdots & \cdots & \cdots & \cdots & \cdots \\ (A_{ML})_{m,1} & (A_{ML})_{m,2} & \cdots & (A_{ML})_{m,11} & (A_{ML})_{m,12} \end{bmatrix} \quad (4\text{-}102)$$

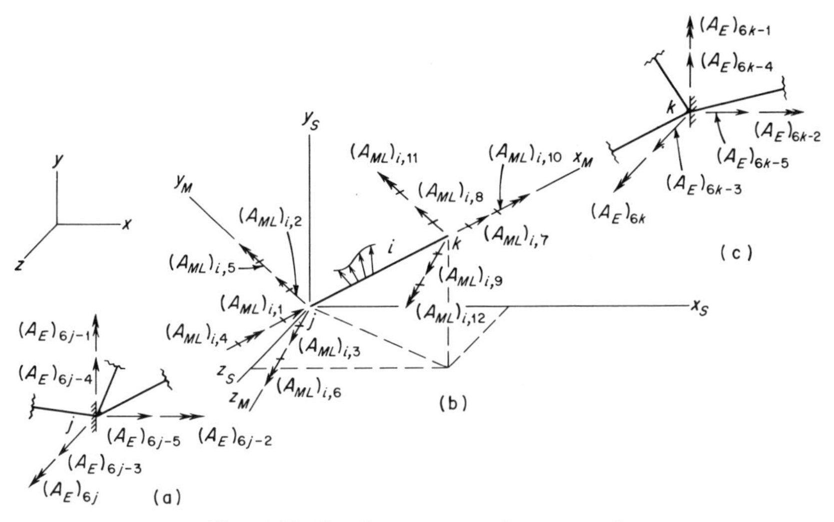

FIG. 4-51. Loads on a space frame member.

The construction of the equivalent load vector $\mathbf{A}_E$ is accomplished through the rotation of axes procedure used previously. Figures 4-51a and 4-51c show the equivalent loads at joints j and k which receive contributions from member i. The negatives of the incremental portions of $\mathbf{A}_E$ contributed by this member are obtained by evaluating the product of $[R'_T]_i$ and $\{A_{ML}\}_i$. The form of the rotation matrix $\mathbf{R}$ used for this purpose is dictated by the category into which the particular member falls. These categories were discussed in Art. 4.24. Following the calculation of $\mathbf{A}_E$, this vector is added to the vector $\mathbf{A}$ to form the combined load vector $\mathbf{A}_C$.

The analysis is completed by calculating joint displacements D (translations and rotations in the x, y, and z directions), using Eq. (4-8) and then expanding them into the vector $\mathbf{D}_J$. Support reactions A_R (forces and couples in the x, y, and z directions) are computed next using Eq. (4-4). Finally, member end-actions for each member are computed by substituting the member stiffness matrix $[S_M]_i$ for member axes (Table 4-1) and the appropriate form of the rotation transformation matrix $[R_T]_i$ into the following equation:

$$\{A_M\}_i = \{A_{ML}\}_i + [S_M]_i[R_T]_i\{D_J\}_i \tag{4-103}$$

This equation is of the same form as Eq. (4-76) given previously for plane frames.

A computer program for the analysis of space frames is given in Art. 5.9, and numerical examples are also included in that article.

Problems

The problems for Art. 4.9 are to be solved in the manner described in Arts. 4.8 and 4.9. In each problem all of the joint displacements, support reactions, and member end-actions are to be obtained, unless stated otherwise. Use the arbitrary numbering system shown in Fig. 4-7b when obtaining the over-all joint stiffness matrix.

4.9-1. Analyze the continuous beam in Fig. 2-16a, assuming that the flexural rigidity EI is constant for all spans. Assume also that $wL = P$ and $M = PL$.

4.9-2. Make an analysis of the continuous beam in Fig. 2-17a if the flexural rigidity EI is the same for both spans and $P_1 = 2P$, $P_2 = P$.

4.9-3. Analyze the three-span beam shown in the figure for Prob. 2.9-7 if $L_1 = L_3 = L$, $L_2 = 2L$, $P_1 = P_2 = P_3 = P$, $M = PL$, and $wL = P$. The flexural rigidity for members AB and CD is EI and for member BC is $2EI$.

4.9-4. Analyze the beam shown in the figure for Prob. 2.9-14 assuming that $EI_1 = 2EI$ and $EI_2 = EI$.

4.9-5. Analyze the overhanging beam shown in the figure, taking points A, B, and C as joints. The beam has constant flexural rigidity EI.

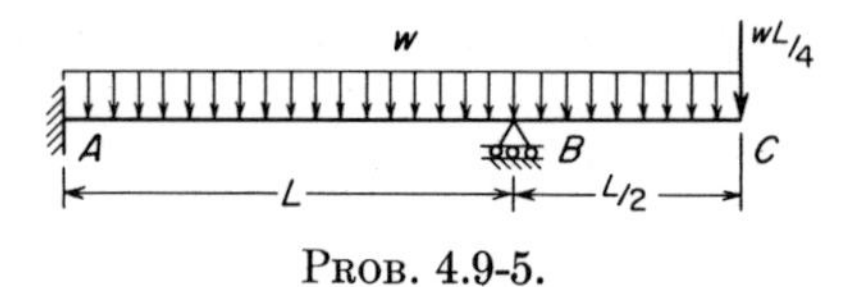

Prob. 4.9-5.

4.9-6. Analyze the three-span beam having constant flexural rigidity EI (see figure).

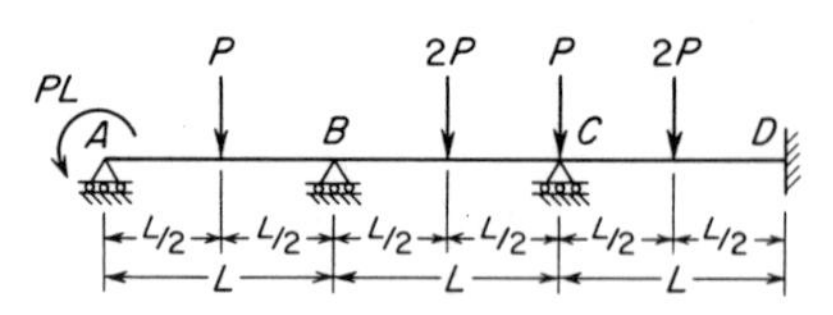

Prob. 4.9-6.

4.9-7. Analyze the beam shown in the figure, taking points A, B, C, and D as joints. The segment AB has a flexural rigidity of $2EI$, and the portion from B to D has a constant flexural rigidity of EI.

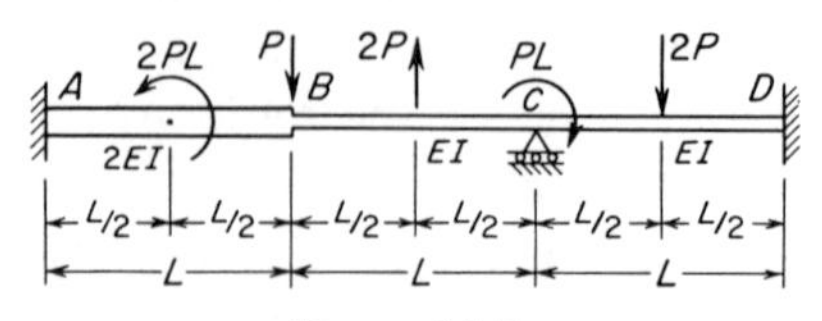

Prob. 4.9-7.

4.9-8. Obtain the joint stiffness matrix $\mathbf{S}_J$ for the beam shown in Prob. 2.9-8; also, rearrange and partition $\mathbf{S}_J$ into the form given by Eq. (4-1). Assume that the beam has constant EI.

4.9-9. Obtain the joint stiffness matrix $\mathbf{S}_J$ for the beam shown in Prob. 2.9-11, assuming that EI is constant. Also, rearrange and partition the matrix into the form given by Eq. (4-1).

4.9-10. Find the joint stiffness matrix $\mathbf{S}_J$ for the beam of Prob. 2.9-12, assuming that the flexural rigidity of the middle span is twice that of the end spans. Also, rearrange and partition the matrix into the form given by Eq. (4-1).

4.9-11. Find the joint stiffness matrix $\mathbf{S}_J$ for a continuous beam on five simple supports having four identical spans, each of length L. Also, rearrange and partition the matrix into the form given by Eq. (4-1). Assume that EI is constant for all spans.

4.9-12. Obtain the joint stiffness matrix $\mathbf{S}_J$ for the beam shown in the figure. Also, rearrange and partition the matrix into the form given by Eq. (4-1). The beam has a flexural rigidity of $2EI$ from A to C and EI from C to E.

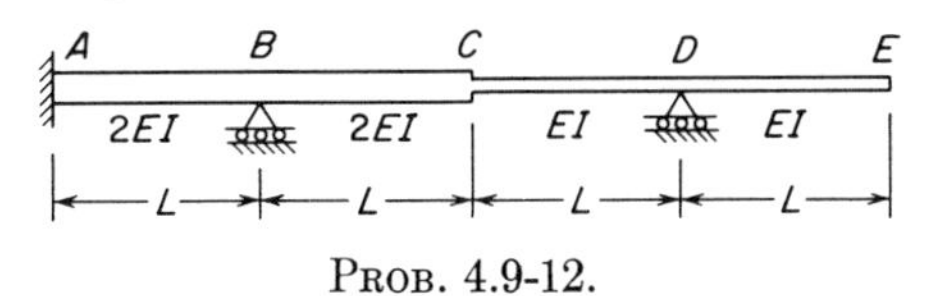

PROB. 4.9-12.

The problems for Art. 4.12 are to be solved in the manner described in Arts. 4.11 and 4.12. In each problem all of the joint displacements, support reactions, and member end-actions are to be obtained unless otherwise stated. Use the arbitrary numbering systems shown in the figures which accompany the problems.

4.12-1. Analyze the plane truss structure in the figure for Prob. 2.9-16 under the effect of the vertical load P shown in the figure plus a horizontal load of $2P$ applied at the mid-height of member AC and directed to the right. Each bar of the truss has the same axial rigidity EA_X. Number the members as shown in the figure, and number the joints in the same sequence as the letters in the figure.

4.12-2. Analyze the plane truss shown in Fig. 4-23 for the loads shown, but with a hinge support which prevents translation at point B (joint 2). Otherwise, the data are the same as in the example problem.

4.12-3. Analyze the plane truss shown in the figure for Prob. 2.9-20 due to the weights of the bars. Each bar has the same axial rigidity EA_X and the same weight w per unit length. Number the joints in the following sequence: D, C, A, and B. Number the members in the following order: DC, AB, AD, BC, DB, and AC.

4.12-4. Analyze the plane truss shown in Fig. 4-23 for the effect of its own weight, assuming a roller support exists at joint B which prevents translation in the x direction. Assume that the weight of each member is w per unit length. Otherwise, the structure data are the same as in the example problem.

4.12-5. Analyze the plane truss for the loads shown (see figure). All members have the same axial rigidity EA_X.

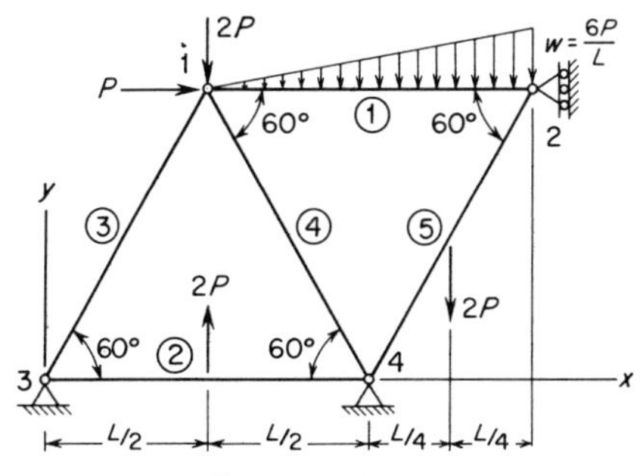

PROB. 4.12-5.

4.12-6. Analyze the plane truss shown in the figure for the effect of its own weight. Assume that the axial rigidity of each member is EA_X, and that each has the same weight w per unit length. (Hint: Take advantage of the symmetry of the structure and loading by working with half the structure as indicated in the second part of the figure. Note that both the area and the weight of the middle bar must be halved.)

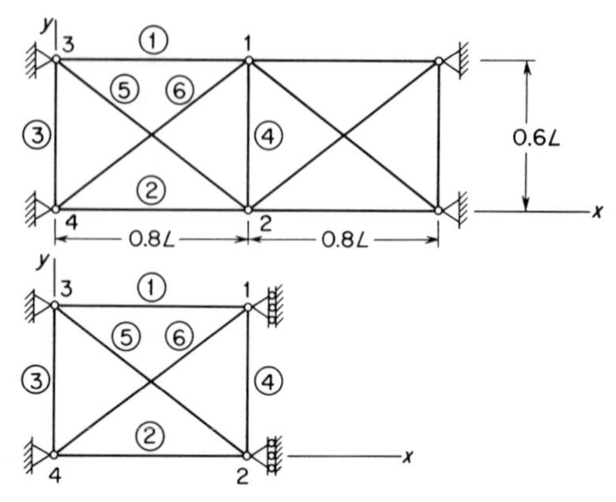

PROB. 4.12-6.

4.12-7. Obtain the joint stiffness matrix $\mathbf{S}_J$ for the plane truss shown in the figure. Assume that all bars have the same axial rigidity EA_X.

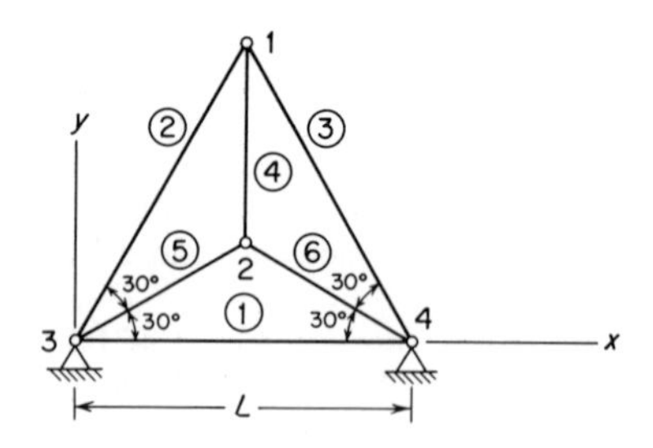

PROB. 4.12-7.

4.12-8. Determine the joint stiffness matrix $\mathbf{S}_J$ for the plane truss shown in the figure. Assume that all bars have the same axial rigidity EA_X.

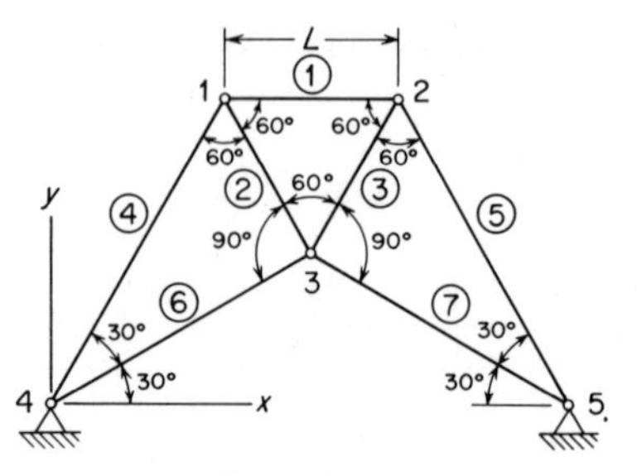

PROB. 4.12-8.

4.12-9. Obtain the joint stiffness matrix $\mathbf{S}_J$ for the plane truss structure shown in the figure. Rearrange and partition the matrix into the form given by Eq. (4-1). Assume that the horizontal and vertical members have cross-sectional areas A_X, and that the diagonal members have cross-sectional areas $2A_X$.

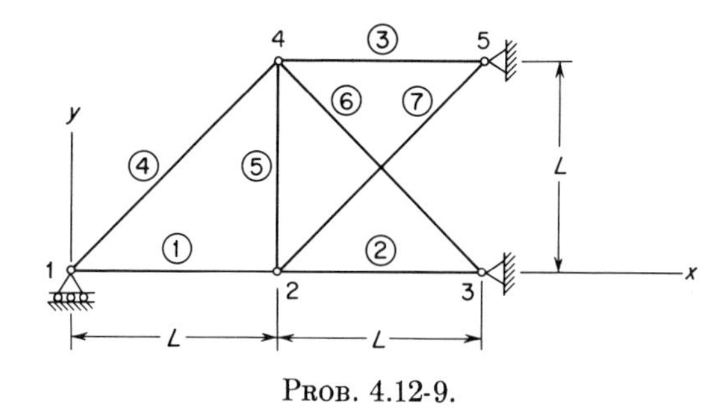

PROB. 4.12-9.

4.12-10. Determine the joint stiffness matrix $\mathbf{S}_J$ for the plane truss structure given in the figure. Assume that the cross-sectional areas of members numbered one through five are A_X, and that the cross-sectional areas of members numbered six through ten are $2A_X$.

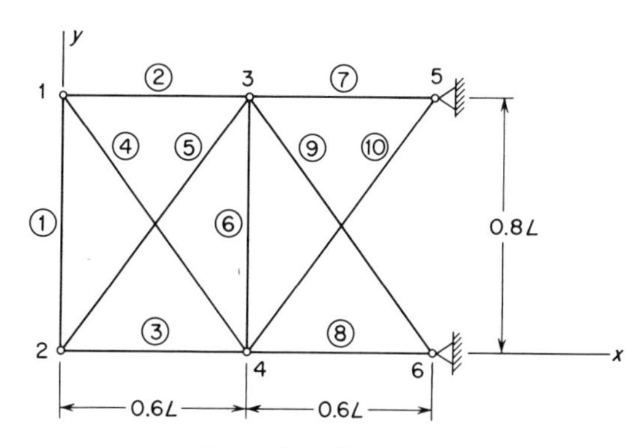

PROB. 4.12-10.

The problems for Art. 4.18 are to be solved in the manner described in Arts. 4.17 and 4.18. In each problem all of the joint displacements, support reactions, and member end-actions are to be obtained unless otherwise stated. Use the numbering systems shown in the figures which accompany the problems. In the calculations use units of kips and inches.

4.18-1. Analyze the plane frame shown in Fig. 2-22a, assuming that both members have the same cross-sectional properties. Use the following numerical data: $P = 10$ kips, $L = H = 12$ ft, $E = 30{,}000$ ksi, $I_Z = 200$ in.^{4}, and $A_X = 10$ in.^{2} Number the joints in the sequence B, A, C, and number the members in the sequence AB, BC.

4.18-2. Analyze the plane frame shown in the figure for Probs. 2.9-25 and 2.9-27 if P = 6 kips, L = 24 ft, H = 16 ft, E = 30,000 ksi, I_Z = 350 in.^{4}, and A_X = 16 in.^{2} Number the joints in the sequence B, A, C, D, and number the members in the sequence AB, BC, DB.

4.18-3. Analyze the plane frame shown in the figure, assuming that both members have the same cross-sectional properties. Use the following data: P = 10 kips, L = 20 ft, E = 30,000 ksi, I_Z = 500 in.^{4}, and A_X = 12 in.^{2}

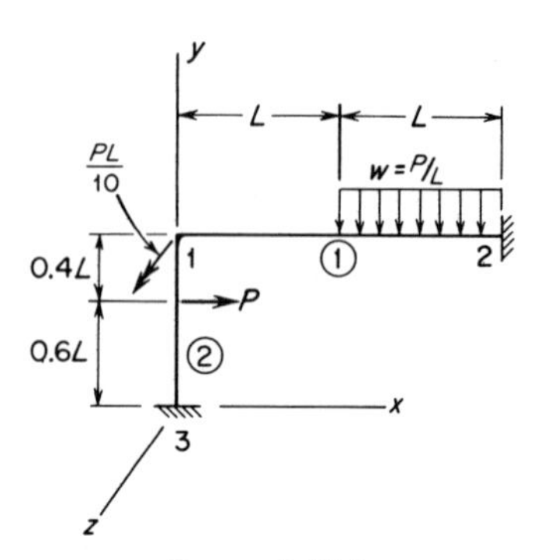

PROB. 4.18-3.

4.18-4. Analyze the plane frame shown in the figure, assuming that P = 5 kips, L = 5 ft., and E = 10,000 ksi. The moment of inertia and cross-sectional area of members 1 and 3 are 80 in.^{4} and 4 in.^{2}, respectively, and for member 2 are 150 in.^{4} and 5 in.^{2}, respectively.

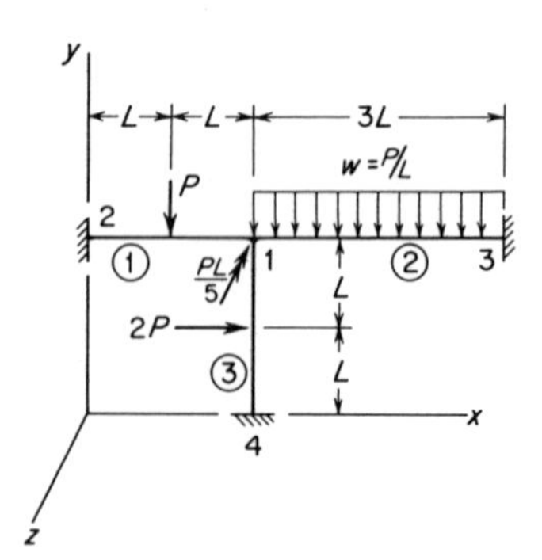

PROB. 4.18-4.

4.18-5. Analyze the plane frame shown in figure (a) for the effect of the loads shown. Assume that all members have E = 10,500 ksi, I_Z = 26 in.^{4}, A_X = 8 in.^{2}, and also assume that L = 60 in. and P = 2000 lb. (Hint: Take advantage of the symmetry of the structure and loading by analyzing the structure shown in figure (b). Note that the area and moment of inertia of member 1 must be halved. The imaginary restraint at joint 1 prevents both x translation and z rotation.)

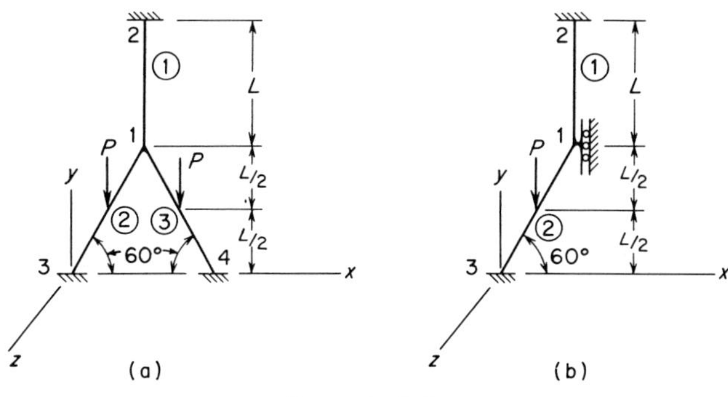

PROB. 4.18-5.

4.18-6. Analyze the plane frame shown in the figure assuming that (for both members) $E = 30{,}000$ ksi, $I_Z = 450$ in.^{4}, and $A_X = 11$ in.^{2} Also, assume $L = 300$ in. and $P = 2$ kips.

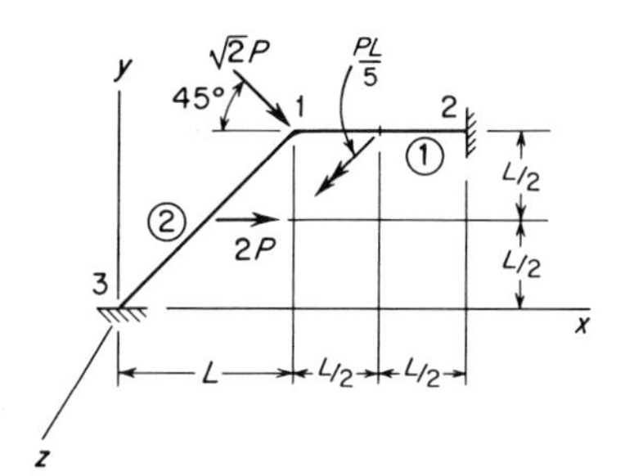

PROB. 4.18-6.

4.18-7. The plane frame shown in the figure is to be analyzed on the basis of the following numerical data: $E = 30{,}000$ ksi, $P = 400$ lb, $L = 30$ in.; for members 1 and 2, $I_Z = 12$ in.^{4}, $A_X = 4$ in.^{2}; for member 3, $I_Z = 14$ in.^{4}, $A_X = 6$ in.^{2}

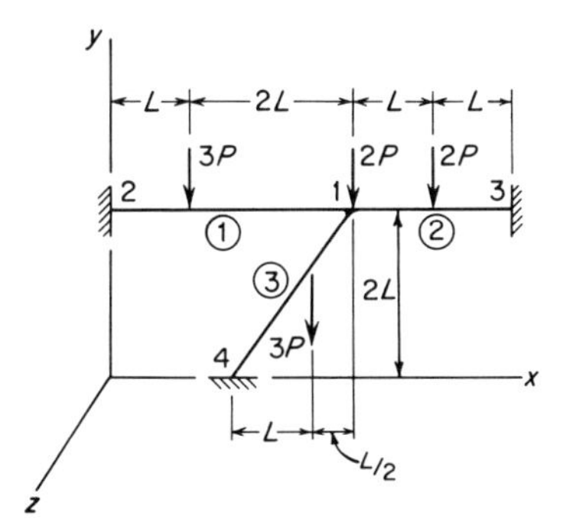

PROB. 4.18-7.

4.18-8. Determine the joint stiffness matrix $\mathbf{S}_J$ for the plane frame shown in the figure if all members have the same cross-sectional properties. Assume the following data: $E = 30{,}000$ ksi, $L = 20$ ft, $I_Z = 1600$ in.^{4}, $A_X = 24$ in.^{2}

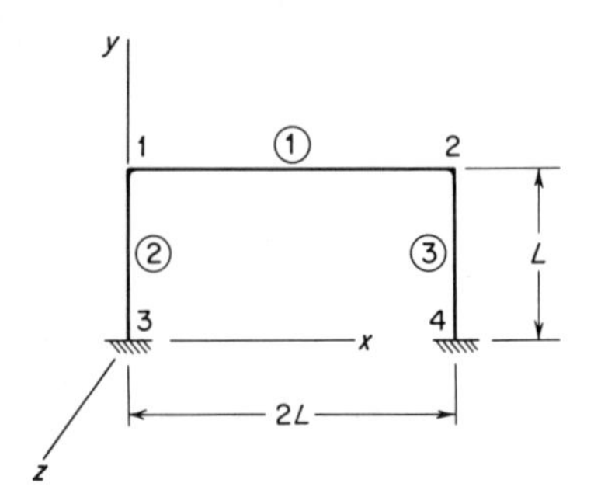

Prob. 4.18-8.

4.18-9. Solve the preceding problem for the plane frame shown in the figure. Also, rearrange and partition the matrix into the form given by Eq. (4-1).

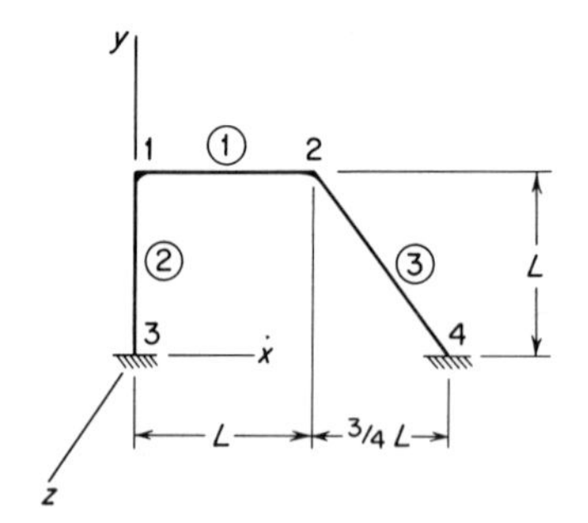

Prob. 4.18-9.

Chapter 5

COMPUTER PROGRAMS FOR FRAMED STRUCTURES

5.1 Introduction. In Chapter 4 the stiffness method of analysis was presented in a form which is suitable for computer programming. The object of this chapter is to present flow charts for the analysis (by the stiffness method) of the six basic types of framed structures. The flow charts contain sufficient detail to permit any person who is familiar with the elements of computer programming to write his own programs for structural analysis.

The programs described in this chapter have been assembled with three primary objectives. The first objective is to apply the power and generality of the stiffness method of analysis to each of the basic categories of framed structures. The second objective is to enable the beginner to interpret the flow charts and to write his own programs, using them as guides. The third objective is to demonstrate to the reader certain programming techniques which can be used later for the purpose of writing more sophisticated programs for structural analysis. Thus, the programs described herein are quite powerful in their own right, but they also serve as an excellent foundation preparatory to writing more advanced programs. Such programs could include modifications of the types discussed in Chapter 6. It is also possible to add refinements for improving the speed of the calculations and for using the storage within the computer more efficiently. In addition, it is possible to combine all six programs into one large program for the analysis of all types of framed structures. These modifications can be considered by the programmer after he has mastered the elementary programs given here.*

In the programs of this chapter all of the structures to be analyzed are assumed to consist of straight prismatic members. The material properties for a given structure are taken to be constant throughout the structure. Only the effects of loads are considered, and no other influences, such as temperature change, are taken into account. For each type of structure the analysis follows the steps described in Chapter 4.

The reader will appreciate the fact that there are many ways of carry-

* More advanced programs are presented in the book by W. Weaver, Jr., *Computer Programs for Structural Analysis*, 1966, D. Van Nostrand Co., Inc., Princeton, N.J.

ing out the detailed steps involved in structural calculations. Also, there are many possible ways that a computer program can be organized, including variations in the manner of reading data into the program and printing the results. Thus, it is apparent that the programs in this chapter, while representing a satisfactory method of calculation, do not represent the only method. The programs have been tested in classroom and consulting work and have proven to be practically useful and easily mastered by beginning programmers. Once they are understood, the programmer can proceed on his own to the development of more sophisticated programs.

5.2 Computer Programming and Flow Charts. It is assumed that the reader has an acquaintance with computer programming to the extent that he is familiar with an algorithmic language, such as ALGOL or FORTRAN.* He should also have some programming experience with elementary problems, not necessarily related to structural engineering. If the reader has this rudimentary programming ability as well as a fundamental understanding of the material in the previous chapters of this book, he should encounter no difficulty in writing the programs described in this chapter. The programs are presented in the form of flow charts so that they are independent of a particular language or a particular type of computer. The flow charts are sufficiently detailed so that every step in the calculations is covered. A person who wishes to write these programs in the language with which he is familiar should be able to do so by working step by step through the flow charts.

Some of the essential features of an algorithmic language, as well as the symbols used in the flow charts, are summarized in this article. The discussion is not intended to be complete, but covers only those features which are required in the flow charts that follow.

Numbers in a computer program may be either decimal numbers or integer numbers. The decimal numbers are also referred to as real numbers or floating-point numbers. Integer numbers are whole numbers which do not include a decimal point. Both types will be used in the programs, but they must be distinguished from each other because of the fact that arithmetic operations are not always the same for the two types of numbers. Most of the calculations in the programs are made with decimal numbers. Integer numbers are used primarily as indexes (or subscripts). At the beginning of each flow chart, a list of the variables that are integers is given.

Identifiers are names for variables, constants, or other entities which are used in the program. For example, the symbol E is used as an identifier for the modulus of elasticity of the material. In general, the symbols

* For detailed discussions of these algorithmic languages, see D. D. McCracken, *A Guide to ALGOL Programming*, John Wiley and Sons, Inc., New York, 1962, and E. I. Organick, *A FORTRAN Primer*, Addison-Wesley Publishing Co., Inc., Reading, Mass., 1963.

used previously in Chapter 4 are used in the programs as identifiers, although some minor modifications are required because of the limited number of characters available on keypunching machines. All alphabetical characters must be capital letters, and all symbols must be on the same line. For example, the symbol for the number of joints n_j becomes the identifier NJ in a computer program. Identifiers of this type require only one storage location in the memory of a computer and are said to represent *simple variables*.

There are other variables, called *subscripted variables*, that require more than one storage location. Thus, an identifier for a subscripted variable represents an array of numbers, as in the case of a vector or a matrix. A particular element in the array is denoted by the subscripts which follow the symbol. The subscripts are enclosed in square brackets in the programs given later. For example, an element of the vector $\mathbf{A}_{\mathrm{E}}$, such as A_{E4}, becomes $AE[4]$ in the program. Similarly, an element of the matrix $\mathbf{A}_{\mathrm{ML}}$, such as A_{ML23}, becomes $AML[2,3]$. It is necessary to reserve in advance a block of storage locations in the computer memory for each subscripted variable, as explained later. Hence, at the beginning of each flow chart, a list is given of the subscripted variables for which this must be done.

Operators used in computer programming are of several types, and those used in the programs of this chapter are summarized in Table 5-1.

TABLE 5-1

Operators Used in Computer Programs

Operator	Symbol
Arithmetic Operators:	
Addition	+
Subtraction	−
Multiplication	x
Division	/
Exponentiation	*
Replacement Operator	←
Relational Operators:	
Less	$<$
Less or equal	$\leq$
Equal	$=$
Greater or equal	$\geq$
Greater	$>$
Not equal	$\neq$

Arithmetic operators are used to signify the operations of addition, subtraction, multiplication, division, and exponentiation. All of these operations are denoted by the usual algebraic symbols except exponentiation. Since all symbols must be on the same line, exponentiation is denoted by an asterisk followed by the exponent. For example, A^4 is represented in a program by $A*4$.

The *replacement operator* is an arrow directed to the left, and it represents the operation of replacing an existing number in a storage location by a new number. The new number is then said to be the *current value* of the variable. The *relational operators* are also listed in Table 5-1. These consist of the relations less, less or equal, equal, greater or equal, greater, and not equal. All of the relational operators are represented by the usual algebraic symbols.

Computer programs consist of *declarations* and *statements*. The purpose of a declaration is to define the properties of one or more identifiers used in the program. For example, a declaration may be used to specify which identifiers represent integer numbers and which represent decimal numbers. The former are said to be identifiers of integer type, and the latter are identifiers of floating (or real) type. Another important kind of declaration specifies the amounts of storage to be reserved for arrays in the program (array declarations). Declarations are also used to specify the items in sets of input or output data, as well as the details of format for the output of information. Statements, on the other hand, define operations to be performed by the computer. The order in which statements are written is an essential feature in computer programming because they are executed one by one in the sequence in which they appear in the program, except when the statements themselves cause control to pass to another statement that is not in sequence.

A statement may be preceded by a *label*. Such statement labels serve to identify particular statements and are useful for the purpose of proceeding from one statement to another in a sequence other than that in which they are listed.

The types of statements used in the programs in this chapter are summarized in Table 5-2. The flow chart symbols denoting these statements also appear in the table. Each of the statements listed in Table 5-2 is described briefly in the following discussion.

The *input statement* causes information to be transferred into the memory unit of the computer from some external source, such as data cards or tape. An input statement is represented in a flow chart by a box in the shape of a data card (see Table 5-2). As an example, suppose that the numbers A, B, and C are to be brought into the computer as input data. This may be represented in the flow chart as shown in Fig. 5-1a. (The arrows denote the flow of steps in a chart.)

The *output statement* causes information in the computer to be trans-

TABLE 5-2

Statements Used in Computer Programs

	Type of Statement	Flow Chart Symbol
(a)	Input	
(b)	Output	
(c)	Assignment	
(d)	Unconditional control	
(e)	Conditional control	T / F
(f)	Iterative control	

ferred outside and communicated to the programmer by means of a printer, a card punch, or other device. An output statement appears in a flow chart as a box with a diagonal line at the base, as shown in Table 5-2. Figure 5-1b demonstrates the appearance in a flow chart of an output statement that causes the numbers A, B, and C to be printed (or otherwise communicated to the programmer).

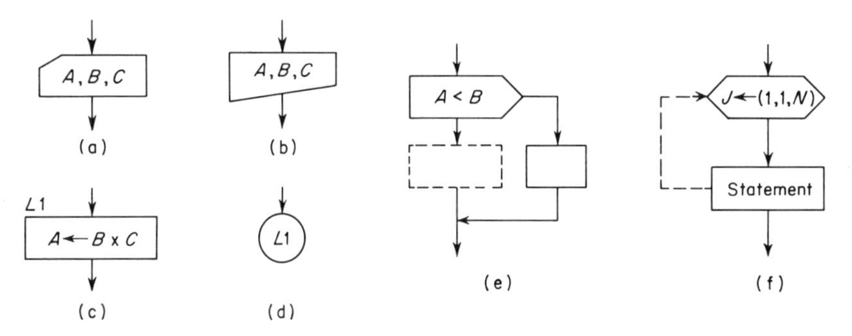

FIG. 5-1. Flow chart symbols for statements: (a) input statement, (b) output statement, (c) assignment statement, (d) unconditional control statement, (e) conditional control statement, and (f) iteration control statement.

The *assignment statement* represents the replacement of the current value of a variable by a new value, which then becomes the current value of the variable. For example, the statement

$$A \leftarrow B \times C$$

means that the product $B \times C$ (the term on the right) is to be evaluated and put into the storage location for the variable A (the term on the left) Therefore, the current value of A is replaced by the product of B times C, which then becomes the current value of A. The replacement operator is an integral part of the assignment statement. In a flow chart the assignment statement is enclosed in a rectangle (see Table 5-2). The assignment statement described above is shown in Fig. 5-1c as it would appear in a flow chart. The manner of showing a statement label in a flow chart is also illustrated in Fig. 5-1c. The label is written at the upper left-hand corner of the rectangle, and in this example the label is assumed to be the identifier $L1$.

The *unconditional control statement* causes a change in the order of execution of statements in a program. By means of this statement, control may be transferred to any other statement in the program by referring to its statement label. The unconditional control statement is represented in a program by a circle, as shown in Table 5-2. The label of the statement to which control is transferred is indicated inside the circle. For example, in Fig. 5-1d the statement label $L1$ in the circle indicates that control is to be transferred to the statement in the program having the label $L1$.

The *conditional control statement* provides the ability to make decisions based upon a test, which may be done with a relational expression. Control is transferred to one statement or another in the program, depending upon whether the expression is true or false. The flow chart symbol for a conditional control statement is shown in Table 5-2 with the true and false branches indicated by a T and an F, respectively. These letters are not required in a flow chart if the true and false branches are always arranged in the manner shown. Figure 5-1e shows a conditional control statement which tests to determine whether A is less than B, using the relational operator $<$. If the condition is true, the branch at the right is followed; but if the condition is false, the program continues along the branch indicated by the lower arrow. In some instances the false branch may contain a statement (indicated by the box with dashed lines in Fig. 5-1e) which is to be executed as an alternative to the statement on the true branch.

The *iterative control statement* is a useful device for executing the same statement or set of statements a specified number of times. A common type of iterative control statement contains a list of three elements which define the initial value, an incremental value, and the final value of a controlling parameter. In a flow chart this list is enclosed in a hexagonal figure of the type shown in Table 5-2. The dashed line and arrowhead associated with the symbol indicate that control is returned to this location in cyclic fashion. Figure 5-1f shows an iterative control statement which causes the statement within the rectanulgar box to be executed N times. The iterative control statement contains the list of three elements (in parentheses), as shown:

$$J \leftarrow (1, 1, N)$$

This expression means that the quantity J takes the value 1 the first time that the statement in the box is executed, and that J is incremented by 1 in each cycle of execution, until finally J is set equal to N and the statement is executed for the last time. As an example, suppose that A, B, and C are all subscripted variables of N elements each. If the corresponding elements of B and C are to be multiplied together and assigned to A, the statement in the rectangular box would be

$$A[J] \leftarrow B[J] \times C[J]$$

After N repetitions of the calculation, all elements of the arrays have been processed, and control is transferred to the next statement in the program.

In any statement in a program it may be necessary to evaluate a *function* in the course of executing the statement. For example, the assignment statement

$$A \leftarrow B \times COS(C)$$

involves the evaluation of the cosine of C. In this example, COS is the function designator, and its use in a program implies that a specific set of operations are to be performed automatically, which operations result in the evaluation of the cosine of C. Certain frequently used functions, such as sine, cosine, square root, logarithm, and absolute value (designated as SIN, COS, $SQRT$, LOG, and ABS), are incorporated into algorithmic languages for the convenience of the programmer.

Another important feature of algorithmic languages enables the programmer to use a series of statements as a *procedure* (or *subroutine* in FORTRAN), which is a type of subprogram. A procedure is a block of statements which may be used at several points in the same program. Procedures may also be used in different programs without alteration. In the programs of this chapter it is assumed that a procedure is available for inverting a square, nonsingular matrix. In this procedure the inverse matrix is assumed to be stored in the same storage locations as the original matrix. Thus, at the stage in each program where it is desired to invert the stiffness matrix **S**, a statement appears which calls for the inversion to be performed, but the details of performing the inversion are omitted.

Although the foregoing discussion of some of the features of an algorithmic language is brief and incomplete, its purpose has been to explain the conventions used in the flow charts of this chapter. With these conventions in mind, the programmer can follow the charts and write his own programs.

5.3 Outline of Programs. The general outline of all the programs in this chapter follows directly from the sequence of operations described in Chapter 4. However, only a part of the over-all joint stiffness matrix $\mathbf{S}_J$ is generated in the programs. Since only the effects of loads are taken into account, the required portions of $\mathbf{S}_J$ are $\mathbf{S}$ and $\mathbf{S}_{RD}$ (see Eq. 4-1). The other two portions ($\mathbf{S}_{DR}$ and $\mathbf{S}_{RR}$) are not generated in these programs. However, it is only a slight additional step to generate them, and this could be done in a more extensive program that included the effects of support displacements (see Art. 6.5).

Rotation transformation matrices are used only in the program for space frame analysis, because otherwise the calculations for a space frame become very complex. For all other structures, however, the calculations are carried out directly by means of the formulas developed in Chapter 4.

Since the detailed steps in a computer program are related to the manner in which data are prepared, each program is preceded by a summary which explains the preparation of the necessary data. In addition, some of the variables in a program will be of integer type while others are of floating-point type. Therefore, a listing of the variables of integer type precedes each flow chart. The variables which are subscripted in the program (and which require blocks of storage in the computer) are also listed at the beginning of each chart.

The general outline of all the programs is shown in the following five steps, each of which is explained in more detail in connection with the flow charts:

1. Input and Print Structure Data
   (a) Structure parameters and elastic moduli
   (b) Joint coordinates (except continuous beams)
   (c) Member designations, properties, and orientations
   (d) Joint restraint list; cumulative restraint list
2. Structure Stiffness Matrix
   (a) Generation of stiffness matrix
   (b) Inversion of stiffness matrix
3. Input and Print Load Data
   (a) Numbers of loaded joints and members
   (b) Actions applied at joints
   (c) Actions at ends of restrained members due to loads
4. Construction of Vectors Associated with Loads

   (a) Equivalent joint loads
   (b) Combined joint loads
5. Calculation and Output of Results
   (a) Joint displacements and support reactions
   (b) Member end-actions

The above outline of steps is repeated in the flow charts in order to identify the nature of the various parts of the program. Explanatory notes for each step in the flow charts accompany the flow charts themselves. These notes are identified by numbers in parentheses on the left-hand sides of the flow charts.

In the outline above and in the programs, it is assumed that the input data and certain descriptive headings will be printed, as well as the final results. Such information is usually needed to identify the structure being analyzed and to provide the programmer with the results which constitute the goal of the analysis. Of course, it is always possible for a programmer to add additional output statements at any intermediate stage of the calculations if it is desired to print some of the intermediate results. For example, under certain conditions it might be desirable to print the stiffness matrix for the structure.

The first program given in this chapter is presented in complete form. However, some of the later programs contain steps which are the same as those in the earlier programs. When this overlap occurs, the steps are not repeated; instead, the reader is referred back to the earlier program in which those steps appear.

After preparing his own computer program for a given type of structure, the reader will want to verify it by solving at least one example problem having a known solution. For this purpose, each program in this chapter is followed by one or more example problems. These problems consist of computer solutions for the examples given in Chapter 4 as well as some additional problems of a more complex nature. These problems were solved using the computer programs of this chapter, and the final results are given for each case.

Table 5-3 summarizes most of the notation of this chapter, with the items listed approximately in the order of their appearance in the programs. Further notation, required for space frames only, is given at the beginning of Art. 5.9. A symbol in the list which is followed by square brackets [ ] denotes a subscripted variable with a single subscript, that is, a vector. A symbol followed by square brackets enclosing a comma [,] denotes a subscripted variable (or a matrix) with two subscripts.

A consistent system of units must be used for the input data, which involve units of force and length only. The results of the calculations will have the same units as the input data, with the addition that all angles will automatically be in radians. One example of a consistent

TABLE 5-3

Identifiers Used in Computer Programs

Identifier	*Definition*
M	Number of members
N	Number of degrees of freedom
NJ	Number of joints
NR	Number of support restraints
NRJ	Number of restrained joints
E	Elastic modulus for tension or compression
G	Elastic modulus for shear
I	Member index
J, K	Joint indexes
$X[\ ], Y[\ ], Z[\ ]$	x, y, and z coordinates of joint
$JJ[\ ]$	Designation for j end of member (joint j)
$JK[\ ]$	Designation for k end of member (joint k)
$L[\ ]$	Length of member
$AX[\ ]$	Cross-sectional area of member
$IX[\ ]$	Torsion constant of member
$IY[\ ], IZ[\ ]$	Moments of inertia about the y and z axes
XCL, YCL, ZCL	x, y, and z components of length of member
$CX[\ ], CY[\ ], CZ[\ ]$	x, y, and z direction cosines of member
$RL[\ ]$	Joint restraint list
$CRL[\ ]$	Cumulative restraint list
$J1, \ldots, J6$	Indexes for displacements at j end of member
$K1, \ldots, K6$	Indexes for displacements at k end of member
SCM	Stiffness constant of member
$SM[\ ,\ ]$	Member stiffness matrix for member-oriented axes
$SMD[\ ,\ ]$	Member stiffness matrix for structure-oriented axes
$S[\ ,\ ]$	Joint stiffness matrix ($\mathbf{S}_J$)
NLJ	Number of loaded joints
NLM	Number of loaded members
$A[\ ]$	Actions (loads) applied at joints (in directions of structure axes)
$AML[\ ,\ ]$	Actions at ends of restrained members (in directions of member axes) due to loads
$AE[\ ]$	Equivalent joint loads (in directions of structure axes)
$AC[\ ]$	Combined joint loads (in directions of structure axes)
$D[\ ]$	Joint displacements (in directions of structure axes)
$AR[\ ]$	Support reactions (in directions of structure axes)
JE, KE	Indexes for expanded vectors

system of units is to give loads in kips, lengths in inches, areas in square inches, modulus of elasticity in kips per square inch, and so on. In such a case all of the final results will be in units of kips, inches, and radians. On the other hand, if units of feet and pounds are used for the input data, the final results will be in terms of feet, pounds, and radians.

5.4 Continuous Beam Program. The program described in this article performs the analysis of continuous beams by the techniques described in Art. 4.8. The required input data for the program are discussed first, since the form of the program itself depends to some extent upon the form in which the data are supplied. Next, the flow chart for the program is presented, and this chart is followed by explanatory comments. Finally, the continuous beam example of Art. 4.9 is analyzed again, using the computer program.

The input data required in the program are summarized in Table 5-4.

TABLE 5-4

Preparation of Data for Continuous Beam Program

	Data	Number of Cards	Items on Data Cards
Structure Data	(a) Structure parameters and elastic modulus	1	M NR NRJ E
	(b) Member designations and properties	M	I L[I] IZ[I]
	(c) Joint restraint list	NRJ	K RL[2K – 1] RL[2K]
Load Data	(a) Numbers of loaded joints and members	1	NLJ NLM
	(b) Actions applied at joints	NLJ	K A[2K – 1] A[2K]
	(c) Actions at ends of restrained members due to loads	NLM	I AML[I, 1] AML[I, 2] AML[I, 3] AML[I, 4]

This table specifies the number of data cards required for each category of information as well as the items which are to be punched on each card. The first card listed in the table contains the number of members M, the number of restraints NR, the number of restrained joints NRJ, and the elastic modulus E of the material. The reason for including NRJ is that the number of data cards required to fill the joint restraint list (see line 3 of the table) is minimized if the number of restrained joints is known.

Each data card containing member information (line 2 of Table 5-4) includes for one member the member number I, the length $L[I]$, and the moment of inertia $IZ[I]$ about the z axis. A total of M cards is required.

Each of the cards in the next series (a total of NRJ cards) contains a joint number K and two code numbers which indicate the conditions of restraint at that joint. The term $RL[2K-1]$ denotes the restraint against translation in the y direction at joint k, and the term $RL[2K]$ denotes the restraint against rotation in the z sense at joint k. The convention adopted in this program is the following. If the restraint exists, the integer 1 is assigned as the value of RL, and if there is no restraint a value of zero is assigned.

The first card in the load data contains two numbers. These are the number of loaded joints NLJ and the number of loaded members NLM. One reason for introducing these numbers is that they serve to minimize the number of subsequent load data cards required. Another reason is that certain calculations in the program can be bypassed if either NLJ or NLM is equal to zero; for example, if NLM is equal to zero the calculation of the vector $\mathbf{A}_E$ can be omitted.

Each joint load card (a total of NLJ cards) contains a joint number K and the two actions applied at that joint. These actions are the applied force and couple in the y direction and the z sense. Finally, each card for member loads (NLM cards total) contains a member number I and the four fixed-end actions for that member. The fixed-end actions consist of a force in the y direction and a couple in the z sense at each end of the member.

The preparation of data for a particular structure is illustrated in the example problem at the end of this article.

At the beginning of a computer program there are various preliminary matters which should be taken into account. It is good practice to introduce a program with a description of the content of the program. This includes the scope of the program, an outline of the steps in the program, the notation used, and the data required. Since this is done primarily for the convenience of the programmer, it is not included in the flow charts in this chapter.

Another preliminary matter is to declare those identifiers in a program which are to be integer numbers as opposed to those which are to be decimal numbers. The following identifiers of type integer appear in the continuous beam program:

$$I, J, JE, J1, J2, K, KE, K1, K2, M, N,$$

$$NJ, NR, NRJ, NLJ, NLM, RL, CRL$$

and an appropriate declaration to this effect must be made.

In addition, storage in the memory unit of the computer must be reserved for the subscripted variables in the program. For the continuous beam program the subscripted variables are the following:

$$L[\ ], IZ[\ ], RL[\ ], CRL[\ ], A[\ ], AML[\ ,\ ],$$

$$AE[\ ], AC[\ ], SM[\ ,\ ], S[\ ,\ ], D[\ ], AR[\ ]$$

Finally, any procedures to be used in the program (such as the procedure for matrix inversion) must be inserted at the beginning of the program in order to have them available for later use.

The flow chart for the continuous beam follows. The outline of the program is interspersed in the chart, and comments noted by the numbers in parentheses are given at the end of the chart.

FLOW CHART FOR CONTINUOUS BEAM PROGRAM

1. Input and Print Structure Data

a. Structure parameters and elastic modulus

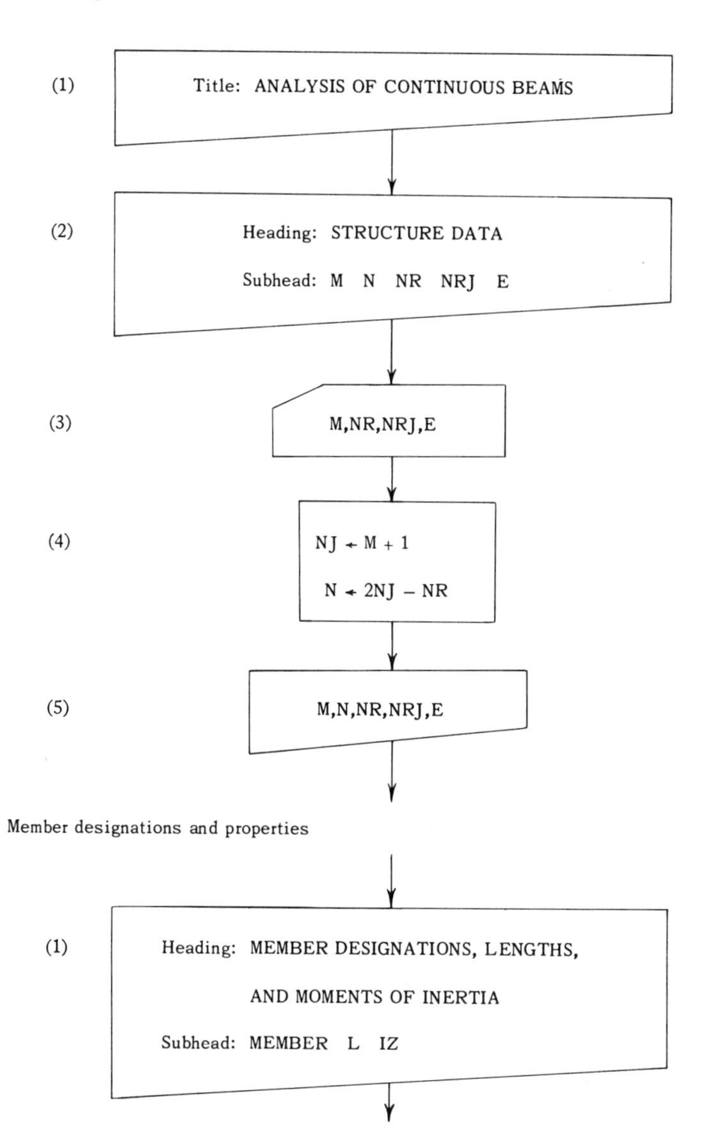

(1)
Title: ANALYSIS OF CONTINUOUS BEAMS
(2)
Heading: STRUCTURE DATA
Subhead: M N NR NRJ E
(3)
M,NR,NRJ,E
(4)
NJ ← M + 1
N ← 2NJ − NR
(5)
M,N,NR,NRJ,E
b. Member designations and properties
(1)
Heading: MEMBER DESIGNATIONS, LENGTHS, AND MOMENTS OF INERTIA
Subhead: MEMBER L IZ

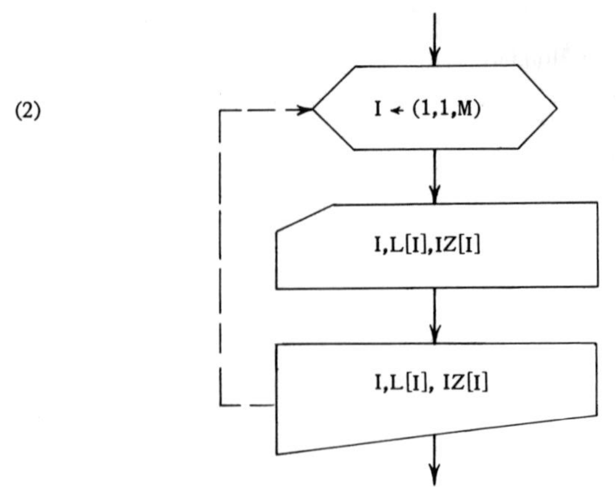

c. Joint restraint list; cumulative restraint list

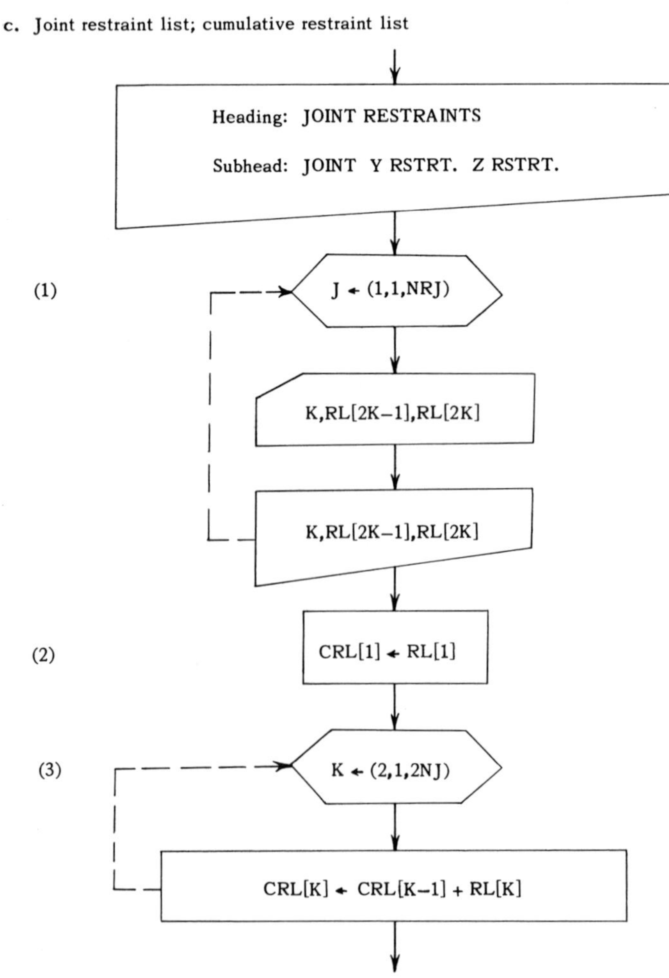

2. Structure Stiffness Matrix

a. Generation of stiffness matrix

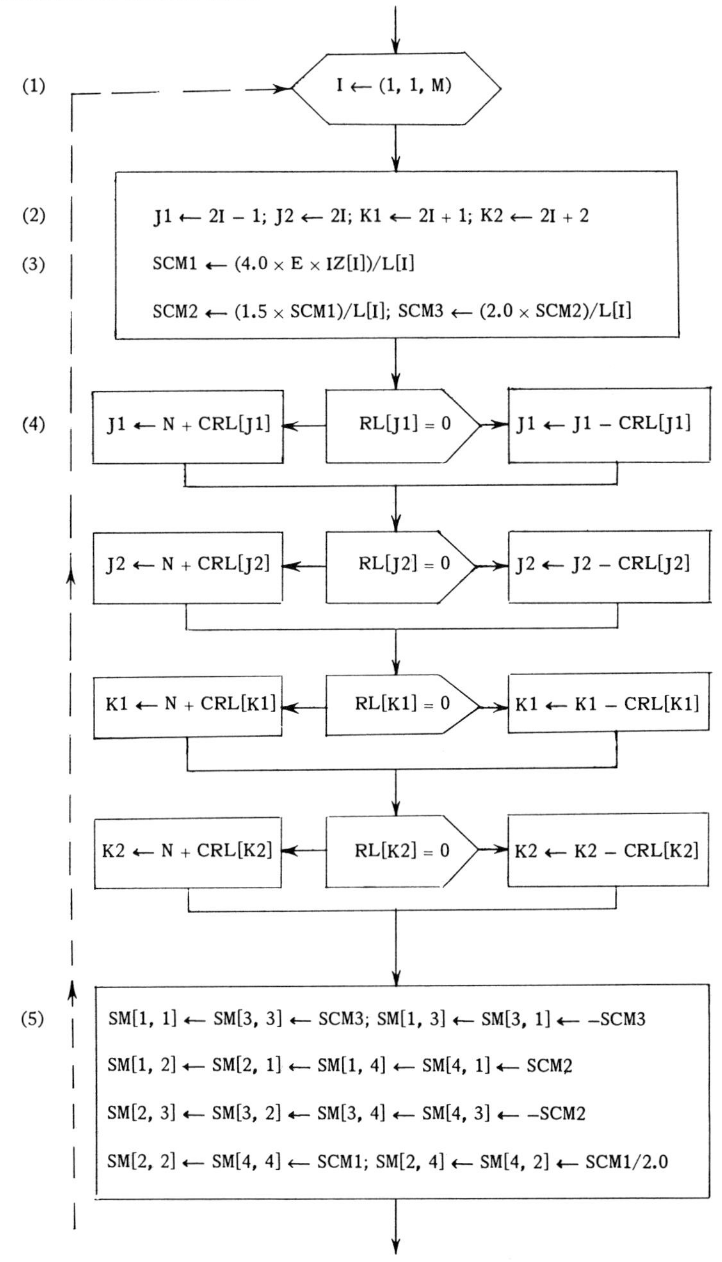

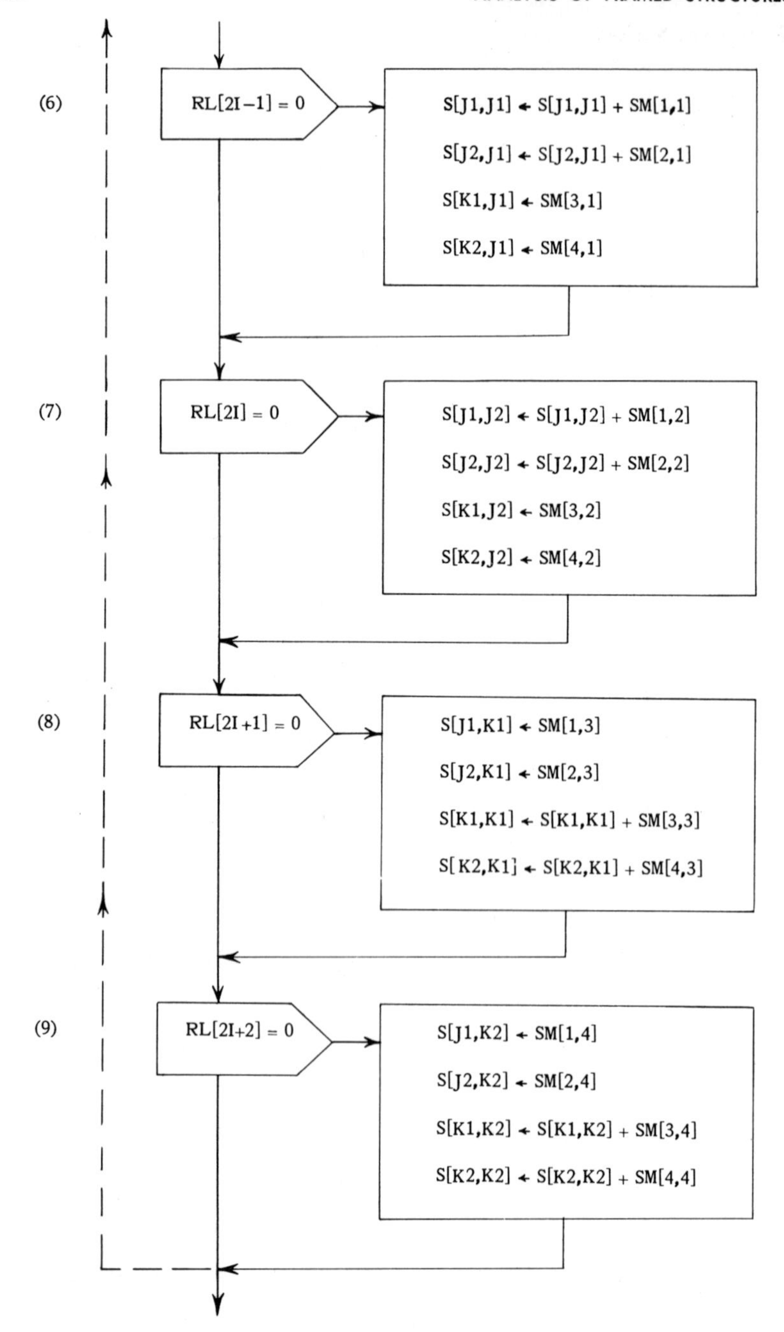
(6)
RL[2I−1] = 0
S[J1,J1] ← S[J1,J1] + SM[1,1]
S[J2,J1] ← S[J2,J1] + SM[2,1]
S[K1,J1] ← SM[3,1]
S[K2,J1] ← SM[4,1]
(7)
RL[2I] = 0
S[J1,J2] ← S[J1,J2] + SM[1,2]
S[J2,J2] ← S[J2,J2] + SM[2,2]
S[K1,J2] ← SM[3,2]
S[K2,J2] ← SM[4,2]
(8)
RL[2I+1] = 0
S[J1,K1] ← SM[1,3]
S[J2,K1] ← SM[2,3]
S[K1,K1] ← S[K1,K1] + SM[3,3]
S[K2,K1] ← S[K2,K1] + SM[4,3]
(9)
RL[2I+2] = 0
S[J1,K2] ← SM[1,4]
S[J2,K2] ← SM[2,4]
S[K1,K2] ← S[K1,K2] + SM[3,4]
S[K2,K2] ← S[K2,K2] + SM[4,4]

b. Inversion of stiffness matrix

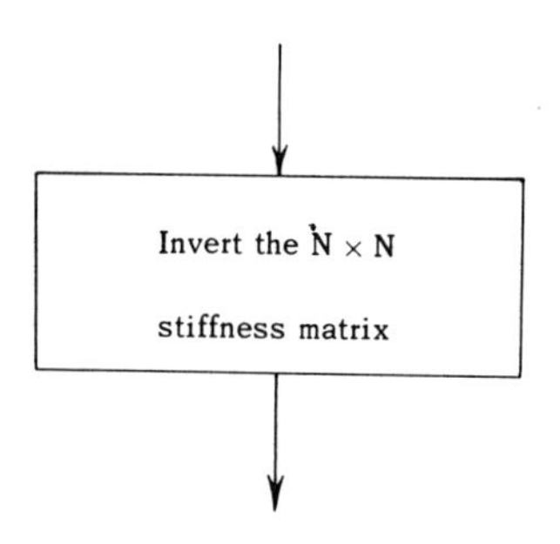

3. Input and Print Load Data

a. Numbers of loaded joints and members

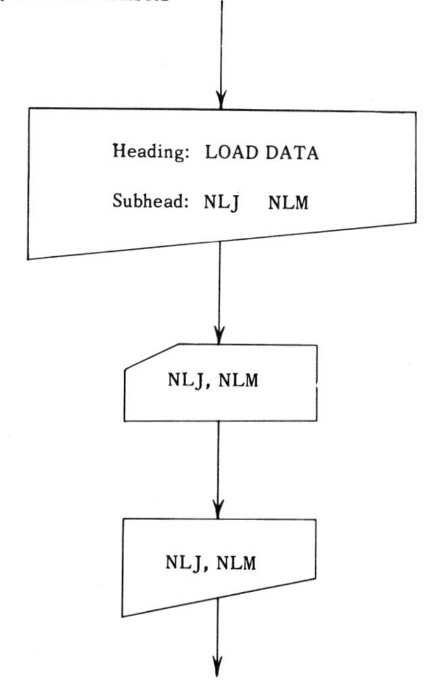

b. Actions applied at joints

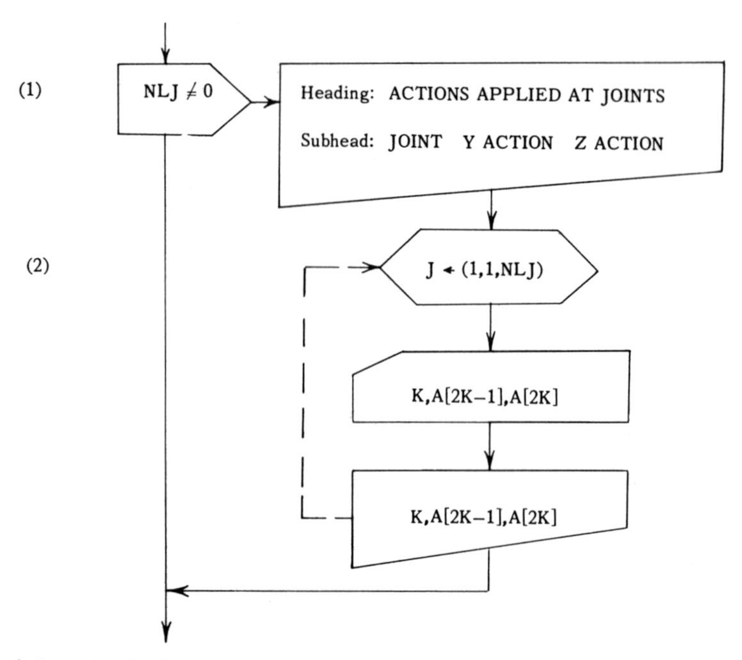

c. Actions at ends of restrained members due to loads

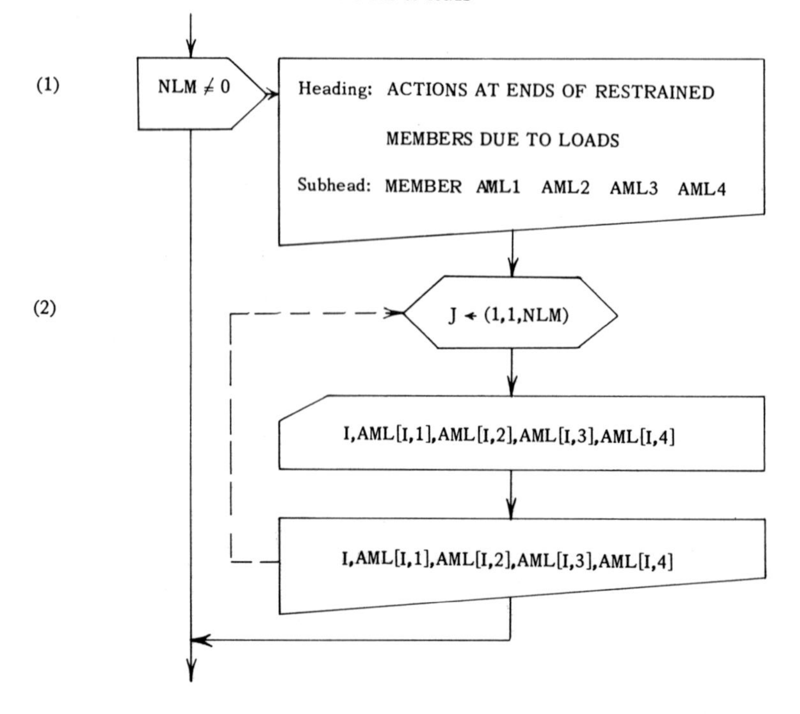

4. Construction of Vectors Associated with Loads

a. Equivalent joint loads

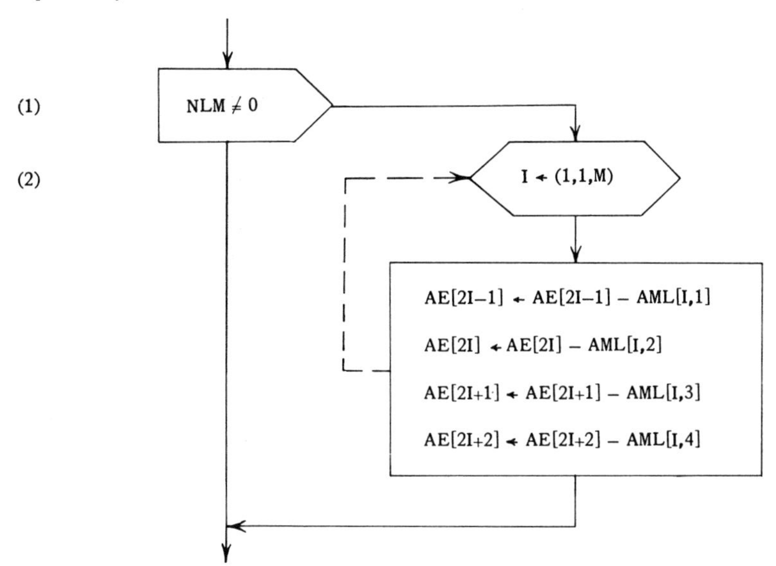

b. Combined joint loads

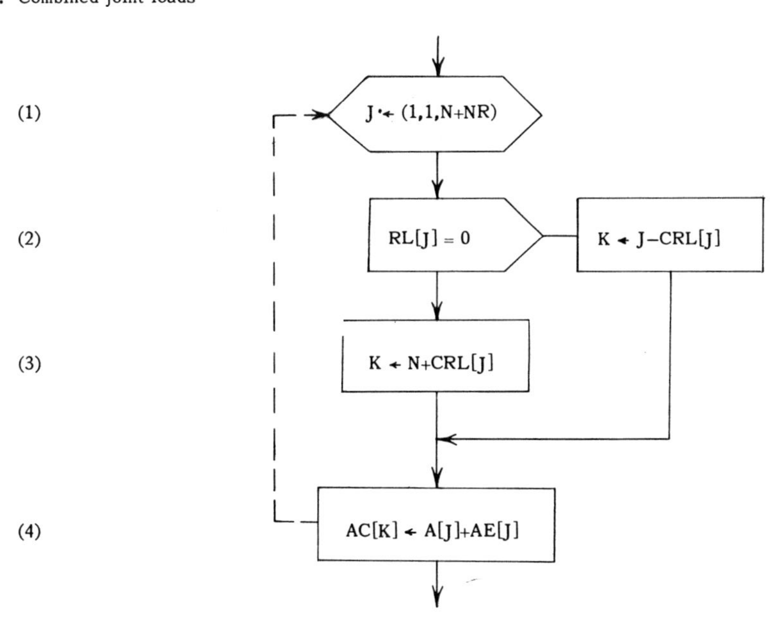

5. Calculation and Output of Results

a. Joint displacements and support reactions

(1)

(2)

(3)

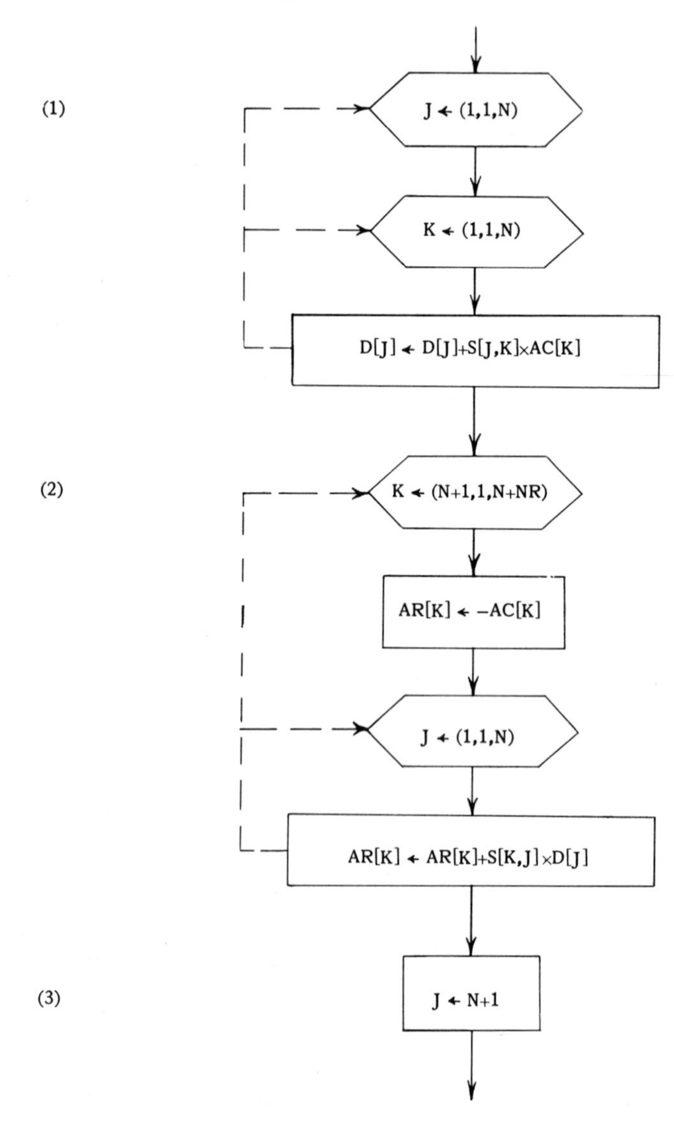

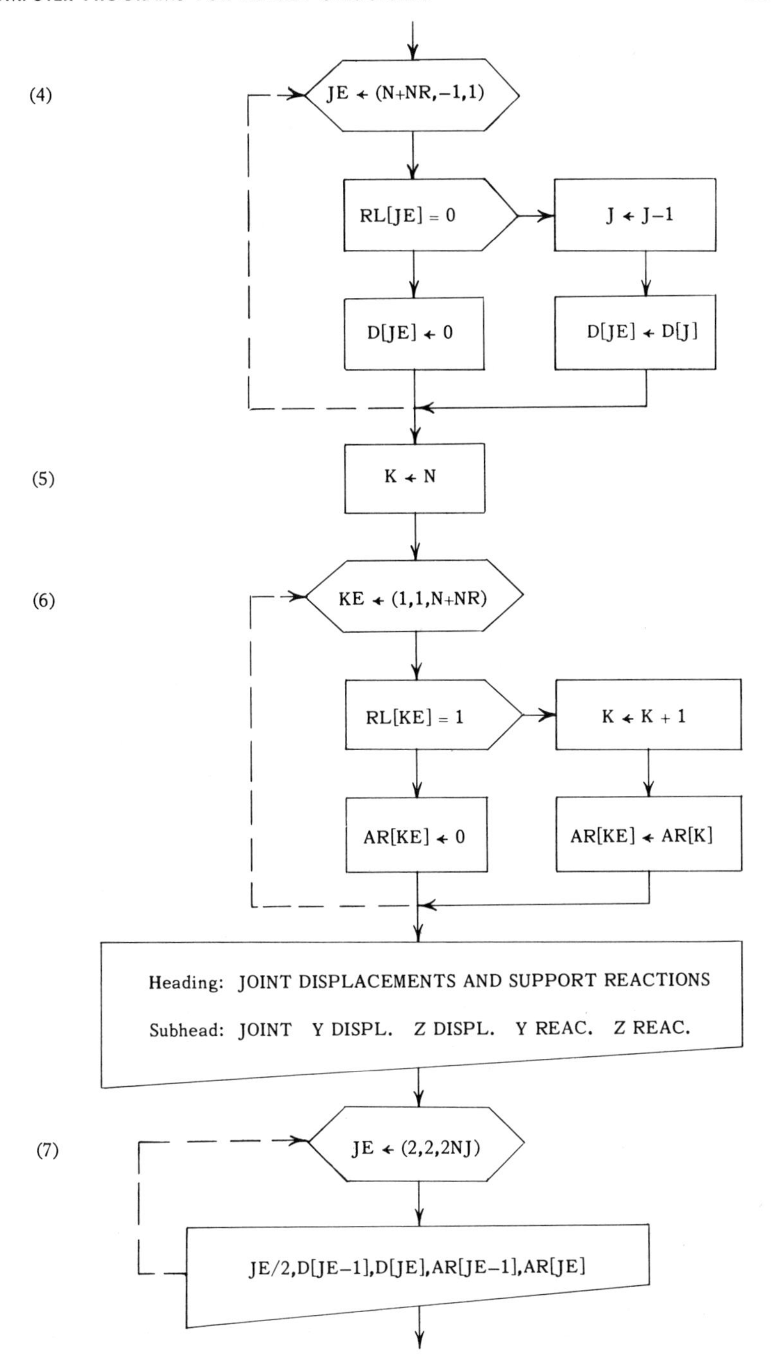

(4)
JE ← (N+NR,−1,1)
RL[JE] = 0
J ← J−1
D[JE] ← 0
D[JE] ← D[J]
(5)
K ← N
(6)
KE ← (1,1,N+NR)
RL[KE] = 1
K ← K + 1
AR[KE] ← 0
AR[KE] ← AR[K]
Heading: JOINT DISPLACEMENTS AND SUPPORT REACTIONS
Subhead: JOINT Y DISPL. Z DISPL. Y REAC. Z REAC.
(7)
JE ← (2,2,2NJ)
JE/2,D[JE−1],D[JE],AR[JE−1],AR[JE]

b. Member end-actions

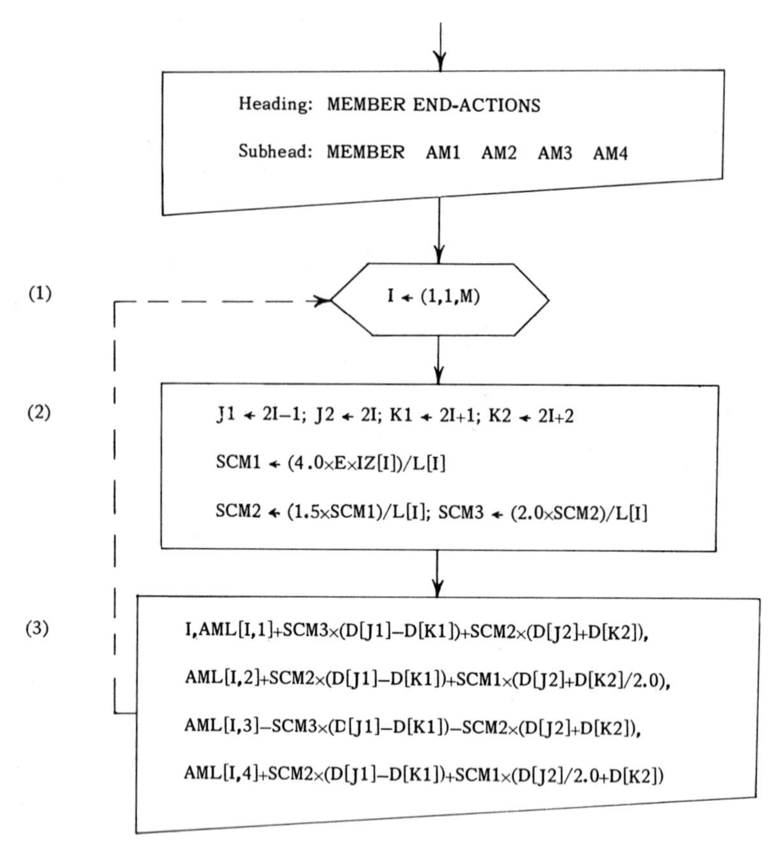

Explanation of Flow Chart for the Continuous Beam Program

Item No. *Remarks*

1. a. (1) An appropriate title for the program output is printed.
   (2) A heading and a subheading for structure data are printed.
   (3) The first data card, containing the structure parameters and elastic modulus, is read into storage (see Table 5-4).
   (4) The number of joints and the number of degrees of freedom are calculated.
   (5) The input data, augmented by N, are printed under the appropriate subheadings.
   b. (1) A heading and a subheading for member information are printed.
   (2) In an iteration control statement, M cards containing the member designations and properties are read (see Table 5-4). The same information is also printed under the appropriate subheadings.
   c. (1) In an iteration control statement, NRJ cards containing the joint designations and joint restraints are read (see Table 5-4). The same information is also printed.
   (2) In this portion of the program the cumulative restraint list CRL is

computed. This list consists of the cumulative sums of the numbers in the restraint list RL. As an example, the restraint list for the problem of Art. 4.9 is shown in Table 4-8. The corresponding cumulative restraint list consists of the following:

$$CRL = \{1, 2, 2, 2, 3, 3, 4, 5\}$$

Note that the first element of CRL is always equal to the first element of RL, and the last element of CRL is numerically equal to the total number of restraints NR. The first step in generating CRL is to assign $RL[1]$ to $CRL[1]$.

(3) In an iteration control statement, all of the other elements of CRL are calculated in sequence.

2\. a. (1) Iteration control statement on members, for M members.

(2) The two possible displacements for the joint to the left ($J1$ and $J2$) and the joint to the right ($K1$ and $K2$) of the member I are indexed in ascending order according to the numbering system for members.

(3) Three stiffness constants for member I are calculated and stored in order to avoid repetition of these expressions when generating the member stiffness matrix. These constants are:

$$SCM1 = \frac{4EI_Z}{L} \qquad SCM2 = \frac{6EI_Z}{L^2} \qquad SCM3 = \frac{12EI_Z}{L^3}$$

(4) Using the joint restraint list as a guide, the two possible displacements for the joints at each end of the member are re-indexed according to whether the joints are actually free to move or not. That is, if a certain displacement is free to occur, the index is decreased by the corresponding element of CRL, placing it in the proper order among the degrees of freedom. On the other hand, if a displacement is not free to occur, the index is reset to N plus the corresponding element of CRL, placing it in the proper order among the support restraints.

(5) The 4×4 member stiffness matrix SM for member I is generated (see Table 4-2, Art. 4.3) using the stiffness constants previously evaluated.

(6) If the index $2I-1$ (the original value of $J1$) corresponds to a degree of freedom, the first column of SM is transferred to the joint stiffness matrix for the structure (see Eqs. 4-13). However, if $2I-1$ corresponds to a restraint, the first column of SM is bypassed. By this device, only the required portions of the over-all joint stiffness matrix are generated. (The result is a rectangular array consisting of the matrix $\mathbf{S}$ in the upper portion and the matrix $\mathbf{S}_{RD}$ in the lower portion. Nevertheless, the whole array is identified as S in the program.) Elements of S involving the coupling of $J1$ and $J2$ are cumulative because, in general, they receive contributions from more than one member. On the other hand, elements of S involving the coupling of $J1$ with $K1$ and $K2$ are single values.

(7) Similar to (6) but for $2I$, which is the original value of $J2$ (see Eqs. 4-14).

(8) Similar to (6) but for $2I+1$, which is the original value of $K1$ (see Eqs. 4-15).

(9) Similar to (6) but for $2I+2$, which is the original value of $K2$ (see Eqs. 4-16).

b. The upper part of the rectangular array (of order $N \times N$) is inverted, using a procedure.

3. a. The numbers of loaded joints and loaded members are read (see Table 5-4) and printed.
   b. (1) If the number of loaded joints is not equal to zero, actions applied at joints must be read and printed.
   (2) In an iteration control statement, NLJ cards (see Table 5-4) containing the joint designations and the actions applied at joints (see Fig. 4-11) are read and printed.
   c. (1) If the number of loaded members is not equal to zero, data concerning loads on members must be read and printed.
   (2) In an iteration control statement, NLM cards (see Table 5-4) containing the member designations and the actions at the ends of restrained members due to loads (see Fig. 4-12) are read and printed.
4. a. (1) If NLM is not equal to zero, equivalent joint loads must be calculated.
   (2) In an iteration control statement for M members, the contributions to AE from each member are identified and transferred (see Eqs. 4-20). Note that the calculation of equivalent joint loads involves taking the negatives of fixed-end actions for members. Each of the statements for generating an element of AE consists of a summation of contributions from more than one member, because only the end points of the continuous beam receive contributions from a single member.
   b. (1) Iteration control statement on all possible joint displacements ($N+NR$ displacements).
   (2) If the J-th possible displacement is unrestrained, the index K for an element of AC is set to the proper position in the first part of the vector: $K \leftarrow J-CRL[J]$
   (3) Otherwise, the J-th possible displacement is restrained; thus, the index K is set to the proper position in the latter part of the vector: $K \leftarrow N+CRL[J]$
   (4) Elements of AC are calculated as the sums of elements of A and AE. Because of the fact that the indexes have been reset as described above, the vector AC is generated in the form given by Eq. (4-6).
5. a. (1) Within two iteration control statements, the displacements are calculated by matrix multiplication, using Eq. (4-8).
   (2) Within two iteration control statements, the support reactions are calculated by matrix algebra, using Eq. (4-4).
   (3) In preparation for expanding the displacement vector, J is initialized to $N+1$.
   (4) In an iteration control statement with descending values of the variable JE, the displacement vector is expanded in accordance with the degrees of freedom denoted by the joint restraint list. J is the index for the condensed form of D, and JE is the index for the expanded form of D (defined as $\mathbf{D}_J$ in Chapter 4).
   (5) In preparation for expanding the vector of support reactions, K is initialized to N.
   (6) In an iteration control statement, the vector of support reactions is expanded in accordance with the restraints denoted by the joint restraint list. K is the index for the condensed form of AR, and KE is the index for the expanded form of AR.
   (7) In an iteration control statement for all joints, the joint number, the joint displacements, and the support reactions are printed in sequence.
   b. (1) Iteration control statement on members, for M members.
   (2) The same member information is computed as in steps 2a(2) and (3) given previously.

(3) For each member, the member number is printed, and the four member end-actions are computed and printed. The calculations are made on the basis of one member at a time, using Eqs. (4-23). Thus, all results have been printed, and the program is completed.

Example: The continuous beam structure analyzed in Art. 4.9 (see Fig. 4-13) is presented as an example of the use of the computer program. For the purposes of machine calculations the following numerical values are assumed:

$$E = 10{,}000 \text{ ksi} \qquad L = 100 \text{ in.} \qquad P = 10 \text{ kips} \qquad IZ = 1000 \text{ in.}^4$$

The input data required by the computer program are summarized in Table 5-5.

TABLE 5-5

Data Cards for Continuous Beam Example

Type of Data		Numerical Data on Card					Card No.
Structure Data	(a)	3	5	3	10000.0		1
	(b)	1	100.0	1000.0			2
		2	100.0	2000.0			3
		3	200.0	2000.0			4
	(c)	1	1	1			5
		3	1	0			6
		4	1	1			7
Load Data	(a)	2	3				8
	(b)	2	−10.0	1000.0			9
		3	−10.0	0			10
	(c)	1	10.0	250.0	10.0	−250.0	11
		2	10.0	250.0	10.0	−250.0	12
		3	10.0	333.333	10.0	−333.333	13

The format of the cards follows that described in Table 5-4, with the first seven cards containing structure data and the last six cards containing load data.

The output of the computer program is given in Table 5-6.

5.5 Plane Truss Program. The program presented in this article executes the analysis of plane trusses by the method given in Art. 4.11. The steps in this program are very similar to those in the continuous beam program (see Art. 5.4) because of the fact that both types of structures have two possible displacements at each joint. Therefore, in

TABLE 5-6

Output of Computer Program for Continuous Beam Example

ANALYSIS OF CONTINUOUS BEAMS

STRUCTURE DATA

M	N	NR	NRJ	E
3	3	5	3	10000.0

MEMBER DESIGNATIONS, LENGTHS, AND MOMENTS OF INERTIA

MEMBER	L	IZ
1	100.0	1000.0
2	100.0	2000.0
3	200.0	2000.0

JOINT RESTRAINTS

JOINT	Y RSTRT.	Z RSTRT.
1	1	1
3	1	0
4	1	1

LOAD DATA

NLJ	NLM
2	3

ACTIONS APPLIED AT JOINTS

JOINT	Y ACTION	Z ACTION
2	−10.0	1000.0
3	−10.0	0

ACTIONS AT ENDS OF RESTRAINED MEMBERS DUE TO LOADS

MEMBER	AML1	AML2	AML3	AML4
1	10.0	250.0	10.0	−250.0
2	10.0	250.0	10.0	−250.0
3	10.0	333.333	10.0	−333.333

JOINT DISPLACEMENTS AND SUPPORT REACTIONS

JOINT	Y DISPL.	Z DISPL.	Y REAC.	Z REAC.
1	0	0	33.06	1281.75
2	−0.1316	0.001210	0	0
3	0	0.000843	39.47	0
4	0	0	7.47	−164.68

MEMBER END-ACTIONS

MEMBER	AM1	AM2	AM3	AM4
1	33.06	1281.75	−13.06	1023.81
2	3.06	−23.81	16.94	−670.64
3	12.53	670.64	7.47	−164.68

order to avoid repetition, the following discussion emphasizes those parts of the program which are different from the continuous beam program.

The input data required for the plane truss program are shown in Table 5-7, which has the same form as Table 5-4 for continuous beams.

TABLE 5-7

Preparation of Data for Plane Truss Program

	Data	Number of Cards	Items on Data Cards
Structure Data	(a) Structure parameters and elastic modulus	1	M NJ NR NRJ E
	(b) Joint coordinates	NJ	J X[J] Y[J]
	(c) Member designations and properties	M	I JJ[I] JK[I] AX[I]
	(d) Joint restraint list	NRJ	K RL[2K – 1] RL[2K]
Load Data	(a) Numbers of loaded joints and members	1	NLJ NLM
	(b) Actions applied at joints	NLJ	K A[2K – 1] A[2K]
	(c) Actions at ends of restrained members due to loads	NLM	I AML[I, 1] AML[I, 2] AML[I, 3] AML[I, 4]

The first card listed in the table contains the structure parameters (that is, the numbers of members, joints, restraints, and restrained joints) and the modulus of elasticity. Next, a set of cards (NJ cards total) is required to specify the coordinates of the joints. Each card in this set contains a joint number J, the x coordinate $X[J]$ of the joint, and the y coordinate $Y[J]$ of the joint.

On each of the cards pertaining to the member designations and properties, the member number I is listed first, and this is followed by the joint j number $JJ[I]$ and the joint k number $JK[I]$ for the two ends of the member. The choice of which end of the member is to be the j end and which is to be the k end is made arbitrarily by the programmer. The last item on each card is the cross-sectional area $AX[I]$ of the member.

The data cards for the joint restraint list are similar to those in the continuous beam program, except that the types of restraints are different. For the plane truss, the terms $RL[2K-1]$ and $RL[2K]$ denote the restraints against translations in the x and y directions, respectively, at joint k. Similarly, the cards for load data are symbolically the same as in the continuous beam program, but the applied actions A consist of forces in the x and y directions, and the actions AML consist of forces in the x_M and y_M directions at the ends of the members.

The identifiers of type integer in the plane truss program are the same as in the continuous beam program, with the addition of the joint designations JJ and JK. The subscripted variables for which storage must be reserved are the following:

$$X[\ ],\ Y[\ ],\ JJ[\ ],\ JK[\ ],\ AX[\ ],\ L[\ ],\ CX[\ ],\ CY[\ ],\ RL[\ ],\ CRL[\ ],$$

$$A[\ ],\quad AML[\ ,\ ],\quad AE[\ ],\quad AC[\ ],\quad SMD[\ ,\ ],\quad S[\ ,\ ],\quad D[\ ],\quad AR[\ ]$$

Explanation of Flow Chart for the Plane Truss Program

Item No.	*Remarks*
1. b.	In an iteration control statement, NJ cards containing the joint numbers and the x and y coordinates of joints are read (see Table 5-7). The same information is also printed.
c. (1)	In an iteration control statement, a series of M cards are read, each of which contains the member designations and cross-sectional area for a given member (see Table 5-7).
(2)	The x component of the length of the member (called XCL in the program) is computed by taking the difference between the x coordinates of the joints at the ends of the member. Similarly, the y component of the length (YCL) is computed as the difference between y coordinates.
(3)	The true length of the member is computed as the square root of the sum of the squares of the x and y components of the length.
(4)	The x and y direction cosines of the member are computed by dividing the x and y components of the length by the length of the member itself.
(5)	The input information plus the length and the direction cosines are printed for each member.
2. a. (1)	Iteration control statement on members, for M members.
(2)	The two possible displacements are indexed for the joint J ($J1$ and $J2$) and the joint K ($K1$ and $K2$) of the member I.
(3)	The stiffness constant for axial deformation is calculated and stored in order to avoid repetition of this expression when generating the member stiffness matrix.
(4)	The two possible displacements for the joints at each end of the member are re-indexed by the same method as that used in the continuous beam program.
(5)	The 4×4 member stiffness matrix SMD for member I is generated (see Table 4-15, Art. 4.10).
4. a. (1)	If NLM is not equal to zero, equivalent joint loads must be calculated.
(2)	In an iteration control statement for M members, the contributions to AE from each member are identified and transferred (see Eqs. 4-33).
5. b. (1)	For each member, the same information is computed as in Secs. 2a(2) and (3) given previously.
(2)	The member number is printed, and the four member end-actions are computed and printed, using Eqs. (4-34).

FLOW CHART FOR PLANE TRUSS PROGRAM

1. Input and Print Structure Data

a. Structure parameters and elastic modulus

b. Joint coordinates

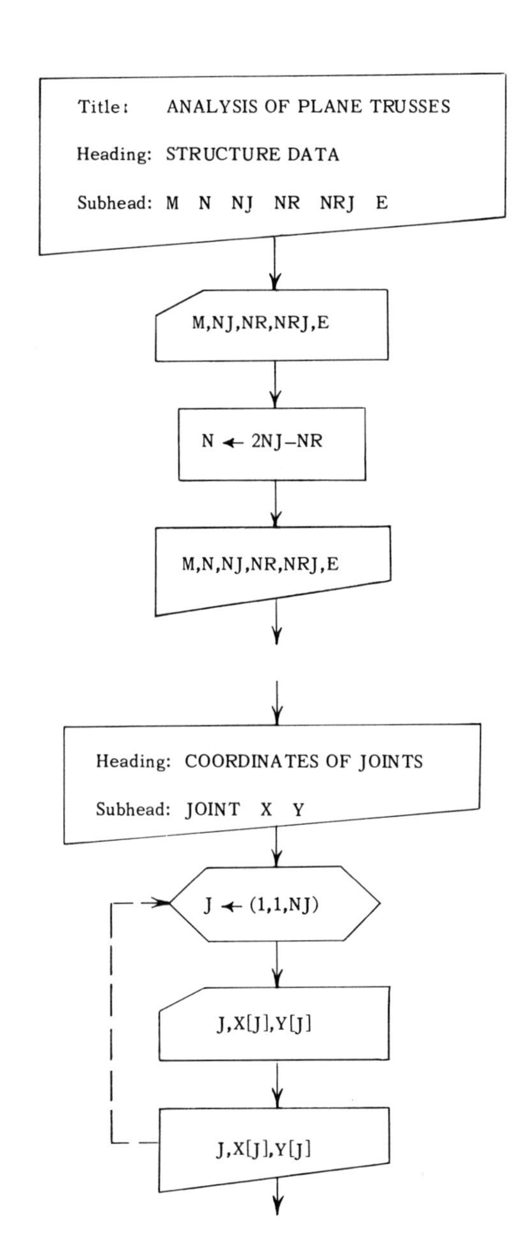

c. Member designations and properties

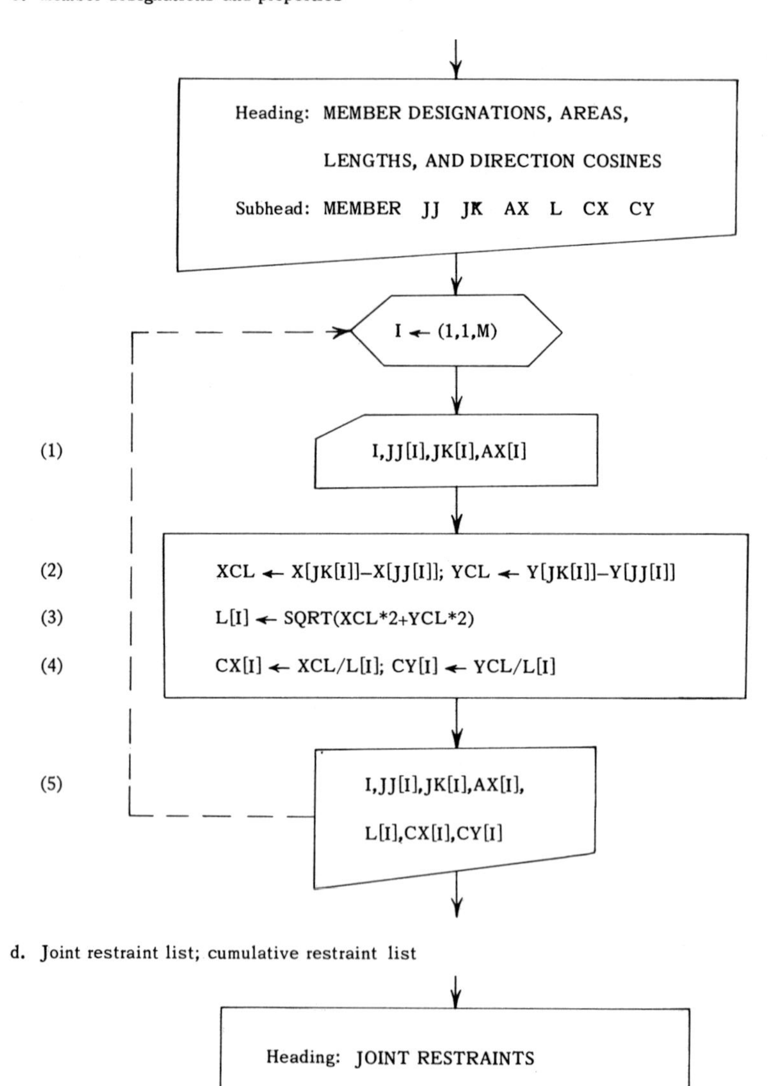

d. Joint restraint list; cumulative restraint list

Heading: JOINT RESTRAINTS

Subhead: JOINT X RSTRT. Y RSTRT.

The remaining part of this section is the same as in Sec. 1c of the continuous beam program.

2. Structure Stiffness Matrix

a. Generation of stiffness matrix

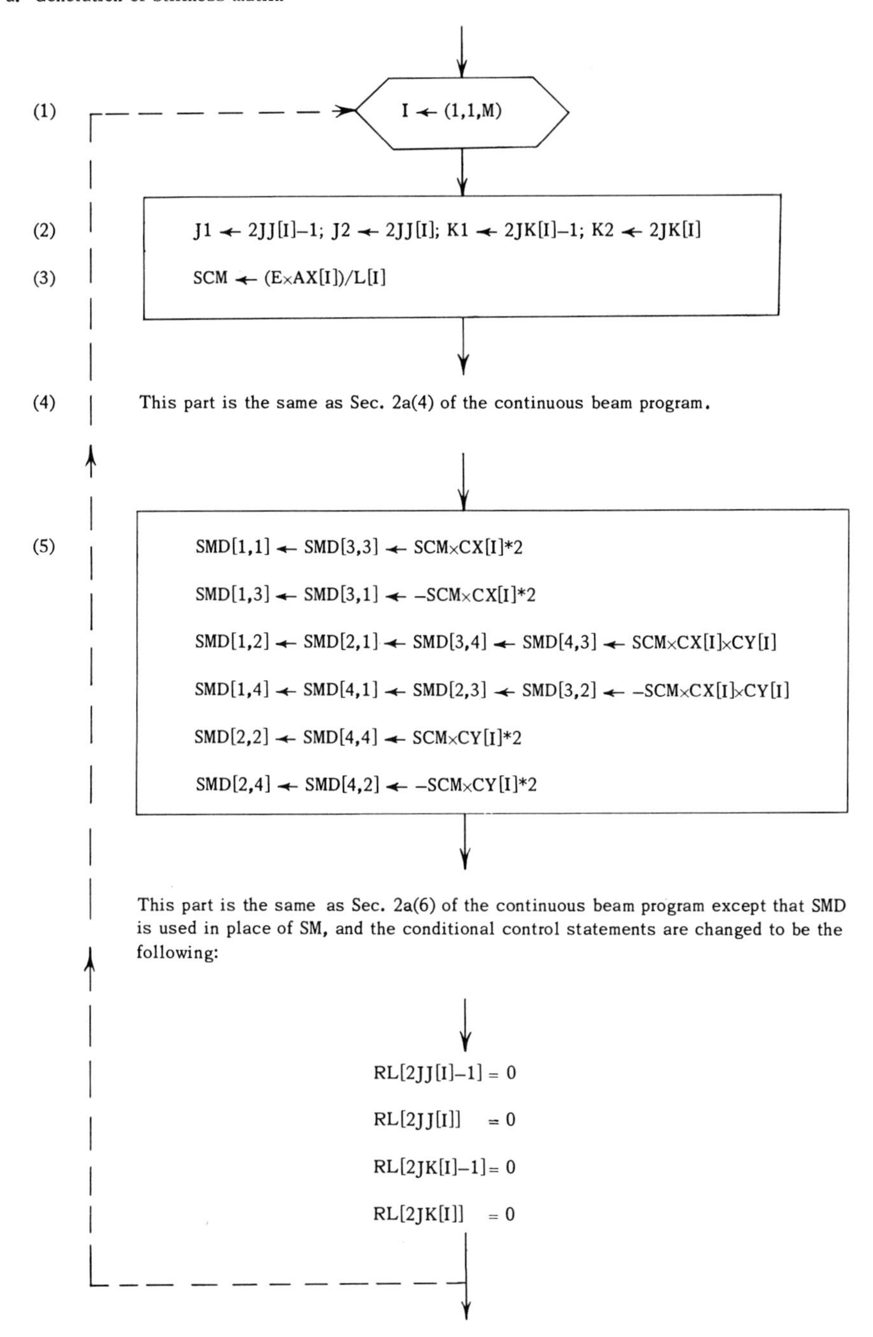

(1)
I ← (1,1,M)
(2)
J1 ← 2JJ[I]–1; J2 ← 2JJ[I]; K1 ← 2JK[I]–1; K2 ← 2JK[I]
(3)
SCM ← (E×AX[I])/L[I]
(4)
This part is the same as Sec. 2a(4) of the continuous beam program.
(5)
SMD[1,1] ← SMD[3,3] ← SCM×CX[I]*2
SMD[1,3] ← SMD[3,1] ← –SCM×CX[I]*2
SMD[1,2] ← SMD[2,1] ← SMD[3,4] ← SMD[4,3] ← SCM×CX[I]×CY[I]
SMD[1,4] ← SMD[4,1] ← SMD[2,3] ← SMD[3,2] ← –SCM×CX[I]×CY[I]
SMD[2,2] ← SMD[4,4] ← SCM×CY[I]*2
SMD[2,4] ← SMD[4,2] ← –SCM×CY[I]*2
This part is the same as Sec. 2a(6) of the continuous beam program except that SMD is used in place of SM, and the conditional control statements are changed to be the following:
RL[2JJ[I]–1] = 0
RL[2JJ[I]] = 0
RL[2JK[I]–1] = 0
RL[2JK[I]] = 0

Sections 2b and 3a, b, c are the same as the corresponding sections in the continuous beam program, except that the subheadings Y ACTION and Z ACTION are changed to X ACTION and Y ACTION, respectively.

4. Construction of Vectors Associated with Loads

a. Equivalent joint loads

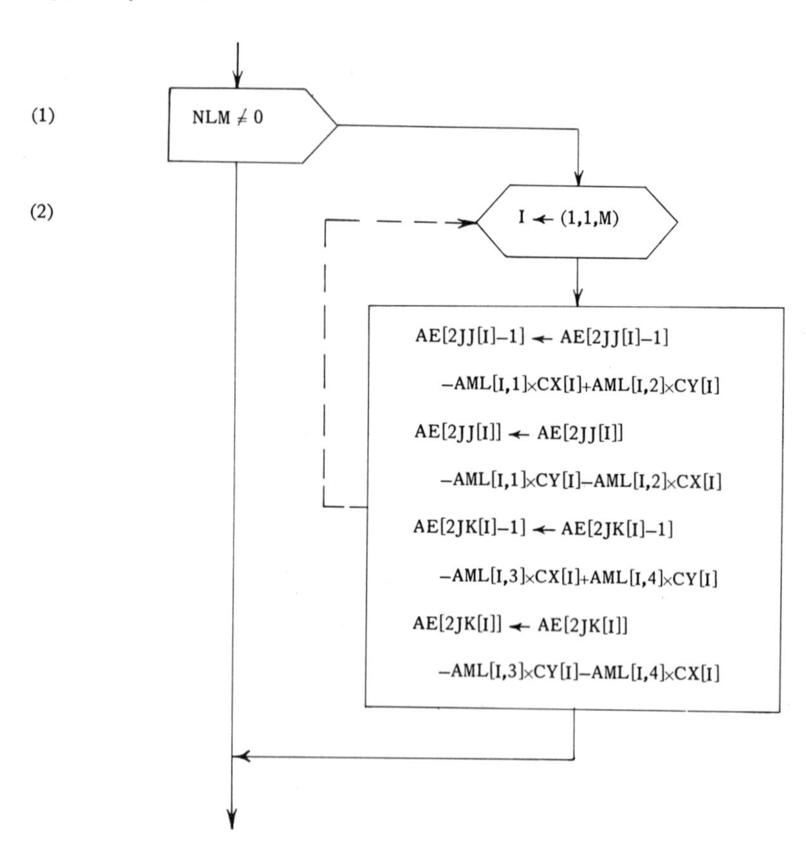

Sections 4b and 5a are the same as in the continuous beam program, except that the subheadings Y DISPL., Z DISPL., Y REAC., and Z REAC. are changed to X DISPL., Y DISPL., X REAC., and Y REAC., respectively.

5b. Member end-actions

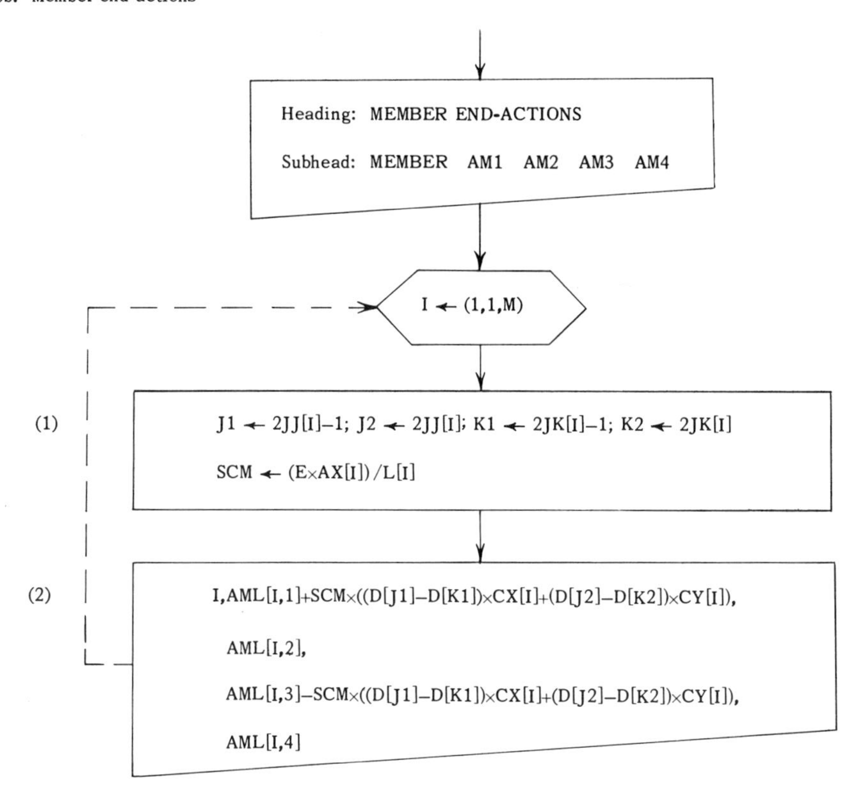

Example: The following example demonstrates the analysis of the plane truss structure of Art. 4.12 (see Fig. 4-23), using the computer program. The following numerical values are assumed for the parameters in the problem:

$$E = 10{,}000 \text{ ksi} \qquad L = 100 \text{ in.} \qquad P = 10 \text{ kips} \qquad AX = 10 \text{ in.}^2$$

Table 5-8 contains the input data required for the program, and Table 5-9 shows the output of the computer program.

5.6 Plane Frame Program. This article contains a program for the analysis of plane rigid frames using the concepts described in Art. 4.17. Since this type of structure has three possible displacements at each joint, the detailed steps in the program will be somewhat different from the programs for continuous beams and plane trusses. However, the over-all methodology is the same, and the reader will notice that the only change in certain statements is that the integer 2, representing the number of possible displacements at a joint, is replaced by the integer 3.

Table 5-10 summarizes the input data required for the plane frame program. This table contains information which is very similar to that in Table 5-7 for plane trusses. However, an additional member property

TABLE 5-8

Data Cards for Plane Truss Example

Type of Data		Numerical Data on Card						Card No.
Structure Data	(a)	6	4	4	2	10000.0		1
	(b)	1	0	80.0				2
		2	60.0	80.0				3
		3	0	0				4
		4	60.0	0				5
	(c)	1	1	2	6.0			6
		2	3	4	6.0			7
		3	3	1	8.0			8
		4	4	2	8.0			9
		5	1	4	10.0			10
		6	3	2	10.0			11
	(d)	3	1	1				12
		4	1	1				13
Load Data	(a)	1	4					14
	(b)	2	20.0	10.0				15
	(c)	1	0	−20.0	0	20.0		16
		2	0	10.0	0	10.0		17
		3	0	10.0	0	10.0		18
		4	5.0	5.0	5.0	5.0		19

IZ (moment of inertia of the cross-section) is required on each of the cards pertaining to member designations and properties.

Each of the cards in the joint restraint series contains a joint number K and three code numbers which indicate the conditions of restraint at that joint. The terms $RL[3K-2]$, $RL[3K-1]$, and $RL[3K]$ denote the restraints against translations in the x and y directions and rotation in the z sense, respectively, at joint k.

Each joint load card contains a joint number K and the three actions applied at that joint. These actions are the applied forces in the x and y directions and the couple in the z sense. In addition, each card for member loads contains a member number I and the six actions at the ends

TABLE 5-9

Output of Computer Program for Plane Truss Example

ANALYSIS OF PLANE TRUSSES

STRUCTURE DATA

M	N	NJ	NR	NRJ	E
6	4	4	4	2	10000.0

COORDINATES OF JOINTS

JOINT	X	Y
1	0	80.0
2	60.0	80.0
3	0	0
4	60.0	0

MEMBER DESIGNATIONS, AREAS, LENGTHS, AND DIRECTION COSINES

MEMBER	JJ	JK	AX	L	CX	CY
1	1	2	6.0	60.0	1.0	0
2	3	4	6.0	60.0	1.0	0
3	3	1	8.0	80.0	0	1.0
4	4	2	8.0	80.0	0	1.0
5	1	4	10.0	100.0	0.6	–0.8
6	3	2	10.0	100.0	0.6	0.8

JOINT RESTRAINTS

JOINT	X RSTRT.	Y RSTRT.
3	1	1
4	1	1

LOAD DATA

NLJ	NLM
1	4

ACTIONS APPLIED AT JOINTS

JOINT	X ACTION	Y ACTION
2	20.0	10.0

ACTIONS AT ENDS OF RESTRAINED MEMBERS DUE TO LOADS

MEMBER	AML1	AML2	AML3	AML4
1	0	–20.0	0	20.0
2	0	10.0	0	10.0
3	0	10.0	0	10.0
4	5.0	5.0	5.0	5.0

JOINT DISPLACEMENTS AND SUPPORT REACTIONS

JOINT	X DISPL.	Y DISPL.	X REAC.	Y REAC.
1	0.1000	0.04147	0	0
2	0.1061	–0.04020	0	0
3	0	0	–28.90	–56.67
4	0	0	–21.10	76.67

(Table 5-9 continued on next page.)

(TABLE 5-9, continued)

MEMBER END-ACTIONS

MEMBER	AM1	AM2	AM3	AM4
1	-6.10	-20.00	6.10	20.00
2	0	10.00	0	10.00
3	-41.47	10.00	41.47	10.00
4	45.20	5.00	-35.20	5.00
5	26.83	0	-26.83	0
6	-31.50	0	31.50	0

of the restrained member. These actions consist of forces in the x_M and y_M directions and a couple in the z_M (or z) sense at the ends j and k, respectively.

The identifiers of type integer in the plane frame program are the same as those in the plane truss program, with the addition of the displacement indexes $J3$ and $K3$. Moreover, the subscripted variables are also the same, except that $IZ[\ ]$ is added to the list.

TABLE 5-10

Preparation of Data for Plane Frame Program

	Data	Number of Cards	Items on Data Cards
Structure Data	(a) Structure parameters and elastic modulus	1	M NJ NR NRJ E
	(b) Joint coordinates	NJ	J X[J] Y[J]
	(c) Member designations and properties	M	I JJ[I] JK[I] AX[I] IZ[I]
	(d) Joint restraint list	NRJ	K RL[3K – 2] RL[3K – 1] RL[3K]
Load Data	(a) Numbers of loaded joints and members	1	NLJ NLM
	(b) Actions applied at joints	NLJ	K A[3K – 2] A[3K – 1] A[3K]
	(c) Actions at ends of restrained members due to loads	NLM	I AML[I, 1] AML[I, 2] AML[I, 3] AML[I, 4] AML[I, 5] AML[I, 6]

Explanation of Flow Chart for the Plane Frame Program

Item No. *Remarks*

2. a. (1) Iteration control statement on members, for M members.
   (2) The three possible displacements are indexed for the joint J ($J1$, $J2$, and $J3$) and the joint K ($K1$, $K2$, and $K3$) of the member I.

(*Remarks continued on p. 354.*)

FLOW CHART FOR PLANE FRAME PROGRAM

1. Input and Print Structure Data

a. Structure parameters and elastic modulus

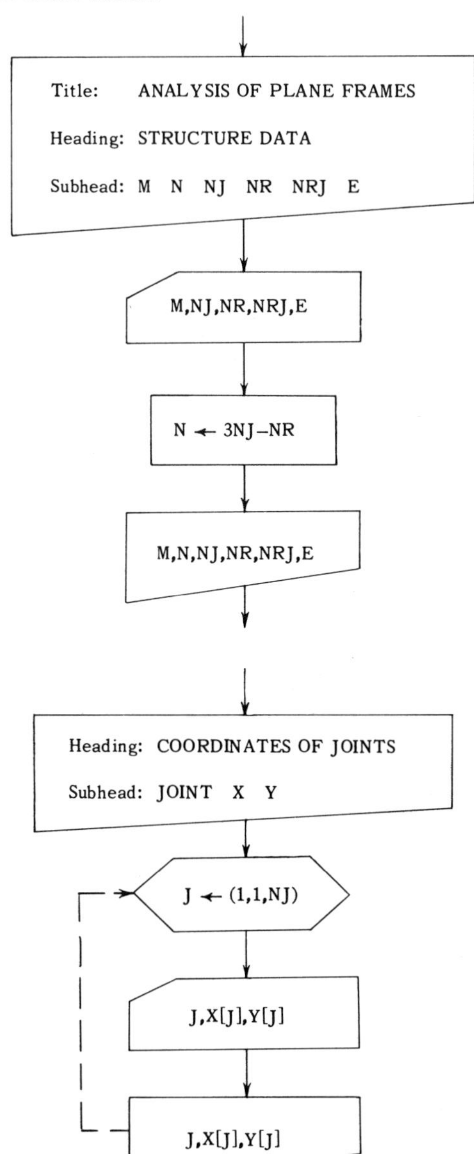

b. Joint coordinates

c. Member designations and properties

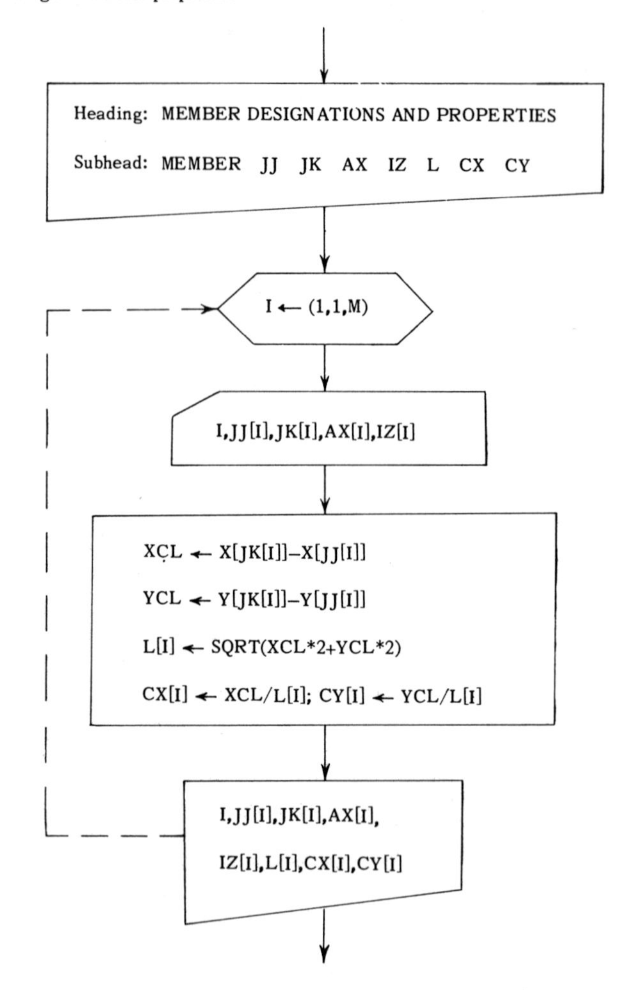

d. Joint restraint list; cumulative restraint list

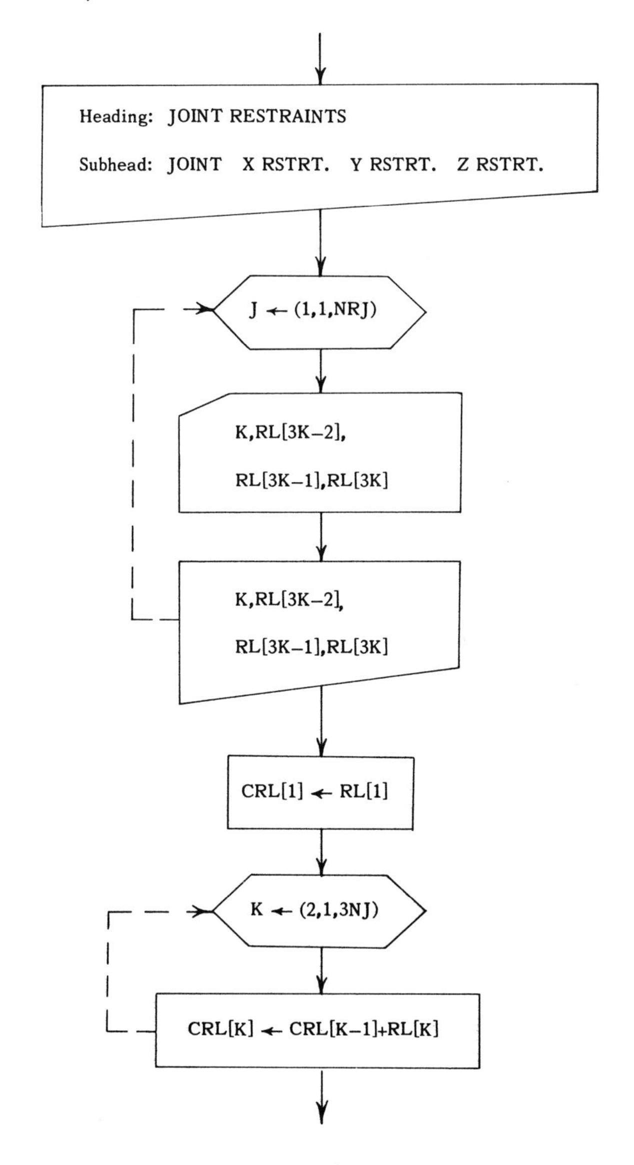

2. Structure Stiffness Matrix

a. Generation of stiffness matrix

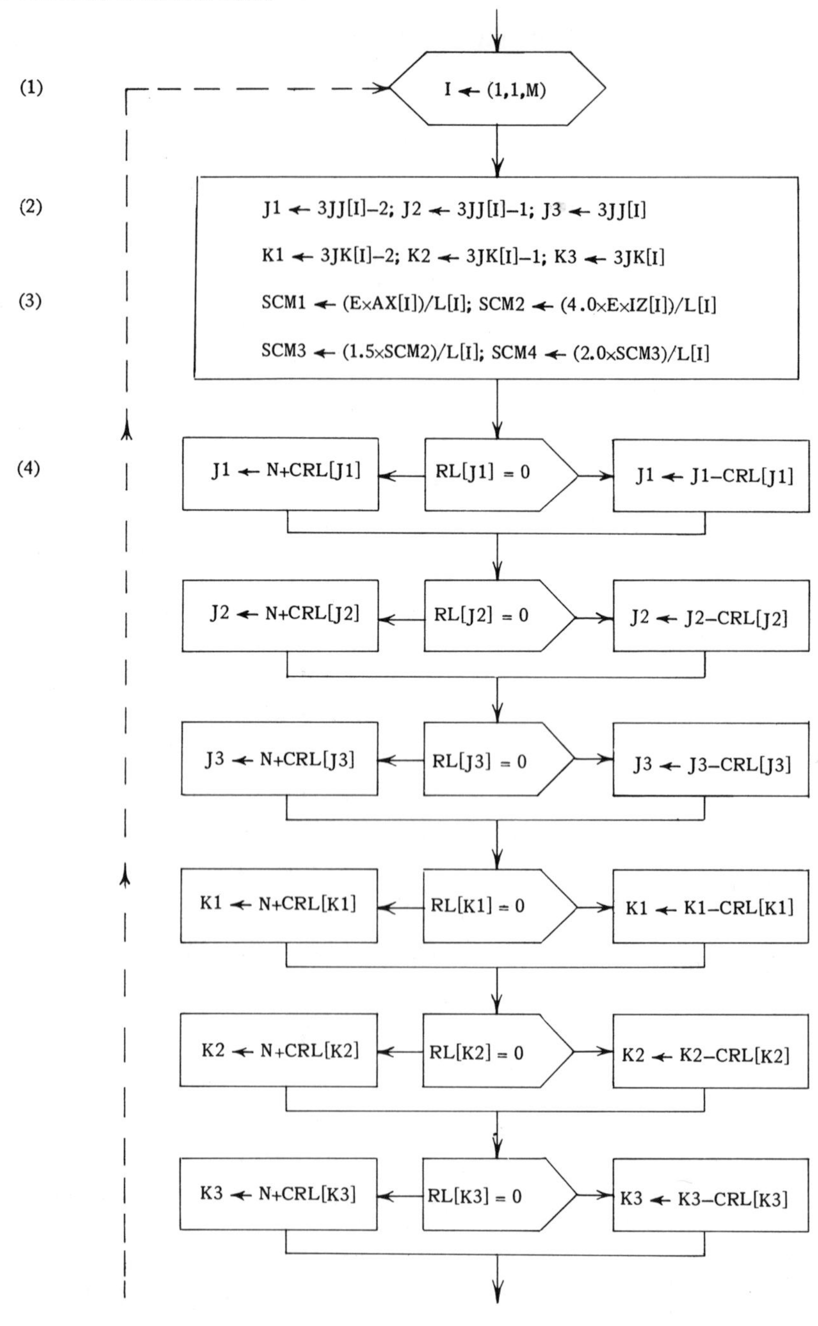

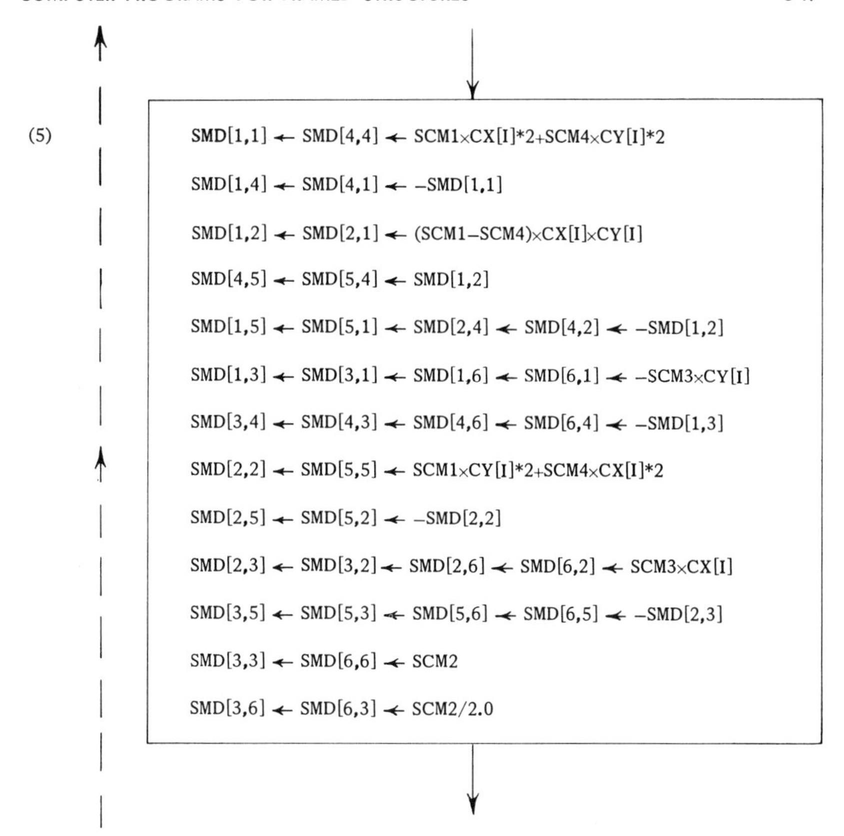
(5)
SMD[1,1] ← SMD[4,4] ← SCM1×CX[I]*2+SCM4×CY[I]*2
SMD[1,4] ← SMD[4,1] ← –SMD[1,1]
SMD[1,2] ← SMD[2,1] ← (SCM1–SCM4)×CX[I]×CY[I]
SMD[4,5] ← SMD[5,4] ← SMD[1,2]
SMD[1,5] ← SMD[5,1] ← SMD[2,4] ← SMD[4,2] ← –SMD[1,2]
SMD[1,3] ← SMD[3,1] ← SMD[1,6] ← SMD[6,1] ← –SCM3×CY[I]
SMD[3,4] ← SMD[4,3] ← SMD[4,6] ← SMD[6,4] ← –SMD[1,3]
SMD[2,2] ← SMD[5,5] ← SCM1×CY[I]*2+SCM4×CX[I]*2
SMD[2,5] ← SMD[5,2] ← –SMD[2,2]
SMD[2,3] ← SMD[3,2] ← SMD[2,6] ← SMD[6,2] ← SCM3×CX[I]
SMD[3,5] ← SMD[5,3] ← SMD[5,6] ← SMD[6,5] ← –SMD[2,3]
SMD[3,3] ← SMD[6,6] ← SCM2
SMD[3,6] ← SMD[6,3] ← SCM2/2.0

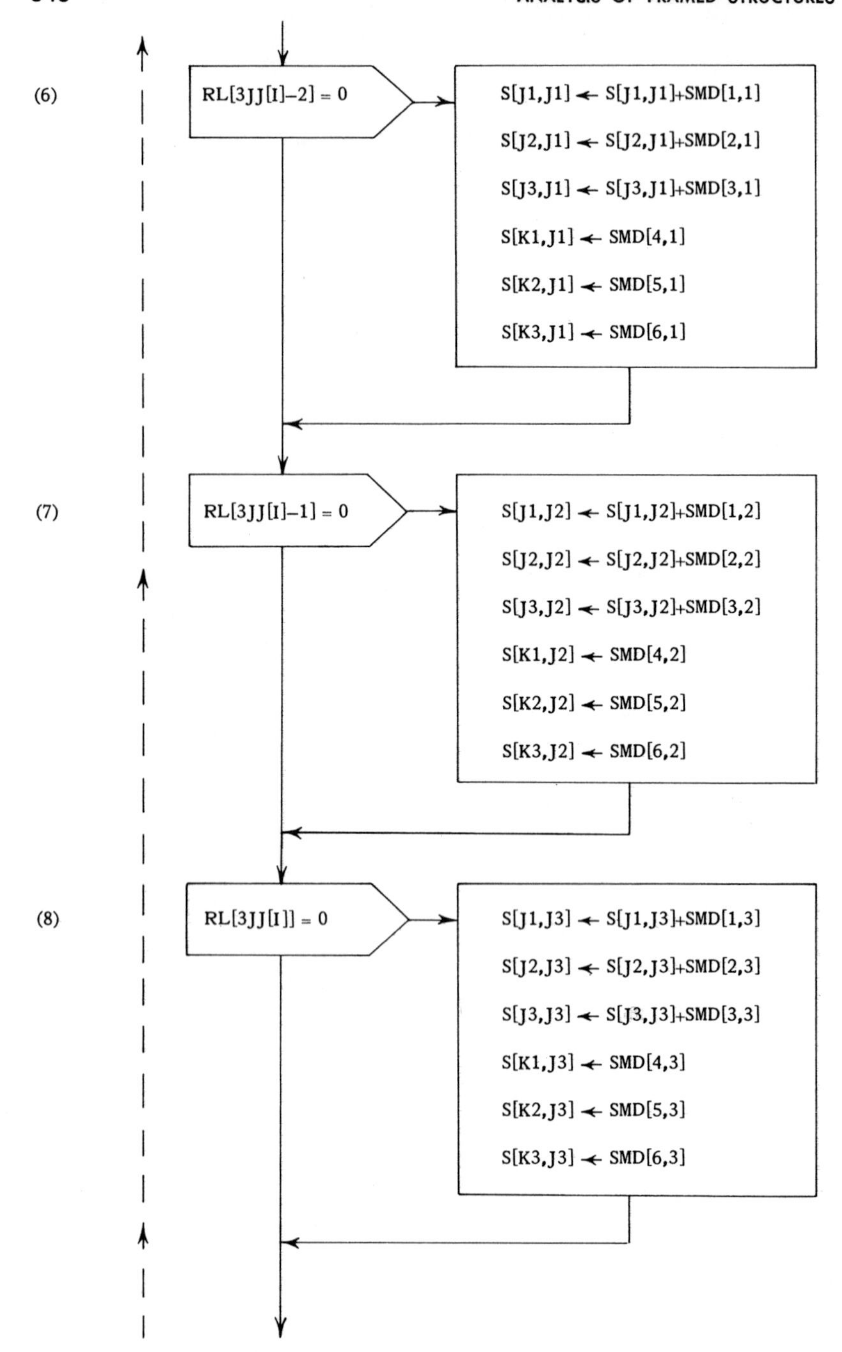

(6)
RL[3JJ[I]–2] = 0
S[J1,J1] ← S[J1,J1]+SMD[1,1]
S[J2,J1] ← S[J2,J1]+SMD[2,1]
S[J3,J1] ← S[J3,J1]+SMD[3,1]
S[K1,J1] ← SMD[4,1]
S[K2,J1] ← SMD[5,1]
S[K3,J1] ← SMD[6,1]
(7)
RL[3JJ[I]–1] = 0
S[J1,J2] ← S[J1,J2]+SMD[1,2]
S[J2,J2] ← S[J2,J2]+SMD[2,2]
S[J3,J2] ← S[J3,J2]+SMD[3,2]
S[K1,J2] ← SMD[4,2]
S[K2,J2] ← SMD[5,2]
S[K3,J2] ← SMD[6,2]
(8)
RL[3JJ[I]] = 0
S[J1,J3] ← S[J1,J3]+SMD[1,3]
S[J2,J3] ← S[J2,J3]+SMD[2,3]
S[J3,J3] ← S[J3,J3]+SMD[3,3]
S[K1,J3] ← SMD[4,3]
S[K2,J3] ← SMD[5,3]
S[K3,J3] ← SMD[6,3]

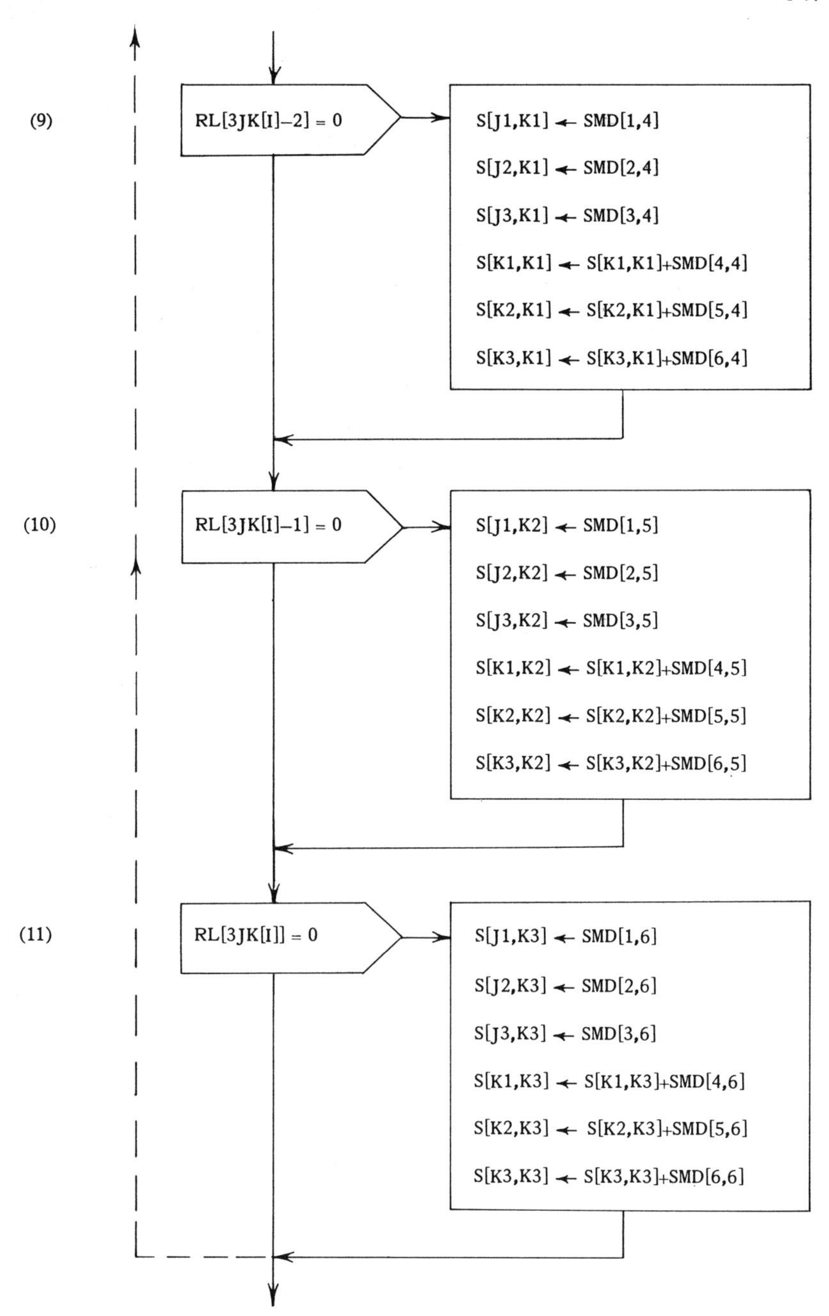
(9)
RL[3JK[I]–2] = 0
S[J1,K1] ← SMD[1,4]
S[J2,K1] ← SMD[2,4]
S[J3,K1] ← SMD[3,4]
S[K1,K1] ← S[K1,K1]+SMD[4,4]
S[K2,K1] ← S[K2,K1]+SMD[5,4]
S[K3,K1] ← S[K3,K1]+SMD[6,4]
(10)
RL[3JK[I]–1] = 0
S[J1,K2] ← SMD[1,5]
S[J2,K2] ← SMD[2,5]
S[J3,K2] ← SMD[3,5]
S[K1,K2] ← S[K1,K2]+SMD[4,5]
S[K2,K2] ← S[K2,K2]+SMD[5,5]
S[K3,K2] ← S[K3,K2]+SMD[6,5]
(11)
RL[3JK[I]] = 0
S[J1,K3] ← SMD[1,6]
S[J2,K3] ← SMD[2,6]
S[J3,K3] ← SMD[3,6]
S[K1,K3] ← S[K1,K3]+SMD[4,6]
S[K2,K3] ← S[K2,K3]+SMD[5,6]
S[K3,K3] ← S[K3,K3]+SMD[6,6]

Sections 2b and 3a are the same as in the continuous beam program.

3b. Actions applied at joints

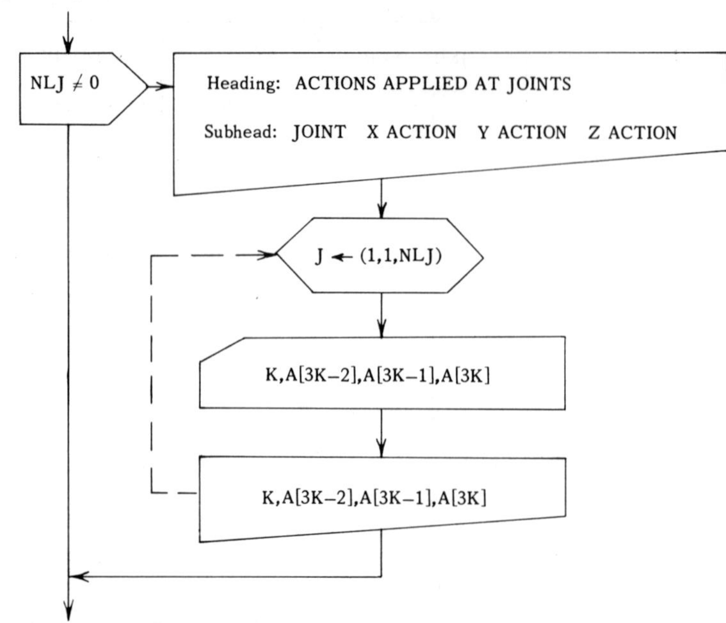

c. Actions at ends of restrained members due to loads

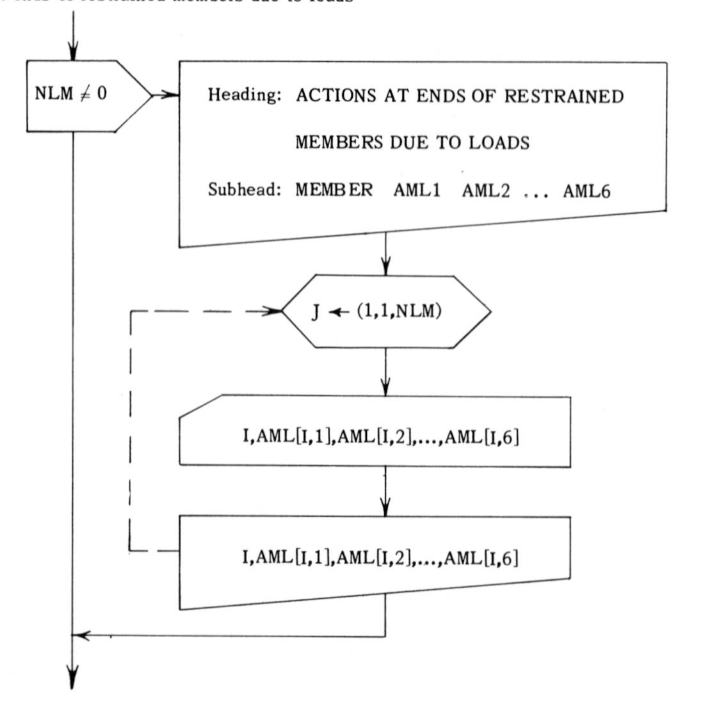

4. Construction of Vectors Associated with Loads

a. Equivalent joint loads

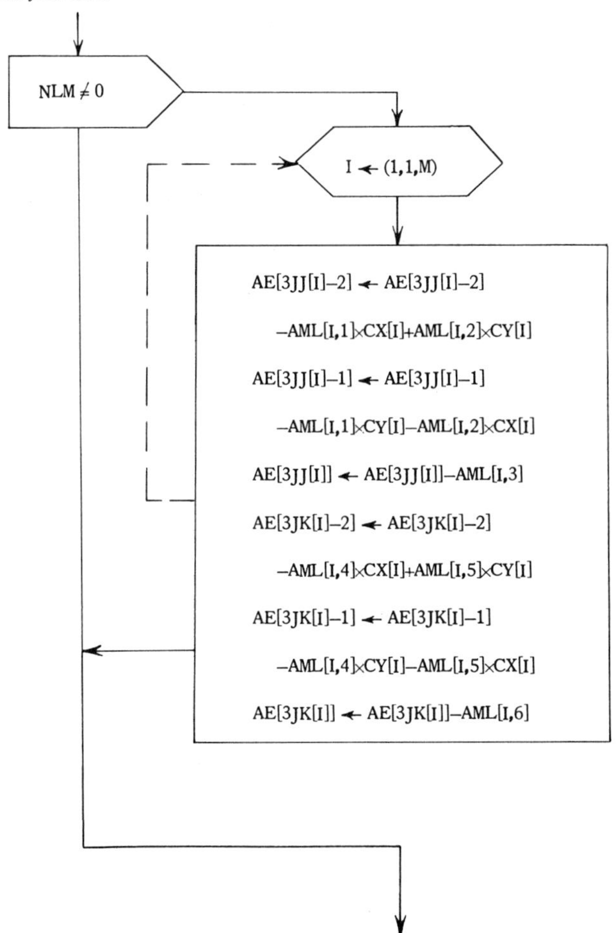

Sections 4b and 5a are the same as in the continuous beam program, except that the following printing is done in Sec. 5a:

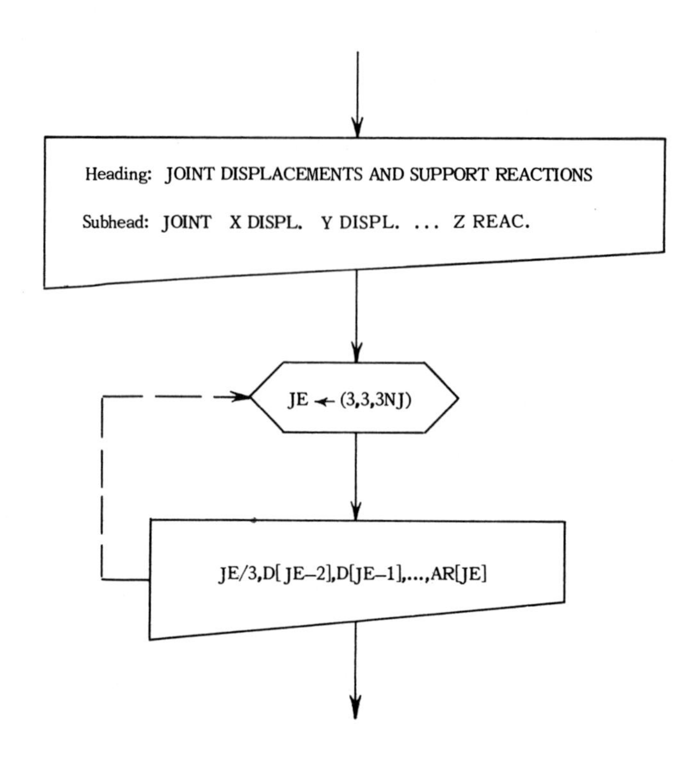

5b. Member end-actions

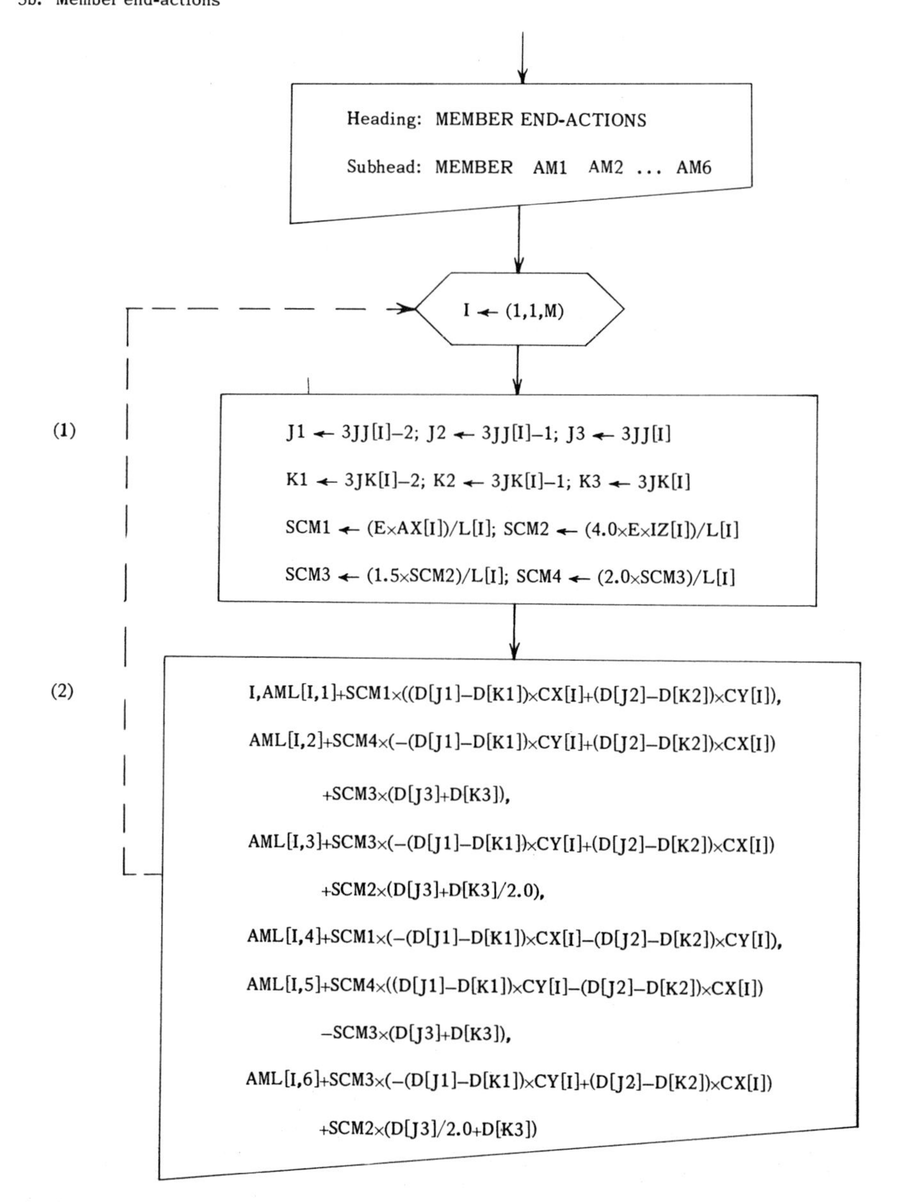

Heading: MEMBER END-ACTIONS
Subhead: MEMBER AM1 AM2 ... AM6
I ← (1,1,M)
(1)
J1 ← 3JJ[I]–2; J2 ← 3JJ[I]–1; J3 ← 3JJ[I]
K1 ← 3JK[I]–2; K2 ← 3JK[I]–1; K3 ← 3JK[I]
SCM1 ← (E×AX[I])/L[I]; SCM2 ← (4.0×E×IZ[I])/L[I]
SCM3 ← (1.5×SCM2)/L[I]; SCM4 ← (2.0×SCM3)/L[I]
(2)
I,AML[I,1]+SCM1×((D[J1]–D[K1])×CX[I]+(D[J2]–D[K2])×CY[I]),
AML[I,2]+SCM4×(–(D[J1]–D[K1])×CY[I]+(D[J2]–D[K2])×CX[I])
+SCM3×(D[J3]+D[K3]),
AML[I,3]+SCM3×(–(D[J1]–D[K1])×CY[I]+(D[J2]–D[K2])×CX[I])
+SCM2×(D[J3]+D[K3]/2.0),
AML[I,4]+SCM1×(–(D[J1]–D[K1])×CX[I]–(D[J2]–D[K2])×CY[I]),
AML[I,5]+SCM4×((D[J1]–D[K1])×CY[I]–(D[J2]–D[K2])×CX[I])
–SCM3×(D[J3]+D[K3]),
AML[I,6]+SCM3×(–(D[J1]–D[K1])×CY[I]+(D[J2]–D[K2])×CX[I])
+SCM2×(D[J3]/2.0+D[K3])

(*Continued from p. 342.*)

(3) Four stiffness constants for member I are calculated and stored in order to avoid repetition of these expressions when generating the member stiffness matrix.

(4) Using the joint restraint list as a guide, the three possible displacements for the joints at each end of the member are re-indexed according to whether the joints are actually free to move or not. The program statements required for this purpose are similar to those in the continuous beam program.

(5) The 6×6 member stiffness matrix SMD for member I is generated (see Table 4-27, Art. 4.16).

(6) If the index $3JJ[I]-2$ (the original value of $J1$) corresponds to a degree of freedom, the first column of SMD is transferred to the joint stiffness matrix for the structure (see Eqs. 4-67). However, if $3JJ[I]-2$ corresponds to a restraint, the first column of SMD is bypassed. Elements of S involving the coupling of $J1$ with $J2$ and $J3$ are cumulative because, in general, they receive contributions from more than one member. On the other hand, elements of S involving the coupling of $J1$ with $K1$, $K2$, and $K3$ are single values.

(7) Similar to (6) but for $3JJ[I]-1$, which is the original value of $J2$ (see Eqs. 4-68).

(8) Similar to (6) but for $3JJ[I]$, which is the original value of $J3$ (see Eqs. 4-69).

(9) Similar to (6) but for $3JK[I]-2$, which is the original value of $K1$ (see Eqs. 4-70).

(10) Similar to (6) but for $3JK[I]-1$, which is the original value of $K2$ (see Eqs. 4-71).

(11) Similar to (6) but for $3JK[I]$, which is the original value of $K3$ (see Eqs. 4-72).

4. a. If NLM is not equal to zero, the contributions to AE from each member are identified and transferred (see Eqs. 4-75).

5. b. (1) For each member, the same information is computed as in Secs. 2a(2) and (3) given previously.

(2) The member number is printed, and the six member end-actions are computed and printed, using Eqs. (4-77).

Example: The plane frame analyzed in Art. 4.18 (see Fig. 4-31) is given as an example of the use of the computer program. The numerical values used in the example are the following:

$$E = 10{,}000 \text{ ksi} \quad L = 100 \text{ in.} \quad P = 10 \text{ kips} \quad AX = 10 \text{ in.}^2 \quad IZ = 1000 \text{ in.}^4$$

The input data for the problem are listed in Table 5-11, and the output of the computer program appears in Table 5-12.

5.7 Grid Program. A program for the analysis of grid structures is given in this article, using the methods described earlier in Art. 4.20. The analysis of grids is symbolically similar to the analysis of plane frames. Therefore, the following discussion emphasizes only those parts of the grid program which are different from the plane frame program.

The input data required for the grid program are nearly the same as that shown in Table 5-10 for plane frames. However, an additional material property (the shear modulus of elasticity G) is required on the first data card. Also, on the cards containing member designations and properties, the cross-sectional area AX and moment of inertia IZ must be replaced by the torsion constant IX and moment of inertia IY, respectively.

TABLE 5-11

Data Cards for Plane Frame Example

Type of Data		Numerical Data on Card							Card No.
Structure Data	(a)	2	3	6	2	10000.0			1
	(b)	1	100.0		75.0				2
		2	0		75.0				3
		3	200.0		0				4
	(c)	1	2	1	10.0	1000.0			5
		2	1	3	10.0	1000.0			6
	(d)	2	1	1	1				7
		3	1	1	1				8
Load Data	(a)	1	2						9
	(b)	1	0	–10.0	–1000.0				10
	(c)	1	0	12.0	200.0	0	12.0	–200.0	11
		2	–6.0	8.0	250.0	–6.0	8.0	–250.0	12

The data cards for the joint restraint list are similar to those in the plane frame program, except that the nature of the restraints is different. For the grid program, the terms $RL[3K-2]$, $RL[3K-1]$, and $RL[3K]$ denote the restraints against rotations in the x and y senses and translation in the z direction, respectively, at joint k.

Similarly, the cards for load data are symbolically the same as in the plane frame program, but the meanings are different. For a grid structure, the actions applied at joints consist of couples in the x and y senses and a force in the z direction. Finally, the actions AML at the ends of restrained members due to loads are couples in the x_M and y_M senses and a force in the z_M (or z) direction at the ends j and k, respectively.

The identifiers of type integer in the grid program are exactly the same as in the plane frame program. The list of subscripted variables is almost the same, but $AX[\ ]$ and $IZ[\ ]$ are replaced by $IX[\ ]$ and $IY[\ ]$. The flow chart begins on p. 357.

TABLE 5-12

Output of Computer Program for Plane Frame Example

ANALYSIS OF PLANE FRAMES

STRUCTURE DATA

M	N	NJ	NR	NRJ	E
2	3	3	6	2	10000.0

COORDINATES OF JOINTS

JOINT	X	Y
1	100.0	75.0
2	0	75.0
3	200.0	0

MEMBER DESIGNATIONS AND PROPERTIES

MEMBER	JJ	JK	AX	IZ	L	CX	CY
1	2	1	10.0	1000.0	100.0	1.0	0
2	1	3	10.0	1000.0	125.0	0.8	−0.6

JOINT RESTRAINTS

JOINT	X RSTRT.	Y RSTRT.	Z RSTRT.
2	1	1	1
3	1	1	1

LOAD DATA

NLJ	NLM
1	2

ACTIONS APPLIED AT JOINTS

JOINT	X ACTION	Y ACTION	Z ACTION
1	0	−10.00	−1000.00

ACTIONS AT ENDS OF RESTRAINED MEMBERS DUE TO LOADS

MEMBER	AML1	AML2	AML3	AML4	AML5	AML6
1	0	12.0	200.0	0	12.0	−200.0
2	−6.0	8.0	250.0	−6.0	8.0	−250.0

JOINT DISPLACEMENTS AND SUPPORT REACTIONS

JOINT	X DISPL.	Y DISPL.	Z DISPL.	X REAC.	Y REAC.	Z REAC.
1	−0.02026	−0.09936	−0.001797	0	0	0
2	0	0	0	20.26	13.14	436.6
3	0	0	0	−20.26	40.86	−889.5

MEMBER END-ACTIONS

MEMBER	AM1	AM2	AM3	AM4	AM5	AM6
1	20.26	13.14	436.6	−20.26	10.86	−322.9
2	28.72	−4.53	−677.1	−40.73	20.53	−889.5

FLOW CHART FOR GRID PROGRAM

1. Input and Print Structure Data

a. Structure parameters and elastic moduli

Title: ANALYSIS OF GRIDS

Heading: STRUCTURE DATA

Subhead: M N NJ NR NRJ E G

Section 1 of this program is similar to the plane frame program except that the shear modulus G is added and the member properties AX and IZ are replaced by IX and IY.

2. Structure Stiffness Matrix

a. Generation of stiffness matrix

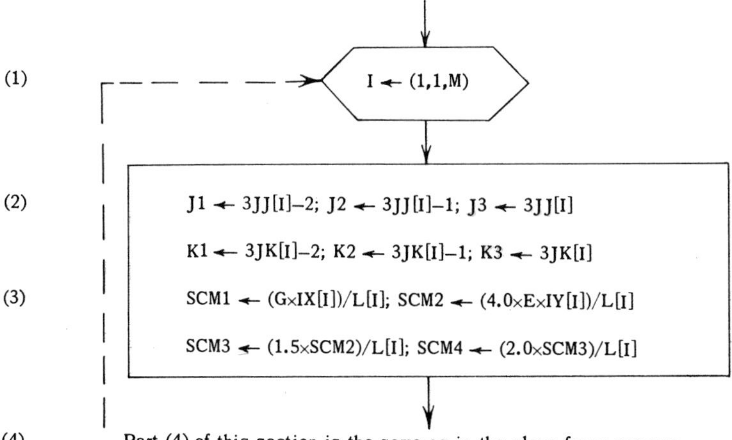
(1)
I ← (1,1,M)
(2)
J1 ← 3JJ[I]–2; J2 ← 3JJ[I]–1; J3 ← 3JJ[I]
K1 ← 3JK[I]–2; K2 ← 3JK[I]–1; K3 ← 3JK[I]
(3)
SCM1 ← (G×IX[I])/L[I]; SCM2 ← (4.0×E×IY[I])/L[I]
SCM3 ← (1.5×SCM2)/L[I]; SCM4 ← (2.0×SCM3)/L[I]

(4) Part (4) of this section is the same as in the plane frame program.

(5)

```
SMD[1,1] ← SMD[4,4] ← SCM1×CX[I]*2+SCM2×CY[I]*2

SMD[1,2] ← SMD[2,1] ← (SCM1–SCM2)×CX[I]×CY[I]

SMD[4,5] ← SMD[5,4] ← SMD[1,2]

SMD[1,3] ← SMD[3,1] ← SMD[3,4] ← SMD[4,3] ← SCM3×CY[I]

SMD[1,6] ← SMD[6,1] ← SMD[4,6] ← SMD[6,4] ← –SMD[1,3]

SMD[1,4] ← SMD[4,1] ← –SCM1×CX[I]*2+(SCM2/2.0)×CY[I]*2

SMD[1,5] ← SMD[5,1] ← –(SCM1+SCM2/2.0)×CX[I]×CY[I]

SMD[2,4] ← SMD[4,2] ← SMD[1,5]

SMD[2,2] ← SMD[5,5] ← SCM1×CY[I]*2+SCM2×CX[I]*2

SMD[2,3] ← SMD[3,2] ← SMD[3,5] ← SMD[5,3] ← –SCM3×CX[I]

SMD[2,6] ← SMD[6,2] ← SMD[5,6] ← SMD[6,5] ← –SMD[2,3]

SMD[2,5] ← SMD[5,2] ← –SCM1×CY[I]*2+(SCM2/2.0)×CX[I]*2

SMD[3,3] ← SMD[6,6] ← SCM4

SMD[3,6] ← SMD[6,3] ← –SCM4
```

The remainder of Section 2 is the same as in the plane frame program.

Sections 3a through 5a are the same as in the plane frame program.

5b. Member end-actions

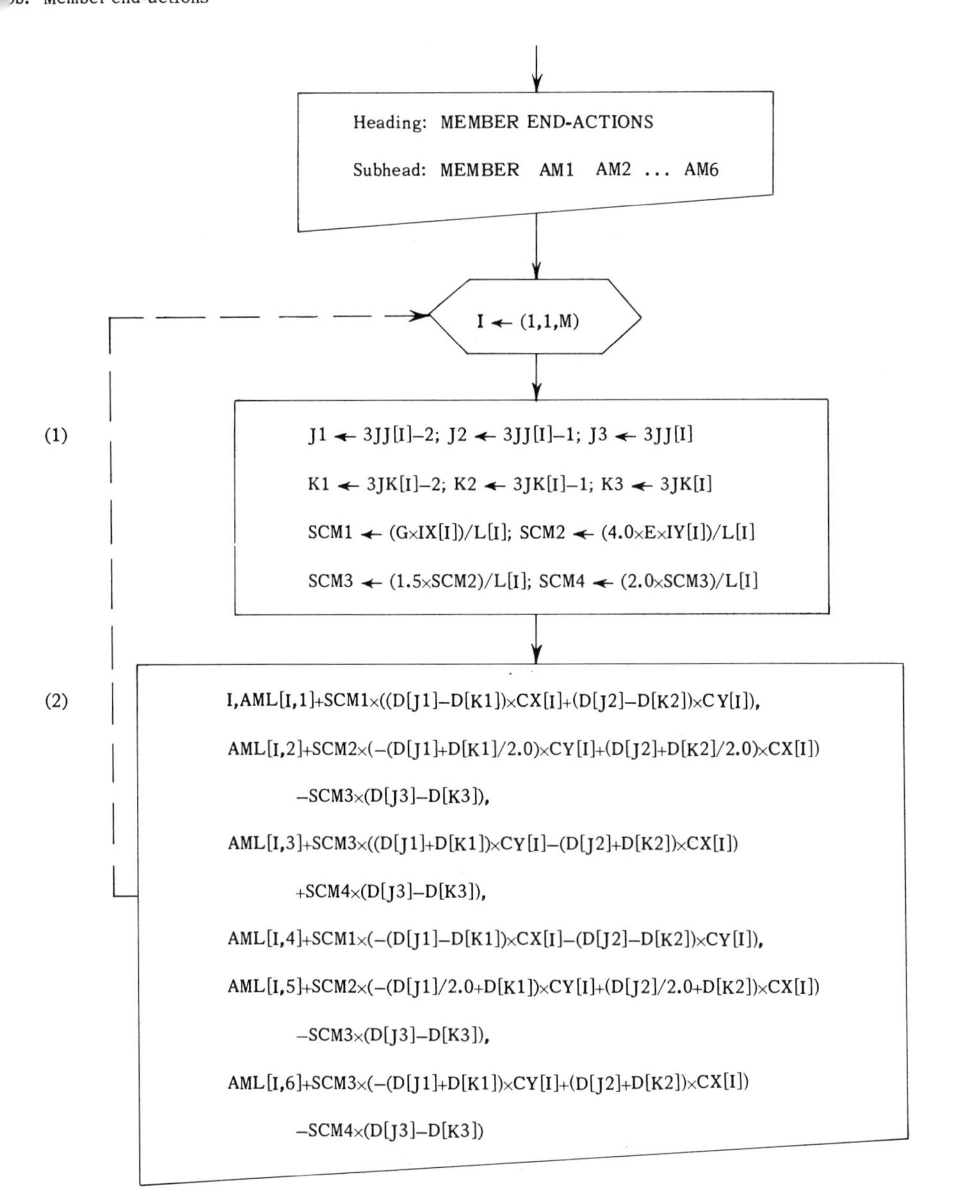
Heading: MEMBER END-ACTIONS
Subhead: MEMBER AM1 AM2 ... AM6
I ← (1,1,M)
(1)
J1 ← 3JJ[I]–2; J2 ← 3JJ[I]–1; J3 ← 3JJ[I]
K1 ← 3JK[I]–2; K2 ← 3JK[I]–1; K3 ← 3JK[I]
SCM1 ← (G×IX[I])/L[I]; SCM2 ← (4.0×E×IY[I])/L[I]
SCM3 ← (1.5×SCM2)/L[I]; SCM4 ← (2.0×SCM3)/L[I]
(2)
I,AML[I,1]+SCM1×((D[J1]–D[K1])×CX[I]+(D[J2]–D[K2])×CY[I]),
AML[I,2]+SCM2×(–(D[J1]+D[K1]/2.0)×CY[I]+(D[J2]+D[K2]/2.0)×CX[I])
–SCM3×(D[J3]–D[K3]),
AML[I,3]+SCM3×((D[J1]+D[K1])×CY[I]–(D[J2]+D[K2])×CX[I])
+SCM4×(D[J3]–D[K3]),
AML[I,4]+SCM1×(–(D[J1]–D[K1])×CX[I]–(D[J2]–D[K2])×CY[I]),
AML[I,5]+SCM2×(–(D[J1]/2.0+D[K1])×CY[I]+(D[J2]/2.0+D[K2])×CX[I])
–SCM3×(D[J3]–D[K3]),
AML[I,6]+SCM3×(–(D[J1]+D[K1])×CY[I]+(D[J2]+D[K2])×CX[I])
–SCM4×(D[J3]–D[K3])

Explanation of Flow Chart for the Grid Program

Item No. *Remarks*

2. a. (1) Iteration control statement on members, for M members.
   (2) The three possible displacements are indexed for the joints J and K of the member I.
   (3) Four stiffness constants for member I are calculated and stored in order to avoid repetition of these expressions when generating the member stiffness matrix.
   (4) The three possible displacements for the joints at each end of the member are re-indexed by the same method as that used in the plane frame program.
   (5) The 6×6 member stiffness matrix SMD for member I is generated (see Table 4-38, Art. 4.19).
5. b. (1) For each member the same information is computed as in Secs. 2a(2) and (3) given previously.
   (2) The member number is printed, and the six member end-actions are computed and printed, using Eqs. (4-78).

Example 1: Figure 5-2a shows a grid structure having geometrical properties and loads similar to the plane frame problem in Art. 4.18 (see Fig. 4-31a). Assume that the cross-sectional properties of both members of the grid are the same and that the following numerical values apply:

$$E = 10{,}000 \text{ ksi} \qquad P = 10 \text{ kips} \qquad IX = 1000 \text{ in.}^4$$

$$G = 4{,}000 \text{ ksi} \qquad L = 100 \text{ in.} \qquad IY = 1000 \text{ in.}^4$$

A numbering system for members and joints is given in Fig. 5-2b, which shows the restrained structure. This numbering system is the same as in the previous example for the plane frame.

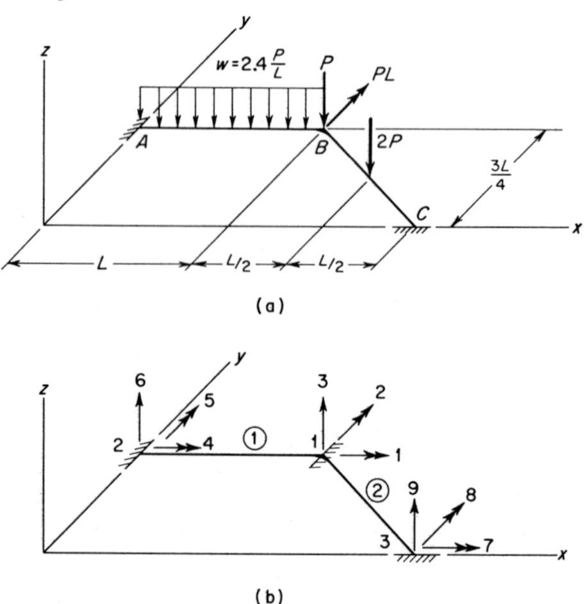

FIG. 5-2. Example 1 (grid).

TABLE 5-13

Data Cards for Grid Example 1

Type of Data		Numerical Data on Card									Card No.
Structure Data	(a)	2	3	6	2	10000.0	4000.0				1
	(b)	1	100.0		75.0						2
		2	0		75.0						3
		3	200.0		0						4
	(c)	1	2	1	1000.0	1000.0					5
		2	1	3	1000.0	1000.0					6
	(d)	2	1	1	1						7
		3	1	1	1						8
Load Data	(a)	1	2								9
	(b)	1	0	1000.0	−10.0						10
	(c)	1	0	−200.0	12.0	0	200.0	12.0			11
		2	0	−312.5	10.0	0	312.5	10.0			12

The numerical values to be supplied on data cards are shown in Table 5-13. (Note the similarity of the data in this table to the data in Table 5-11 for the plane frame.) The output of the computer program is given in Table 5-14.

Example 2: A second problem consisting of a rectangular grid is shown in Fig. 5-3a. This structure has five members, six joints, twelve restraints, and six degrees of freedom. The cross-sectional properties of all of the members are assumed to be the same, and the numerical constants in the problem are the following:

$$E = 30{,}000 \text{ ksi} \qquad P = 16 \text{ kips} \qquad IX = 2000 \text{ in.}^4$$

$$G = 12{,}000 \text{ ksi} \qquad L = 60 \text{ in.} \qquad IY = 1000 \text{ in.}^4$$

Figure 5-3b shows a numbering system for members and joints in the restrained structure. As in previous problems, the joints which are free to displace are numbered first.

Table 5-15 contains the input data for this problem, and the solution is given in Table 5-16, which shows only the final results from the computer program.

5.8 Space Truss Program. The program in this article performs the analysis of space trusses by the methods given in Art. 4.23. The steps in this program are similar to those in the plane frame and grid programs (Arts. 5.6 and 5.7) because all three types of structures have three possible displacements at each joint. The space truss program also bears a great similarity to the plane truss program, but the three-dimensional nature of space trusses causes some additional complications.

The input data required for the space truss program are similar to that shown in Table 5-10 for plane frames, with some exceptions. For

TABLE 5-14

Output of Computer Program for Grid Example 1

ANALYSIS OF GRIDS

STRUCTURE DATA

M	N	NJ	NR	NRJ	E	G
2	3	3	6	2	10000.0	4000.0

COORDINATES OF JOINTS

JOINT	X	Y
1	100.0	75.0
2	0	75.0
3	200.0	0

MEMBER DESIGNATIONS AND PROPERTIES

MEMBER	JJ	JK	IX	IY	L	CX	CY
1	2	1	1000.0	1000.0	100.0	1.0	0
2	1	3	1000.0	1000.0	125.0	0.8	–0.6

JOINT RESTRAINTS

JOINT	X RSTRT.	Y RSTRT.	Z RSTRT.
2	1	1	1
3	1	1	1

LOAD DATA

NLJ	NLM
1	2

ACTIONS APPLIED AT JOINTS

JOINT	X ACTION	Y ACTION	Z ACTION
1	0	1000.0	–10.0

ACTIONS AT ENDS OF RESTRAINED MEMBERS DUE TO LOADS

MEMBER	AML1	AML2	AML3	AML4	AML5	AML6
1	0	–200.0	12.0	0	200.0	12.0
2	0	–312.5	10.0	0	312.5	10.0

JOINT DISPLACEMENTS AND SUPPORT REACTIONS

JOINT	X DISPL.	Y DISPL.	Z DISPL.	X REAC.	Y REAC.	Z REAC.
1	–0.007599	0.005095	–0.3551	0	0	0
2	0	0	0	303.9	–1311.5	24.04
3	0	0	0	1193.1	1103.5	29.96

MEMBER END-ACTIONS

MEMBER	AM1	AM2	AM3	AM4	AM5	AM6
1	303.9	–1311.5	24.04	–303.9	107.5	–0.0397
2	–292.3	896.4	– 9.96	292.3	1598.7	29.96

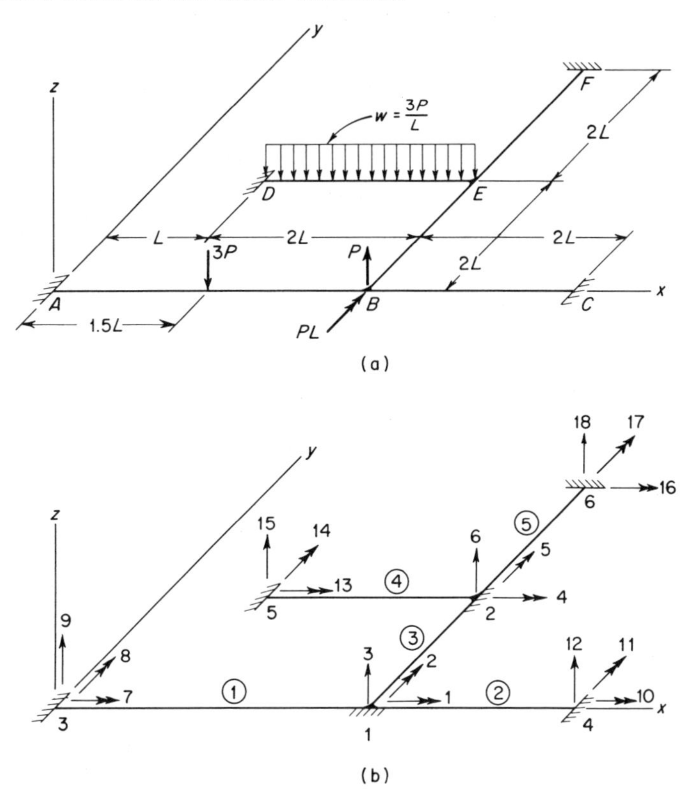

FIG. 5-3. Example 2 (grid).

example, each of the cards containing joint coordinates must provide not only the x and y coordinates but also the z coordinate for the joint. Also, on the cards containing member designations and properties, the moment of inertia IZ is not required.

The data cards for the joint restraint list are similar to those in the plane frame program, except that the nature of the restraints is different. For the space truss program, the terms $RL[3K-2]$, $RL[3K-1]$, and $RL[3K]$ denote the restraints against translations in the x, y, and z directions, respectively, at joint k.

Similarly, the cards for load data are symbolically the same as in the plane frame program, but the meanings are different. Actions applied at joints consist of forces in the x, y, and z directions, and end-actions AML consist of forces in the x_M, y_M, and z_M directions at each end of a loaded member.

The identifiers of type integer in the space truss program are exactly the same as in the plane frame program. However, the list of subscripted variables requires the deletion of $IZ[\ ]$ and the addition of $Z[\ ]$ and $CZ[\ ]$. In addition, the identifier Q is used in the program for temporary storage of a square root expression (see Sec. 4a(2)). The flow chart begins on p. 365.

TABLE 5-15

Data Cards for Grid Example 2

TYPE OF DATA		NUMERICAL DATA ON CARD	CARD NO.
Structure Data	(a)	5 6 12 4 30 000.0 12 000.0	1
	(b)	1 180.0 0	2
		2 180.0 120.0	3
		3 0 0	4
		4 300.0 0	5
		5 60.0 120.0	6
		6 180.0 240.0	7
	(c)	1 3 1 2000.0 1000.0	8
		2 1 4 2000.0 1000.0	9
		3 1 2 2000.0 1000.0	10
		4 5 2 2000.0 1000.0	11
		5 2 6 2000.0 1000.0	12
	(d)	3 1 1 1	13
		4 1 1 1	14
		5 1 1 1	15
		6 1 1 1	16
Load Data	(a)	1 2	17
	(b)	1 0 960.0 16.0	18
	(c)	1 0 −1080.0 24.0 0 1080.0 24.0	19
		4 0 −960.0 48.0 0 960.0 48.0	20

TABLE 5-16

Final Results for Grid Example 2

JOINT DISPLACEMENTS AND SUPPORT REACTIONS

JOINT	X DISPL.	Y DISPL.	Z DISPL.	X REAC.	Y REAC.	Z REAC.
1	−0.000684	−0.000288	−0.07017	0	0	0
2	0.000554	0.000353	−0.12092	0	0	0
3	0	0	0	91.13	−1565.7	29.93
4	0	0	0	136.70	733.4	11.02
5	0	0	0	−110.81	−2295.1	68.78
6	0	0	0	−1234.4	−70.56	18.27

MEMBER END-ACTIONS

MEMBER	AM1	AM2	AM3	AM4	AM5	AM6
1	91.13	−1565.7	29.93	−91.13	498.5	18.07
2	−136.70	589.6	−11.02	136.70	733.4	11.02
3	−128.07	−227.8	8.95	128.07	−846.6	−8.95
4	−110.81	−2295.1	68.78	110.81	−198.6	27.22
5	70.56	957.4	−18.27	−70.57	1234.4	18.27

FLOW CHART FOR SPACE TRUSS PROGRAM

1. Input and Print Structure Data

a. Structure parameters and elastic modulus

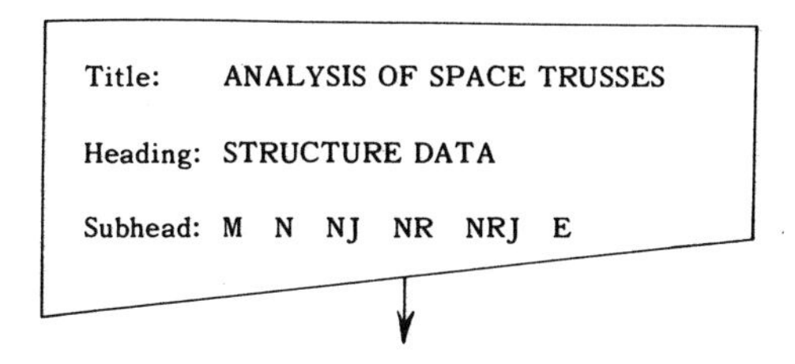

Sections 1a and 1b of this program are similar to the plane frame program, except that Z coordinates of joints are also required

c. Member designations and properties

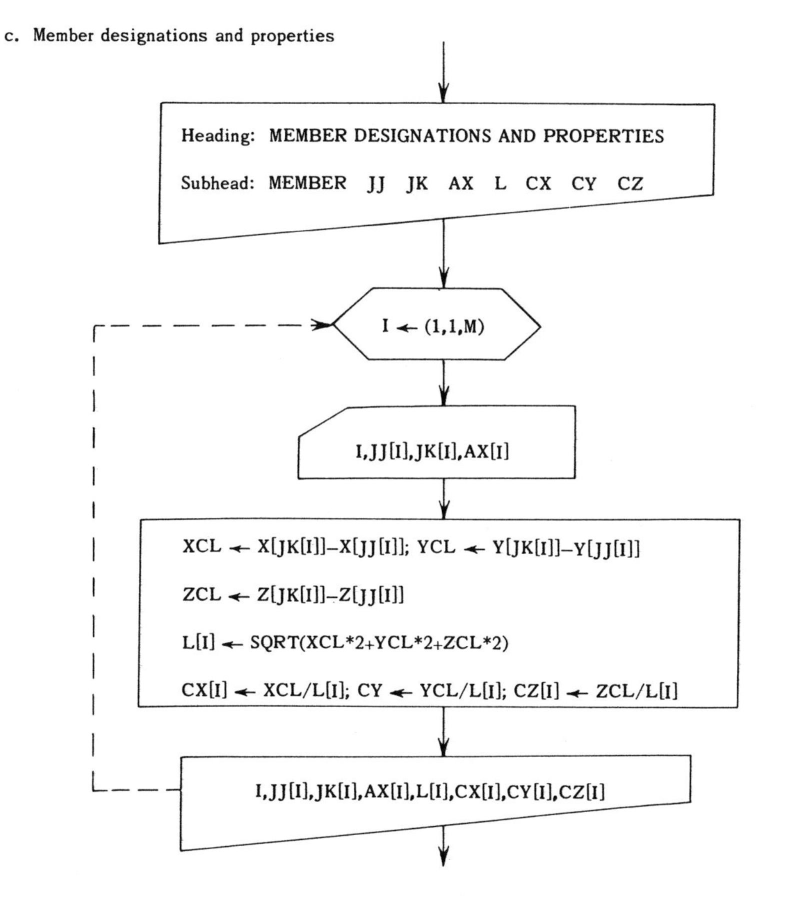

Section 1d is the same as in the plane frame program.

2. Structure Stiffness Matrix

a. Generation of stiffness matrix

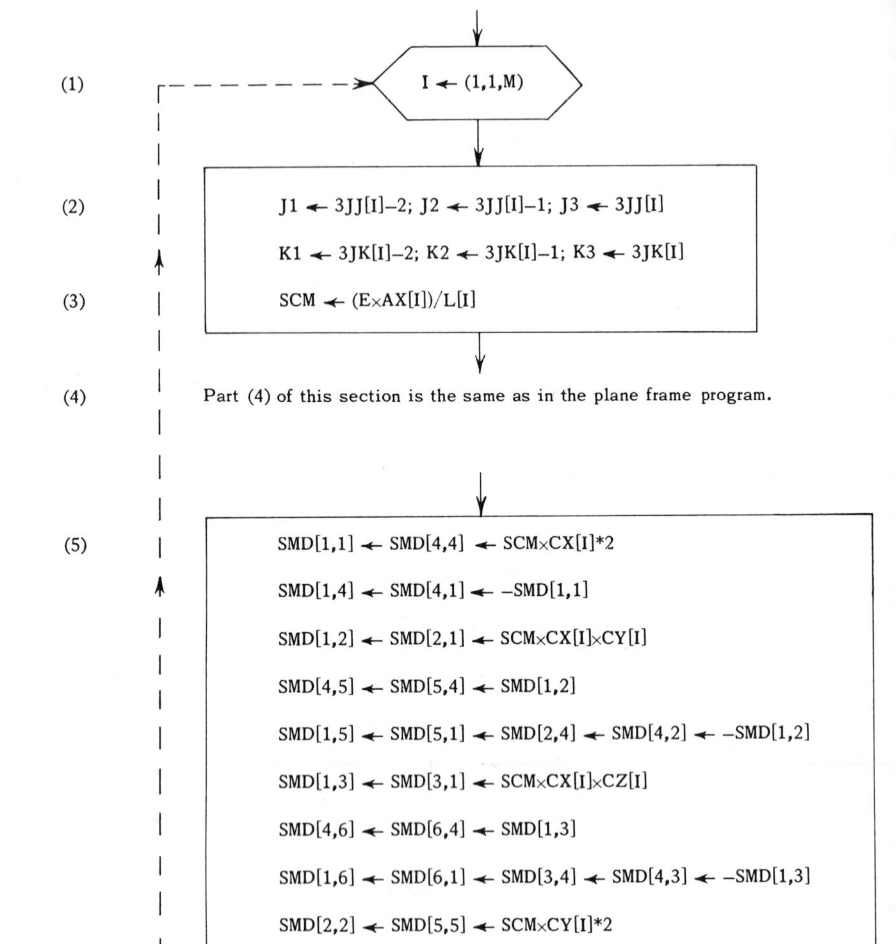

(1)
I ← (1,1,M)
(2)
J1 ← 3JJ[I]–2; J2 ← 3JJ[I]–1; J3 ← 3JJ[I]
K1 ← 3JK[I]–2; K2 ← 3JK[I]–1; K3 ← 3JK[I]
(3)
SCM ← (E×AX[I])/L[I]
(4)
Part (4) of this section is the same as in the plane frame program.
(5)
SMD[1,1] ← SMD[4,4] ← SCM×CX[I]*2
SMD[1,4] ← SMD[4,1] ← –SMD[1,1]
SMD[1,2] ← SMD[2,1] ← SCM×CX[I]×CY[I]
SMD[4,5] ← SMD[5,4] ← SMD[1,2]
SMD[1,5] ← SMD[5,1] ← SMD[2,4] ← SMD[4,2] ← –SMD[1,2]
SMD[1,3] ← SMD[3,1] ← SCM×CX[I]×CZ[I]
SMD[4,6] ← SMD[6,4] ← SMD[1,3]
SMD[1,6] ← SMD[6,1] ← SMD[3,4] ← SMD[4,3] ← –SMD[1,3]
SMD[2,2] ← SMD[5,5] ← SCM×CY[I]*2
SMD[2,5] ← SMD[5,2] ← –SMD[2,2]
SMD[2,3] ← SMD[3,2] ← SCM×CY[I]×CZ[I]
SMD[5,6] ← SMD[6,5] ← SMD[2,3]
SMD[2,6] ← SMD[6,2] ← SMD[3,5] ← SMD[5,3] ← –SMD[2,3]
SMD[3,3] ← SMD[6,6] ← SCM×CZ[I]*2
SMD[3,6] ← SMD[6,3] ← –SMD[3,3]

The remainder of this section is the same as in the plane frame program.

Section 3 is the same as in the plane frame program.

4. Construction of Vectors Associated with Loads

a. Equivalent joint loads

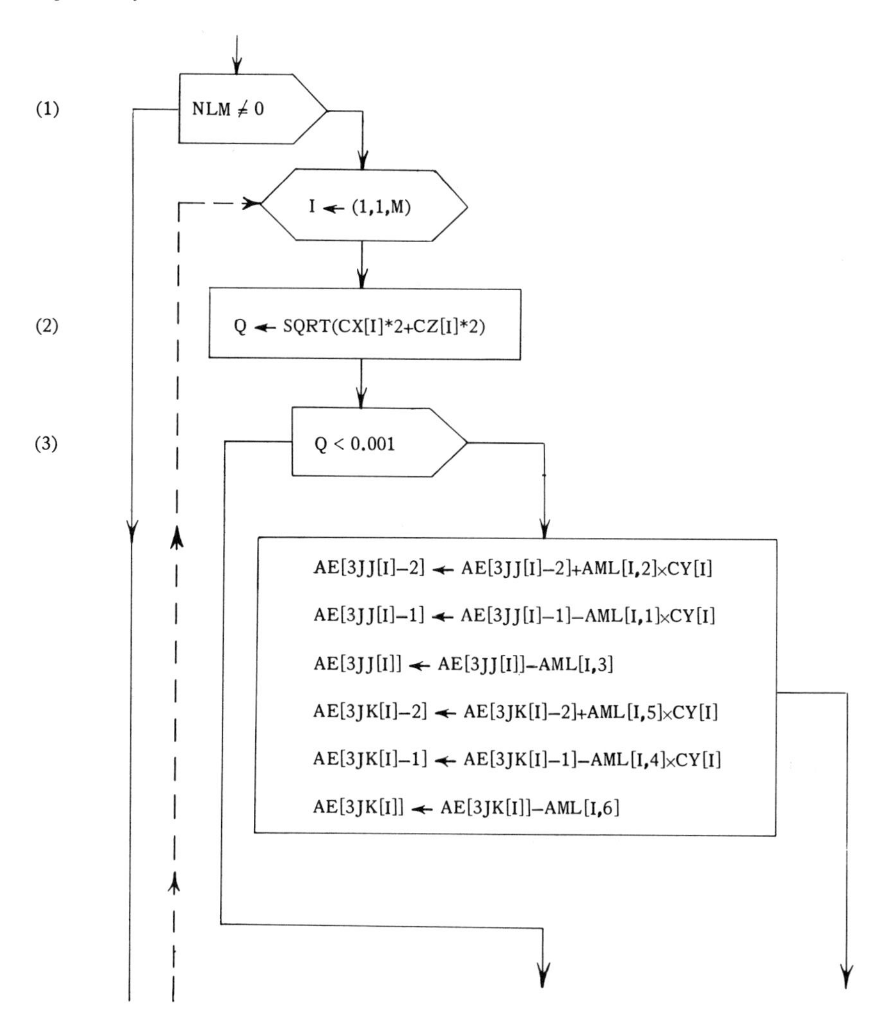

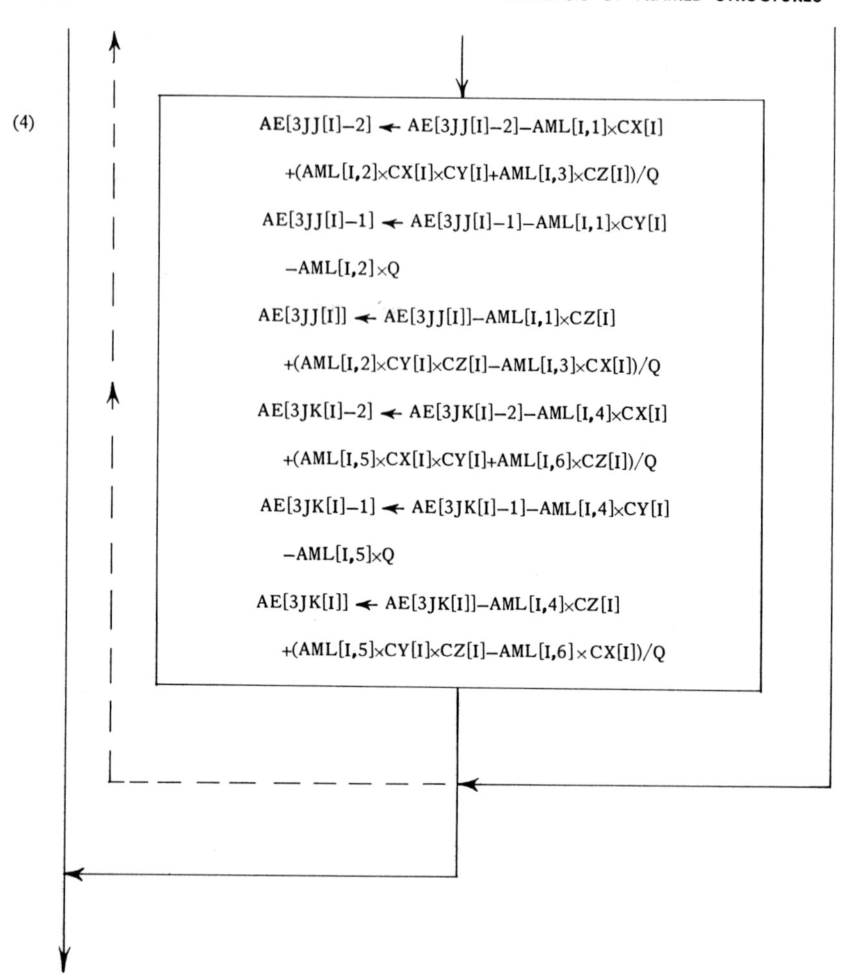

(4)
AE[3JJ[I]−2] ← AE[3JJ[I]−2]−AML[I,1]×CX[I]
+(AML[I,2]×CX[I]×CY[I]+AML[I,3]×CZ[I])/Q
AE[3JJ[I]−1] ← AE[3JJ[I]−1]−AML[I,1]×CY[I]
−AML[I,2]×Q
AE[3JJ[I]] ← AE[3JJ[I]]−AML[I,1]×CZ[I]
+(AML[I,2]×CY[I]×CZ[I]−AML[I,3]×CX[I])/Q
AE[3JK[I]−2] ← AE[3JK[I]−2]−AML[I,4]×CX[I]
+(AML[I,5]×CX[I]×CY[I]+AML[I,6]×CZ[I])/Q
AE[3JK[I]−1] ← AE[3JK[I]−1]−AML[I,4]×CY[I]
−AML[I,5]×Q
AE[3JK[I]] ← AE[3JK[I]]−AML[I,4]×CZ[I]
+(AML[I,5]×CY[I]×CZ[I]−AML[I,6]×CX[I])/Q

Sections 4b and 5a are the same as in the plane frame program.

5b. Member end-actions

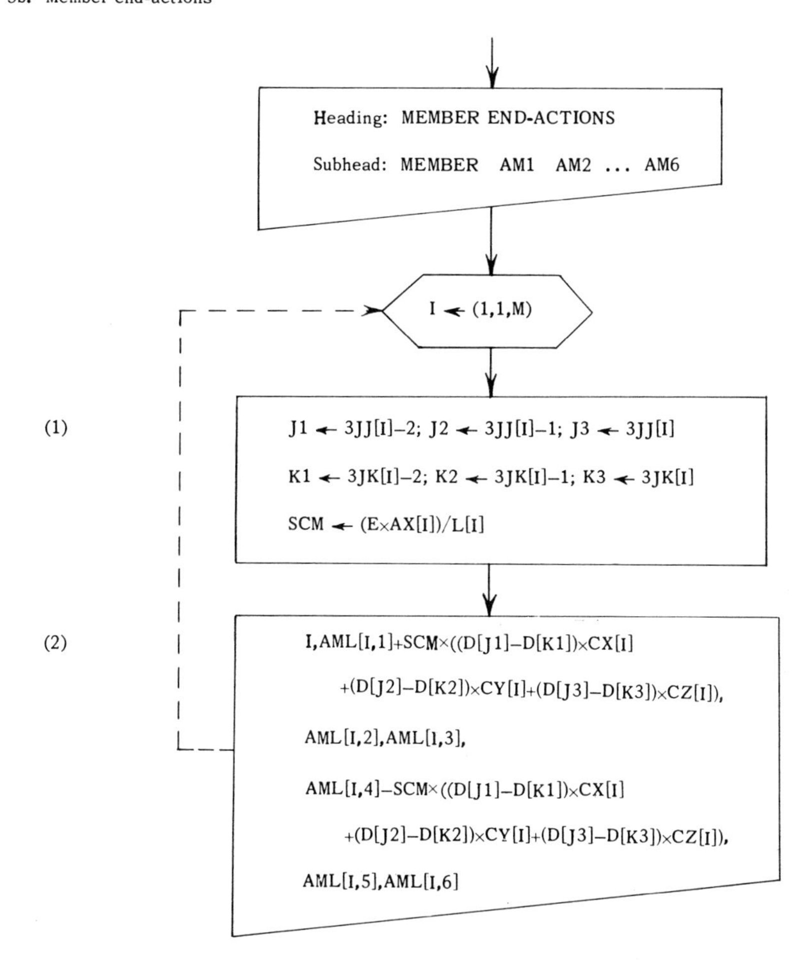
Heading: MEMBER END-ACTIONS
Subhead: MEMBER AM1 AM2 ... AM6
I ← (1,1,M)
(1)
J1 ← 3JJ[I]−2; J2 ← 3JJ[I]−1; J3 ← 3JJ[I]
K1 ← 3JK[I]−2; K2 ← 3JK[I]−1; K3 ← 3JK[I]
SCM ← (E×AX[I])/L[I]
(2)
I,AML[I,1]+SCM×((D[J1]−D[K1])×CX[I]
+(D[J2]−D[K2])×CY[I]+(D[J3]−D[K3])×CZ[I]),
AML[I,2],AML[I,3],
AML[I,4]−SCM×((D[J1]−D[K1])×CX[I]
+(D[J2]−D[K2])×CY[I]+(D[J3]−D[K3])×CZ[I]),
AML[I,5],AML[I,6]

Explanation of Flow Chart for the Space Truss Program

Item No.	*Remarks*
2. a. (1)	Iteration control statement on members, for M members.
(2)	The three possible displacements are indexed for the joints J and K of the member I.
(3)	The stiffness constant for axial deformation of the member is calculated.
(4)	The three possible displacements for the joints at each end of the member are re-indexed by the same method as that used in the plane frame program.
(5)	The 6×6 member stiffness matrix SMD for member I is generated (see Table 4-40, Art. 4.21).
4. a. (1)	If NLM is not equal to zero, the contributions from AML to AE for each member must be calculated and transferred.

(2) The quantity $Q = \sqrt{C_X^2 + C_Z^2}$ is computed and stored in order to avoid repetitious calculations later.

(3) If Q is zero or very small, the member is taken to be vertical, and the contributions to AE from AML are drawn from Table 4-42 in Art. 4.23.

(4) On the other hand, if Q is not small, the member is taken to be skew, and the contributions to AE from AML are drawn from Table 4-41 in Art. 4.23.

5. b. (1) For each member the same information is computed as in Secs. 2a(2) and (3) given previously.

(2) The member number is printed, and the six member end-actions are computed and printed using Eqs. (4-87).

Example 1: The space truss structure shown in Fig. 5-4a has six members, four joints, nine restraints, and three degrees of freedom. In the figure an arrow with a short cross line indicates that a support restraint exists at that location and prevents motion in the direction of the arrow. Loads, on the other hand, are represented by plain arrows. Assume that all members have the same cross-sectional area and that the numerical values in the problem are as follows:

$$E = 10{,}000 \text{ ksi} \qquad L = 25 \text{ in.} \qquad P = 10 \text{ kips} \qquad AX = 10 \text{ in.}^2$$

Figure 5-4b shows the restrained structure with a numbering system for members and joints. Because of the locations of restraints, rearrangement of terms in the matrices is required in the analysis.

Table 5-17 shows the data cards required as input for this problem, and Table 5-18 gives the output of the computer program.

Example 2: Figure 5-5a shows a second example having seven members, five joints, nine restraints, and six degrees of freedom. The cross-sectional areas of all members are assumed to be the same, and the following numerical values apply:

$$E = 30{,}000 \text{ ksi} \qquad L = 120 \text{ in.} \qquad P = 20 \text{ kips} \qquad AX = 10 \text{ in.}^2$$

Figure 5-5b shows the restrained structure and a numbering system which requires rearrangement of matrices in the analysis.

The input data required for the problem are listed in Table 5-19, and the final results from the computer program are given in Table 5-20.

5.9 Space Frame Program. This article contains a program for the analysis of space frames by the methods given in Art. 4.25. The steps in this program are considerably different from those in the previous programs for two principal reasons. First, the analysis is complicated by the fact that each joint has six possible displacements instead of three, and, second, the program incorporates the use of rotation matrices. It was not necessary to use rotation matrices in the previous programs, but the complex nature of space frames makes their use almost essential.

Several additional identifiers are required in the space frame program that were not needed previously. These identifiers are listed in Table 5-21.

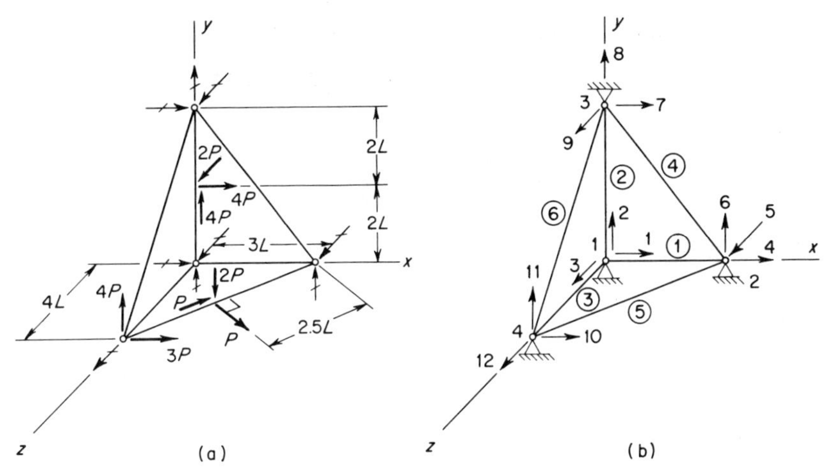

FIG. 5-4. Example 1 (space truss).

TABLE 5-17

Data Cards for Space Truss Example 1

Type of Data		Numerical Data on Card							Card No.
Structure Data	(a)	6	4	9	4	10000.0			1
	(b)	1	0	0	0				2
		2	75.0	0	0				3
		3	0	100.0	0				4
		4	0	0	100.0				5
	(c)	1	1	2	10.0				6
		2	1	3	10.0				7
		3	1	4	10.0				8
		4	2	3	10.0				9
		5	2	4	10.0				10
		6	3	4	10.0				11
	(d)	1	1	1	1				12
		2	0	1	1				13
		3	1	1	1				14
		4	0	0	1				15
Load Data	(a)	1	2						16
	(b)	4	30.0	40.0	0				17
	(c)	2	−20.0	20.0	−10.0	−20.0	20.0	−10.0	18
		5	5.0	10.0	5.0	5.0	10.0	5.0	19

TABLE 5-18

Output of Computer Program for Space Truss Example 1

ANALYSIS OF SPACE TRUSSES

STRUCTURE DATA

M	N	NJ	NR	NRJ	E
6	3	4	9	4	10000.0

COORDINATES OF JOINTS

JOINT	X	Y	Z
1	0	0	0
2	75.0	0	0
3	0	100.0	0
4	0	0	100.0

MEMBER DESIGNATIONS AND PROPERTIES

MEMBER	JJ	JK	AX	L	CX	CY	CZ
1	1	2	10.0	75.0	1.000	0	0
2	1	3	10.0	100.0	0	1.000	0
3	1	4	10.0	100.0	0	0	1.000
4	2	3	10.0	125.0	–0.600	0.800	0
5	2	4	10.0	125.0	–0.600	0	0.800
6	3	4	10.0	141.4	0	–0.707	0.707

JOINT RESTRAINTS

JOINT	X RSTRT.	Y RSTRT.	Z RSTRT.
1	1	1	1
2	0	1	1
3	1	1	1
4	0	0	1

LOAD DATA

NLJ	NLM
1	2

ACTIONS APPLIED AT JOINTS

JOINT	X ACTION	Y ACTION	Z ACTION
4	30.0	40.0	0

ACTIONS AT ENDS OF RESTRAINED MEMBERS DUE TO LOADS

MEMBER	AML1	AML2	AML3	AML4	AML5	AML6
2	–20.0	20.0	–10.0	–20.0	20.0	–10.0
5	5.0	10.0	5.0	5.0	10.0	5.0

JOINT DISPLACEMENTS AND SUPPORT REACTIONS

JOINT	X DISPL.	Y DISPL.	Z DISPL.	X REAC.	Y REAC.	Z REAC.
1	0	0	0	–56.18	–20.00	–10.00
2	0.02714	0	0	0	–0.42	50.33
3	0	0	0	–27.82	–39.58	20.00
4	0.1556	0.08485	0	0	0	–78.33

(TABLE 5-18, continued)

MEMBER END-ACTIONS

MEMBER	AM1	AM2	AM3	AM4	AM5	AM6
1	−36.18	0	0	36.18	0	0
2	−20.00	20.00	−10.00	−20.00	20.00	−10.00
3	0	0	0	0	0	0
4	−13.03	0	0	13.03	0	0
5	66.67	10.00	5.00	−56.67	10.00	5.00
6	42.43	0	0	−42.43	0	0

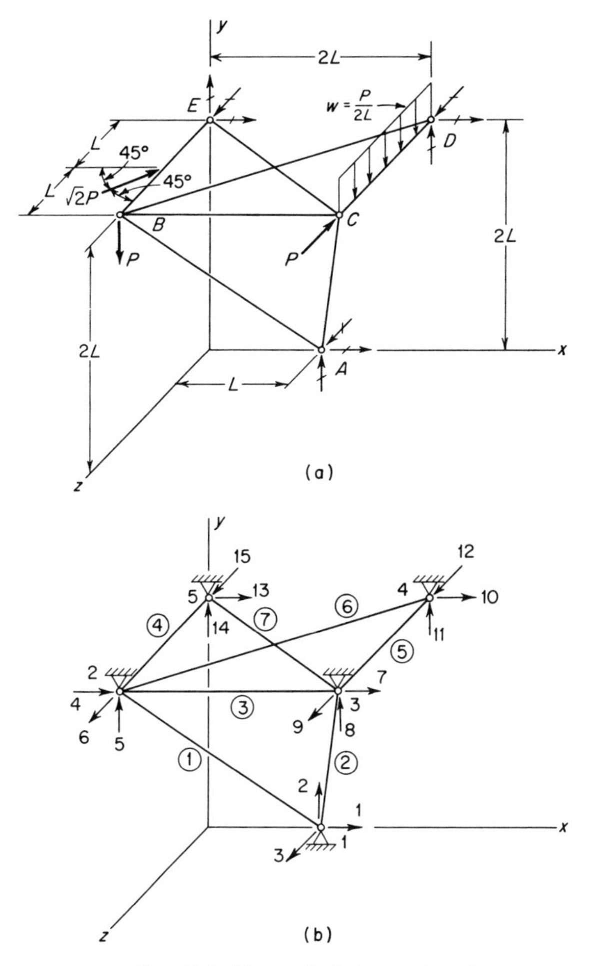

FIG. 5-5. Example 2 (space truss).

TABLE 5-19

Data Cards for Space Truss Example 2

Type of Data		Numerical Data on Card							Card No.
Structure Data	(a)	7	5	9	3	30000.0			1
	(b)	1	120.0	0	0				2
		2	0	240.0	240.0				3
		3	240.0	240.0	240.0				4
		4	240.0	240.0	0				5
		5	0	240.0	0				6
	(c)	1	1	2	10.0				7
		2	1	3	10.0				8
		3	2	3	10.0				9
		4	2	5	10.0				10
		5	3	4	10.0				11
		6	2	4	10.0				12
		7	3	5	10.0				13
	(d)	1	1	1	1				14
		4	1	1	1				15
		5	1	1	1				16
Load Data	(a)	2	2						17
	(b)	2	0	−20.0	0				18
		3	0	0	−20.0				19
	(c)	4	−10.0	0	−10.0	−10.0	0	−10.0	20
		5	0	10.0	0	0	10.0	0	21

TABLE 5-20

Final Results for Space Truss Example 2

JOINT DISPLACEMENTS AND SUPPORT REACTIONS

JOINT	X DISPL.	Y DISPL.	Z DISPL.	X REAC.	Y REAC.	Z REAC.
1	0	0	0	−5.00	30.00	30.00
2	0.01477	−0.05638	0.009769	0	0	0
3	0.01654	−0.02504	−0.01023	0	0	0
4	0	0	0	−2.21	10.00	15.00
5	0	0	0	−12.79	0	−5.00

MEMBER END-ACTIONS

MEMBER	AM1	AM2	AM3	AM4	AM5	AM6
1	30.00	0	0	−30.00	0	0
2	15.00	0	0	−15.00	0	0
3	−2.21	0	0	2.21	0	0
4	−22.21	0	−10.0	2.21	0	−10.0
5	12.79	10.0	0	−12.79	10.0	0
6	3.13	0	0	−3.13	0	0
7	−3.94	0	0	3.94	0	0

TABLE 5-21

Additional Identifiers Used in Space Frame Program

Identifier	*Definition*
AA	Identifier used to indicate whether or not the angle α is zero
XP, YP, ZP	x, y, and z coordinates of point p
XPS, YPS, ZPS	x_S, y_S, and z_S coordinates of point p
YPG, ZPG	y_γ and z_γ coordinates of point p
Q, SQ	Identifiers used for temporary storage of square root expressions
SINA, COSA	Sine and cosine of the angle α
R[,]	Rotation matrix
SMR[,]	Identifier used for temporary storage of the product $\mathbf{S}_M\mathbf{R}_T$

The input data required for the space frame program are summarized in Table 5-22, which is very similar to Table 5-10 for plane frames.

TABLE 5-22

Preparation of Data for Space Frame Program

		Data	Number of Cards	Items on Data Cards
Structure Data	(a)	Structure parameters and elastic moduli	1	M NJ NR NRJ E G
	(b)	Joint coordinates	NJ	J X[J] Y[J] Z[J]
	(c)	Member designations, properties, and orientations	M	I JJ[I] JK[I] AX[I] IX[I] IY[I] IZ[I] AA
		Coordinates of point p (required when AA = 1)		I XP YP ZP
	(d)	Joint restraint list	NRJ	K RL[6K−5] RL[6K−4] RL[6K−3] RL[6K−2] RL[6K−1] RL[6K]
Load Data	(a)	Numbers of loaded joints and members	1	NLJ NLM
	(b)	Actions applied at joints	NLJ	K A[6K − 5] A[6K − 4] A[6K − 3] A[6K − 2] A[6K − 1] A[6K]
	(c)	Actions at ends of restrained members due to loads	2NLM	I AML[I,1] AML[I,2] AML[I,3] AML[I,4] AML[I,5] AML[I,6]
				AML[I,7] AML[I,8] AML[I,9] AML[I,10] AML[I,11] AML[I,12]

However, the first data card includes the shear modulus G as well as the modulus E, and each of the cards containing joint coordinates must provide the z coordinate as well as the x and y coordinates of a joint.

The cards pertaining to member designations, properties, and orientations must include the cross-sectional properties AX, IX, IY, and IZ. In addition, an identifier AA is provided in order to indicate whether the rotation angle α is zero or not (see Art. 4.24 for the meaning of the angle α). If the angle α is zero for a given member, the value of zero is assigned to AA. On the other hand, if the angle α is not zero, the integer 1 is assigned. This rule is an arbitrary convention which serves to indicate that the member has its principal axes in skew directions. Under these conditions the principal axes are located by means of the coordinates of a point p in the x_M-y_M plane (see Figs. 4-45 and 4-46) but not on the axis of the member. Therefore, each data card for which AA is equal to 1 must be followed by an additional data card on which the member index I and the three coordinates XP, YP, and ZP are given. These coordinates are used to obtain the rotation matrix as described in Art. 4.24.

An alternate method of programming would be to provide for the contingency that the angle α is known, in which case it would not be necessary to give the coordinates of a point p. The actual value of α could be assigned to the identifier AA, after which the rotation matrix can be calculated directly from Eq. (4-90) or Eq. (4-96).

Each of the cards in the joint restraint series contains a joint number K and six code numbers which indicate the conditions of restraint at that joint. The terms $RL[6K-5]$ through $RL[6K]$ denote the restraints against translations and rotations in the x, y, and z directions at joint k.

Each joint load card contains a joint number K and the six actions applied at that joint. These actions are the components of the applied force vector and the applied moment vector in the x, y, and z directions.

In the last series, two cards must be provided for each member to which loads are applied because the information may not fit on one card. The first card contains a member number I and the six actions at the j end of the restrained member. The second card provides the six actions at the k end of the member. The actions at each end are the forces and couples in the x_M, y_M, and z_M directions.

The identifiers of type integer in the space frame program are the same as those in the plane frame program, with the addition of the indexes $J4$, $J5$, $J6$, $K4$, $K5$, and $K6$ and the identifier AA. The list of subscripted variables requires the deletion of $CX[\ ]$ and $CY[\ ]$ and the addition of $Z[\ ]$, $IX[\ ]$, $IY[\ ]$, $SM[\ ,\ ]$, $SMR[\ ,\ ]$, and $R[\ ,\ ]$.

Explanation of Flow Chart for the Space Frame Program

Item No. *Remarks*

1. c. (1) In addition to the member indexes and properties, the identifier AA is read. This identifier denotes whether the angle α is zero ($AA = 0$) or is not zero ($AA = 1$).

(2) Components of length, true length, and direction cosines are computed for each member. (*Continued on p. 390.*)

FLOW CHART FOR SPACE FRAME PROGRAM

1. Input and Print Structure Data

a. Structure parameters and elastic moduli

b. Joint coordinates

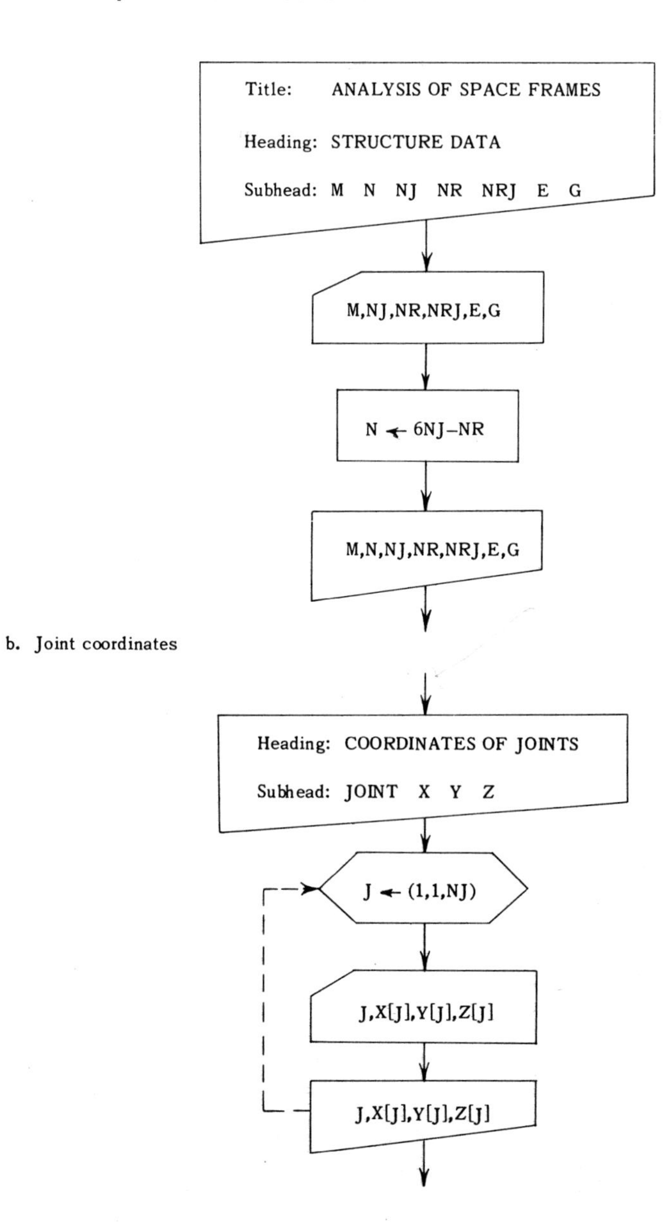

c. Member designations, properties, and orientations

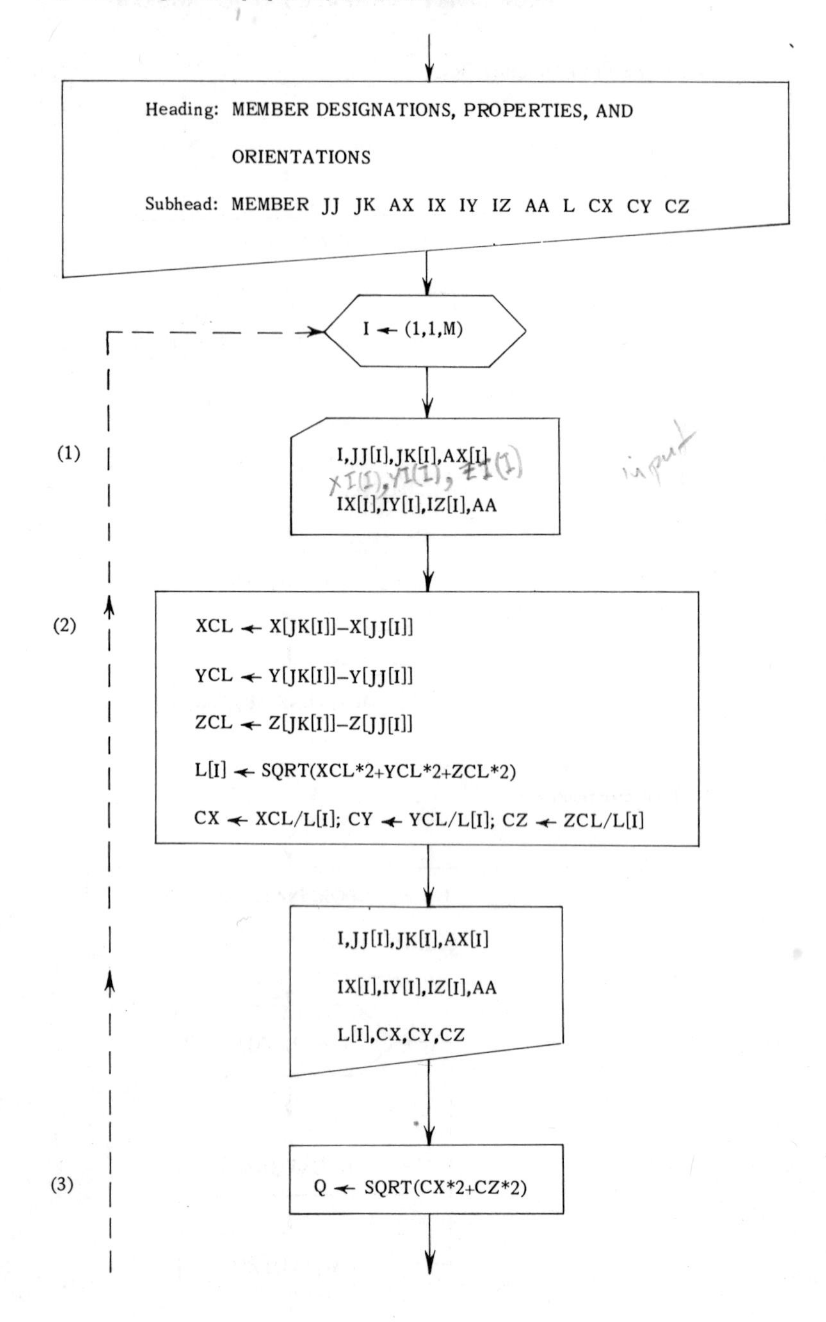
Heading: MEMBER DESIGNATIONS, PROPERTIES, AND ORIENTATIONS
Subhead: MEMBER JJ JK AX IX IY IZ AA L CX CY CZ
I ← (1,1,M)
(1)
I,JJ[I],JK[I],AX[I]
IX[I],IY[I],IZ[I],AA
(2)
XCL ← X[JK[I]]–X[JJ[I]]
YCL ← Y[JK[I]]–Y[JJ[I]]
ZCL ← Z[JK[I]]–Z[JJ[I]]
L[I] ← SQRT(XCL*2+YCL*2+ZCL*2)
CX ← XCL/L[I]; CY ← YCL/L[I]; CZ ← ZCL/L[I]
I,JJ[I],JK[I],AX[I]
IX[I],IY[I],IZ[I],AA
L[I],CX,CY,CZ
(3)
Q ← SQRT(CX*2+CZ*2)

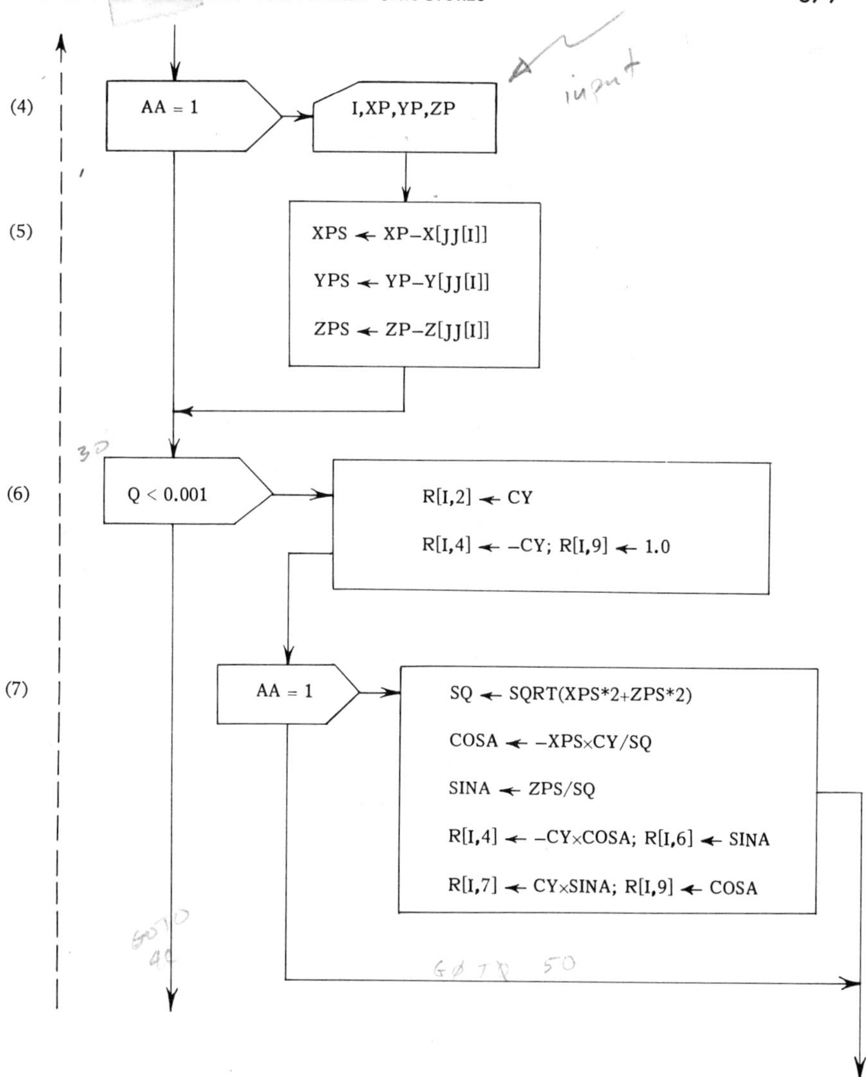
(4)
AA = 1
I,XP,YP,ZP
input
(5)
XPS ← XP−X[JJ[I]]
YPS ← YP−Y[JJ[I]]
ZPS ← ZP−Z[JJ[I]]
30
(6)
Q < 0.001
R[I,2] ← CY
R[I,4] ← −CY; R[I,9] ← 1.0
(7)
AA = 1
SQ ← SQRT(XPS*2+ZPS*2)
COSA ← −XPS×CY/SQ
SINA ← ZPS/SQ
R[I,4] ← −CY×COSA; R[I,6] ← SINA
R[I,7] ← CY×SINA; R[I,9] ← COSA
GOTO 40
GØTO 50

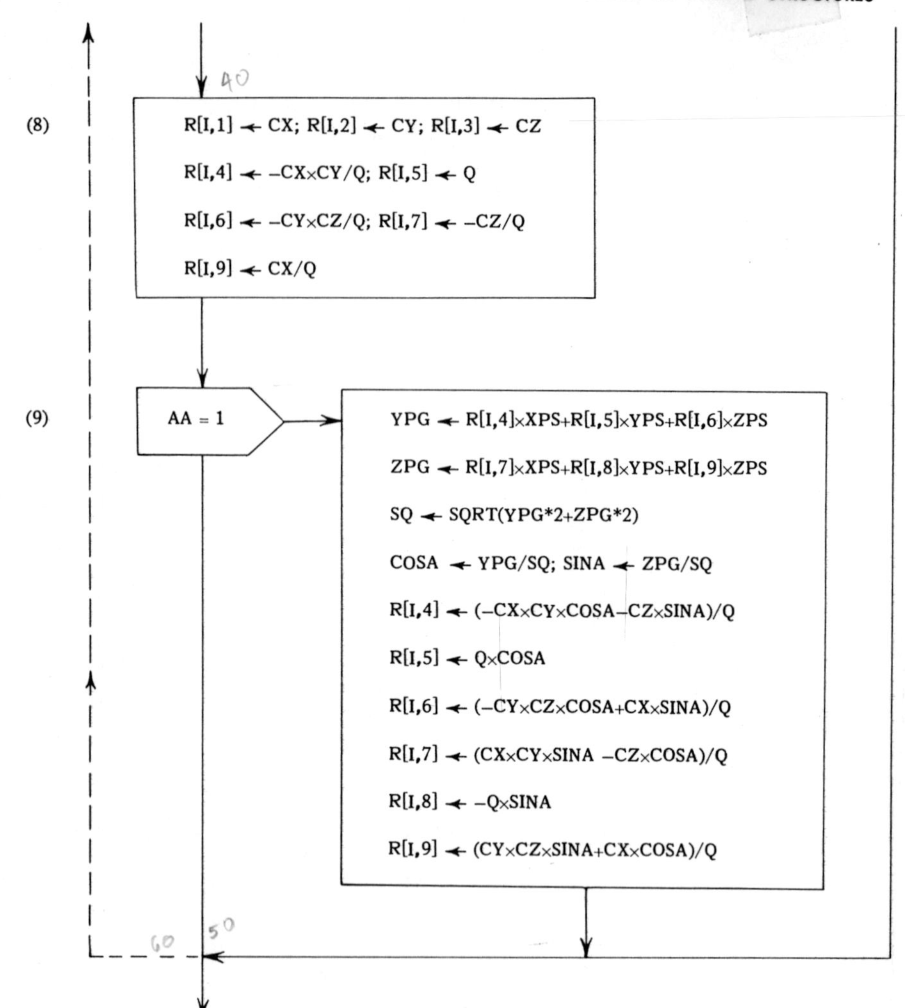
40
(8)
R[I,1] ← CX; R[I,2] ← CY; R[I,3] ← CZ
R[I,4] ← −CX×CY/Q; R[I,5] ← Q
R[I,6] ← −CY×CZ/Q; R[I,7] ← −CZ/Q
R[I,9] ← CX/Q
(9)
AA = 1
YPG ← R[I,4]×XPS+R[I,5]×YPS+R[I,6]×ZPS
ZPG ← R[I,7]×XPS+R[I,8]×YPS+R[I,9]×ZPS
SQ ← SQRT(YPG*2+ZPG*2)
COSA ← YPG/SQ; SINA ← ZPG/SQ
R[I,4] ← (−CX×CY×COSA−CZ×SINA)/Q
R[I,5] ← Q×COSA
R[I,6] ← (−CY×CZ×COSA+CX×SINA)/Q
R[I,7] ← (CX×CY×SINA −CZ×COSA)/Q
R[I,8] ← −Q×SINA
R[I,9] ← (CY×CZ×SINA+CX×COSA)/Q
60
50

d. Joint restraint list; cumulative restraint list

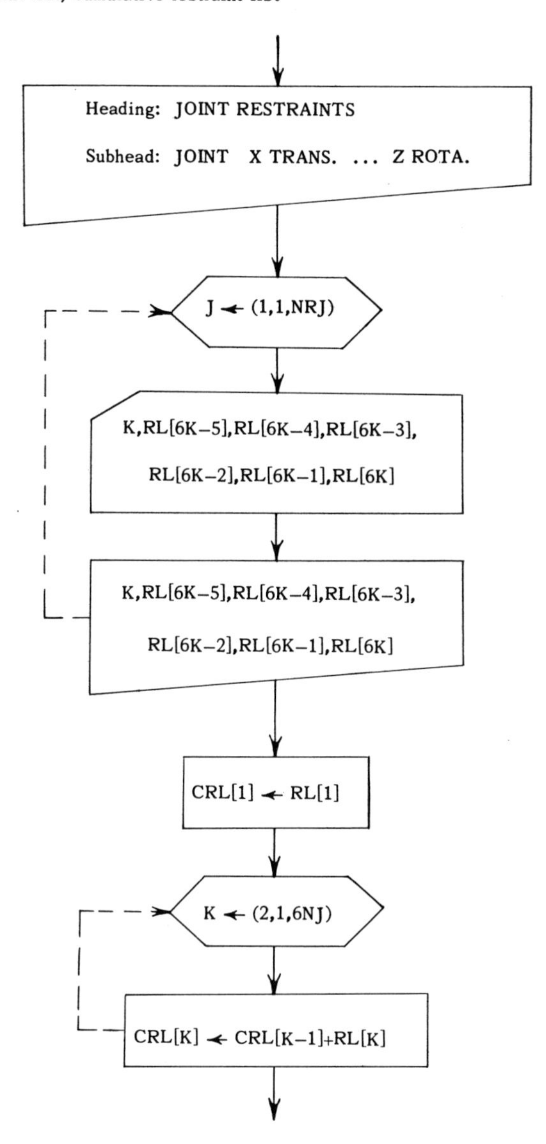

2. Structure Stiffness Matrix

a. Generation of stiffness matrix

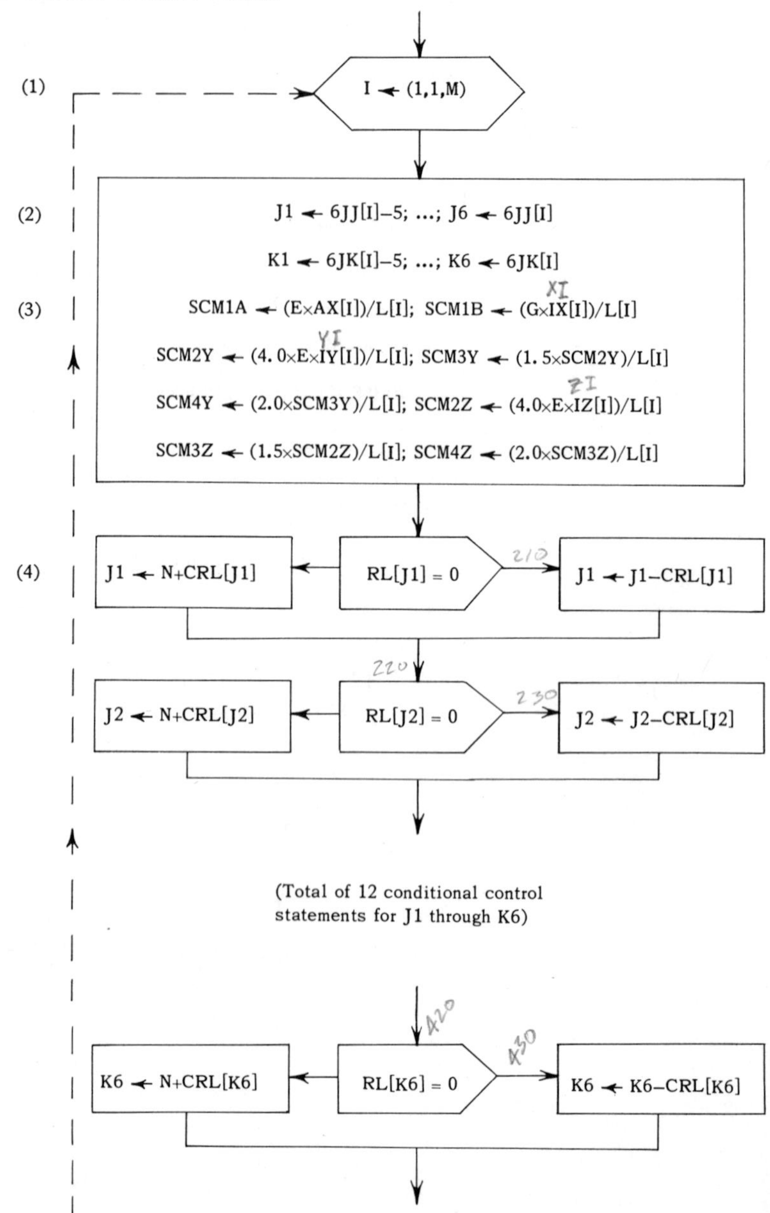

(5)

```
SM[1,1] ← SM[7,7] ← SCM1A; SM[1,7] ← SM[7,1] ← –SCM1A
SM[2,2] ← SM[8,8] ← SCM4Z; SM[2,8] ← SM[8,2] ← –SCM4Z
SM[2,6] ← SM[6,2] ← SM[2,12] ← SM[12,2] ← SCM3Z
SM[6,8] ← SM[8,6] ← SM[8,12] ← SM[12,8] ← –SCM3Z
SM[3,3] ← SM[9,9] ← SCM4Y; SM[3,9] ← SM[9,3] ← –SCM4Y
SM[3,5] ← SM[5,3] ← SM[3,11] ← SM[11,3] ← –SCM3Y
SM[4,4] ← SM[10,10] ← SCM1B; SM[4,10] ← SM[10,4] ← –SCM1B
SM[5,5] ← SM[11,11] ← SCM2Y; SM[5,11] ← SM[11,5] ← SCM2Y/2.0
SM[5,9] ← SM[9,5] ← SM[9,11] ← SM[11,9] ← SCM3Y
SM[6,6] ← SM[12,12] ← SCM2Z; SM[6,12] ← SM[12,6] ← SCM2Z/2.0
```

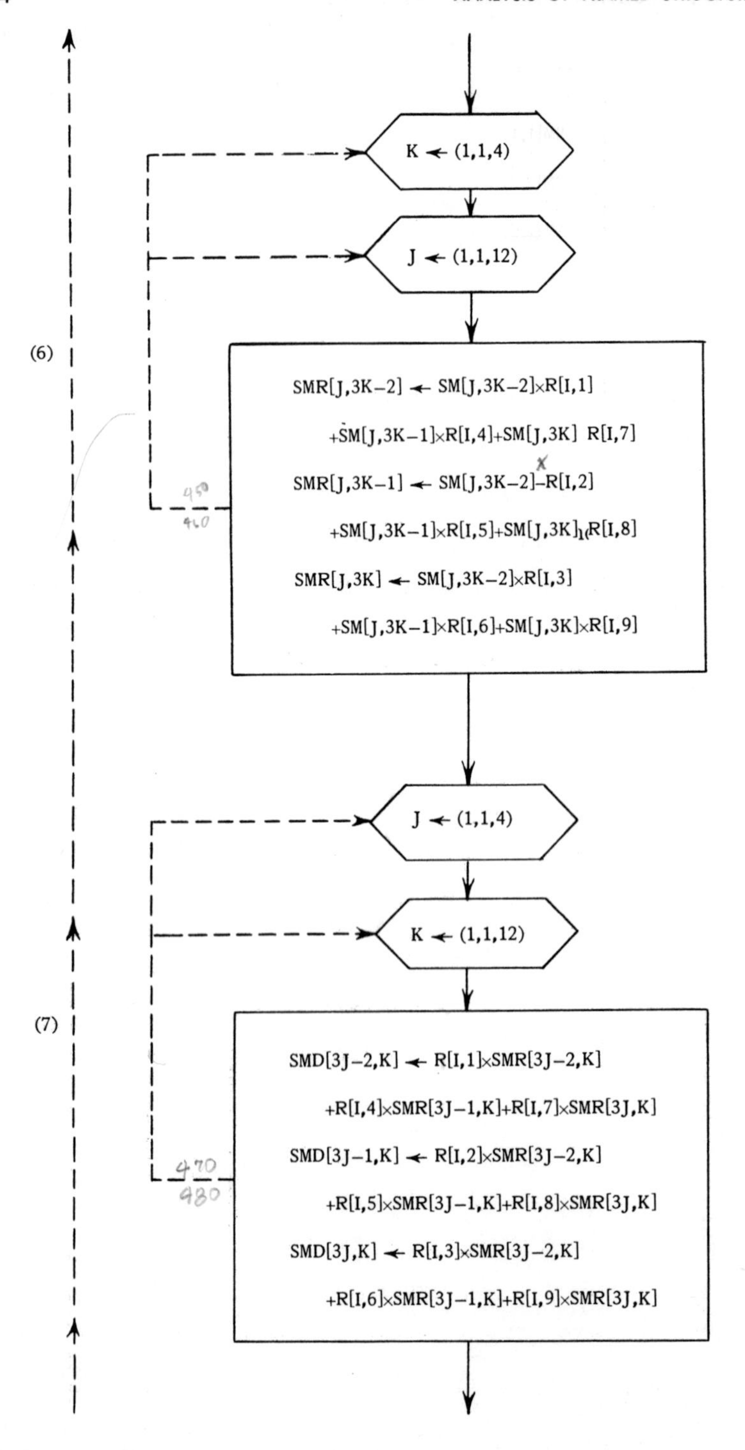
K ← (1,1,4)
J ← (1,1,12)
(6)
SMR[J,3K−2] ← SM[J,3K−2]×R[I,1]
+SM[J,3K−1]×R[I,4]+SM[J,3K] R[I,7]
SMR[J,3K−1] ← SM[J,3K−2]−R[I,2]
+SM[J,3K−1]×R[I,5]+SM[J,3K]×R[I,8]
SMR[J,3K] ← SM[J,3K−2]×R[I,3]
+SM[J,3K−1]×R[I,6]+SM[J,3K]×R[I,9]
J ← (1,1,4)
K ← (1,1,12)
(7)
SMD[3J−2,K] ← R[I,1]×SMR[3J−2,K]
+R[I,4]×SMR[3J−1,K]+R[I,7]×SMR[3J,K]
SMD[3J−1,K] ← R[I,2]×SMR[3J−2,K]
+R[I,5]×SMR[3J−1,K]+R[I,8]×SMR[3J,K]
SMD[3J,K] ← R[I,3]×SMR[3J−2,K]
+R[I,6]×SMR[3J−1,K]+R[I,9]×SMR[3J,K]

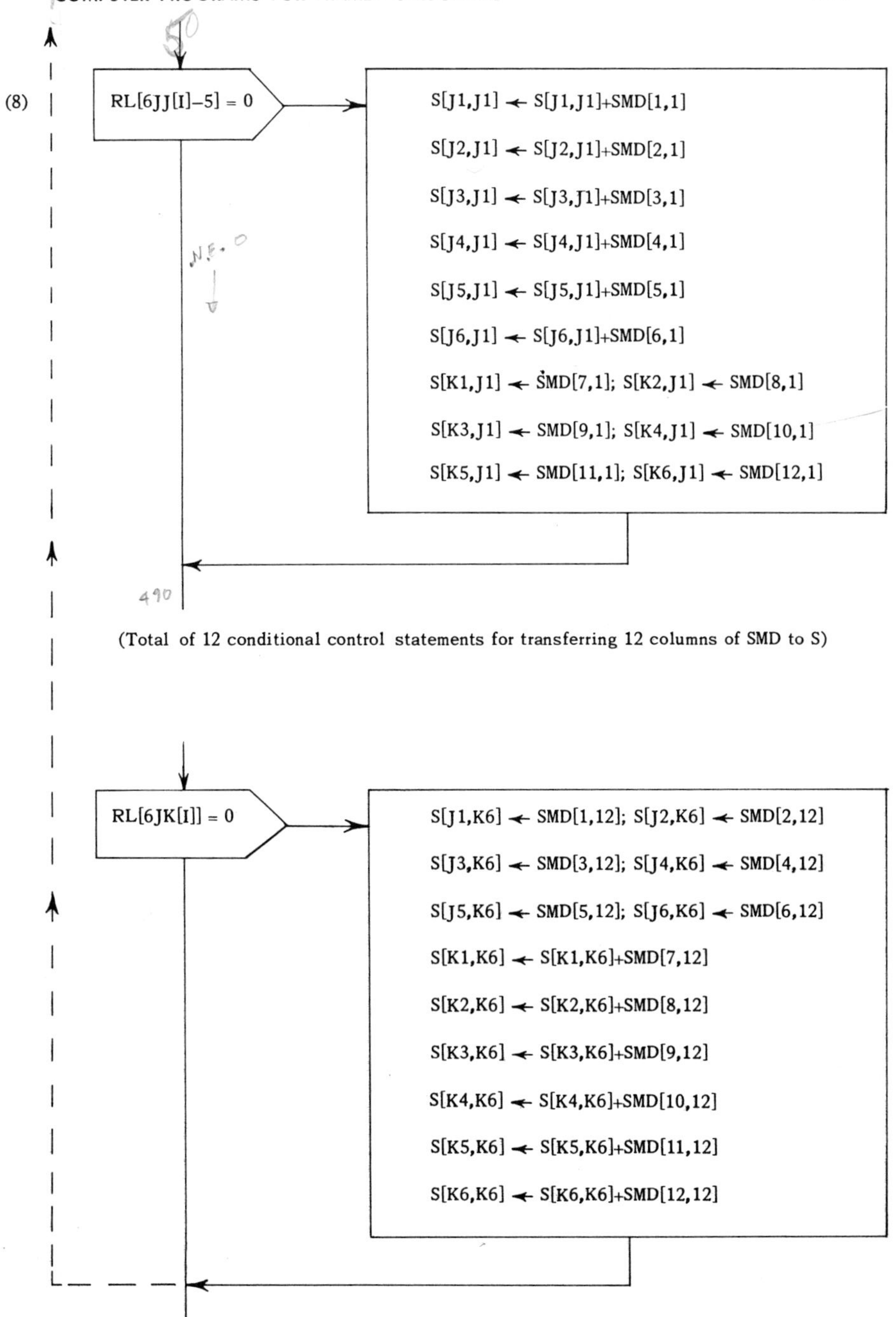

Sections 2b and 3a are the same as in the continuous beam program.

3b. Actions applied at joints

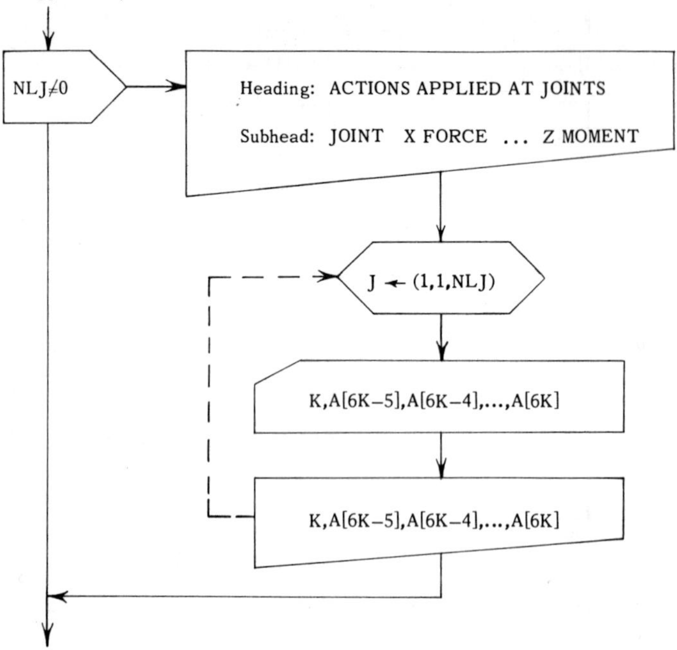

c. Actions at ends of restrained members due to loads

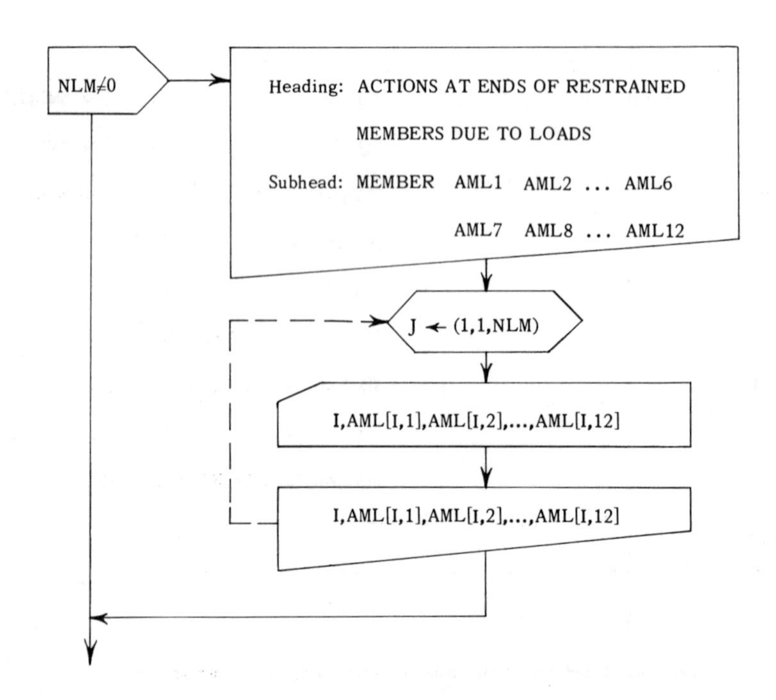

4. Construction of Vectors Associated with Loads

a. Equivalent joint loads

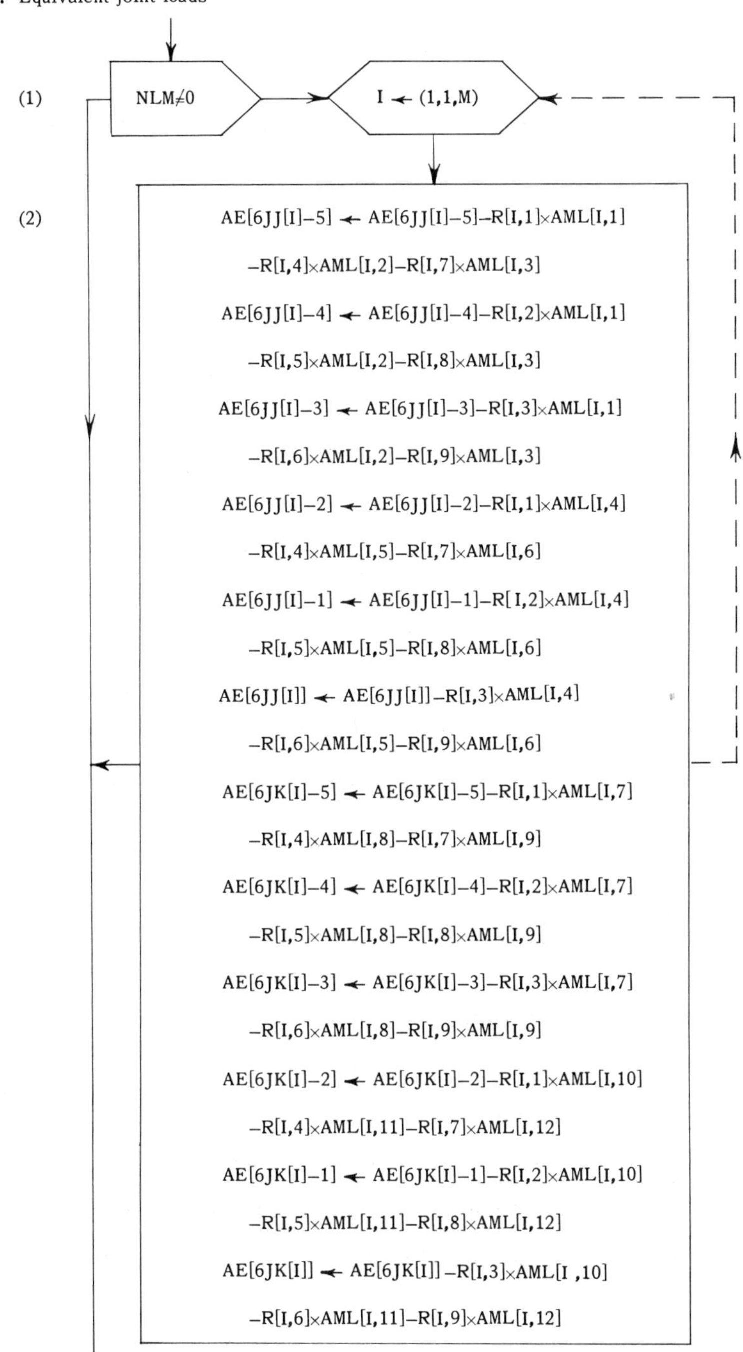

Sections 4b and 5a are the same as in the plane frame program, except the following printing is done in Sec. 5a:

5b. Member end-actions

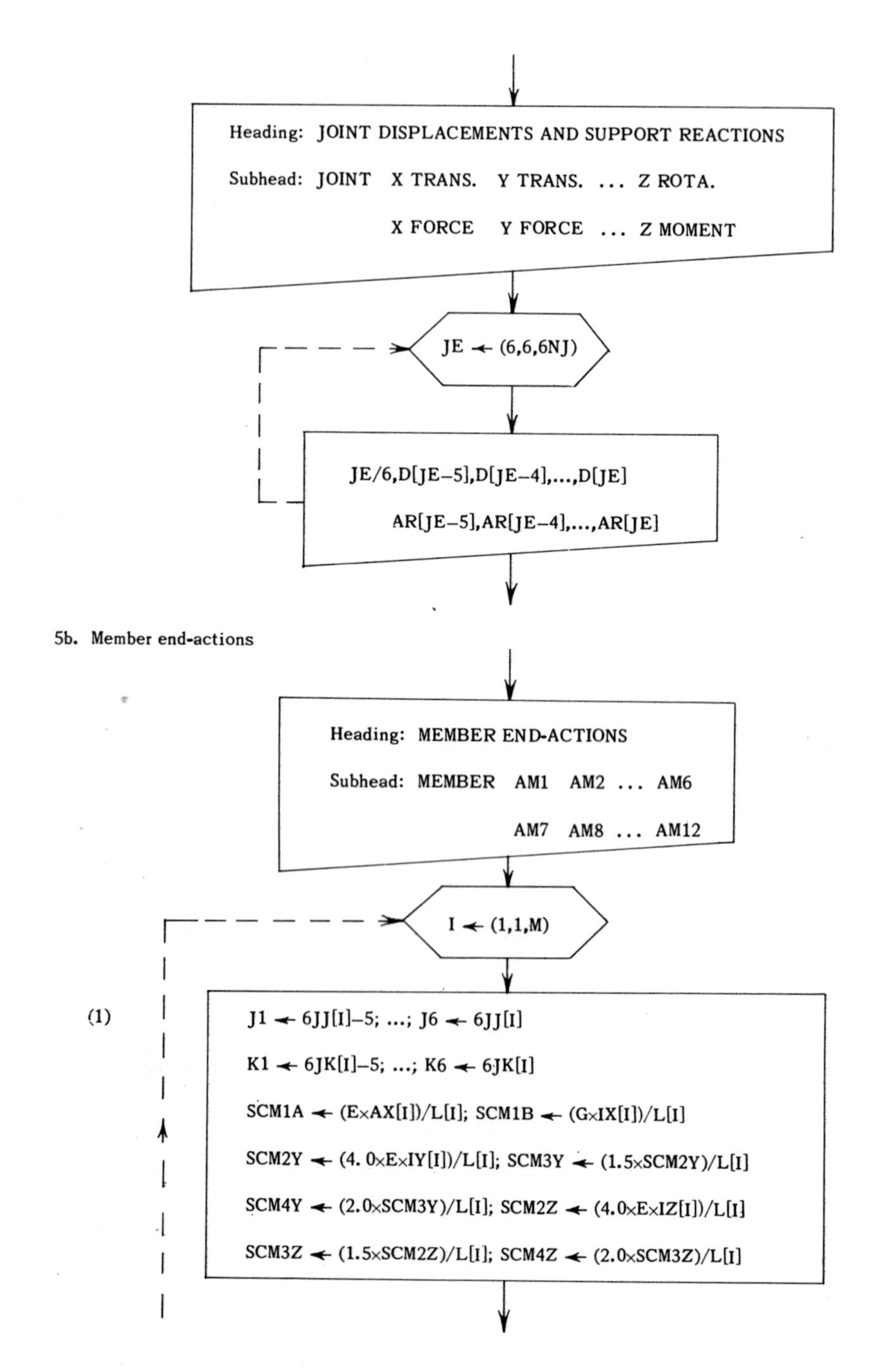

(2) (Generate SM as in Section 2a, then do the following:)

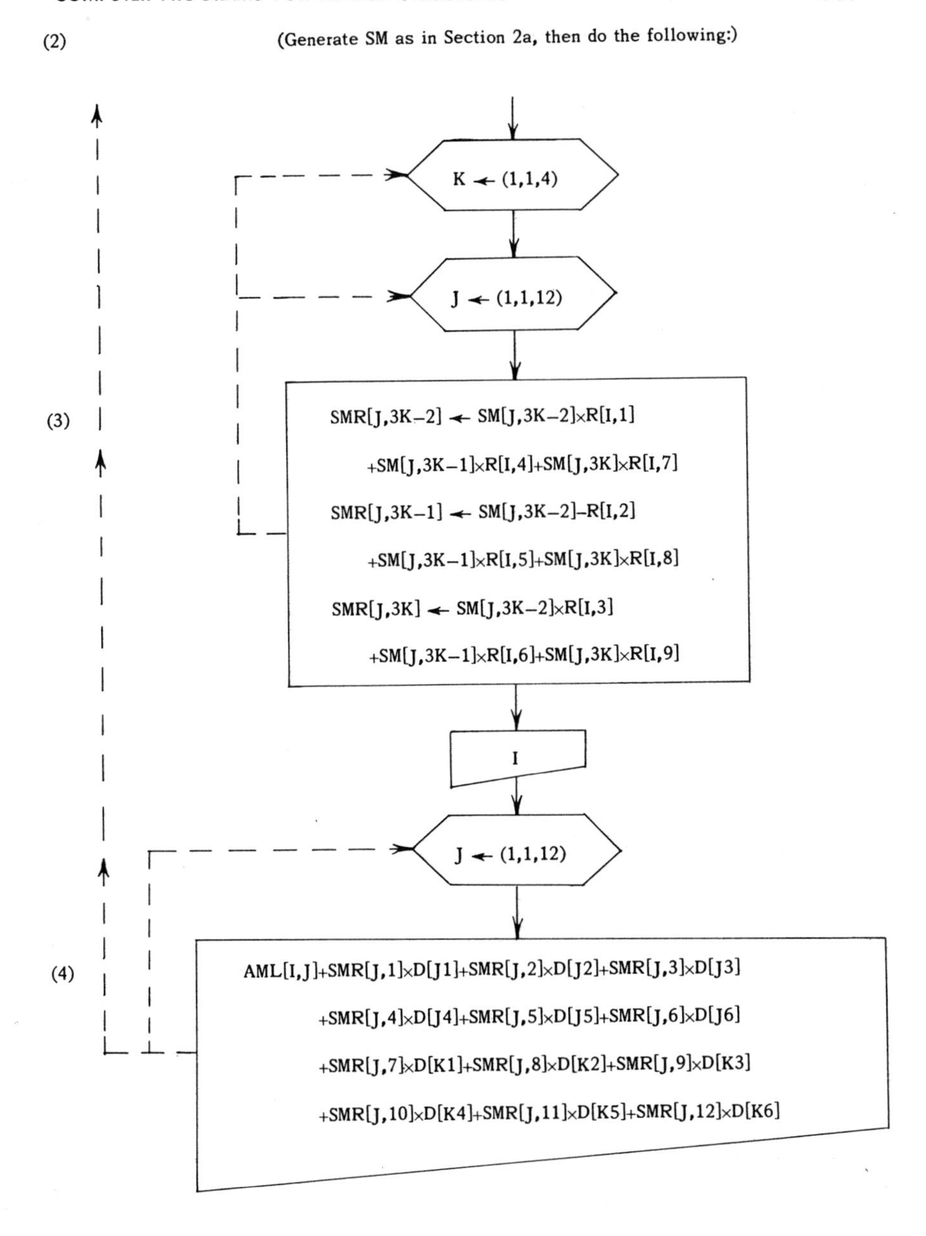

K ← (1,1,4)
J ← (1,1,12)
(3)
SMR[J,3K−2] ← SM[J,3K−2]×R[I,1]
+SM[J,3K−1]×R[I,4]+SM[J,3K]×R[I,7]
SMR[J,3K−1] ← SM[J,3K−2]−R[I,2]
+SM[J,3K−1]×R[I,5]+SM[J,3K]×R[I,8]
SMR[J,3K] ← SM[J,3K−2]×R[I,3]
+SM[J,3K−1]×R[I,6]+SM[J,3K]×R[I,9]
I
J ← (1,1,12)
(4)
AML[I,J]+SMR[J,1]×D[J1]+SMR[J,2]×D[J2]+SMR[J,3]×D[J3]
+SMR[J,4]×D[J4]+SMR[J,5]×D[J5]+SMR[J,6]×D[J6]
+SMR[J,7]×D[K1]+SMR[J,8]×D[K2]+SMR[J,9]×D[K3]
+SMR[J,10]×D[K4]+SMR[J,11]×D[K5]+SMR[J,12]×D[K6]

(*Continued from p. 376*)

(3) The value of Q is computed and stored in order to avoid repetitious calculations later.
(4) If the angle α is nonzero, a second data card for the member I is read. This card contains the x, y, and z coordinates for a point p which lies in principal plane x_M-y_M of the member but not on the member axis.
(5) The x_S, y_S, and z_S coordinates are computed for the point p.
(6) If Q is zero or very small, the member is taken to be vertical, and the nonzero elements of the rotation matrix **R** are calculated for the case of a vertical member having $\alpha = 0$ (see Eq. 4-86). The 3×3 rotation matrix **R** is stored as a vector $R[I,1], \ldots, R[I,9]$, which is the I-th row of a rectangular matrix.
(7) If the angle α is nonzero, R is revised to correspond to the case of a vertical member with $\alpha \neq 0$ (see Eq. 4-96). The identifier SQ is used as temporary storage for the square root term.
(8) On the other hand, if Q is not small (hence, the member is not vertical), the matrix R is generated on the basis of $\alpha = 0$ for a skew member (see Eq. 4-85).
(9) If α is nonzero, R is revised to correspond to the case of a member of completely general orientation in space (see Eq. 4-90). This operation involves the calculation of the y_γ and z_γ coordinates of point p.

2. a. (1) Iteration control statement on members, for M members.
(2) The six possible displacements are indexed for the joints J ($J1$ through $J6$) and K ($K1$ through $K6$) of the member I.
(3) Eight stiffness constants for axial, torsional, and flexural deformations of the member are calculated.
(4) Using the joint restraint list as a guide, the six possible displacements for the joints at each end of the member are re-indexed according to whether the joints are actually free to move or not.
(5) The 12×12 member stiffness matrix SM for member I is generated for principal axes of the member (see Table 4-1, Art. 4.3).
(6) A matrix SMR is calculated as an intermediate step in the transformation of the matrix SM to the matrix SMD, using Eq. (4-92). This intermediate matrix is calculated by postmultiplying the matrix SM by the rotation transformation matrix RT as follows:

$$\mathbf{S}_{\mathrm{MR}} = \mathbf{S}_{\mathrm{M}}\mathbf{R}_{\mathrm{T}}$$

In the program, the matrix RT is not actually formed, but the rotation matrix R (which makes up RT) is used.
(7) Next, the matrix SMD is calculated by premultiplication of the matrix SMR by the transpose of RT:

$$\mathbf{S}_{\mathrm{MD}} = \mathbf{R}_{\mathrm{T}}'\mathbf{S}_{\mathrm{MR}}$$

(8) If the index $6JJ[I]$-5 (the original value of $J1$) corresponds to a degree of freedom, the first column of SMD is transferred to the joint stiffness matrix for the structure (see Eqs. 4-100). On the other hand, if $6JJ[I]$-5 corresponds to a restraint, the first column of SMD is bypassed. Elements of S involving the coupling of $J1$ with $J2$ through $J6$ are cumulative because, in general, they receive contributions from more than one member. However, elements of S involving the coupling of $J1$ with $K1$ through $K6$ are single values. This transfer process is repeated for all 12 columns of SMD.

4. a. (1) If NLM is not equal to zero, the contributions to AE from each member must be calculated and transferred.
   (2) Contributions to AE are computed from the product of the matrix $[R'_T]_i$ and the vector $\{A_{ML}\}_i$.
5. a. (1) For each member the same information is computed as in Secs. 2a(2) and (3) given previously.
   (2) The matrix SM is generated exactly as before in Sec. 2a(5).
   (3) The matrix SMR is calculated as in Sec. 2a(6).
   (4) The member number is printed, and the twelve member end-actions are computed and printed, using Eq. (4-103).

Example 1: A space frame having three members and four joints is shown in Fig. 5-6a. Inspection of the figure shows that there are twelve degrees of freedom (six at each of the joints B and C) and twelve restraints (six at each of the points A and D). The joint loads on the frame consist of a force $2P$ in the positive x direction at point B, a force P in the negative y direction at point C, and a couple PL in the negative z sense at C. Member BC is subjected to a force $4P$ in the positive z direction applied at the mid-length of the member.

The x-y plane is a principal plane for members AB and BC; and member CD

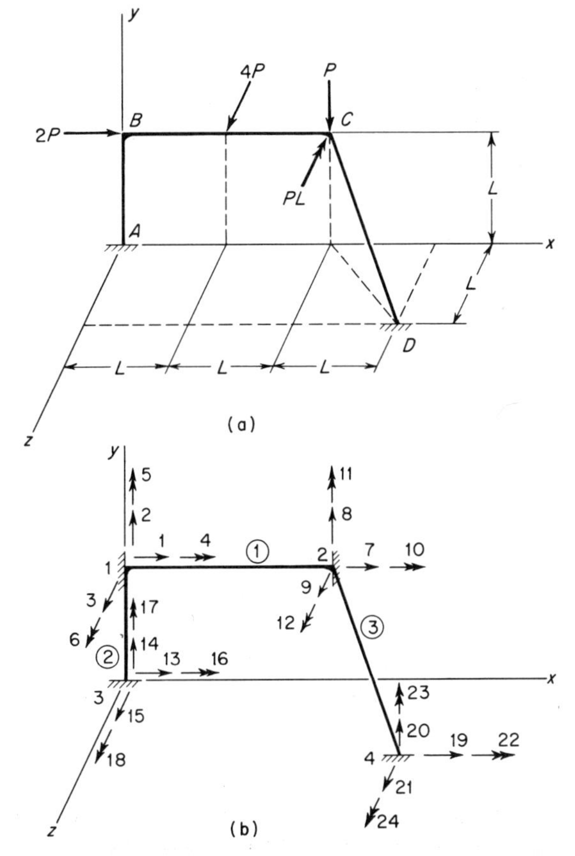

FIG. 5-6. Example 1 (space frame)

has a principal plane parallel to the y axis. Therefore, all members have the angle α equal to zero. It is assumed that all members have the same cross-sectional properties and that the numerical values in the problem are the following:

$$E = 30{,}000 \text{ ksi} \qquad P = 1 \text{ kip} \qquad AX = 11 \text{ in.}^2$$

$$G = 12{,}000 \text{ ksi} \qquad L = 120 \text{ in.} \qquad IX = 83 \text{ in.}^4$$

$$IY = IZ = 56 \text{ in.}^4$$

Figure 5-6b shows the restrained structure with a numbering system for members and joints. The numbering system for the displacements derives, of course, from the numbering system for joints.

The data cards required as input to the computer program are given in Table 5-23. Note that the identifier AA is zero on each of the cards containing member information. Table 5-24 shows the output of the computer program.

TABLE 5-23

Data Cards for Space Frame Example 1

Type of Data		Numerical Data on Card	Card No.
Structure Data	(a)	3 4 12 2 30000.0 12000.0	1
	(b)	1 0 120.0 0	2
		2 240.0 120.0 0	3
		3 0 0 0	4
		4 360.0 0 120.0	5
	(c)	1 1 2 11.0 83.0 56.0 56.0 0	6
		2 3 1 11.0 83.0 56.0 56.0 0	7
		3 2 4 11.0 83.0 56.0 56.0 0	8
	(d)	3 1 1 1 1 1 1	9
		4 1 1 1 1 1 1	10
Load Data	(a)	2 1	11
	(b)	1 2.0 0 0 0 0 0	12
		2 0 −1.0 0 0 0 −120.0	13
	(c)	1 0 0 −2.0 0 120.0 0	14
		0 0 −2.0 0 −120.0 0	15

Example 2: Figure 5-7a shows a space frame having three members, four joints, six degrees of freedom (at joint A), and eighteen restraints (six at each of the points B, C, and D). The frame is loaded at joint A with a force P in the negative

z direction and a couple $PL/4$ in the negative x sense. In addition, a force of magnitude P acting in the negative z direction is applied at the middle of member AB, and a force P acting in the negative y direction is applied at the middle of member AD. All members in the frame have a principal plane x_M-y_M which

TABLE 5-24

Output of Computer Program for Space Frame Example 1

ANALYSIS OF SPACE FRAMES

STRUCTURE DATA

M	N	NJ	NR	NRJ	E	G
3	12	4	12	2	30000.0	12000.0

COORDINATES OF JOINTS

JOINT	X	Y	Z
1	0	120.0	0
2	240.0	120.0	0
3	0	0	0
4	360.0	0	120.0

MEMBER DESIGNATIONS, PROPERTIES, AND ORIENTATIONS

MEMBER	JJ	JK	AX	IX	IY	IZ	AA	L	CX	CY	CZ
1	1	2	11.0	83.0	56.00	56.0	0	240.0	1.000	0	0
2	3	1	11.0	83.0	56.00	56.0	0	120.0	0	1.000	0
3	2	4	11.0	83.0	56.00	56.0	0	207.8	0.577	−0.577	0.577

JOINT RESTRAINTS

JOINT	X TRANS.	Y TRANS.	Z TRANS.	X ROTA.	Y ROTA.	Z ROTA.
3	1	1	1	1	1	1
4	1	1	1	1	1	1

LOAD DATA

NLJ	NLM
2	1

ACTIONS APPLIED AT JOINTS

JOINT	X FORCE	Y FORCE	Z FORCE	X MOMENT	Y MOMENT	Z MOMENT
1	2.0	0	0	0	0	0
2	0	−1.0	0	0	0	−120.0

ACTIONS AT ENDS OF RESTRAINED MEMBERS DUE TO LOADS

MEMBER	AML1	AML2	AML3	AML4	AML5	AML6
	AML7	AML8	AML9	AML10	AML11	AML12
1	0	0	−2.0	0	120.0	0
	0	0	−2.0	0	−120.0	0

JOINT DISPLACEMENTS AND SUPPORT REACTIONS

JOINT	X TRANS.	Y TRANS.	Z TRANS.	X ROTA.	Y ROTA.	Z ROTA.
	X FORCE	Y FORCE	Z FORCE	X MOMENT	Y MOMENT	Z MOMENT
1	−0.1528	−0.0002436	0.6263	0.007536	−0.005463	0.002674
	0	0	0	0	0	0
2	−0.1542	0.4562	0.6139	0.003584	0.005748	−0.002701
	0	0	0	0	0	0

(Continued on next page.)

(TABLE 5-24, continued)

3	0	0	0	0	0	0
	-0.09	-0.67	-2.03	-227.41	45.34	-32.11
4	0	0	0	0	0	0
	-1.91	1.67	-1.97	-52.22	-44.54	30.99

MEMBER END-ACTIONS

MEMBER	AM1	AM2	AM3	AM4	AM5	AM6
	AM7	AM8	AM9	AM10	AM11	AM12
1	1.91	-0.67	-2.03	16.40	45.34	-42.75
	-1·91	0.67	-1.97	-16.40	-37.71	-118.00
2	-0.67	0.09	-2.03	45.34	227.41	-32.11
	0.67	-0.09	2.03	-45.34	16.40	42.75
3	3.20	0.22	0.04	-13.46	36.67	-13.01
	-3.20	-0.22	-0.04	13.46	-45.03	58.84

contains the point p indicated in Fig. 5-7a. Point p is located in the plane containing joints B, C, and D (that is, the x-y plane). Any point on the line pA (except point A) would suffice to define the principal planes in this problem.

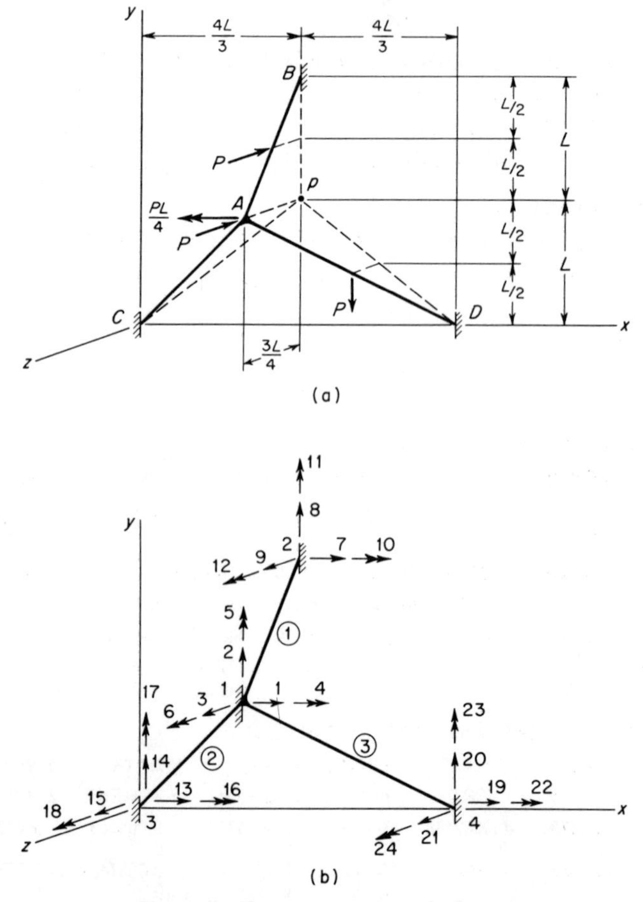

FIG. 5-7. Example 2 (space frame).

Since a principal plane for member AB is parallel to the y axis, the angle α is taken to be zero for this member. For members AC and AD the principal planes are located by giving the coordinates of point p.

The cross-sectional properties of all members are assumed to be the same, and the numerical constants in the problem are as follows:

$$E = 10{,}000 \text{ ksi} \qquad P = 5 \text{ kips} \qquad AX = 9 \text{ in.}^2 \qquad IY = 28 \text{ in.}^4$$

$$G = 4{,}000 \text{ ksi} \qquad L = 96 \text{ in.} \qquad IX = 64 \text{ in.}^4 \qquad IZ = 80 \text{ in.}^4$$

Figure 5-7b shows the structure with all joints restrained. A numbering system for members, joints, and displacements appears in the figure also.

The input data for this problem are listed in Table 5-25. Note that the cards containing member information for members 2 and 3 are each followed by a card which gives the coordinates of point p.

TABLE 5-25

Data Cards for Space Frame Example 2

Type of Data		Numerical Data on Card								Card No.
Structure Data	(a)	3	4	18	3	10000.0	4000.0			1
	(b)	1	128.0	96.0	72.0					2
		2	128.0	192.0	0					3
		3	0	0	0					4
		4	256.0	0	0					5
	(c)	1	1	2	9.0	64.0	28.0	80.0	0	6
		2	3	1	9.0	64.0	28.0	80.0	1	7
		2	128.0	96.0	0					8
		3	1	4	9.0	64.0	28.0	80.0	1	9
		3	128.0	96.0	0					10
	(d)	2	1	1	1	1	1	1		11
		3	1	1	1	1	1	1		12
		4	1	1	1	1	1	1		13
Load Data	(a)	1	2							14
	(b)	1	0	0	−5.0	−120.0	0	0		15
	(c)	1	−1.5	2.0	0	0	0	60.0		16
			−1.5	2.0	0	0	0	−60.0		17
		3	−1.368	0.616	2.0	0	−87.727	27.0		18
			−1.368	0.616	2.0	0	87.727	−27.0		19

The calculation of actions A_{ML} at the ends of the restrained members due to loads (the last four data cards) requires further explanation in this example. The end-actions A_{ML} for member 1 are obtained easily because the load P applied at the middle lies in the principal plane for the member. Figure 5-8 shows all of the end-actions for this member. Note that the orientation of the member axes in the figure results from the sequence of β and γ rotations depicted in Fig. 4-38 (see Art. 4.22).

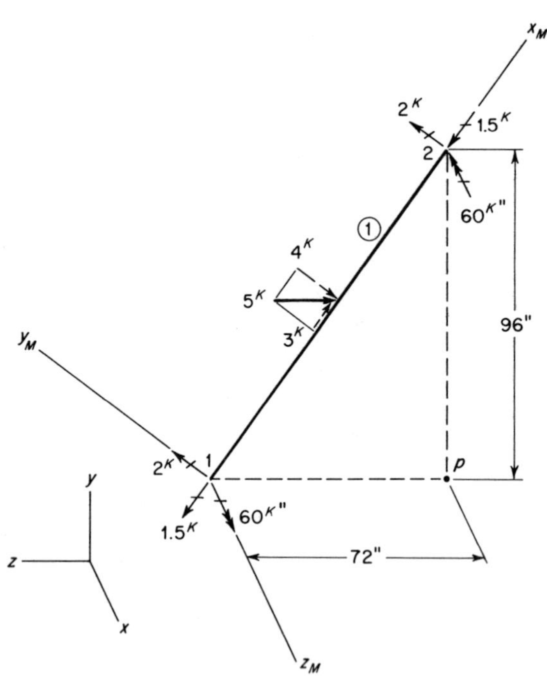

FIG. 5-8. Restraint actions at ends of member 1.

On the other hand, the load P applied to member 3 does not lie in a principal plane of the member. In order to calculate the end-actions A_{ML} for this member, the components of the load P in the directions of member axes must be ascertained. One method for doing this is to use the rotation of axes technique represented by Eq. (4-57) in Art. 4.15; thus:

$$\mathbf{A}_{\mathrm{M}} = \mathbf{R}\mathbf{A}_{\mathrm{S}} \tag{4-57}$$

repeated

Figure 5-9 shows the member axes for member 3 as well as the load P applied in the negative y direction. The rotation matrix $\mathbf{R}$ for the member axes may be obtained either from the geometry of the figure or from Eq. (4-90) in Art. 4.24 (using the coordinates of point p if desired). This rotation matrix is the following:

$$\mathbf{R} = \begin{bmatrix} \frac{16}{\sqrt{481}} & -\frac{12}{\sqrt{481}} & -\frac{9}{\sqrt{481}} \\ -\frac{36}{5\sqrt{481}} & \frac{27}{5\sqrt{481}} & -\frac{20}{\sqrt{481}} \\ \frac{3}{5} & \frac{4}{5} & 0 \end{bmatrix} \tag{a}$$

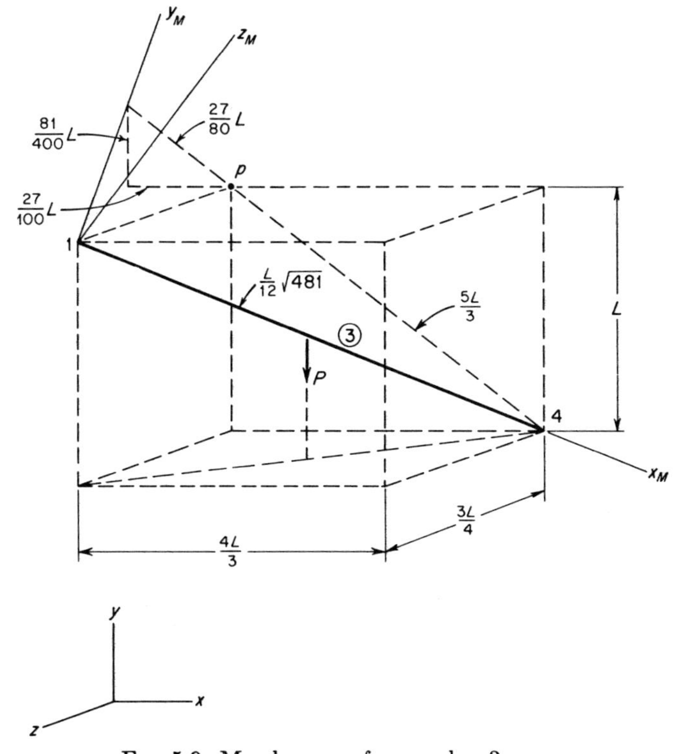

FIG. 5-9. Member axes for member 3.

Since the applied load P is 5 kips, the vector $\mathbf{A}_S$ becomes:

$$\mathbf{A}_S = \begin{bmatrix} 0 \\ -5 \\ 0 \end{bmatrix} \tag{b}$$

Substitution of expressions (a) and (b) into Eq. (4-57) produces the vector $\mathbf{A}_M$ as follows:

$$\mathbf{A}_M = \begin{bmatrix} \dfrac{60}{\sqrt{481}} \\ -\dfrac{27}{\sqrt{481}} \\ -4 \end{bmatrix} = \begin{bmatrix} 2.736 \\ -1.231 \\ -4.000 \end{bmatrix} \qquad \text{(c)}$$

The elements in this vector are the components of the applied load in the directions of member axes. The end-actions A_{ML} caused by these forces acting on member 3 are listed in the last two lines of Table 5-25.

Table 5-26 shows the final results from the computer program.

TABLE 5-26

Final Results for Space Frame Example 2

JOINT DISPLACEMENTS AND SUPPORT REACTIONS

JOINT	X TRANS. X FORCE	Y TRANS. Y FORCE	Z TRANS. Z FORCE	X ROTA. X MOMENT	Y ROTA. Y MOMENT	Z ROTA. Z MOMENT
1	−0.0002176	−0.004062	−0.01674	−0.005202	0.0001870	−0.004495
	0	0	0	0	0	0
2	0	0	0	0	0	0
	−0.41	−2.98	7.01	−134.64	−14.60	−9.34
3	0	0	0	0	0	0
	3.70	2.61	1.51	−8.18	28.39	−3.62
4	0	0	0	0	0	0
	−3.29	5.37	1.48	−63.93	−16.89	−98.43

MEMBER END-ACTIONS

MEMBER	AM1 AM7	AM2 AM8	AM3 AM9	AM4 AM10	AM5 AM11	AM6 AM12
1	3.59	0.18	0.41	6.07	−32.49	−84.00
	−6.59	3.82	−0.41	−6.07	−16.23	−134.64
2	4.74	0.48	−0.13	8.08	7.61	27.62
	−4.74	−0.48	0.13	−8.08	15.39	57.44
3	3.21	0.18	1.68	−3.00	−50.18	−24.96
	−5.95	1.05	2.32	3.00	106.59	−51.86

Chapter 6

ADDITIONAL TOPICS FOR THE STIFFNESS METHOD

6.1 Introduction. The techniques described in Chapter 4 for implementing the stiffness method, and the corresponding computer programs (presented in Chapter 5), apply to framed structures having prismatic members and subjected to the effects of loads only. However, the techniques can be extended readily in order to incorporate many other considerations into the computer programs. The purpose of this chapter is to describe how some of the more important considerations may be treated. First, the effects of temperature changes and prestrains, which were described in Chapter 2 from the standpoint of a manual analysis, are discussed again in the next article for the purpose of including them in a computer program. The next two articles (6.3 and 6.4) deal with the effects of loads between joints and methods for determining internal actions and displacements at intermediate locations along the lengths of members. Then several topics relating to the supports of framed structures are discussed (Arts. 6.5 through 6.7); these topics include support displacements, inclined supports, and elastic supports. The remaining articles in the chapter are devoted to subjects whose primary effects are to alter the stiffnesses and fixed-end actions for individual members. Covered under this category are nonprismatic members, discontinuities in members, elastic connections, shear deformations, and axial-flexural interactions.

The material in this chapter is intended to be supplementary to the subject matter of previous chapters. Therefore, there is no attempt to present complete and detailed explanations; rather, the intention is to suggest additional topics which the interested reader may wish to pursue further.

6.2 Temperature and Prestrain. The effects of temperature and prestrain were considered in the hand solutions of Chapter 2. These effects may be readily incorporated into a computer-oriented analysis by treating them in a manner analogous to the procedure described in Chapter 4 (Art. 4.5) for loads on members, that is, by replacing them with equivalent joint loads. Both temperature changes and prestrain effects will cause fixed-end actions in the members of a restrained structure. Formulas for fixed-end actions of this nature are listed in Tables

B-2 and B-3 of Appendix B. Such end-actions may be handled in the same manner as those due to loads on members; in other words, they may be listed in matrices $\mathbf{A}_{MT}$ and $\mathbf{A}_{MP}$, respectively, analogous to the listing of fixed-end actions due to loads in the matrix $\mathbf{A}_{ML}$. Then the sum of all three effects can be combined into a single matrix $\mathbf{A}_{MS}$ (see Eq. 2-30). Of course, in computer programming it may be desirable to place all terms directly into the matrix $\mathbf{A}_{MS}$, without the necessity of forming separately the matrices $\mathbf{A}_{ML}$, $\mathbf{A}_{MT}$, and $\mathbf{A}_{MP}$. Following this step, the fixed-end actions in the matrix $\mathbf{A}_{MS}$ are converted to equivalent joint loads, and the analysis proceeds as described in Chapter 4 from then on. All joint displacements D, reactions A_R, and member end-actions A_M are calculated in exactly the same manner as for the case of loads only.

6.3 Loads Between Joints. In Chapters 4 and 5 the effects of loads acting on members were handled indirectly by making use of the fixed-end actions caused by such loads. It is also possible to extend the computer programs of Chapter 5 to accept these loads directly as input data, and some of the feasible techniques for doing so are covered in the following discussion.

From the previous descriptions of member-oriented and structure-oriented axes, it is apparent that loads acting on members may be resolved into components parallel to either set of axes. In general, it is desirable to resolve the member loads into components parallel to the member axes (x_M, y_M, and z_M). The loads on beam and grid structures are usually aligned with the member axes by their very nature, but the orientations of loads on plane and space frames may be quite general. Therefore, some preliminary effort may be required to resolve the loads into components parallel to member axes, as demonstrated for a space frame member in Example 2 of Art. 5.9.

It frequently happens that the loads on truss members are expressed with respect to structure-oriented axes, and it may be more convenient to handle them in that form than to resolve them into components parallel to member axes. The reason for this conclusion is that the pinned-end actions in a restrained truss structure are easily calculated regardless of the orientations of the members (see Table B-5 of Appendix B). In such a case the negatives of the pinned-end actions may be taken as equivalent joint loads acting in the directions of structure axes, and the analysis may be carried out as usual from this point on. The drawback to this approach is the fact that the final member end-actions will not include the initial pinned-end actions A_{ML} in the directions of member axes. Thus, the only member end-actions A_M that are calculated are the axial forces in the members due to the equivalent joint loads. This limitation is not a serious disadvantage in the analysis of trusses, because their principal features relate to the effects of loads (actual or equivalent) at the joints.

Now consider again those structures for which the member loads have been resolved into components parallel to member axes. The task of incorporating the loads directly into the stiffness method of analysis will be facilitated if they are expressed in discrete form (that is, as concentrated forces and couples). Any distributed loadings can be approximated by a series of concentrated loads which cause essentially the same

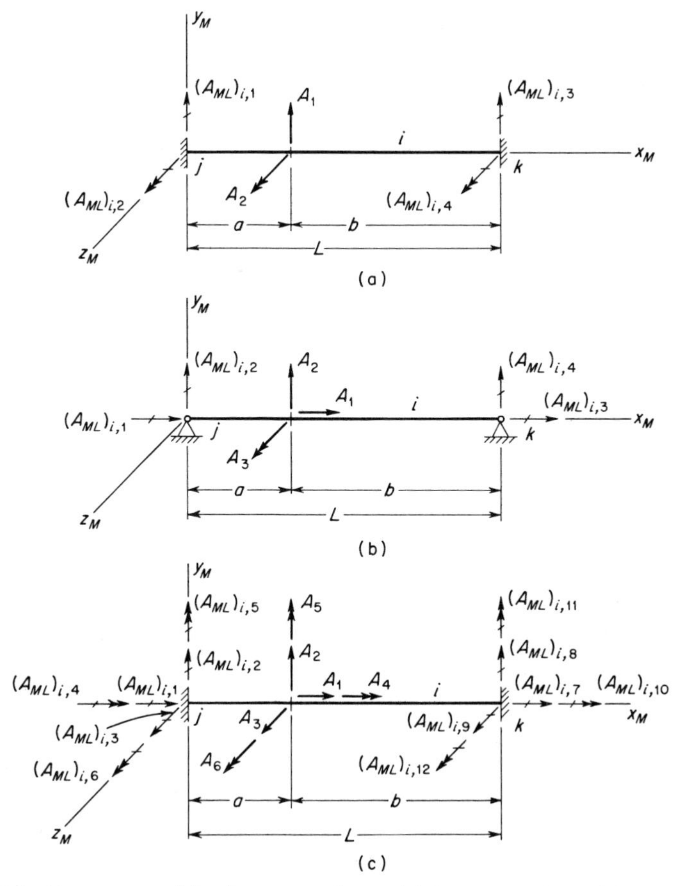

FIG. 6-1. Concentrated loads on members: (a) beam member, (b) plane truss member, and (c) space frame member.

actions and displacements in the structure. If this approach is followed, only a certain number of concentrated loads will be possible at any point on a member. For example, at a point on a beam there may exist either a concentrated force A_1 in the y_M direction or a couple A_2 in the z_M sense, as shown in Fig. 6-1a. Each of these loads will cause four fixed-end actions of the types indicated in the figure. Similarly, there are three possible concentrated forces at any point on a plane truss member (see

Fig. 6-1b), and four possible pinned-end actions. The most general category of member (a space frame member) is illustrated in Fig. 6-1c. Six load components may exist at any point on such a member, causing twelve types of fixed-end actions, as shown in the figure.

If the loads on a member are discretized as described above, there are several methods by which they may be incorporated directly into the analysis of a structure. The simplest approach is to assume that a joint exists at every point of loading on the member, and then to analyze the structure as if it were subjected to joint loads only. This method is suitable for structures with rigid joints but cannot be adapted easily to trusses; moreover, the loads must be structure-oriented instead of member-oriented in this approach. The principal disadvantage of this method is that the number of joints in the structure may become excessive if there are loads on many members.

A second method consists of analyzing each loaded member as a separate structure, using the stiffness method of analysis, and then incorporating the results of each such subanalysis into the over-all analysis of the structure. For example, the restrained beam in Fig. 6-1a could be analyzed as a structure consisting of two members of lengths a and b, respectively, and having two degrees of freedom at the point of application of the loads. The fixed-end actions A_{ML} for the member i will be the reactions determined in the subanalysis of the beam considered as two members. This line of reasoning may be generalized to include any type of framed structure and any number of load points on a member.

A third approach to the problem involves the use of transfer matrices for fixed-end actions due to unit loads on the members. The relationships between end-actions and loads acting at one point on a member may be expressed by the following equation:

$$\mathbf{A}_{\mathrm{ML}i} = \mathbf{T}_{\mathrm{ML}i}\mathbf{A}_i \tag{6-1}$$

In Eq. (6-1), $\mathbf{A}_i$ is a column vector of concentrated loads which may exist at any point on the member i. For example, the beam pictured in Fig. 6-1a has the concentrated loads A_1 and A_2 associated with an arbitrary point on the member. Thus, for this type of structure:

$$\mathbf{A}_i = \{A_1, A_2\} \tag{6-2}$$

The matrix $\mathbf{A}_{\mathrm{ML}i}$ in Eq. (6-1) consists of a column vector of fixed-end actions caused by the loads A_i. For a beam, this vector is the following (see Fig. 6-1a):

$$\mathbf{A}_{\mathrm{ML}i} = \{(A_{ML})_{i,1}, (A_{ML})_{i,2}, (A_{ML})_{i,3}, (A_{ML})_{i,4}\} \tag{6-3}$$

The matrix $\mathbf{T}_{\mathrm{ML}i}$ is a transfer matrix for the actions $\mathbf{A}_{\mathrm{ML}i}$ at the ends of the restrained member i due to unit values of the actions $\mathbf{A}_i$. The elements of such a transfer matrix may be obtained from the formulas given in Table B-1 of Appendix B, and for a beam the matrix will be the following 4×2 array (omitting the subscript i):

$$\mathbf{T}_{\mathrm{ML}} = \begin{bmatrix} -\frac{b^2}{L^3}(3a+b) & \frac{6ab}{L^3} \\ -\frac{ab^2}{L^2} & \frac{b}{L^2}(2a-b) \\ -\frac{a^2}{L^3}(a+3b) & -\frac{6ab}{L^3} \\ \frac{a^2b}{L^2} & \frac{a}{L^2}(2b-a) \end{bmatrix} \tag{6-4}$$

The final vector of member end-actions $\mathbf{A}_{\mathrm{ML}}$ for a member with loads acting at several load points is obtained by applying Eq. (6-1) for every point on the member at which a load is applied and then summing the results.

Transfer matrices similar to Eq. (6-4) may be developed for all types of framed structures. For a plane truss member subjected to loads as indicated in Fig. 6-1b, the transfer matrix $\mathbf{T}_{\mathrm{ML}}$ for pinned-end actions will be of order 4×3, as follows:

$$\mathbf{T}_{\mathrm{ML}} = \begin{bmatrix} -\frac{b}{L} & 0 & 0 \\ 0 & -\frac{b}{L} & \frac{1}{L} \\ -\frac{a}{L} & 0 & 0 \\ 0 & -\frac{a}{L} & -\frac{1}{L} \end{bmatrix} \tag{6-5}$$

Similarly, the transfer matrices for the other types of framed structures may be easily derived. Such transfer matrices may be incorporated into a computer program for the purpose of automatically calculating the actions A_{ML} at the ends of restrained members due to loads at a discrete number of points. After this is done, the calculation of equivalent joint loads and the remaining part of the analysis proceed as before.

6.4 Actions and Displacements Between Joints. End-displacements D_M and end-actions A_M for all members of a structure may be obtained by the methods of Chapter 4. In addition, actions and displacements at intermediate locations along the lengths of members may also be of interest. There are several methods by which such actions and displacements may be calculated. For instance, as mentioned in Art. 2.8, they may be determined by hand calculations after the actions and displacements at the ends have been obtained. The internal actions are determined by applying static equilibrium principles to a portion of the member taken as a free body, and displacements may be calculated using elementary deflection theories (such as the moment-area theorems).

Matrix methods are also suitable for this purpose, and the simplest modification of the stiffness method of analysis is to assume that a joint exists at every point where actions and displacements are desired. The joint displacements and member end-actions resulting from the analysis will include displacements and internal actions at the points of interest. In this approach the displacements will automatically be in the directions of structure axes. As mentioned previously, the disadvantage of this method is that the number of joints may become too large to be handled successfully.

Another matrix approach is to analyze a given member as a subproblem within the total analysis of the whole structure. The restrained member may be divided into segments and analyzed by the stiffness method, treating the displacements of the ends of the segmented member as if they were support displacements. This subanalysis will produce the same results as the preceding method.

A third matrix approach involves the use of transfer matrices for actions and displacements at intermediate locations. This approach will be described for beams, but the method is generally applicable to all types of framed structures. The task of computing actions and displacements at intermediate locations may be divided into two steps. The first step consists of calculating the actions and displacements in the restrained member due to the loads (or other effects) acting on the member. The second step involves calculating the additional actions and displacements due to the displacements of the ends of the member. In both steps the actions and displacements would ordinarily be computed with respect to member axes, but these quantities may always be converted to structure axes by an appropriate rotation of axes. The calculation of actions in beams will be considered first, and then the matter of determining displacements will be discussed.

Figure 6-2a shows a restrained beam member i with a point of load (where A_1 and A_2 act) at the distance a from the j end. A second point, at which actions and displacements are to be calculated, is located at the distance x_M from the j end. The vector of fixed-end actions $\mathbf{A}_{\mathrm{ML}i}$ due to the actions A_1 and A_2 may be calculated as described in the previous article (see Eq. 6-1), using the transfer matrix $\mathbf{T}_{\mathrm{ML}}$ (Eq. 6-4). The first two elements of the vector $\mathbf{A}_{\mathrm{ML}i}$ are the actions at the j end of the member, and the last two elements are the actions at the k end (see Fig. 6-2a). Figure 6-2b shows a free-body diagram for the segment of length x_M, assuming that $x_M \leq a$. The figure also shows the internal actions A_{MXL1} and A_{MXL2} at the point of interest, acting in their positive senses. The sign convention adopted here is that the internal actions on the end of a segment that is toward the k end of the member are positive when acting in the positive directions of the member axes.

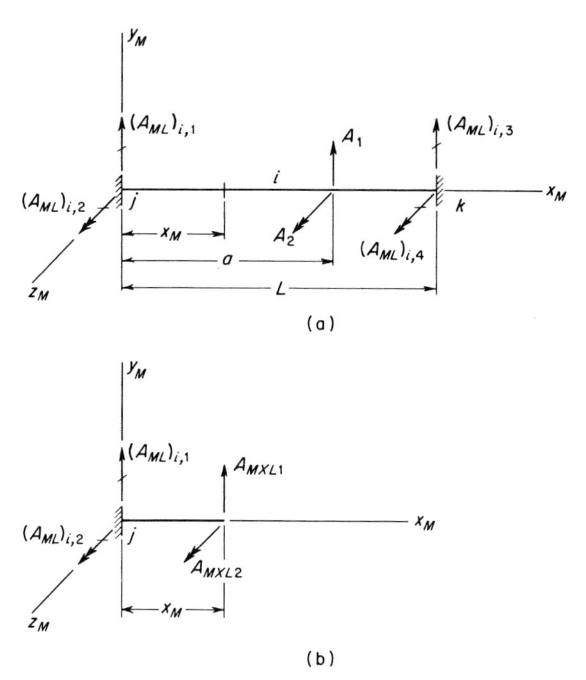

FIG. 6-2. Actions in a beam: (a) applied actions and (b) internal actions.

The internal actions $\mathbf{A}_{\text{MXL}}$ may be computed from the 4×1 vector of fixed-end actions $\mathbf{A}_{\text{ML}}$ using a transfer matrix $\mathbf{T}_{\text{AXJ}}$ as follows (omitting the subscript i):

$$\mathbf{A}_{\text{MXL}} = \mathbf{T}_{\text{AXJ}}\mathbf{A}_{\text{ML}} \tag{6-6}$$

The elements of this transfer matrix may be obtained from Fig. 6-2b by inspection:

$$\mathbf{T}_{\text{AXJ}} = \begin{bmatrix} -1 & 0 & 0 & 0 \\ x_M & -1 & 0 & 0 \end{bmatrix} \tag{6-7}$$

The left-hand half of the transfer matrix $\mathbf{T}_{\text{AXJ}}$ consists of actions at the point of interest due to unit restraint actions at the j end of the member. The right-hand half consists of zeros inasmuch as the actions A_{ML} at the k end of the member do not enter into the calculations. The matrix $\mathbf{T}_{\text{AXJ}}$ is valid only for points on the beam at which $x_M \leq a$. If the point of interest is located such that $x_M \geq a$, the fixed-end actions at the k end may be used instead, and the actions $\mathbf{A}_{\text{MXL}}$ are computed, using a transfer matrix $\mathbf{T}_{\text{AXK}}$:

$$\mathbf{A}_{\text{MXL}} = \mathbf{T}_{\text{AXK}}\mathbf{A}_{\text{ML}} \tag{6-8}$$

in which the transfer matrix $\mathbf{T}_{AXK}$ is the following:

$$\mathbf{T}_{AXK} = \begin{bmatrix} 0 & 0 & 1 & 0 \\ 0 & 0 & L - x_M & 1 \end{bmatrix} \tag{6-9}$$

The elements in the right-hand half of this matrix are actions at the point of interest due to unit restraint actions at the k end of the member.

Equations (6-6) and (6-8) may be combined into one equation for the purpose of calculating internal actions on both sides of the load point simultaneously; thus,

$$\mathbf{A}_{MXL} = \mathbf{T}_{AX}\mathbf{A}_{ML} \tag{6-10}$$

The vector $\mathbf{A}_{MXL}$ in Eq. (6-10) consists of the actions at two points of interest, one on each side of the load point, as indicated by the points 1 and 2 in Fig. 6-3a. The composite transfer matrix $\mathbf{T}_{AX}$ is, therefore, constituted as follows:

$$\mathbf{T}_{AX} = \begin{bmatrix} \mathbf{T}_{AXJ} \\ \mathbf{T}_{AXK} \end{bmatrix} = \left[\begin{array}{cccc} -1 & 0 & 0 & 0 \\ x_{M1} & -1 & 0 & 0 \\ \hdashline 0 & 0 & 1 & 0 \\ 0 & 0 & L - x_{M2} & 1 \end{array}\right] \tag{6-11}$$

in which x_{M1} is the distance from the j end to point 1, and x_{M2} is the distance to point 2. Thus, the actions to one side of the load point or the other may be computed, using either Eq. (6-6) or Eq. (6-8); alternatively, they may be computed simultaneously, using Eq. (6-10).

Substitution of Eq. (6-1) into Eq. (6-10) gives the internal actions in terms of the transfer matrix $\mathbf{T}_{ML}$ and the load vector $\mathbf{A}$, as follows:

$$\mathbf{A}_{MXL} = \mathbf{T}_{AX}\mathbf{T}_{ML}\mathbf{A} \tag{6-12}$$

If there are several points at which loads are applied, the expression for $\mathbf{A}_{MXL}$ becomes

$$\mathbf{A}_{MXL} = \sum (\mathbf{T}_{AX}\mathbf{T}_{ML}\mathbf{A}) \tag{6-13}$$

in which the summation is performed for all load points.

Equation (6-13) gives the actions at points of interest caused by loads acting on the restrained member. Additional actions occur due to the displacements of the ends of the member. These actions may be calculated as follows:

$$\mathbf{A}_{MXD} = \mathbf{T}_{AX}\mathbf{S}_{M}\mathbf{D}_{M} \tag{6-14}$$

In this equation the matrix $\mathbf{A}_{MXD}$ is a vector of actions at two points of interest caused by the displacements $\mathbf{D}_M$ at the ends of the member, $\mathbf{S}_M$ is the member stiffness matrix, and $\mathbf{T}_{AX}$ is given by Eq. (6-11).

The addition of Eqs. (6-13) and (6-14) produces the total action vector $\mathbf{A}_{MX}$ at the two points of interest:

$$\mathbf{A}_{MX} = \mathbf{A}_{MXL} + \mathbf{A}_{MXD} \tag{6-15}$$

or,

$$\mathbf{A}_{MX} = \sum (\mathbf{T}_{AX}\mathbf{T}_{ML}\mathbf{A}) + \mathbf{T}_{AX}\mathbf{S}_{M}\mathbf{D}_{M} \tag{6-16}$$

For the special case of only one point of load, the summation process is not required, and the matrix $\mathbf{T}_{\mathrm{AX}}$ can be factored from the right-hand side of Eq. (6-16) as follows:

$$\begin{aligned}\mathbf{A}_{\mathrm{MX}} &= \mathbf{T}_{\mathrm{AX}}(\mathbf{T}_{\mathrm{ML}}\mathbf{A} + \mathbf{S}_{\mathrm{M}}\mathbf{D}_{\mathrm{M}}) \\ &= \mathbf{T}_{\mathrm{AX}}(\mathbf{A}_{\mathrm{ML}} + \mathbf{S}_{\mathrm{M}}\mathbf{D}_{\mathrm{M}}) \\ &= \mathbf{T}_{\mathrm{AX}}\mathbf{A}_{\mathrm{M}} \end{aligned} \tag{6-17}$$

in which $\mathbf{A}_{\mathrm{M}}$ is the vector of final end-actions in the member.

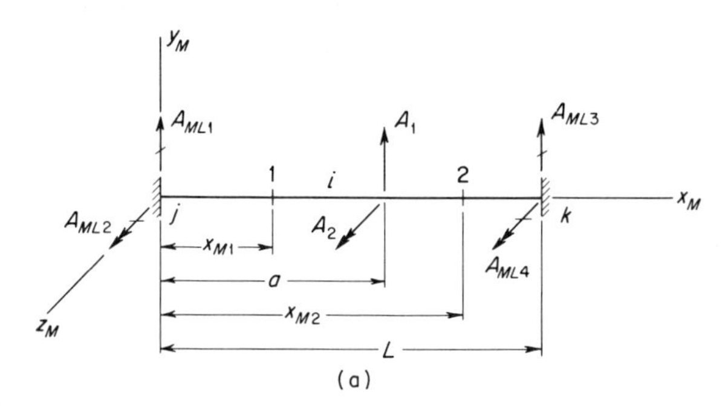

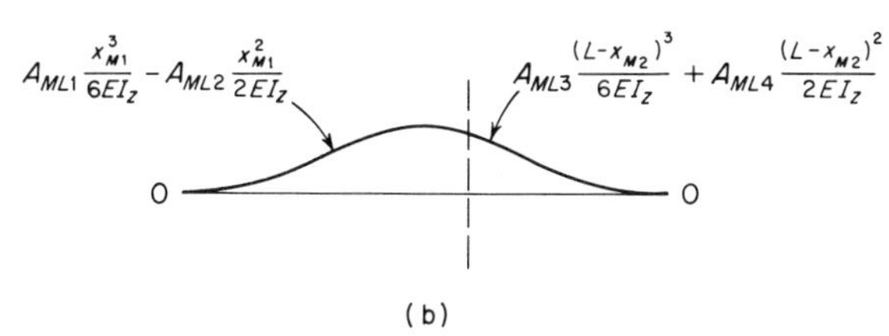

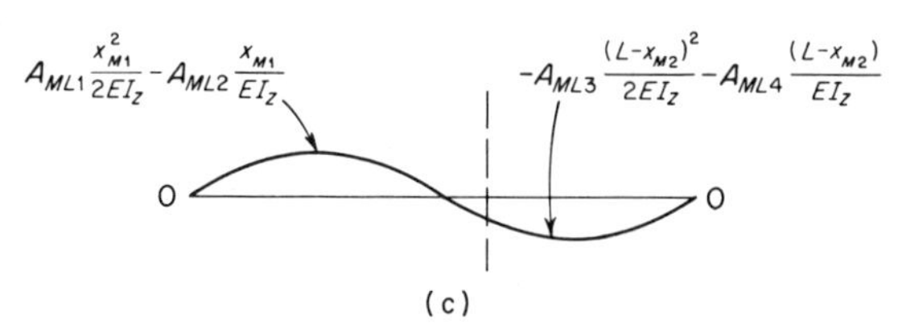

FIG. 6-3. Displacements in a beam: (a) applied actions, (b) translations, and (c) rotations.

Consider next the task of calculating displacements at intermediate locations between joints. The general formulation for obtaining such displacements is similar to that for actions, in that they consist of the sums of displacements due to loads on the restrained member plus the effects of displacements of the ends. Thus,

$$\mathbf{D}_{\mathrm{MX}} = \mathbf{D}_{\mathrm{MXL}} + \mathbf{D}_{\mathrm{MXD}} \tag{6-18}$$

The vector $\mathbf{D}_{\mathrm{MX}}$ in Eq. (6-18) contains the displacements at two different points at which $x_{M1} \leq a$ and $x_{M2} \geq a$, respectively (see Fig. 6-3a). The vectors $\mathbf{D}_{MXL}$ and $\mathbf{D}_{MXD}$ are the displacements at these points caused by the member loads and the end-displacements, respectively. Elements of the vector $\mathbf{D}_{MXL}$ for a beam are given by the formulas shown in Figs. 6-3b and c. The formula in Fig. 6-3b to the left of the load point (delineated by the vertical dashed line) is for translations in terms of the actions at the j end of the member, and that to the right of the load point is in terms of the actions at the k end. In a similar manner, Fig. 6-3c shows the rotations along the length of the beam in terms of the actions at the j and k ends. The vector $\mathbf{D}_{MXL}$ can be factored into the matrix products:

$$\mathbf{D}_{MXL} = \sum (\mathbf{T}_{DX}\mathbf{A}_{ML}) = \sum (\mathbf{T}_{DX}\mathbf{T}_{ML}\mathbf{A}) \tag{6-19}$$

in which $\mathbf{T}_{DX}$ is the following transfer matrix:

$$\mathbf{T}_{DX} = \begin{bmatrix} \mathbf{T}_{DXJ} \\ \mathbf{T}_{DXK} \end{bmatrix}$$

$$= \left[\begin{array}{cccc} \dfrac{x_{M1}^3}{6EI_Z} & -\dfrac{x_{M1}^2}{2EI_Z} & 0 & 0 \\ \dfrac{x_{M1}^2}{2EI_Z} & -\dfrac{x_{M1}}{EI_Z} & 0 & 0 \\ \hline 0 & 0 & \dfrac{(L - x_{M2})^3}{6EI_Z} & \dfrac{(L - x_{M2})^2}{2EI_Z} \\ 0 & 0 & -\dfrac{(L - x_{M2})^2}{2EI_Z} & -\dfrac{(L - x_{M2})}{EI_Z} \end{array}\right] \tag{6-20}$$

It can also be shown that the vector $\mathbf{D}_{MXD}$ in Eq. (6-18) may be obtained by the following calculation:

$$\mathbf{D}_{MXD} = (\mathbf{T}_{DB} + \mathbf{T}_{DX}\mathbf{S}_M)\mathbf{D}_M \tag{6-21}$$

Elements of the first column of the matrix enclosed in parentheses in Eq. (6-21) are drawn from the formulas for translations and rotations in Figs. 6-4a and 6-4b, respectively. These figures represent the effects of a unit translation in the y_M direction at the j end. Similarly, formulas may be deduced for translations and rotations due to the three other types of unit displacements of the ends of the member. These formulas may be considered to be composed of two parts, as indicated in Eq. (6-21). The first part (matrix $\mathbf{T}_{DB}$) consists of the effects of the displacements of the ends of the member as a rigid body, and the second part involves the displacements of a point of interest relative to the tangents to the elastic curve at the displaced ends. Either end of the member is suitable as a reference point, and, for consistency, points of interest which lie to the left of a load point (see point 1, Fig. 6-5a) are

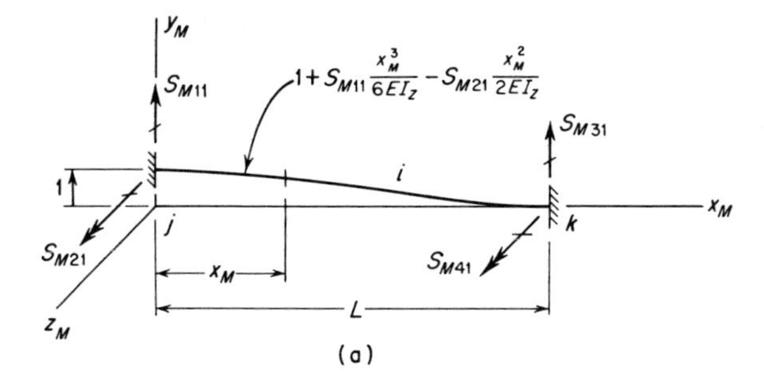

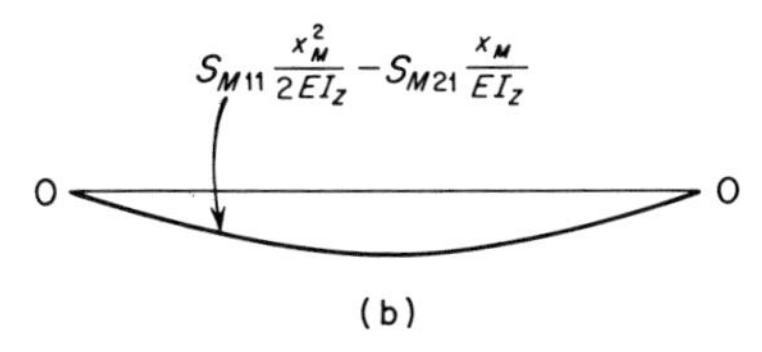

FIG. 6-4. Displacements in a beam due to a unit translation of the j end: (a) translations and (b) rotations.

referred to the j end. On the other hand, points to the right of a load point (see point 2, Fig. 6-5a) are referred to the k end.

Thus, the matrix $\mathbf{T}_{DB}$ is a transfer matrix for displacements at the points of interest due to the rigid body motions of the joints at the ends

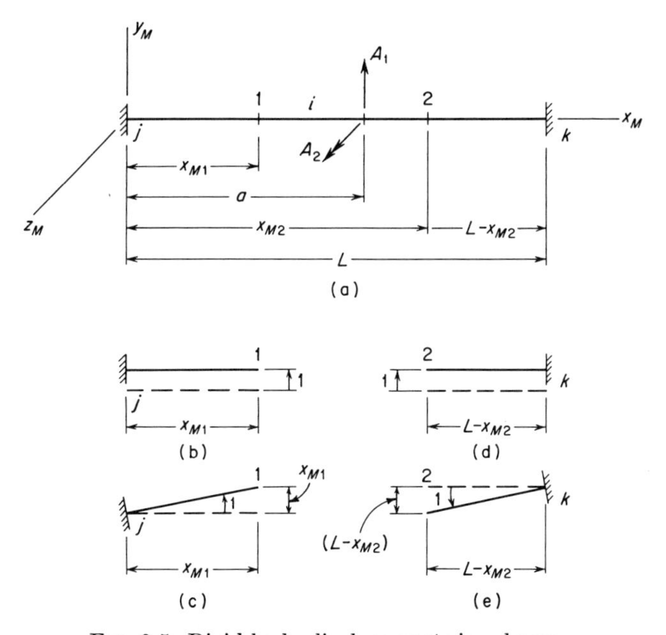

FIG. 6-5. Rigid body displacements in a beam.

of the member. The elements of this matrix may be obtained by inspection of Figs. 6-5b through 6-5e:

$$\mathbf{T}_{DB} = \begin{bmatrix} \mathbf{T}_{DBJ} \\ \mathbf{T}_{DBK} \end{bmatrix} = \left[\begin{array}{cccc} 1 & x_{M1} & 0 & 0 \\ 0 & 1 & 0 & 0 \\ \hline 0 & 0 & 1 & -(L - x_{M2}) \\ 0 & 0 & 0 & 1 \end{array}\right] \tag{6-22}$$

Substitution of Eqs. (6-19) and (6-21) into Eq. (6-18) produces the expanded equation:

$$\mathbf{D}_{MX} = \sum (\mathbf{T}_{DX}\mathbf{T}_{ML}\mathbf{A}) + (\mathbf{T}_{DB} + \mathbf{T}_{DX}\mathbf{S}_{M})\mathbf{D}_{M} \tag{6-23}$$

For the special case of only one point of load, the summation of terms is not required, and the matrix $\mathbf{T}_{DX}$ can be factored as follows:

$$\begin{aligned} \mathbf{D}_{MX} &= \mathbf{T}_{DB}\mathbf{D}_{M} + \mathbf{T}_{DX}(\mathbf{T}_{ML}\mathbf{A} + \mathbf{S}_{M}\mathbf{D}_{M}) \\ &= \mathbf{T}_{DB}\mathbf{D}_{M} + \mathbf{T}_{DX}\mathbf{A}_{M} \end{aligned} \tag{6-24}$$

A simple example will be used now to demonstrate the calculation of actions and displacements at a pair of intermediate points on a beam. The above discussion has been oriented toward calculating actions and displacements at two points simultaneously in order to make use of the full matrices $\mathbf{T}_{ML}$ and $\mathbf{S}_{M}$. However, if only one point is of interest, the appropriate halves of these and other required matrices may be used in the calculations.

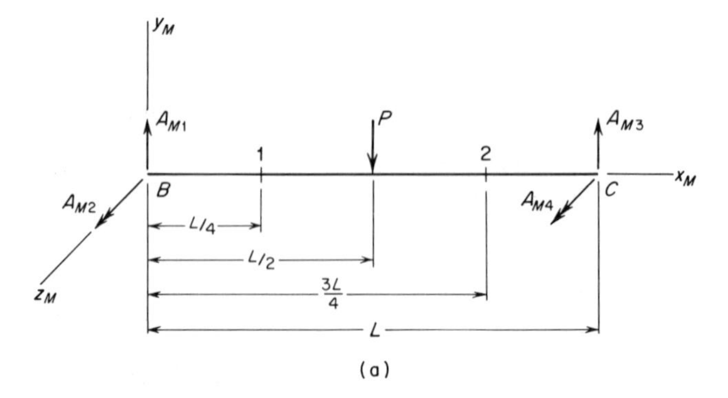

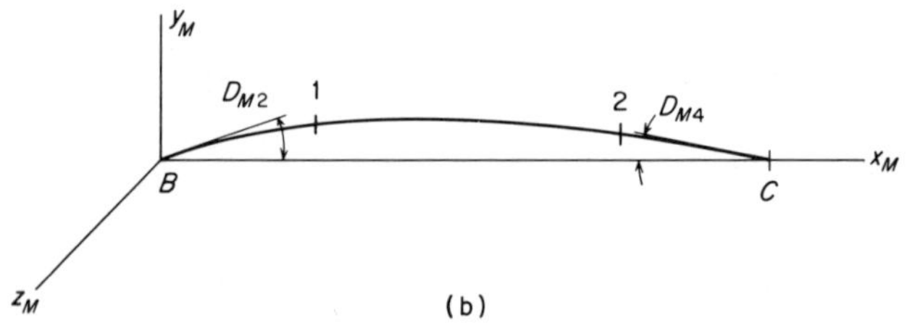

FIG. 6-6. Example.

Example. Assume that the internal actions and displacements are to be calculated at two points of member BC in the continuous beam analyzed in Art. 4.6 and illustrated in Fig. 4-5. From the solution of this structure by the stiffness method the following vectors are available for member BC:

$$\mathbf{A}_\mathrm{M} = \frac{P}{56}\{64, 36L, -8, 0\} \tag{a}$$

$$\mathbf{D}_\mathrm{M} = \frac{PL^2}{112EI_Z}\{0, 17, 0, -5\} \tag{b}$$

The numbering system for the elements of the vector $\mathbf{A}_\mathrm{M}$ is shown in Fig. 6-6a, which also indicates the load P acting on the member and the locations of the two points of interest. These points are labeled 1 and 2, and the following dimensions are assumed:

$$x_{M1} = \frac{L}{4} \quad \text{and} \quad x_{M2} = \frac{3L}{4}$$

Substitution of these values into Eqs. (6-11), (6-20), and (6-22) yields the following transfer matrices:

$$\mathbf{T}_\mathrm{AX} = \left[\begin{array}{cc|cc} -1 & 0 & 0 & 0 \\ \frac{L}{4} & -1 & 0 & 0 \\ \hline 0 & 0 & 1 & 0 \\ 0 & 0 & \frac{L}{4} & 1 \end{array}\right] \tag{c}$$

$$\mathbf{T}_\mathrm{DX} = \frac{1}{384EI_Z}\left[\begin{array}{cccc} L^2 & -12L & 0 & 0 \\ 12L & -96 & 0 & 0 \\ \hline 0 & 0 & L^2 & 12L \\ 0 & 0 & -12L & -96 \end{array}\right] \tag{d}$$

$$\mathbf{T}_\mathrm{DB} = \left[\begin{array}{cccc} 1 & \frac{L}{4} & 0 & 0 \\ 0 & 1 & 0 & 0 \\ \hline 0 & 0 & 1 & -\frac{L}{4} \\ 0 & 0 & 0 & 1 \end{array}\right] \tag{e}$$

When Eqs. (a) and (c) are substituted into Eq. (6-17), the following action vector is obtained:

$$\mathbf{A}_\mathrm{MX} = \begin{bmatrix} \mathbf{A}_\mathrm{MX1} \\ \mathbf{A}_\mathrm{MX2} \end{bmatrix} = \frac{P}{56}\left[\begin{array}{c} -64 \\ -20L \\ \hline -8 \\ -2L \end{array}\right]$$

Figure 6-6b shows approximately the deformed shape of member BC, and the displacements at points 1 and 2 are calculated by substituting Eqs. (a), (b), (d), and (e) into Eq. (6-24). Thus,

$$\mathbf{D}_{\mathrm{MX}} = \begin{bmatrix} \mathbf{D}_{\mathrm{MX1}} \\ \mathbf{D}_{\mathrm{MX2}} \end{bmatrix} = \frac{PL^2}{2688EI_Z} \begin{bmatrix} 56L \\ 72 \\ \hline 29L \\ -108 \end{bmatrix}$$

The methods described above for determining actions and displacements in beams may be extended to all types of framed structures. However, the calculation of displacements at intermediate points in trusses is somewhat more complicated than for structures with rigid joints, because the ends of all members in trusses are free to rotate independently of one another. In addition, the flexural properties of truss members are usually not specified and are not of general interest in the analysis. Therefore, only the intermediate actions in truss members are apt to receive the attention of the analyst.

6.5 Support Displacements. Support displacements consist of known translations or rotations of support restraints. Two alternative methods will be described for including the effects of support displacements in the stiffness method of analysis.

The first approach requires the calculation of the actions at the ends of members in the restrained structure due to the displacements of the supports. (This technique was used in Chapter 2 for support displacements and in Art. 6.2 for temperature and prestrain effects.) These fixed-end actions are then placed in the matrix $\mathbf{A}_{\mathrm{MR}}$ (or $\mathbf{A}_{\mathrm{MS}}$; see Eq. 2-30) and treated in the same manner as those due to loads. The conversion of these quantities to equivalent joint loads and the subsequent analysis then proceed as described in Chapter 4.

The second approach involves the generation of the over-all joint stiffness matrix $\mathbf{S}_{\mathrm{J}}$ given in Eq. (4-1) and repeated below:

$$\mathbf{S}_{\mathrm{J}} = \begin{bmatrix} \mathbf{S} & \mathbf{S}_{\mathrm{DR}} \\ \mathbf{S}_{\mathrm{RD}} & \mathbf{S}_{\mathrm{RR}} \end{bmatrix} \tag{4-1 repeated}$$

This matrix may be incorporated into an over-all action equation for all joints in a structure, including the supports; thus,

$$\mathbf{A}_{\mathrm{J}} = \mathbf{A}_{\mathrm{JL}} + \mathbf{S}_{\mathrm{J}}\mathbf{D}_{\mathrm{J}} \tag{6-25}$$

or, in expanded form:

$$\begin{bmatrix} \mathbf{A}_{\mathrm{D}} \\ \mathbf{A}_{\mathrm{R}} \end{bmatrix} = \begin{bmatrix} \mathbf{A}_{\mathrm{DL}} \\ \mathbf{A}_{\mathrm{RL}} \end{bmatrix} + \begin{bmatrix} \mathbf{S} & \mathbf{S}_{\mathrm{DR}} \\ \mathbf{S}_{\mathrm{RD}} & \mathbf{S}_{\mathrm{RR}} \end{bmatrix} \begin{bmatrix} \mathbf{D} \\ \mathbf{D}_{\mathrm{R}} \end{bmatrix} \tag{6-26}$$

All of the submatrices in Eq. (6-26) have been defined before except the matrix $\mathbf{D}_{\mathrm{R}}$, which is a vector of displacements of the support restraints.

Equation (6-26) may be written as two sets of equations by performing the indicated matrix multiplication:

$$\mathbf{A}_{\mathrm{D}} = \mathbf{A}_{\mathrm{DL}} + \mathbf{SD} + \mathbf{S}_{\mathrm{DR}}\mathbf{D}_{\mathrm{R}} \tag{6-27}$$

$$\mathbf{A}_{\mathrm{R}} = \mathbf{A}_{\mathrm{RL}} + \mathbf{S}_{\mathrm{RD}}\mathbf{D} + \mathbf{S}_{\mathrm{RR}}\mathbf{D}_{\mathrm{R}} \tag{6-28}$$

The first of these equations represents the action equation corresponding to the actual degrees of freedom in the structure, and the second relates to the support restraints. If the support displacements $\mathbf{D}_{\mathrm{R}}$ are zero, Eqs. (6-27) and (6-28) will be the same as Eqs. (2-23) and (4-4), respectively.

Equation (6-27) may be solved for the displacement vector $\mathbf{D}$ as follows:

$$\mathbf{D} = \mathbf{S}^{-1}(\mathbf{A}_{\mathrm{D}} - \mathbf{A}_{\mathrm{DL}} - \mathbf{S}_{\mathrm{DR}}\mathbf{D}_{\mathrm{R}}) \tag{6-29}$$

Then the reactions $\mathbf{A}_{\mathrm{R}}$ can be found by substituting the results of Eq. (6-29) into Eq. (6-28). The calculation of member end-actions is the same as described in Chapter 4.

Both of the methods described above for calculating the effects of support displacements may be readily incorporated into a computer program.

6.6 Inclined Supports. The restraint list code used in the computer programs of Chapter 5 will account for the presence or absence of restraints in directions parallel to the structure axes x, y, and z. Thus, for example, a roller support parallel to one of the structure axes can be taken into account. The code, however, will not accommodate roller supports in directions inclined to the structure axes, such as those indicated in Fig. 6-7a for a plane truss and in Fig. 6-7c for a plane frame. There are several methods by which such support conditions may be included in the analysis of structures by the stiffness method. One possible approach is to rotate the structure axes so that the reference planes are either parallel or perpendicular to the inclined planes. This technique is limited to those cases in which there is only one inclined support or in which all inclines are mutually orthogonal. In those cases where this method is applicable, however, it is likely that the joint coordinates and the orientations of members and loads with respect to the structure axes will become more complicated.

Another method of handling the problem is to replace the actual support by a member having a large cross-sectional area A_X and having its longitudinal axis in the direction normal to the inclined support. Such a substitution is indicated in Fig. 6-7b for the support shown in Fig. 6-7a. A similar substitution is also indicated in Fig. 6-7d for the frame support of Fig. 6-7c. In this latter case the cross-sectional area A_X must be made large and the moment of inertia I_Z must be set equal to zero. Since the cross-sectional areas of the substitute members are large compared to those of the other members of the structure, their axial changes of length will be negligible in the over-all analysis. Therefore, such members will create essentially the same effects as the rollers on the inclined planes. Also, the axial forces in the substitute members will be approximately the same as the reactions of the roller supports. The length of a substitute member should be at least the same order of magnitude as the other members in the structure in order to insure that the angle of rotation of

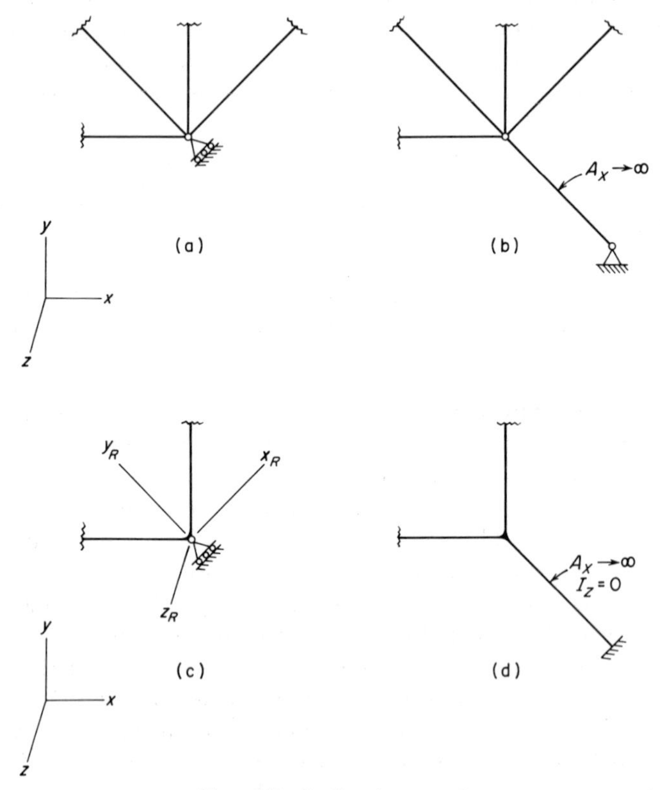

FIG. 6-7. Inclined supports.

the member is small. The advantage of this method of handling inclined supports lies in the fact that the addition of one or two extra members and joints does not require any changes in the programs of Chapter 5.

A third approach to the problem of inclined supports requires a modification of the method of writing action equations in the stiffness method. These equations represent summations of forces and couples in the directions of the structure axes. However, at an inclined support the suitable directions are parallel and perpendicular to the inclined plane. Therefore, the over-all formulation of the action equations for the structure should include some equations which are written for *restraint-oriented axes* instead of structure-oriented axes. A set of restraint-oriented axes x_R, y_R, and z_R is shown in Fig. 6-7c in conjunction with the inclined support. Actions and displacements at an inclined support may be transformed from structure axes to restraint axes (or vice versa), using the type of rotation matrix **R** described in Chapter 4. In this case the matrix **R** will consist of the direction cosines of the restraint axes with respect to the structure axes. Moreover, the over-all joint stiffness matrix $\mathbf{S}_J$ may be generated with respect to structures axes as before, and then the rows and columns associated with the inclined supports may be altered by operations on the matrix using an appropriate rotation transformation matrix $\mathbf{R}_R$. The matrix $\mathbf{R}_R$ contains as submatrices on the principal diagonal either the identity matrix or the rotation matrix **R**, the latter appearing in those positions that correspond to the inclined supports. The required operations may be represented symbolically by premultiplying the over-all action equation (Eq. 6-25 before rearrangement) by the matrix $\mathbf{R}_R$ and inserting $\mathbf{I} = \mathbf{R}_R^{-1}\mathbf{R}_R = \mathbf{R}_R'\mathbf{R}_R$ in the last term as follows:

$$\mathbf{R}_R\mathbf{A}_J = \mathbf{R}_R\mathbf{A}_{JL} + \mathbf{R}_R\mathbf{S}_J\mathbf{R}_R'\mathbf{R}_R\mathbf{D}_J \tag{6-30}$$

Equation (6-30) represents in general a change of coordinates, and quantities in the new coordinate system may be identified with an asterisk and defined as follows:

$$\left.\begin{aligned}\mathbf{R}_R\mathbf{A}_J &= \mathbf{A}_J^* &\quad (a)\\ \mathbf{R}_R\mathbf{A}_{JL} &= \mathbf{A}_{JL}^* &\quad (b)\\ \mathbf{R}_R\mathbf{S}_J\mathbf{R}_R' &= \mathbf{S}_J^* &\quad (c)\\ \mathbf{R}_R\mathbf{D}_J &= \mathbf{D}_J^* &\quad (d)\end{aligned}\right\} \tag{6-31}$$

When Eq. (6-30) is rewritten in these terms, it becomes:

$$\mathbf{A}_J^* = \mathbf{A}_{JL}^* + \mathbf{S}_J^*\mathbf{D}_J^* \tag{6-32}$$

which has the same form as Eq. (6-25). The solution for displacements and reactions in the new coordinate system may now be carried out as before, but the computation of member end-actions requires that $\mathbf{D}_J^*$ be transformed back to the original coordinates, using Eq. (6-31d). Thus,

$$\mathbf{D}_J = \mathbf{R}_R^{-1}\mathbf{D}_J^* = \mathbf{R}_R'\mathbf{D}_J^* \tag{6-33}$$

Among the three methods described above for handling inclined supports, the approach using substitute members is the most convenient because it requires no changes in the programs of Chapter 5. Although the method involving rotation of axes is mathematically more elegant, its application involves a substantial increase in the amount of programming effort.

6.7 Elastic Supports. Cases of support restraint conditions may exist which are intermediate between the extremes of zero restraint and full restraint. If such restraints against either translations or rotations are linearly elastic, they may be included easily within the scope of the stiffness method of analysis.

In order to illustrate conditions of elastic support, consider the continuous beam shown in Fig. 6-8. A comparison of this figure with Fig. 4-7a in Art. 4.8 shows that full restraints and zero restraints at every joint have been replaced by elastic springs having stiffness constants denoted by the symbol S_R. The odd-numbered stiffnesses are restraints against translations in the y direction, and the even-numbered stiffnesses are restraints against rotations in the z sense. If two springs exist at every joint, and if there are m members in the beam, there will be $2(m + 1)$ elastic restraints, and the analysis of the structure by the stiffness method will involve that number of degrees of freedom. The restrained structure for the beam in Fig. 6-8 will be the same as before

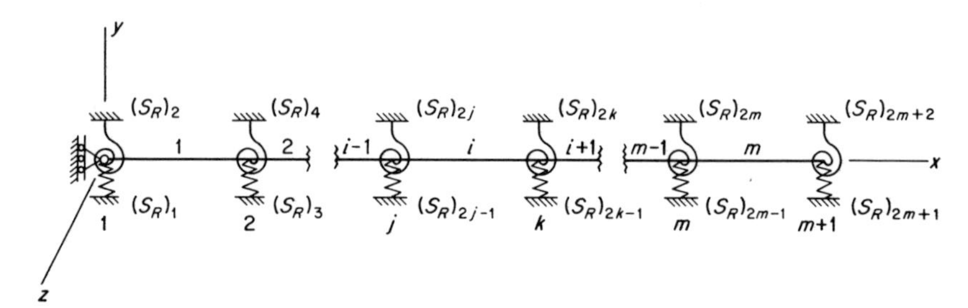

FIG. 6-8. Beam with elastic supports.

(see Fig. 4-7b), but the joint stiffness matrix will be altered because of the restraint stiffnesses S_R. If these restraint stiffnesses are added to the diagonal elements of the over-all joint stiffness matrix $\mathbf{S}_J$, the resulting matrix can then be used in the solution for joint displacements as in Art. 4.8. Member end-actions may also be computed as described in that article. On the other hand, the reactions at the elastic supports will be equal to the actions in the elastic springs and may be calculated as the negatives of the products of the spring constants times the joint displacements, thus:

$$A_R = -S_R D_J \tag{6-34}$$

Although a beam serves as an example in the preceding discussion, this method of handling elastic supports is applicable to all types of framed structures.

An alternative approach to the problem of elastic supports is to substitute equivalent structural members for the restraints. As an example of this technique, consider the problem of handling the joint j of the plane truss shown in Fig. 6-9a. The joint is restrained by two translational springs having stiffnesses $(S_R)_{2j-1}$ and $(S_R)_{2j}$. The two springs can be replaced by the two additional truss members 1 and 2 shown in

Fig. 6-9b. These members can be assigned arbitrary lengths L_1 and L_2, comparable to the lengths of the other members in the truss, and cross-sectional areas A_{X1} and A_{X2} computed as follows:

$$A_{X1} = \frac{L_1}{E}(S_R)_{2j-1} \qquad A_{X2} = \frac{L_2}{E}(S_R)_{2j}$$

The assignment of these cross-sectional areas gives the substitute members axial stiffnesses equal to those of the elastic restraints which they replace.

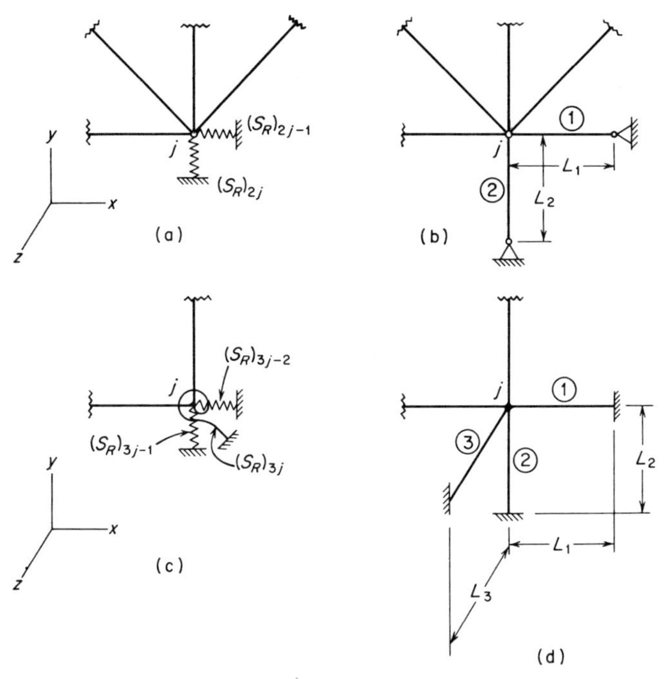

FIG. 6-9. Substitution of members.

A similar example is illustrated in Figs. 6-9c and 6-9d, which show a joint j in a plane frame. The substitution of members 1 and 2 for the translational restraints (see Fig. 6-9d) is similar to that for the plane truss example, except that the moments of inertia of these members must be set equal to zero. A suitable substitution for the rotational restraint of stiffness $(S_R)_{3j}$ consists of a torsion bar, denoted as member 3 in Fig. 6-9d, parallel to the z axis (and hence perpendicular to the plane of the frame). This member would be assigned an arbitrary length L_3 and a torsion constant I_{X3} computed as follows:

$$I_{X3} = \frac{L_3}{G}(S_R)_{3j}$$

The difficulty which arises in this example is that the plane frame is converted into a space frame by the addition of member 3, and the analysis of the structure must proceed on that basis. Therefore, there may be some cases in which the replacement method is disadvantageous compared to the first method discussed above, which involves augmenting the joint stiffness matrix directly.

6.8 Nonprismatic Members. Nonprismatic members are members that do not have the same cross-section from one end to the other. Examples are tapered members and prismatic members having reinforcements over parts of their lengths. In addition, members that do not have straight axes are automatically classified as nonprismatic. The incorporation of nonprismatic members into the stiffness method of analysis requires a recognition of the fact that the member stiffnesses and the fixed-end actions are not the same as for prismatic members. Thus, if the two matrices $\mathbf{S}_{\mathrm{M}}$ and $\mathbf{T}_{\mathrm{ML}}$ are replaced by their counterparts for each nonprismatic member, the stiffness method of analysis will otherwise be unaltered. This same general conclusion, namely, that the only alteration in the stiffness method of analysis is in the determination of member stiffnesses and fixed-end actions, applies also to the topics discussed in the remaining articles of this chapter.*

There are several suitable methods for analyzing nonprismatic members. Sometimes it is possible to obtain the elements of $\mathbf{S}_{\mathrm{M}}$ and $\mathbf{T}_{\mathrm{ML}}$ by direct analytical derivations, as is the case when the variations in section properties may be expressed as suitable functions of x_M. In that event, each element of $\mathbf{S}_{\mathrm{M}}$ and $\mathbf{T}_{\mathrm{ML}}$ is expressible by a formula involving the parameters of the member, just as in the case of a prismatic beam. Otherwise, the elements of $\mathbf{S}_{\mathrm{M}}$ and $\mathbf{T}_{\mathrm{ML}}$ always can be determined in numerical form, either by obtaining the values from previously prepared charts or tables, or by obtaining the values through an appropriate numerical integration procedure.

A feasible approach for analyzing a structure with nonprismatic members consists of assuming joints at intermediate points along the length of each nonprismatic member. A segment between two such joints may be approximated as a prismatic member (if it is not already prismatic) having section properties obtained by averaging those at the two ends of the segment. This approach will cause the stiffness matrix $\mathbf{S}$ for the structure to become very large, but it does not require any changes in the computer programs of Chapter 5.

Another approach involves the determination of the properties of each nonprismatic member by matrix methods. This procedure will be illus-

* Many formulas and derivations for stiffnesses and fixed-end actions can be found in *Moment Distribution*, by J. M. Gere, D. Van Nostrand Company, Inc., Princeton, N.J., 1963. Included are formulas that cover all of the subjects discussed in Arts. 6.8 through 6.12 of this chapter.

trated using a beam as an example. Figure 6-10a shows a restrained nonprismatic beam consisting of two prismatic segments labeled as submembers 1 and 2. This discussion applies for any number of segments, but only two are taken for simplicity. Consider the beam as an isolated structure having two degrees of freedom at point 2 and a total

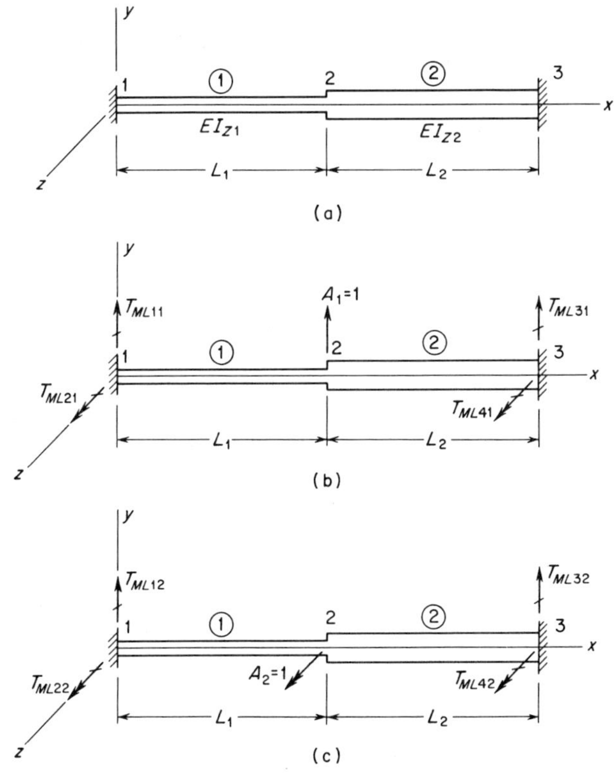

Fig. 6-10. Nonprismatic beam.

of four support restraints at points 1 and 3. Let quantities that pertain to the degrees of freedom in the problem be distinguished by the subscript A and those pertaining to the support restraints by the subscript B. The over-all action-displacement relationships for this structure (see Eq. 6-25) can be written in rearranged and partitioned form as follows:

$$\begin{bmatrix} \mathbf{A}_A \\ \mathbf{A}_B \end{bmatrix} = \begin{bmatrix} \mathbf{S}_{AA} & \mathbf{S}_{AB} \\ \mathbf{S}_{BA} & \mathbf{S}_{BB} \end{bmatrix} \begin{bmatrix} \mathbf{D}_A \\ \mathbf{D}_B \end{bmatrix} \tag{6-35}$$

Performing the indicated multiplication gives

$$\mathbf{A}_A = \mathbf{S}_{AA}\mathbf{D}_A + \mathbf{S}_{AB}\mathbf{D}_B \tag{6-36}$$

$$\mathbf{A}_B = \mathbf{S}_{BA}\mathbf{D}_A + \mathbf{S}_{BB}\mathbf{D}_B \tag{6-37}$$

Solving for $\mathbf{D}_{\mathrm{A}}$ in the first equation results in:

$$\mathbf{D}_{\mathrm{A}} = \mathbf{S}_{\mathrm{AA}}^{-1}(\mathbf{A}_{\mathrm{A}} - \mathbf{S}_{\mathrm{AB}}\mathbf{D}_{\mathrm{B}}) \tag{6-38}$$

Then substituting this expression for $\mathbf{D}_{\mathrm{A}}$ into the second equation and rearranging gives:

$$\mathbf{A}_{\mathrm{B}} - \mathbf{S}_{\mathrm{BA}}\mathbf{S}_{\mathrm{AA}}^{-1}\mathbf{A}_{\mathrm{A}} = (\mathbf{S}_{\mathrm{BB}} - \mathbf{S}_{\mathrm{BA}}\mathbf{S}_{\mathrm{AA}}^{-1}\mathbf{S}_{\mathrm{AB}})\mathbf{D}_{\mathrm{B}} \tag{6-39}$$

The left-hand side of Eq. (6-39) contains the actions of type B minus a term involving the effect of the actions of type A. If the displacements D_B are all equal to zero (fixed-end member), Eq. (6-39) gives the following relation between the actions of types B and A:

$$\mathbf{A}_{\mathrm{B}} = \mathbf{S}_{\mathrm{BA}}\mathbf{S}_{\mathrm{AA}}^{-1}\mathbf{A}_{\mathrm{A}}$$

Therefore, the transfer matrix for actions of type B due to unit actions of type A is

$$\mathbf{T}_{\mathrm{ML}} = \mathbf{S}_{\mathrm{BA}}\mathbf{S}_{\mathrm{AA}}^{-1} \tag{6-40}$$

The elements of this transfer matrix are illustrated in Figs. 6-10b and 6-10c, which show the effects of unit actions at point 2. Note that the transfer matrix $\mathbf{T}_{\mathrm{ML}}$ given by Eq. (6-40) for a point where the section changes is slightly different from the one discussed previously in Art. 6.3, which was a transfer matrix for end-actions due to unit actions at any point along the length of a prismatic member.

If all actions of type A are set equal to zero, Eq. (6-39) becomes

$$\mathbf{A}_{\mathrm{B}} = (\mathbf{S}_{\mathrm{BB}} - \mathbf{S}_{\mathrm{BA}}\mathbf{S}_{\mathrm{AA}}^{-1}\mathbf{S}_{\mathrm{AB}})\mathbf{D}_{\mathrm{B}}$$

Since the actions $\mathbf{A}_{\mathrm{B}}$ and displacements $\mathbf{D}_{\mathrm{B}}$ are corresponding, the term in the parentheses becomes the stiffness matrix for unit displacements of the ends of the nonprismatic member; thus:

$$\mathbf{S}_{\mathrm{M}} = \mathbf{S}_{\mathrm{BB}} - \mathbf{S}_{\mathrm{BA}}\mathbf{S}_{\mathrm{AA}}^{-1}\mathbf{S}_{\mathrm{AB}} \tag{6-41}$$

Substitution of Eq. (6-40) into the last term of Eq. (6-41) gives:

$$\mathbf{S}_{\mathrm{M}} = \mathbf{S}_{\mathrm{BB}} - \mathbf{T}_{\mathrm{ML}}\mathbf{S}_{\mathrm{AB}} \tag{6-42}$$

In summary, the member stiffness matrix $\mathbf{S}_{\mathrm{M}}$ and the transfer matrix $\mathbf{T}_{\mathrm{ML}}$ for a nonprismatic member may be determined by subdividing the member into a number of segments, each of which is treated as an individual member. The over-all joint stiffness matrix is obtained and partitioned as shown in Eq. (6-35) to give the matrices $\mathbf{S}_{\mathrm{AA}}$, $\mathbf{S}_{\mathrm{AB}}$, $\mathbf{S}_{\mathrm{BA}}$, and $\mathbf{S}_{\mathrm{BB}}$. Inversion of the matrix $\mathbf{S}_{\mathrm{AA}}$ and substitution of the appropriate matrices into Eqs. (6-40) and (6-42) produce the transfer matrix $\mathbf{T}_{\mathrm{ML}}$ and the stiffness matrix $\mathbf{S}_{\mathrm{M}}$ for the nonprismatic member. The use of Eqs. (6-40) and (6-42) will be demonstrated by a brief example.

Example: Assume that the nonprismatic beam in Fig. 6-10 has the following properties:

$$L_1 = L_2 = L \qquad I_{Z1} = I \qquad I_{Z2} = 2I$$

Then the rearranged and partitioned over-all stiffness matrix for the structure is:

$$\begin{bmatrix} \mathbf{S}_{AA} & \mathbf{S}_{AB} \\ \mathbf{S}_{BA} & \mathbf{S}_{BB} \end{bmatrix} = \frac{EI}{L^3}\left[\begin{array}{cc|cccc} 36 & 6L & -12 & -6L & -24 & 12L \\ 6L & 12L^2 & 6L & 2L^2 & -12L & 4L^2 \\ \hline -12 & 6L & 12 & 6L & 0 & 0 \\ -6L & 2L^2 & 6L & 4L^2 & 0 & 0 \\ -24 & -12L & 0 & 0 & 24 & -12L \\ 12L & 4L^2 & 0 & 0 & -12L & 8L^2 \end{array}\right]$$

Substitution of the appropriate submatrices from this matrix into Eqs. (6-40) and (6-42) produces the transfer matrix $\mathbf{T}_{ML}$ and the member stiffness matrix $\mathbf{S}_M$ for the nonprismatic member, as follows:

$$\mathbf{T}_{ML} = \frac{1}{33L^2}\begin{bmatrix} -15L^2 & 24L \\ -7L^3 & 9L^2 \\ -18L^2 & -24L \\ 10L^3 & 6L^2 \end{bmatrix}$$

$$\mathbf{S}_M = \frac{4EI}{11L^3}\begin{bmatrix} 6 & 5L & -6 & 7L \\ 5L & 6L^2 & -5L & 4L^2 \\ -6 & -5L & 6 & -7L \\ 7L & 4L^2 & -7L & 10L^2 \end{bmatrix}$$

The method described above for determining the matrices $\mathbf{T}_{ML}$ and $\mathbf{S}_M$ for nonprismatic members was explained using a beam of two segments as an example, but the method is quite general.* Equations (6-40) and (6-42) may be applied to any type of structural member having either a straight or a curved axis and divided into any number of segments.

6.9 Discontinuities in Members. Figure 6-11 shows the types of partial discontinuities which can exist in the members of framed structures. The symbols in Figs. 6-11a, 6-11b, 6-11c, and 6-11d denote the inability to transmit shear, moment, thrust, and torque, respectively. These inabilities to transmit actions result in translational or rotational displacement discontinuities. Combinations of these discontinuities are

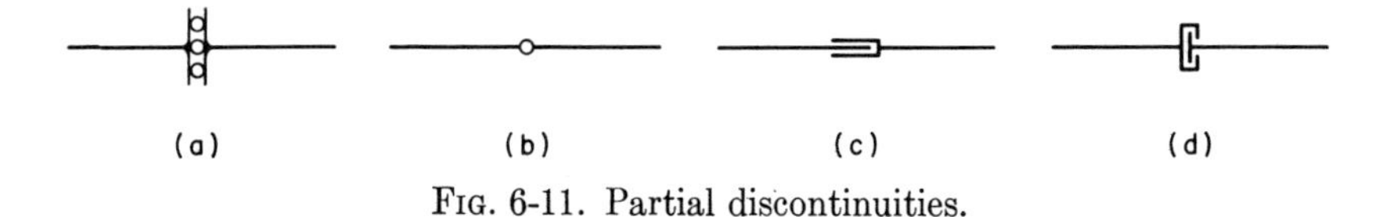

FIG. 6-11. Partial discontinuities.

possible also, and complete discontinuity takes the form of a free end. The general nature of a framed structure determines the types of discontinuities which are important. In beams, for example, only shear and

* The process of eliminating one or more matrices by substituting one matrix equation into another, as was done in deriving Eq. (6-39), is sometimes referred to as *matrix condensation*. Thus, the matrices $\mathbf{S}_M$ and $\mathbf{T}_{ML}$ for the nonprismatic member represent combinations of other matrices obtained by a condensation procedure.

moment discontinuities have significance. The member stiffness matrices and end-action transfer matrices for beams having such discontinuities are presented in this article, and the concepts may be extended to the members of other types of structures as well.

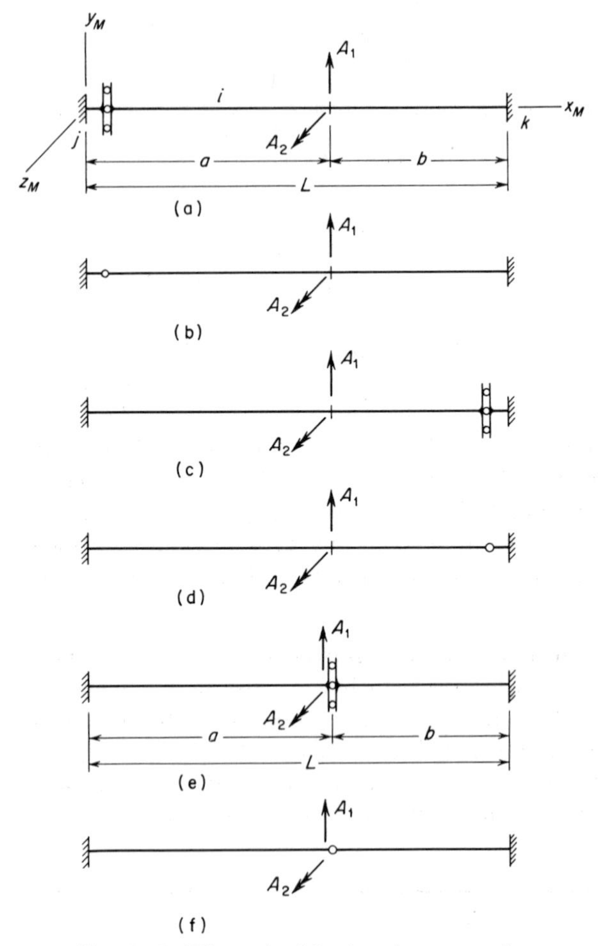

FIG. 6-12. Discontinuities in a beam member.

Consider first the possibility of a shear or moment discontinuity at one end of a prismatic beam. Figures 6-12a and 6-12b show such conditions at a small distance from the j end of a restrained member, and Figs. 6-12c and 6-12d indicate the same conditions at the k end. The member stiffness matrices $\mathbf{S_M}$ for these cases are easily obtained by analysis of the beams shown in the figure, and they are listed in Table 6-1 as matrices (a) through (d). Figures 6-12a through 6-12d also show actions A_1 and A_2 at the distance a from the j end of the member, and the transfer matrices $\mathbf{T}_{ML}$ for the resulting fixed-end actions are given in Table 6-2 as matrices (a) through (d). If the matrices $\mathbf{S}_M$ and $\mathbf{T}_{ML}$ shown in Tables 6-1 and 6-2 are used for a beam member having a shear or moment discontinuity at either end, the analysis may then proceed as before, with no further modifications required.

TABLE 6-1

Member Stiffness Matrices for Beams of Fig. 6-12

$$\frac{EI_Z}{L}\begin{bmatrix} 0 & 0 & 0 & 0 \\ 0 & 1 & 0 & -1 \\ 0 & 0 & 0 & 0 \\ 0 & -1 & 0 & 1 \end{bmatrix} \qquad \frac{3EI_Z}{L^3}\begin{bmatrix} 1 & 0 & -1 & L \\ 0 & 0 & 0 & 0 \\ -1 & 0 & 1 & -L \\ L & 0 & -L & L^2 \end{bmatrix}$$

(a) (b)

$$\frac{EI_Z}{L}\begin{bmatrix} 0 & 0 & 0 & 0 \\ 0 & 1 & 0 & -1 \\ 0 & 0 & 0 & 0 \\ 0 & -1 & 0 & 1 \end{bmatrix} \qquad \frac{3EI_Z}{L^3}\begin{bmatrix} 1 & L & -1 & 0 \\ L & L^2 & -L & 0 \\ -1 & -L & 1 & 0 \\ 0 & 0 & 0 & 0 \end{bmatrix}$$

(c) (d)

$$\frac{EI_Z}{L}\begin{bmatrix} 0 & 0 & 0 & 0 \\ 0 & 1 & 0 & -1 \\ 0 & 0 & 0 & 0 \\ 0 & -1 & 0 & 1 \end{bmatrix} \qquad \frac{3EI_Z}{a^3 + b^3}\begin{bmatrix} 1 & a & -1 & b \\ a & a^2 & -a & ab \\ -1 & -a & 1 & -b \\ b & ab & -b & b^2 \end{bmatrix}$$

(e) (f)

Consider next the determination of the stiffness matrix $\mathbf{S}_M$ for a beam having a shear or moment discontinuity at an intermediate point. Figures 6-12e and 6-12f show such beams with the discontinuities at the distance a from the j end of the member. One method of analyzing these beams consists of treating them as two members by assuming a joint just to the left or right of the discontinuity. Then, one of the submembers will have a discontinuity at one end (as in Figs. 6-12a through 6-12d), and the other submember will have no discontinuities. This approach has the advantage that no new matrices are required other than those discussed above. Alternatively, the stiffness matrices for beams having discontinuities at intermediate points always can be obtained directly from an elementary beam analysis. The results of such analyses for the beams in Figs. 6-12e and 6-12f are given in Table 6-1 as matrices (e) and (f).

Transfer matrices $\mathbf{T}_{ML}$ for fixed-end actions for beams with discontinuities at intermediate points may also be found by the procedures mentioned above. If actions A_1 and A_2 are applied just to the left of the discontinuities (see Figs. 6-12e and 6-12f), the transfer matrices have the forms given in Table 6-2 as matrices (e) and (f). If the actions A_1 and A_2 are applied just to the right of the discontinuities, the matrices are as given in Table 6-2 as matrices (g) and (h).

TABLE 6-2

Transfer Matrices for Beams of Fig. 6-12

$$\frac{1}{2L}\begin{bmatrix} 0 & 0 \\ b^2 & -2b \\ -2L & 0 \\ b(L+a) & -2a \end{bmatrix}$$
(a)

$$\frac{1}{2L^3}\begin{bmatrix} -b^2(2L+a) & 3b(L+a) \\ 0 & 0 \\ -a(3L^2-a^2) & -3b(L+a) \\ abL(L+a) & L(L^2-3a^2) \end{bmatrix}$$
(b)

$$\frac{1}{2L}\begin{bmatrix} -2L & 0 \\ -a(L+b) & -2b \\ 0 & 0 \\ -a^2 & -2a \end{bmatrix}$$
(c)

$$\frac{1}{2L^3}\begin{bmatrix} b(3L^2-b^2) & 3a(L+b) \\ -abL(L+b) & L(L^2-3b^2) \\ -a^2(2L+b) & -3a(L+b) \\ 0 & 0 \end{bmatrix}$$
(d)

$$\frac{1}{2L}\begin{bmatrix} -2L & 0 \\ -a(L+b) & -2b \\ 0 & 0 \\ -a^2 & -2a \end{bmatrix}$$
(e)

$$\frac{1}{2(a^3+b^3)}\begin{bmatrix} -2b^3 & 3a^2 \\ -2ab^3 & a^3-2b^3 \\ -2a^3 & -3a^2 \\ 2a^3b & 3a^2b \end{bmatrix}$$
(f)

$$\frac{1}{2L}\begin{bmatrix} 0 & 0 \\ b^2 & -2b \\ -2L & 0 \\ b(L+a) & -2a \end{bmatrix}$$
(g)

$$\frac{1}{2(a^3+b^3)}\begin{bmatrix} -2b^3 & 3b^2 \\ -2ab^3 & 3ab^2 \\ -2a^3 & -3b^2 \\ 2a^3b & -2a^3+b^3 \end{bmatrix}$$
(h)

Note that the transfer matrices (e) through (h) in Table 6-2 are for end-actions at j and k due to unit actions applied adjacent to the discontinuities. If actions are applied at other points along the length, suitable transfer matrices can be developed for cases in which $x_{M1} \leq a$ and $x_{M2} \geq a$. One way of obtaining these transfer matrices is to make use of the matrices given in Table 6-2. For this purpose, a temporary restraint is imposed at a point adjacent to the discontinuity, and matrices (a) through (d) are used to compute the restraint actions at that point. The negatives of these restraint actions constitute equivalent loads adjacent to the discontinuity, and matrices (e) through (h) may be used to transfer the effects to the ends of the member.

Matrices similar to those in Tables 6-1 and 6-2 can be developed for the members of other types of framed structures. Frames and grids may have such discontinuities, but the topic is of lesser importance in trusses, because the ends of all members are assumed to be pinned, and any additional discontinuities are unlikely.

It is interesting to note that certain discontinuities in members may be taken into account by trivial changes in the analysis. For example, if a plane frame member has hinges at both ends, its member stiffness matrix will be the same as that for a space truss member. Such a member may be handled in the computer program for a plane frame (Art. 5.6) simply by setting its cross-sectional moment of inertia I_Z equal to zero. Such "tricks" can often be used to good advantage without resorting to additional programming.

6.10 Elastic Connections. Joints of framed structures are usually idealized to be either pinned or completely rigid. However, the connections themselves may have a significant degree of flexibility which may be important in the analysis. If such connections are assumed to be linearly elastic, they may be incorporated into the stiffness properties of the individual members as modifications of the idealized cases.

Several types of elastic connections are theoretically possible, according to the relative translations and rotations which can occur at the joints of a structure. Connections for shear, bending moment, thrust, and torque may all possess a certain amount of flexibility, but the most important of these is the rotational type which transmits bending moment. This type of connection is discussed in this article in conjunction with the analysis of beams. The ideas may, of course, be extended to other types of elastic connections and other types of framed structures.

Figure 6-13 shows a prismatic beam with a rotational elastic connec-

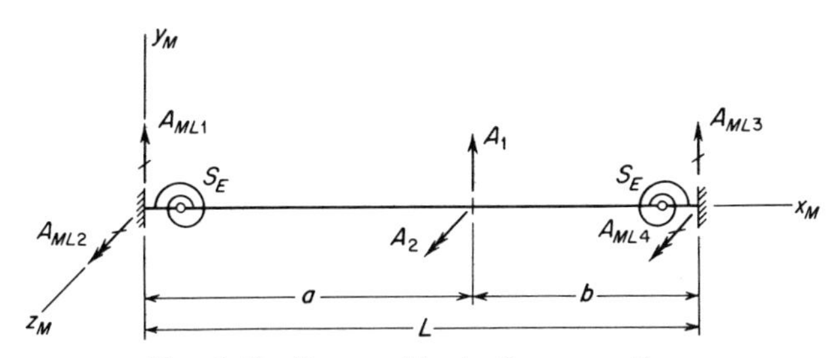

FIG. 6-13. Beam with elastic connections.

tion at a small distance from each end. Let S_E be the stiffness constant for the elastic connection at each end of the member. A constant of this type is defined as the moment per unit relative rotation at the elastic connection. The stiffness constant S_E will be incorporated into the stiffness matrix $\mathbf{S_M}$ and transfer matrix $\mathbf{T_{ML}}$ for the member. For convenience in writing these matrices, the following nondimensional parameters are defined:

$$e = \frac{EI_Z}{LS_E} \tag{6-43}$$

$$e_1 = e + 1 \qquad e_2 = 2e + 1 \qquad e_3 = 3e + 1$$

$$e_4 = 4e + 1 \qquad e_6 = 6e + 1$$

The modified member stiffness matrix $\mathbf{S_M}$ for the beam in Fig. 6-13 may be determined by analyzing the member as a support displacement problem (see Art. 6.5) for a beam on elastic supports (see Art. 6.7). Unit displacements of the supports will produce support reactions which become the elements of $\mathbf{S_M}$. Table 6-3 shows the resulting matrix, and the reader should compare this table with Table 4-2 in Chapter 4. Note that when S_E is allowed to approach infinity (for a rigid connection), the constant e approaches zero, and the matrix in Table 6-3 becomes the same as the one in Table 4-2.

TABLE 6-3

Member Stiffness Matrix for Beam with Elastic Connections

$$\mathbf{S_M} = \frac{EI_Z}{e_2 e_6} \begin{bmatrix} \frac{12}{L^3} e_2 & \frac{6}{L^2} e_2 & -\frac{12}{L^3} e_2 & \frac{6}{L^2} e_2 \\ \frac{6}{L^2} e_2 & \frac{4}{L} e_3 & -\frac{6}{L^2} e_2 & \frac{2}{L} \\ -\frac{12}{L^3} e_2 & -\frac{6}{L^2} e_2 & \frac{12}{L^3} e_2 & -\frac{6}{L^2} e_2 \\ \frac{6}{L^2} e_2 & \frac{2}{L} & -\frac{6}{L^2} e_2 & \frac{4}{L} e_3 \end{bmatrix}$$

Similarly, formulas for the elements of the transfer matrix $\mathbf{T_{ML}}$ may be derived in literal form, and the results are listed in Table 6-4. When e is set equal to zero, the elements of $\mathbf{T_{ML}}$ in this table become the same as those in Eq. (6-4) for a beam with rigid joints (see Art. 6.3).

TABLE 6-4

Transfer Matrix for Beam with Elastic Connections

$$\mathbf{T_{ML}} = \frac{1}{12e^2ab + e_4L^2} \begin{bmatrix} -\frac{b^2}{L}(3ae_2^2 + be_4) & \frac{6ab}{L} e_1e_2 \\ -\frac{ab^2}{L}(ae_2 + be_4) & \frac{b}{L^2}(2a^3e_1 + 3a^2be_2 - b^3e_4) \\ -\frac{a^2}{L}(ae_4 + 3be_2^2) & -\frac{6ab}{L} e_1e_2 \\ \frac{a^2b}{L}(ae_4 + be_2) & \frac{a}{L^2}(-a^3e_4 + 3ab^2e_2 + 2b^3e_1) \end{bmatrix}$$

Matrices similar to those in Tables 6-3 and 6-4 may also be derived for the case of unequal elastic connections at ends j and k of the beam and for the case of an elastic connection at only one end of the beam. In addition, other types of elastic connections and other types of structural members may be considered.

6.11 Shear Deformations. In beams, frames, and grids the deformations of members due to shear forces may be of significant magnitudes. If this is true, the effects of shear deformations should be included in the analyses of such structures by making the appropriate modifications of the matrices $\mathbf{S}_{\mathrm{M}}$ and $\mathbf{T}_{\mathrm{ML}}$. The modified matrices for prismatic beams are presented in this article, and the changes made in the beam matrices may be easily extended to obtain the required changes in the matrices for frames and grids.

All elements of the 4×4 member stiffness matrix $\mathbf{S}_{\mathrm{M}}$ will be modified because of the fact that all unit displacements of the ends of a restrained beam cause shear in the member. Consider, for example, a unit displacement of the j end of the member in the y_M direction as shown in Fig. 6-14. The unit displacement consists of two parts as follows:

$$\frac{S_{M11}L^3}{12EI_Z} + \frac{S_{M11}Lf}{GA_X} = 1 \qquad \text{(a)}$$

The first part is due to the flexural deformations in the member, and the second part is due to shear deformations (see Appendix A). It is convenient to introduce the dimensionless shear constant g for purposes of simplifying Eq. (a):

$$g = \frac{6fEI_Z}{GA_XL^2} \qquad \text{(6-44)}$$

Solving Eq. (a) for S_{M11} in terms of the shear constant g gives:

$$S_{M11} = \frac{12EI_Z}{L^3}\left(\frac{1}{1+2g}\right) \qquad \text{(b)}$$

Expressions for the other restraint actions in Fig. 6-14 are easily obtained from equilibrium principles, and these actions become elements in the first column of the modified member stiffness matrix $\mathbf{S}_{\mathrm{M}}$. The other columns of the stiffness matrix may be obtained in a similar manner, and the results are given in Table 6-5. Note that if the shear constant g is set equal to zero the matrix in Table 6-5 becomes the same as that in Table 4-2.

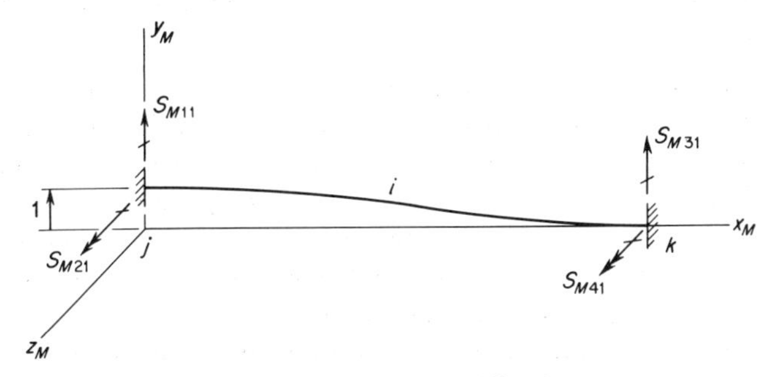

FIG. 6-14. Member stiffnesses.

TABLE 6-5

Member Stiffness Matrix for Beam with Shear Deformations

$$\mathbf{S}_{\mathrm{M}} = \frac{EI_Z}{1+2g}\begin{bmatrix} \frac{12}{L^3} & \frac{6}{L^2} & -\frac{12}{L^3} & \frac{6}{L^2} \\ \frac{6}{L^2} & \frac{4}{L}\left(1+\frac{g}{2}\right) & -\frac{6}{L^2} & \frac{2}{L}(1-g) \\ -\frac{12}{L^3} & -\frac{6}{L^2} & \frac{12}{L^3} & -\frac{6}{L^2} \\ \frac{6}{L^2} & \frac{2}{L}(1-g) & -\frac{6}{L^2} & \frac{4}{L}\left(1+\frac{g}{2}\right) \end{bmatrix}$$

The transfer matrix $\mathbf{T}_{\mathrm{ML}}$ for a beam must also be modified for the effects of shear deformation. This matrix may be developed without difficulty, and Table 6-6 shows the results. If the constant g is set equal to zero, the elements in Table 6-6 become equal to those in Eq. (6-4).

TABLE 6-6

Transfer Matrix for Beam with Shear Deformations

$$\mathbf{T}_{\mathrm{ML}} = \frac{1}{1+2g}\begin{bmatrix} -\frac{b^2}{L^3}[3a+b+2b(L/b)^2g] & \frac{6ab}{L^3} \\ -\frac{ab^2}{L^2}[1+(L/b)g] & \frac{b}{L^2}(2a-b-2gL) \\ -\frac{a^2}{L^3}[a+3b+2a(L/a)^2g] & -\frac{6ab}{L^3} \\ \frac{a^2b}{L^2}[1+(L/a)g] & \frac{a}{L^2}(2b-a-2gL) \end{bmatrix}$$

Thus, the effects of shear deformations in beams may be taken into account in convenient fashion by using the modified matrices $\mathbf{S}_{\mathrm{M}}$ and $\mathbf{T}_{\mathrm{ML}}$ described in this article.

6.12 Axial-Flexural Interaction. Structural members which carry both axial forces and bending moments are subjected to an interaction between these two effects. The lateral deflection of a member causes additional bending moments due to the presence of an axial force. If this axial-flexural interaction is to be taken into account, the member stiffness matrix $\mathbf{S}_{\mathrm{M}}$ and the transfer matrix $\mathbf{T}_{\mathrm{ML}}$ must be altered accordingly. The changes required in these matrices for a beam subjected to both axial and flexural effects are described in this article, and the ideas may also be extended to other types of structures.

Consider first the required changes in the member stiffness matrix $\mathbf{S_M}$ for a beam. Figures 6-15a and 6-15b show a restrained beam which carries an axial compression force P and which is subjected to a unit translation and a unit rotation at the j end (figures showing similar

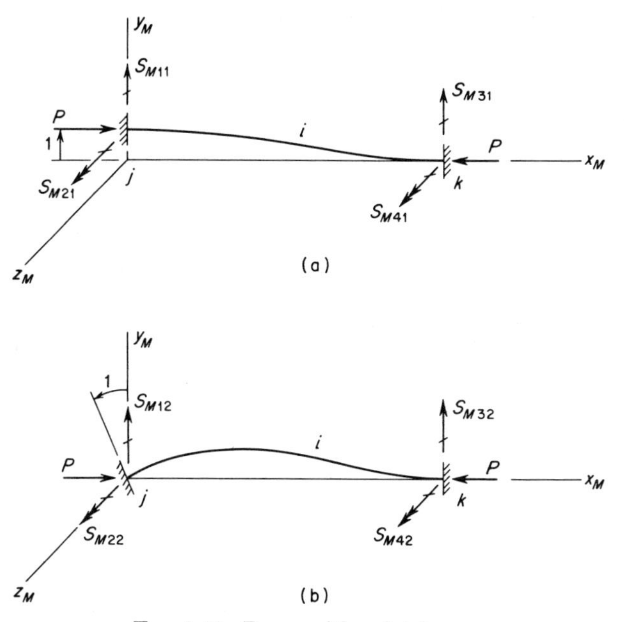

FIG. 6-15. Beam with axial force.

displacements at the k end are omitted). The axial force may be either tension or compression, but compression is usually of greater interest than tension because of the possibility of buckling. Because of the lateral displacements shown in the figures, the bending moments in the beam are modified by the presence of the axial force. The terms in the member stiffness matrix $\mathbf{S_M}$ are indicated in Table 6-7, and are expressed

TABLE 6-7

Member Stiffness Matrix for Beam Subjected to Axial Force

$$\mathbf{S_M} = EI_Z \begin{bmatrix} \frac{12}{L^3}s_1 & \frac{6}{L^2}s_2 & -\frac{12}{L^3}s_1 & \frac{6}{L^2}s_2 \\ \frac{6}{L^2}s_2 & \frac{4}{L}s_3 & -\frac{6}{L^2}s_2 & \frac{2}{L}s_4 \\ -\frac{12}{L^3}s_1 & -\frac{6}{L^2}s_2 & \frac{12}{L^3}s_1 & -\frac{6}{L^2}s_2 \\ \frac{6}{L^2}s_2 & \frac{2}{L}s_4 & -\frac{6}{L^2}s_2 & \frac{4}{L}s_3 \end{bmatrix}$$

in terms of four stiffness functions s_1, s_2, s_3, and s_4. These functions are defined in Table 6-8 for the cases of an axial force which is compression,

TABLE 6-8

Stiffness Functions for Beam Subjected to Axial Force

Function	Condition of Axial Force		
	Compression	Zero	Tension
s_1	$\frac{(kL)^3 \sin kL}{12\phi_C}$	1	$\frac{(kL)^3 \sinh kL}{12\phi_T}$
s_2	$\frac{(kL)^2(1 - \cos kL)}{6\phi_C}$	1	$\frac{(kL)^2(\cosh kL - 1)}{6\phi_T}$
s_3	$\frac{kL(\sin kL - kL \cos kL)}{4\phi_C}$	1	$\frac{kL(kL \cosh kL - \sinh kL)}{4\phi_T}$
s_4	$\frac{kL(kL - \sin kL)}{2\phi_C}$	1	$\frac{kL(\sinh kL - kL)}{2\phi_T}$

$\phi_C = 2 - 2 \cos kL - kL \sin kL \qquad \phi_T = 2 - 2 \cosh kL + kL \sinh kL \qquad k = \sqrt{\frac{P}{EI_Z}}$

tension, or zero. All of these formulas can be derived by elementary beam analysis, considering the presence of the axial force (beam-column analysis).

The transfer matrix $\mathbf{T}_{ML}$ will also be affected by the presence of an axial force, and the elements of this matrix may be obtained also by beam-column analysis.

If axial-flexural interactions are to be taken into account in the analysis of plane or space frames, it is necessary to make other modifications of the stiffness method in addition to those already described. The analysis is complicated by the fact that the axial forces in the members are related to the joint displacements. Therefore, the analysis must be conducted in a cyclic fashion. In the first cycle of analysis the stiffness method is applied as explained in Chapter 4. In the second cycle, the axial forces in the members, as obtained from the first cycle, are used in determining the modified member stiffnesses given in Table 6-7, and also in determining the modified fixed-end actions. The second cycle is then completed, using the modified stiffnesses and fixed-end actions, and new values for the axial forces are obtained. This process is repeated until two successive analyses yield approximately the same results.

The cyclic method of analysis described above may be used to determine the buckling load for a frame. The loads on the frame may be gradually increased until the stiffness matrix **S** becomes singular. This singularity is the criterion for obtaining the magnitude of loading which causes elastic instability in the fundamental buckling mode.

General References on Matrix Methods in Structural Analysis

1. Argyris, J. H., "On the Analysis of Complex Elastic Structures," *Applied Mechanics Reviews*, vol. 11, no. 7, July 1959, pp. 331–338.
2. Argyris, J. H., and Kelsey, S., *Energy Theorems and Structural Analysis*, Butterworths, London, 1960.
3. Gallagher, R. H., *A Correlation Study of Methods of Matrix Structural Analysis*, The Macmillan Company, New York, 1964.
4. Hall, A. S., and Woodhead, R. W., *Frame Analysis*, John Wiley and Sons, Inc., New York, 1961.
5. Matheson, J. A. L., *Hyperstatic Structures*, vol. 1, Academic Press Inc., New York, 1959.
6. McMinn, S. J., *Matrices for Structural Analysis*, John Wiley and Sons, Inc., New York, 1962.
7. Morice, P. B., *Linear Structural Analysis*, The Ronald Press Company, New York, 1959.
8. Pestel, E. C., and Leckie, F. A., *Matrix Methods in Elastomechanics*, McGraw-Hill Book Co., Inc., New York, 1963.
9. Pipes, L. A., *Matrix Methods for Engineering*, Prentice-Hall, Inc., Englewood Cliffs, N.J., 1963.
10. deVeubeke, B. F. (ed.), *Matrix Methods of Structural Analysis*, The Macmillan Company, New York, 1964.

Appendix A

DISPLACEMENTS OF STRUCTURES

A.1 Stresses and Deformations in Bars. Whenever a load is applied to a structure, stresses will be developed within the material, and deformations will occur. Deformation means any change in the shape of some part of the structure, such as a change in shape of an element cut from a member, while stresses refer to the distributed actions that occur internally between adjoining elements of the structure. It is assumed in subsequent analyses that the deformations are very small and that the material is linearly elastic (Hooke's law). Under these conditions the stresses are proportional to the corresponding strains in the material, and the principle of superposition may be used for combining stresses, strains, and deformations due to various load systems.

The principal types of deformations to be considered are axial, flexural, torsional, and shearing deformations. These are caused by stress resultants in the form of axial forces, bending couples, torsional couples, and shearing forces, respectively. In each of these four cases the expressions for the stresses acting on the cross-section, the strains in an element, and the deformation of an element are summarized in this article. In addition, the deformations caused by temperature effects are described.

The calculation of displacements in structures is described in Articles A.2 and A.3. This subject is an important part of the flexibility method of analysis (see Chapters 2 and 3), and is presented in this Appendix for review purposes. Further information on the subject may be found in textbooks on mechanics of materials and elementary theory of structures.

Axial Deformations. The member shown in Fig. A-1a is assumed to be acted upon by a tensile force P at each end. The member will be in pure tension due to these forces, provided each force acts at the centroid of the cross-sectional area. At any distance x from the left end the tension stress σ_x on the cross-section is

$$\sigma_x = \frac{P}{A} \tag{A-1}$$

in which A is the cross-sectional area. The axial strain ϵ_x in the member is equal to the stress divided by the modulus of elasticity E of the material:

$$\epsilon_x = \frac{\sigma_x}{E} = \frac{P}{EA} \tag{A-2}$$

The quantity EA is called the *axial rigidity* of the bar.

The change in length $d\Delta$ of an element of initial length dx is indicated in Fig. A-1b, and is given by the formula

$$d\Delta = \epsilon_x\,dx = \frac{P}{EA}\,dx \tag{A-3}$$

The total elongation Δ of the bar shown in Fig. A-1a is obtained by integration of $d\Delta$ over the length L of the member:

$$\Delta = \int d\Delta = \int_0^L \frac{P}{EA}\,dx \tag{A-4}$$

If the member is prismatic and E is constant, the integration of Eq. (A-4) gives:

$$\Delta = \frac{PL}{EA} \tag{A-5}$$

This equation can be used to calculate the change in length of a prismatic bar subjected to a constant axial force.

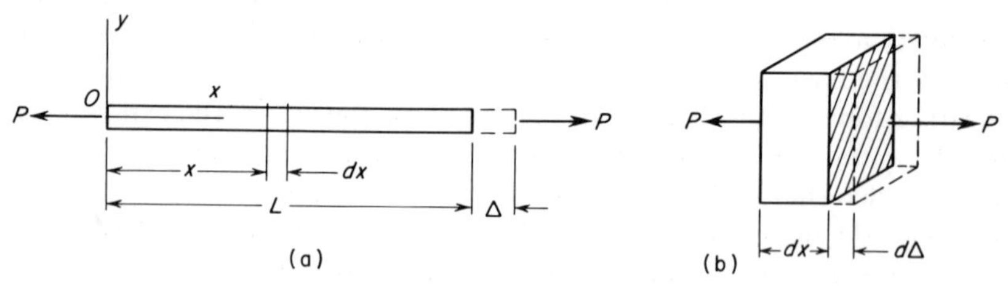

FIG. A-1. Axial deformations.

If the axial force P varies along the length of the bar, Eq. (A-4) can still be used. All that is necessary is to express P as a function of x and then perform the integration. If the bar is tapered slightly, then A must be expressed as a function of x, after which the integration can be carried out. Lastly, it should be noted that all of the above equations are valid for a bar of any cross-sectional shape, provided that P acts through the centroid of the section.

Flexural Deformations. A bar subjected to pure bending moment produced by couples M acting at each end of the bar is shown in Fig. A-2a. It is assumed that the plane of bending (the x-y plane) is a plane of symmetry of the beam, and hence the y axis is an axis of symmetry of the cross-sectional area (see Fig. A-2b). This requirement also means that the y and z axes are principal axes through point O, which is selected at the centroid of the cross-section. With the bending couples M acting as shown in Fig. A-2, it follows that all displacements of the beam will be in the x-y plane.*

At any cross-section of the beam the flexural stress σ_x is given by the formula

$$\sigma_x = -\frac{My}{I_Z} \tag{A-6}$$

in which y is the distance from the neutral axis (the z axis) to any point A in the cross-section (see Fig. A-2b), and I_Z is the moment of

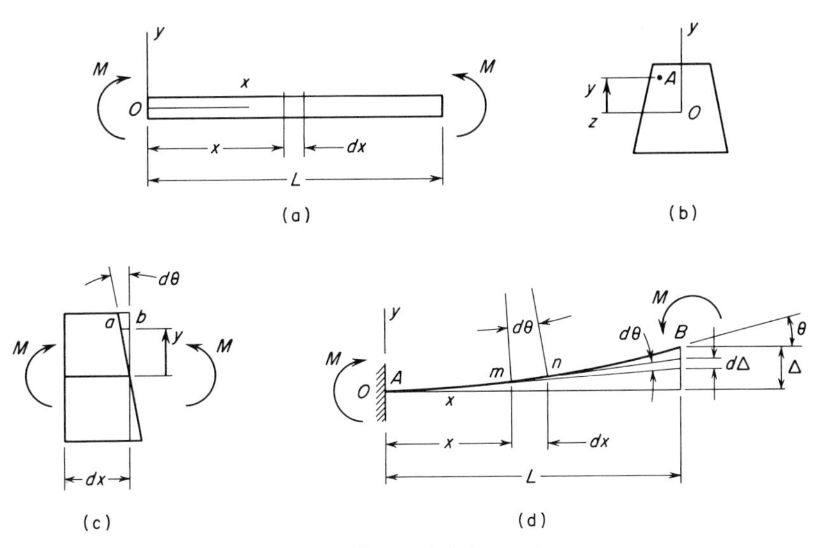

Fig. A-2. Flexural deformations.

inertia of the cross-sectional area with respect to the z axis. The flexural strain ϵ_x at the same point is equal to the flexural stress divided by the modulus of elasticity; therefore,

$$\epsilon_x = \frac{\sigma_x}{E} = -\frac{My}{EI_Z} \tag{A-7}$$

The minus sign appears in Eqs. (A-6) and (A-7) because positive bending moment M produces negative stresses (compression) in the region where y is positive.

The relative angle of rotation $d\theta$ between two cross-sections is shown in Fig. A-2c. For small angles of rotation this angle can be found by

* If the cross-section of the beam is not symmetric about the y axis, the bending analysis becomes more complicated since bending will no longer occur in a single plane. It then becomes necessary to take the origin O through the shear center of the cross-section and to take the y and z axes as principal axes through that point. Then the beam is analyzed for bending in both principal planes, as well as for torsion, and the results combined to give the final stresses and displacements.

dividing the shortening ab of a fiber at distance y from the neutral axis by the distance y itself. Since the distance ab is equal to $-\epsilon_x dx$, the expression for $d\theta$ becomes

$$d\theta = \frac{-\epsilon_x dx}{y}$$

Substitution of Eq. (A-7) into this equation results in:

$$d\theta = \frac{M}{EI_Z} dx \tag{A-8}$$

The quantity EI_Z in the denominator is called the *flexural rigidity* of the beam.

Expression (A-8) can sometimes be used to calculate angles of rotation and displacements of beams. An example of this kind is shown in Fig. A-2d, where it is assumed that the left end A of the bar in pure bending is fixed to a support and does not rotate. The angle of rotation θ of end B may be determined by integration of $d\theta$ (see Eq. A-8) along the length of the member. The expression for θ is

$$\theta = \int d\theta = \int_0^L \frac{M}{EI_Z} dx \tag{A-9}$$

in which dx is the length of the small element mn of the beam. If the member is prismatic and E is constant, integration of Eq. (A-9) gives the following expression for the angle at B for pure bending:

$$\theta = \frac{ML}{EI_Z} \tag{A-10}$$

However, Eq. (A-9) may also be used with good accuracy for cases in which the bending moment varies along the beam or in which the bar is slightly tapered. The procedure is to substitute the appropriate expression for M and I_Z into Eq. (A-9) and then perform the integration.

The deflection Δ at end B of the beam (Fig. A-2d) is seen to consist of the summation of the small distances $d\Delta$, each of which is an intercept on the vertical through B of the tangent lines from points m and n. Thus, for small angles of rotation the intercept $d\Delta$ is

$$d\Delta = (L - x)d\theta$$

or, using Eq. (A-8),

$$d\Delta = (L - x) \frac{M}{EI_Z} dx \tag{A-11}$$

Integration over the length of the member gives the total displacement Δ for a prismatic beam as follows:

$$\Delta = \int d\Delta = \int_0^L (L - x) \frac{M}{EI_Z} dx = \frac{ML^2}{2EI_Z} \tag{A-12}$$

This example for a beam in pure bending requires only very simple calculations in order to find the displacement. The same scheme can be used if either M or the flexural rigidity EI_Z varies along the length.

Under more general conditions it is necessary to use other methods for determining displacements of beams, such as the unit-load method described in Art. A.2.

Torsional Deformations. The deformations caused by pure torsion of a bar having a circular cross-section are illustrated in Fig. A-3. The bar has a length L and is subjected to twisting couples T at its ends (see Fig. A-3a).* The deformation of an element located at distance x from one end of the bar is shown in Fig. A-3b. The deformation consists of a relative rotation about the x axis of one cross-section with respect to

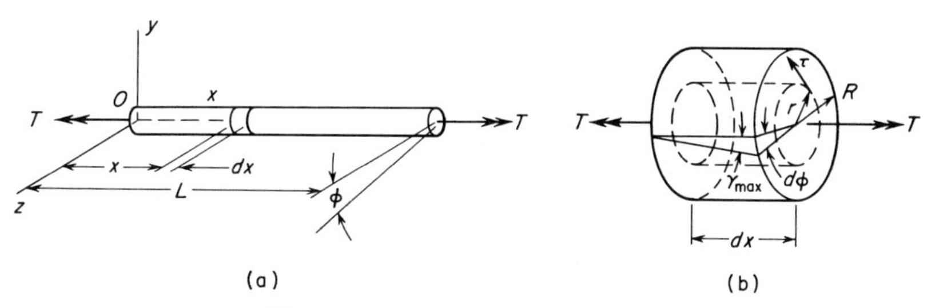

FIG. A-3. Torsional deformations.

another. The relative angle of rotation is denoted $d\phi$ in the figure. Associated with the deformation are shear stresses τ and shear strains γ.

The torsional shearing stresses are directly proportional to the distance from the longitudinal axis, and, at a distance r from the axis (see Fig. A-3b), the stress intensity is given by the formula

$$\tau = \frac{Tr}{J} \tag{A-13}$$

in which J is the polar moment of inertia of the circular cross-section; thus, J is equal to $\pi R^4/2$, where R is the radius of the bar. The maximum shear stress occurs at the outer surface of the bar and is obtained from the formula

$$\tau_{\max} = \frac{TR}{J} \tag{A-14}$$

The shear stresses acting on the circular cross-sections are always in a direction normal to the radius and in the same sense as the torque T.

The shear strain γ at the radius r is equal to the shear stress divided by the shearing modulus of elasticity G of the material; therefore,

$$\gamma = \frac{\tau}{G} = \frac{Tr}{GJ} \tag{A-15}$$

* In this book a double-headed arrow (or vector) is used to represent a couple. The sense of the couple is according to the right-hand rule of signs.

The expression for the maximum shear strain (see Fig. A-3b) is

$$\gamma_{\max} = \frac{TR}{GJ} \tag{A-16}$$

The quantity GJ appearing in the above formula is called the *torsional rigidity* of the bar.

The relative angle of rotation $d\phi$ between the sides of the element in Fig. A-3b is

$$d\phi = \frac{\gamma_{\max}}{R}\, dx$$

as can be seen from the geometry of the figure. This expression takes the following form when Eq. (A-16) is substituted:

$$d\phi = \frac{T}{GJ}\, dx \tag{A-17}$$

From Eq. (A-17) the total angle of twist ϕ (see Fig. A-3a) can be found by integration of $d\phi$ over the length of the member. The result is

$$\phi = \int d\phi = \int_0^L \frac{T}{GJ}\, dx \tag{A-18}$$

which for a cylindrical member with constant torque T becomes

$$\phi = \frac{TL}{GJ} \tag{A-19}$$

All of the formulas given above may be used for either a solid or a tubular circular bar. Of course, in the latter case J must be taken equal to the polar moment of inertia of the annular cross-section.

It should be noted that Eq. (A-18) may be used for a solid or tubular circular bar that is subjected to a torque T which varies along the length of the bar. It may also be used when J varies, provided the variation is gradual. In either of these cases the expression for T or J as a function of x is substituted into Eq. (A-18) before the integration is performed.

If the cross-section of the bar is not circular or annular, the torsion analysis is more complicated than the one described above for a bar with a circular section. However, for pure torsion in which the twisting couple T is constant along the length, the formula for the angle of twist (see Eq. A-19) can still be used with good accuracy, provided that J is taken as the appropriate torsion constant for the particular cross-section. Torsion constants for several shapes of cross-sections are tabulated in Appendix C.

If the cross-section of the bar is not circular there will be warping of the cross-sections. Warping refers to the longitudinal displacement of points in the cross-section, so that the cross-section is no longer a plane surface.

Warping occurs in the case of I beams and channel beams, as well as other sections, and a more complicated analysis is required. However, in such cases it is usually found that the analysis based upon pure torsion alone, with the warping effects neglected, gives conservative results. Torsion in which warping occurs is called *nonuniform torsion.**

Shearing Deformations. There are usually shearing forces as well as bending moments acting on the cross-sections of a beam. For example, at distance x from the fixed support of the cantilever beam shown in Fig. A-4a there will be a bending moment M (assumed positive when the top of the beam is in compression) given by the equation

$$M = -P(L - x) \tag{A-20}$$

The positive direction for M is shown in Fig. A-4b, which shows an element of length dx from the beam. The shearing force V is constant throughout the length of the beam and obtained in this example from the expression

$$V = P \tag{A-21}$$

In this expression it is assumed that positive shearing force is downward on the right-hand side of the element and upward on the left-hand side (see Fig. A-4b). In a more general case, the shearing force V as well as the bending moment M will vary along the length of the beam.

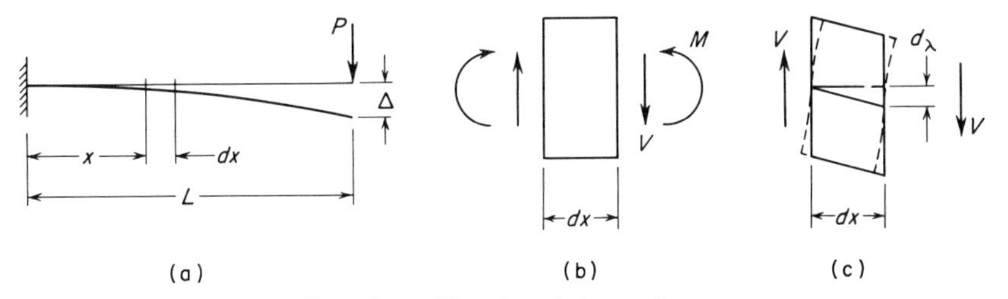

FIG. A-4. Shearing deformations.

The shearing stresses on the cross-section of a beam with rectangular cross-section due to a shear force V can be found from the formula

$$\tau = \frac{VQ}{I_Z b} \tag{A-22}$$

in which Q is the first moment (or static moment) about the neutral axis of the portion of the cross-sectional area which is outside of the section where the shear stress is to be determined; I_Z is the moment of inertia of the cross-sectional area about the neutral axis; and b is the width of the rectangular beam. Equation (A-22) can be used to find

* For the theory of nonuniform torsion of beams, see S. P. Timoshenko, *Strength of Materials*, Part II, 3rd ed., D. Van Nostrand Co., Inc., Princeton, N.J., 1956, pp. 244–273.

shear stresses in a few other shapes of beams; for example, it can be used to calculate the shear stresses in the web of an I beam, provided that b is taken as the thickness of the web of the beam. On the other hand, it cannot be used to calculate stresses in a beam of circular cross-section. The shearing strain γ can be found by dividing the shear stress τ by the shearing modulus of elasticity G.

Previously, the deformation of an element of a beam due solely to the action of bending moments was considered (see Fig. A-2c). In the present discussion, only the deformations caused by the shearing forces V will be taken into account. These deformations consist of a relative displacement $d\lambda$ of one side of the element with respect to the other (see Fig. A-4c). The displacement $d\lambda$ is given by the expression

$$d\lambda = f\frac{V dx}{GA} \tag{A-23}$$

in which A is the area of the cross-section and f is a form factor that is dependent upon the shape of the cross-section. Values of the form factor for several shapes of cross-section are given in Appendix C. The quantity GA/f is called the *shearing rigidity* of the bar.

The presence of shearing deformations $d\lambda$ in the elements of a beam means that the total displacement at any other point along the beam will be influenced by both flexural and shearing deformations. Usually the effects of shear are small compared to the effects of bending and can be neglected; however, if it is desired to include the shearing deformations in the calculation of displacements, it is possible to do so by using the unit-load method, as described in Art. A.2.

In a few elementary cases the deflections due to shearing deformations can be calculated by a direct application of Eq. (A-23). This equation can be used, for example, in calculating the deflection Δ at the end of the cantilever beam in Fig. A-4a. The portion Δ_s of the total deflection that is due solely to the effect of shearing deformations is equal to (see Eqs. A-21 and A-23)

$$\Delta_s = \int d\lambda = \frac{fP}{GA}\int_0^L dx = \frac{fPL}{GA} \tag{A-24}$$

The remaining part Δ_b of the deflection is due to bending and can be found by integrating Eq. (A-11). However, Eq. (A-11) was derived on the basis that upward deflection Δ was positive, whereas in Fig. A-4a the deflection Δ is downward. Thus, it is necessary to change the sign of the expression appearing in Eq. (A-11); the result is:

$$\Delta_b = \int_0^L -(L - x)\frac{M}{EI_Z}dx$$

The bending moment M for the beam is given by Eq. (A-20) and, when this expression is substituted into the above equation, the following expression is obtained:

$$\Delta_b = \int_0^L \frac{(L - x)^2 P}{EI_Z} dx = \frac{PL^3}{3EI_Z} \tag{A-25}$$

Summing the deflections due to both bending and shearing deformations gives the total deflection Δ, as follows:

$$\Delta = \Delta_b + \Delta_s = \frac{PL^3}{3EI_Z} + \frac{fPL}{GA} \tag{A-26}$$

From this equation it is found that the ratio of the shear deflection to the bending deflection is $3fEI_Z/GAL^2$. This ratio is very small compared to unity except in the case of short, deep beams; hence, in most cases it can be neglected.

Temperature Deformations. When the temperature of a structure varies, there is a tendency to produce changes in the shape of the structure. The resulting deformations and displacements may be of considerable importance in the analysis of the structure. In order to obtain formulas for the deformation due to temperature, consider the bar shown in Fig. A-5a. A uniform temperature change throughout the bar results in an increase in the length of the bar by the amount

$$\Delta L = \alpha L(\Delta T) \tag{A-27}$$

in which ΔL is the change in length (positive sign denotes elongation); α is the coefficient of thermal expansion; L is the length of the bar; and ΔT is the change in temperature (positive sign means increase in temperature). In addition, all other dimensions of the bar will be changed proportionately, but only the change in length will be of importance in the analysis of framed structures.

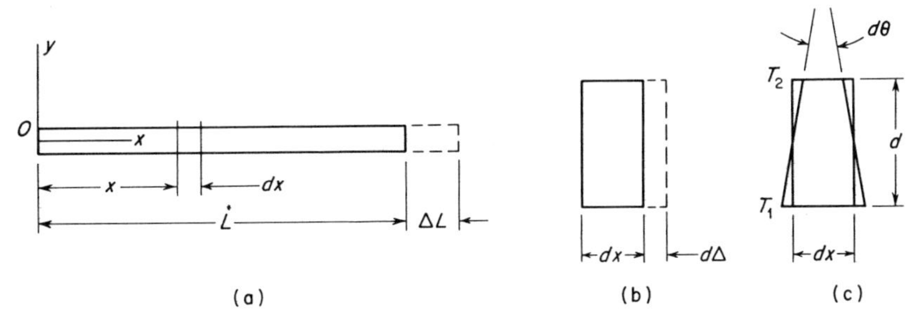

FIG. A-5. Temperature deformations.

The deformation of an element of the bar of length dx (Fig. A-5a) will be analogous to that for the entire bar. The longitudinal deformation of the element is shown in Fig. A-5b, and is seen to be of the same type as that caused by an axial force (see Fig. A-1b). This deformation is given by the expression

$$d\Delta = \alpha(\Delta T)dx \tag{A-28}$$

Equation (A-28) can be used in calculating displacements of structures due to uniform temperature changes, as described in the next article.

A temperature differential between two sides of the bar causes each element of the bar to deform as shown in Fig. A-5c. If the temperature varies linearly between T_2 at the top of the beam and a higher temperature T_1 at the bottom of the beam, the cross-sections of the beam will remain plane, as illustrated in the figure. The relative angle of rotation $d\theta$ between the sides of the element is

$$d\theta = \frac{\alpha(T_1 - T_2)dx}{d} \tag{A-29}$$

in which d is the depth of the beam. The deformation represented by the angle $d\theta$ in Eq. (A-29) is similar to that caused by bending moments acting on the beam (see Fig. A-2c). The use of Eq. (A-29) in finding beam displacements will be shown also in Art. A.2.

The formulas in the preceding paragraphs have been presented for reference use in the solution of problems and examples throughout this book. For a more complete treatment of deformations, a textbook on mechanics of materials should be consulted.

A.2 Unit-Load Method. The unit-load method is a very general and versatile method for calculating displacements of structures. It may be used (in theory) for either determinate or indeterminate structures, although for practical calculations it is used almost exclusively for statically determinate structures because its use requires that the stress resultants be known throughout the structure. The method may be used to determine displacements caused by loads on a structure, as well as displacements caused by temperature changes, misfit of parts, and other effects. The effects of axial, flexural, shearing, and torsional deformations may be included in the calculations. The fundamental ideas of the unit-load method are described in this article, and several numerical examples are solved. It is assumed throughout the discussion that the displacements of the structure are small and that the material is linearly elastic.

The basic equation of the unit-load method is usually derived from the principle of virtual work, and hence the method is sometimes called the *method of virtual work*. It is also known as the *dummy-load method* and the *Maxwell-Mohr method*. The former name arises from the use of a fictitious or dummy load (as described later), and the latter name is used because J. C. Maxwell in 1864 and O. Mohr in 1874 independently derived the method.

Two systems of loading must be considered when using the unit-load method. The first system consists of the structure in its actual condition;

that is, subjected to the actual loads, temperature changes, or other effects. The second system consists of the same structure subjected to a unit load corresponding to the desired displacement in the actual structure. The unit load is a fictitious or dummy load, and is introduced solely for purposes of analysis. By a unit load corresponding to the displacement is meant a load at the particular point of the structure where the displacement is to be determined and acting in the positive direction of that displacement. The term "displacement" is used here in the generalized sense, as discussed in Art. 1.4. Thus, a displacement may be the translation of a point on the structure, the angle of rotation of the axis of a member, or a combination of translations and rotations. If the displacement to be calculated is a translation, then the unit load is a concentrated force at the point where the displacement occurs. In addition, the unit load must be in the same direction as the displacement and have the same positive sense. If the displacement to be calculated is a rotation, then the unit load is a couple at the point where the rotation occurs, and is assumed positive in the same sense as a positive rotation. Similarly, if the displacement is the relative displacement of two points along the line joining them, the unit load consists of two collinear and oppositely directed forces acting at the two points; and if the displacement is a relative rotation between two points, the unit load consists of two equal and oppositely directed couples at the points.

The actions (forces and couples) in the structure due to the unit load, which is the second system of loading, constitute a system of actions in equilibrium. According to the principle of virtual work, if a deformable system in equilibrium is given a small virtual displacement, the total work of the external actions is equal to the total work of the internal actions. The internal actions (or stress resultants) caused by the unit load in the second system will be represented by the symbols N_U, M_U, T_U, and V_U, denoting the axial force, bending moment, twisting couple, and shearing force, respectively, acting at any cross-section in the members of the structure.

The virtual displacements of the second system are taken the same as the actual displacements of the first system; that is, the same as the displacements caused by the actual loads, temperature changes, and so on. The deformations in the first system will be denoted as $d\Delta$ for axial deformation (see Fig. A-1b), $d\theta$ for flexural deformation (see Fig. A-2c), $d\phi$ for torsional deformation (see Fig. A-3b), and $d\lambda$ for shearing deformation (see Fig. A-4c). Thus, the work done in the second system by the internal actions acting on one element of the structure when the deformations $d\Delta$, $d\theta$, $d\phi$, and $d\lambda$ are imposed on that element will be

$$N_U d\Delta + M_U d\theta + T_U d\phi + V_U d\lambda$$

The first term in this expression represents the work done by the axial force N_U (produced by the unit load) when the displacement $d\Delta$ (due to the actual loads or other effects) is imposed on the element. A similar statement can be made about each of the other terms. Then the total work of the internal actions is

$$\int N_U d\Delta + \int M_U d\theta + \int T_U d\phi + \int V_U d\lambda$$

where the integrations are carried out for all elements of the structure.

The work done by the external actions is the same as the work done by the unit load, because the unit load is the only external force (or couple) that moves during the virtual displacement. This work is

$$1 \cdot \Delta$$

in which Δ represents the desired displacement due to the first system of loading. Equating the work of the external and internal actions gives the fundamental equation of the unit-load method:

$$\Delta = \int N_U d\Delta + \int M_U d\theta + \int T_U d\phi + \int V_U d\lambda \qquad \text{(A-30)}$$

In this equation, Δ represents the displacement that is to be calculated; N_U, M_U, T_U, and V_U represent the axial forces, bending moments, torsional couples, and shearing forces caused by a unit load corresponding to Δ; and $d\Delta$, $d\theta$, $d\phi$, and $d\lambda$ represent deformations caused by the actual load system. Because the unit load has been divided from the left-hand side of Eq. (A-30), leaving only the term Δ, it is necessary to consider the quantities N_U, M_U, T_U, and V_U as having the dimensions of force or moment per unit of the applied unit load.

The quantities $d\Delta$, $d\theta$, $d\phi$, and $d\lambda$ appearing in Eq. (A-30) can be expressed in terms of the properties of the structure. The expression for $d\Delta$ when the axial deformations are caused by loads only is (compare with Eq. A-3):

$$d\Delta = \frac{N_L dx}{EA}$$

in which N_L represents the axial force in the member due to the actual loads on the structure. Similarly, if the deformations are caused by a uniform temperature increase, the expression for $d\Delta$ is (see Eq. A-28):

$$d\Delta = \alpha(\Delta T)dx$$

in which α is the coefficient of thermal expansion and ΔT is the temperature change. The expressions for the remaining deformation quantities due to loads (compare with Eqs. A-8, A-17, and A-23) are:

$$d\theta = \frac{M_L dx}{EI} \qquad d\phi = \frac{T_L dx}{GJ} \qquad d\lambda = \frac{fV_L dx}{GA}$$

The quantities M_L, T_L, and V_L represent the bending moment, twisting couple, and shearing force caused by the loads. If there is a temperature differential across the beam, Eq. (A-29) can be used for the deformation $d\theta$.

When the relations given above for the deformations due to loads only are substituted into Eq. (A-30), the equation for the displacement takes the form

$$\Delta = \int \frac{N_U N_L dx}{EA} + \int \frac{M_U M_L dx}{EI} + \int \frac{T_U T_L dx}{GJ} + \int \frac{f V_U V_L dx}{GA} \qquad \text{(A-31)}$$

Each term in this equation represents the effect of one type of deformation on the total displacement Δ that is to be found. In other words, the first term represents the displacement caused by axial deformations; the second term represents the displacement caused by bending deformations; and so forth for the remaining terms. The sign conventions used for the quantities appearing in Eq. (A-31) must be consistent with one another. Thus, the axial forces N_U and N_L must be obtained according to the same convention; for example, tension is positive. Similarly, the bending moments M_U and M_L must have the same sign convention, as also must T_U and T_L, and V_U and V_L. Only if the sign conventions are consistent will the displacement Δ have the same positive sense as the unit load.

The procedure for calculating a displacement by means of Eq. (A-31) is as follows: (1) determine forces and moments in the structure due to the loads (that is, obtain N_L, M_L, T_L, and V_L); (2) place a unit load on the structure corresponding to the displacement Δ that is to be found; (3) determine forces and moments in the structure due to the unit load (that is, find N_U, M_U, T_U, and V_U); (4) form the products shown in Eq. (A-31) and integrate each term for the entire structure; and (5) sum the results to obtain the total displacement.

Usually not all of the terms given in Eq. (A-31) are required for the calculation of displacements. In a truss with hinged joints and with loads acting only at the joints, there will be no bending, torsional, and shearing deformations. Furthermore, if each member of the truss is prismatic the cross-sectional area A will be a constant for each member. In such a case the equation for Δ can be written as

$$\Delta = \Sigma \frac{N_U N_L L}{EA} \qquad \text{(A-32)}$$

in which L represents the length of a member. The summation is carried out for all bars of the truss.

In a beam it is quite likely that only bending deformations are important. Therefore, the equation for the displacements simplifies to

$$\Delta = \int \frac{M_U M_L dx}{EI} \qquad \text{(A-33)}$$

In an analogous manner it is possible to calculate displacements by using any appropriate combination of terms from Eq. (A-31), depending upon the nature of the structuré and the degree of refinement required for the analysis. Other terms can be used when displacements due to temperature changes, prestrains, etc. are to be found. All that is necessary is to substitute into Eq. (A-30) the appropriate expressions for the deformations. Some examples of the use of the unit-load method will now be given.

Example 1. The truss shown in Fig. A-6a is subjected to loads P and $2P$ at joint A. All members of the truss are assumed to have the same axial rigidity

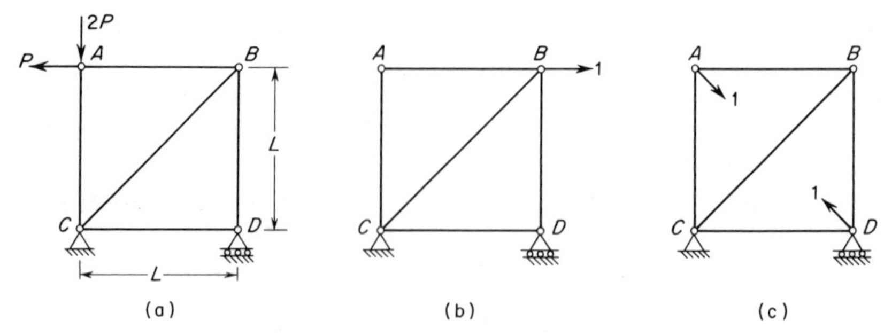

FIG. A-6. Examples 1 and 2.

EA. The horizontal displacement Δ_1 of joint B (positive to the right) is to be calculated.

The calculations for the displacement Δ_1 by the unit-load method are given in Table A-1. The first two columns in the table identify the bars of the truss

TABLE A-1

Bar	Length	N_L	N_U	$N_U N_L L$	N_U	$N_U N_L L$
(1)	(2)	(3)	(4)	(5)	(6)	(7)
AB	L	P	0	0	$-1/\sqrt{2}$	$-PL/\sqrt{2}$
AC	L	$-2P$	0	0	$-1/\sqrt{2}$	$2PL/\sqrt{2}$
BD	L	P	-1	$-PL$	$-1/\sqrt{2}$	$-PL/\sqrt{2}$
CD	L	0	0	0	$-1/\sqrt{2}$	0
CB	$\sqrt{2}\,L$	$-\sqrt{2}\,P$	$\sqrt{2}$	$-2\sqrt{2}\,PL$	1	$-2PL$
				$-3.828PL$		$-2PL$

and their lengths. The axial forces N_L, which are determined by static equilibrium for the truss shown in Fig. A-6a, are listed in column (3) of the table. The unit load corresponding to the displacement Δ_1 is shown in Fig. A-6b, and the resulting axial forces N_U are given in column (4). Finally, the products

$N_U N_L L$ are obtained for each member (column 5), summed, and divided by EA. Thus, the displacement Δ_1 is (see Eq. A-32):

$$\Delta_1 = -3.828\frac{PL}{EA}$$

The negative sign in this result means that Δ_1 is in the opposite direction to the unit load (i.e., to the left).

A similar procedure can be used to find any other displacement of the truss. For example, suppose that it is desired to determine the relative displacement Δ_2 of joints A and D (see Fig. A-6a) along the line joining them. The corresponding unit load consists of two unit forces, as shown in Fig. A-6c. The resulting axial forces N_U in the truss are listed in column (6) of Table A-1, and the products $N_U N_L L$ are given in column (7). Thus, the relative displacement of joints A and D is

$$\Delta_2 = -2\frac{PL}{EA}$$

in which the minus sign indicates that the distance between points A and D has increased (that is, is opposite to the sense of the unit loads).

Example 2. Consider again the truss shown in Fig. A-6a, and assume now that bar BD has been fabricated with a length which is greater by an amount e than the theoretical length L. The horizontal displacement Δ_1 of joint B and the relative displacement Δ_2 between points A and D are to be determined (see Figs. A-6b and A-6c for the corresponding unit loads).

Any displacement of the truss caused by the increased length of member BD can be found by using Eq. (A-30) and retaining only the first term on the right-hand side. For a truss the equation may be expressed in the form

$$\Delta = \sum N_U\,(\Delta L)$$

in which the summation is carried out for all members of the truss and ΔL represents the change in length of any member. In this example the only bar in the truss of Fig. A-6a which has a change in length is bar BD itself, and hence there is only one term in the summation. For bar BD, the term ΔL is

$$\Delta L = e$$

When finding the horizontal displacement of joint B, the force in bar BD due to the unit load shown in Fig. A-6b is

$$N_U = -1$$

as given in the third line of column (4) in Table A-1. Therefore, the displacement of joint B is

$$\Delta_1 = -e$$

and is to the left.

When the decrease in distance between joints A and D due to the lengthening of bar BD is to be found, the value for N_U becomes

$$N_U = -\frac{1}{\sqrt{2}}$$

as given in column (6) of Table A-1; therefore,

$$\Delta_2 = -\frac{e}{\sqrt{2}}$$

The negative sign for Δ_2 shows that joints A and D move apart from one another.

A uniform temperature change in one or more bars of the truss is handled in the same manner as a change in length. The only difference is that the change in length ΔL now is given by Eq. (A-27). Thus, the horizontal displacement of joint B due to a temperature increase of T degrees in member BD becomes

$$\Delta_1 = -\alpha LT$$

and the change in distance between points A and D becomes

$$\Delta_2 = -\frac{\alpha LT}{\sqrt{2}}$$

Example 3. The cantilever beam AE shown in Fig. A-7a is subjected to loads at points B, C, D and E. The translational displacements Δ_1 and Δ_2 of the beam (positive upward) at points C and E, respectively, are to be determined.

The displacements Δ_1 and Δ_2 can be found from Eq. (A-33), which is expressed in terms of the bending moments M_L and M_U. The former moments are due to the actual loads on the beam, and the latter are due to unit loads corresponding to the desired displacements. Expressions for M_L and M_U must be obtained for each segment of the beam between applied loads, and then these expressions are substituted into Eq. (A-33) to obtain the displacements.

The required calculations are shown in Table A-2. The first two columns of

TABLE A-2

Segment	Limits for x	M_L	Unit Load at C		Unit Load at E	
			M_U	$\int M_U M_L dx$	M_U	$\int M_U M_L dx$
(1)	(2)	(3)	(4)	(5)	(6)	(7)
AB	0 to $L/2$	$\frac{P}{2}(4x + L)$	$L - x$	$\frac{17PL^3}{48}$	$2L - x$	$\frac{41PL^3}{48}$
BC	$L/2$ to L	$\frac{3PL}{2}$	$L - x$	$\frac{3PL^3}{16}$	$2L - x$	$\frac{15PL^3}{16}$
CD	L to $\frac{3L}{2}$	$\frac{PL}{2}$	0	0	$2L - x$	$\frac{3PL^3}{16}$
DE	$\frac{3L}{2}$ to $2L$	$P(2L - x)$	0	0	$2L - x$	$\frac{PL^3}{24}$
				$\frac{13PL^3}{24}$		$\frac{97PL^3}{48}$

the table list the segments of the beam and the limits for the distance x, which is measured from the fixed support. Column (3) gives the expressions for the bending moments in the beam due to the actual loads, assuming that compression on top of the beam corresponds to positive bending moment. The moments M_U in column (4) are those caused by a unit load at point C (Fig. A-7b). These moments are evaluated according to the same sign convention that was used in determining the moments M_L. Next, column (5) shows the results of evaluating the integral given in Eq. (A-33), except that the factor EI has been omitted,

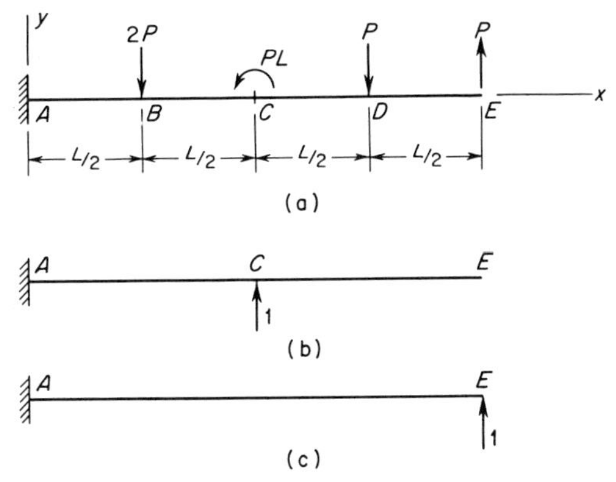

FIG. A-7. Example 3.

since it is assumed to be the same for all segments of the beam. When the expressions in column (5) are summed, and the total is divided by EI, the result is the displacement corresponding to the unit load. Therefore, the displacement at point C in the y direction is

$$\Delta_1 = \frac{13PL^3}{24EI}$$

and is in the positive direction of the y axis (upward).

The calculations for the translation at point E are shown also in Table A-2 (see columns 6 and 7). The unit load used in ascertaining the moments M_U is shown in Fig. A-7c. The result of the calculations is

$$\Delta_2 = \frac{97PL^3}{48EI}$$

which, being positive, shows that the translation is in the direction of the y axis.

Example 4. In this example it is assumed that the beam shown in Fig. A-7a is now subjected to a temperature differential such that the bottom of the beam is at temperature T_1 while the top of the beam is at temperature T_2 (see Fig. A-8). The formula for the displacements is obtained by using only the second term in the right-hand side of Eq. (A-30) and substituting for $d\theta$ the expression given in Eq. (A-29); thus:

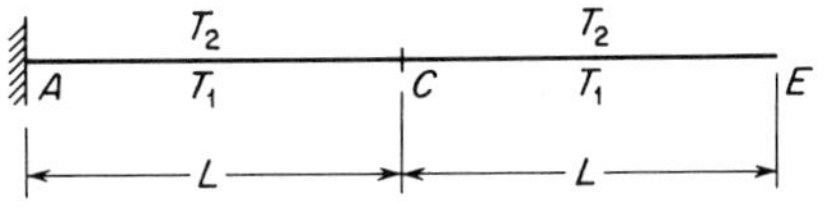

FIG. A-8. Example 4.

$$\Delta = \int M_U \frac{\alpha(T_1 - T_2)dx}{d}$$

TABLE A-3

Displacements of Prismatic Beams

	Beam	Translations (positive downward)	Rotations (positive clockwise)
1 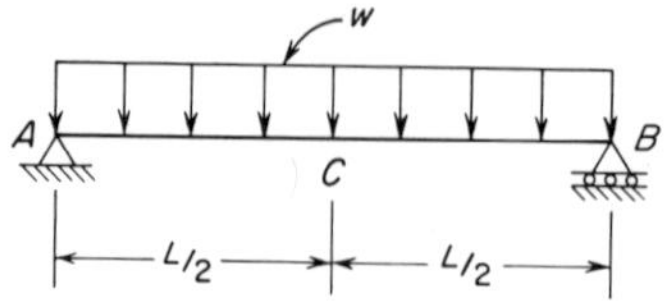	w; A, C, B; $L/2$, $L/2$	$\Delta_C = \dfrac{5wL^4}{384EI}$	$\theta_A = -\theta_B = \dfrac{wL^3}{24EI}$
2 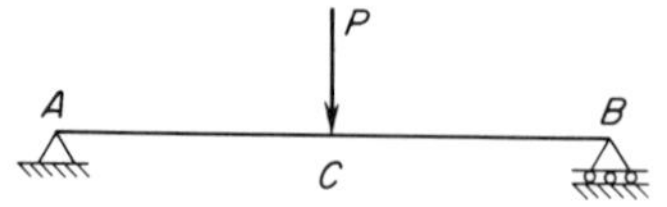	P; A, C, B	$\Delta_C = \dfrac{PL^3}{48EI}$	$\theta_A = -\theta_B = \dfrac{PL^2}{16EI}$
3 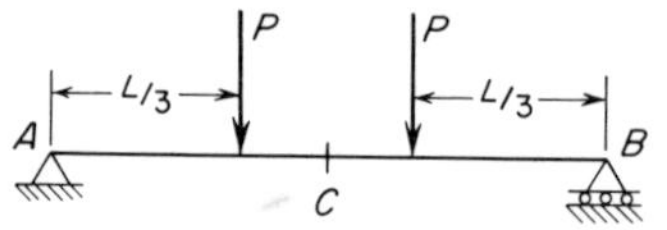	P, P; $L/3$, $L/3$; A, C, B	$\Delta_C = \dfrac{23PL^3}{648EI}$	$\theta_A = -\theta_B = \dfrac{PL^2}{9EI}$
4 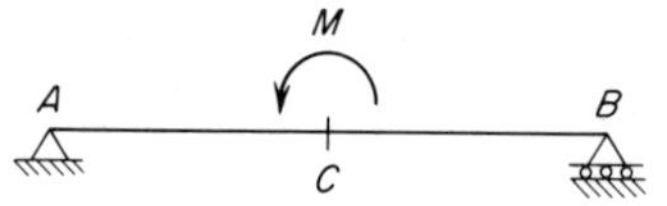	M; A, C, B	$\Delta_C = 0$	$\theta_A = \theta_B = \dfrac{ML}{24EI}$
5 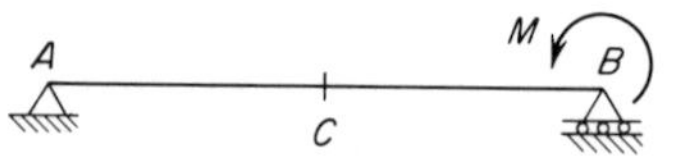	M; A, C, B	$\Delta_C = \dfrac{ML^2}{16EI}$	$\theta_A = \dfrac{ML}{6EI}$ $\theta_B = -\dfrac{ML}{3EI}$
6 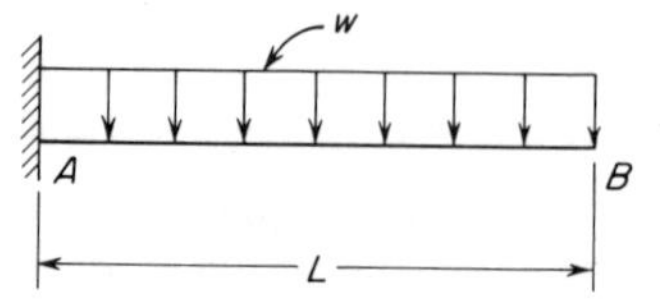	w; A, B; L	$\Delta_B = \dfrac{wL^4}{8EI}$	$\theta_B = \dfrac{wL^3}{6EI}$
7 	P; A, B	$\Delta_B = \dfrac{PL^3}{3EI}$	$\theta_B = \dfrac{PL^2}{2EI}$
8 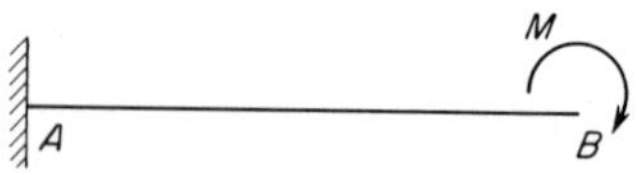	M; A, B	$\Delta_B = \dfrac{ML^2}{2EI}$	$\theta_B = \dfrac{ML}{EI}$

The expressions for M_U that are to be substituted into this equation are given in columns (4) and (6) of Table A-2, assuming that the vertical translations Δ_1 and Δ_2 at points C and E are to be found. The calculations become as follows:

$$\Delta_1 = \frac{\alpha(T_1 - T_2)}{d} \int_0^L (L - x)dx = \frac{\alpha(T_1 - T_2)L^2}{2d}$$

$$\Delta_2 = \frac{\alpha(T_1 - T_2)}{d} \int_0^{2L} (2L - x)dx = \frac{2\alpha(T_1 - T_2)L^2}{d}$$

These results show that if T_1 is greater than T_2, the beam deflects upward.

The preceding examples illustrate the determination of translational displacements in trusses and beams due to various causes. Other types of structures can be analyzed in an analogous manner. Also, techniques that are similar to those illustrated can be used to find rotations at a point (the unit load corresponding to a rotation is a unit couple), as well as to obtain displacements due to shearing and torsional deformations, all of which are included in Eq. (A-30).

A.3 Displacements of Beams. In many of the problems and examples given in Chapters 1, 2, and 3, it is necessary to determine displacements of beams. Such displacements can be found in all cases by the unit-load method, although other standard methods (including integration of the differential equation for displacements of a beam, and the moment-area method) may be suitable also. In most of the examples, however, the desired displacements can be obtained with the aid of the formulas given in Table A-3 for prismatic beams.

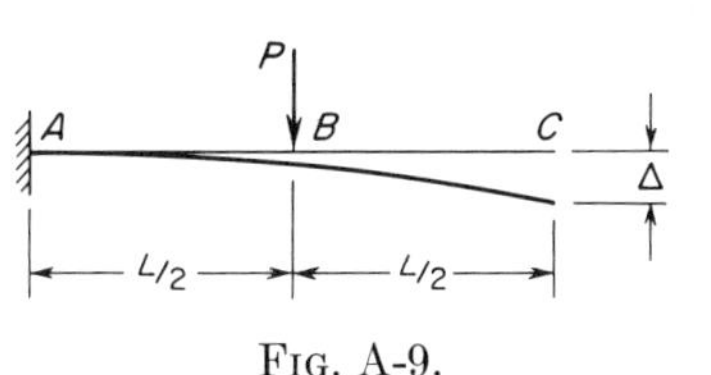

FIG. A-9.

As an illustration of the use of the formulas, consider a cantilever beam with constant EI that is subjected to a concentrated load P at its midpoint (Fig. A-9). The displacement Δ at the end of the beam can be obtained readily by making the following observation: the displacement Δ is equal to the displacement at B plus the rotation at B times the distance from B to C. Thus, using Case 7 in Table A-3, the following expression is obtained:

$$\Delta = \Delta_B + \theta_B \frac{L}{2} = P\left(\frac{L}{2}\right)^3 \frac{1}{3EI} + P\left(\frac{L}{2}\right)^2 \frac{1}{2EI}\frac{L}{2} = \frac{5PL^3}{48EI}$$

Techniques of this kind can be very useful for finding displacements in beams and plane frames.

Appendix B

END-ACTIONS FOR RESTRAINED MEMBERS

A restrained member is one whose ends are restrained against displacement (translation and rotation), as in the case of a fixed-end beam. The end-actions for a restrained member are the reactive actions (forces and couples) developed at the ends when the member is subjected to loads, temperature changes, or other effects. Restrained members are encountered in the stiffness method of analysis (see Art. 2.8) and also in the determination of equivalent joint loads (see Arts. 3.2 and 4.5).

In this appendix, formulas are given for end-actions in restrained members due to various causes. It is assumed in each case that the member is prismatic.

Table B-1 gives end-actions in fixed-end beams that are subjected to various conditions of loading. As shown in the figure at the top of the table, the length of the beam is L, the reactive moments at the left- and right-hand ends are denoted M_A and M_B, respectively, and the reactive forces are denoted R_A and R_B, respectively. The moments are positive when counterclockwise, and the forces are positive when upward. Formulas for these quantities are given in Cases 1, 2, 5, 6, 7, and 8. However, Cases 3 and 4 differ slightly because of the special nature of the loads. In Case 3 the load is an axial force P, and therefore the only reactions are the two axial forces shown in the figure. In Case 4 the load is a twisting couple T, which produces reactions in the form of twisting couples only.

All of the formulas given in Table B-1 can be derived by standard methods of mechanics of materials. For instance, many of the formulas for beams can be obtained by integration of the differential equation for bending of a beam. The flexibility method, as described in Art. 2.2, can also be used to obtain the formulas. Furthermore, the more complicated cases of loading frequently can be obtained from the simpler cases by using the principle of superposition.*

Fixed-end actions due to temperature changes are listed in Table B-2. Case 1 of this table is for a beam subjected to a uniform temperature increase of T degrees. The resulting end-actions consist of axial compressive forces that are independent of the length of the member. The second

* A more extensive table of fixed-end actions caused by loads is given in *Moment Distribution*, by J. M. Gere, D. Van Nostrand Co., Inc., Princeton, N.J., 1963. Also included are formulas and charts for fixed-end actions in nonprismatic members.

TABLE B-1

Fixed-End Actions Caused by Loads

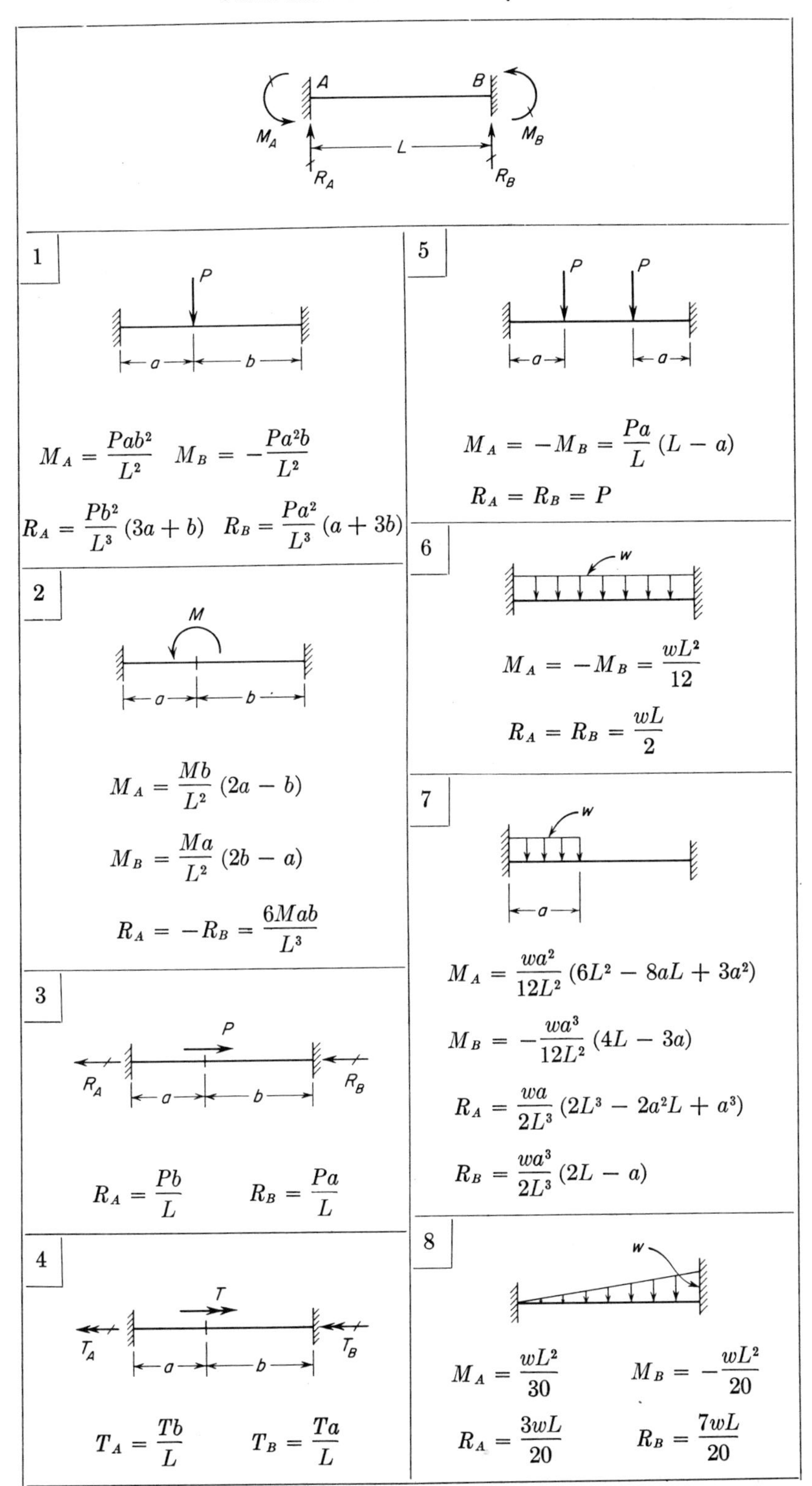

1

$$M_A = \frac{Pab^2}{L^2} \qquad M_B = -\frac{Pa^2b}{L^2}$$

$$R_A = \frac{Pb^2}{L^3}(3a + b) \qquad R_B = \frac{Pa^2}{L^3}(a + 3b)$$

2

$$M_A = \frac{Mb}{L^2}(2a - b)$$

$$M_B = \frac{Ma}{L^2}(2b - a)$$

$$R_A = -R_B = \frac{6Mab}{L^3}$$

3

$$R_A = \frac{Pb}{L} \qquad R_B = \frac{Pa}{L}$$

4

$$T_A = \frac{Tb}{L} \qquad T_B = \frac{Ta}{L}$$

5

$$M_A = -M_B = \frac{Pa}{L}(L - a)$$

$$R_A = R_B = P$$

6

$$M_A = -M_B = \frac{wL^2}{12}$$

$$R_A = R_B = \frac{wL}{2}$$

7

$$M_A = \frac{wa^2}{12L^2}(6L^2 - 8aL + 3a^2)$$

$$M_B = -\frac{wa^3}{12L^2}(4L - 3a)$$

$$R_A = \frac{wa}{2L^3}(2L^3 - 2a^2L + a^3)$$

$$R_B = \frac{wa^3}{2L^3}(2L - a)$$

8

$$M_A = \frac{wL^2}{30} \qquad M_B = -\frac{wL^2}{20}$$

$$R_A = \frac{3wL}{20} \qquad R_B = \frac{7wL}{20}$$

case is a beam subjected to a temperature differential such that the top of the beam is at temperature T_2 while the lower side is at temperature T_1. The temperature at the centroidal axis is assumed to remain unchanged, so that there is no tendency for the beam to change in length. In this case the end-actions consist of moments only.

TABLE B-2

Fixed-End Actions Caused by Temperature Changes

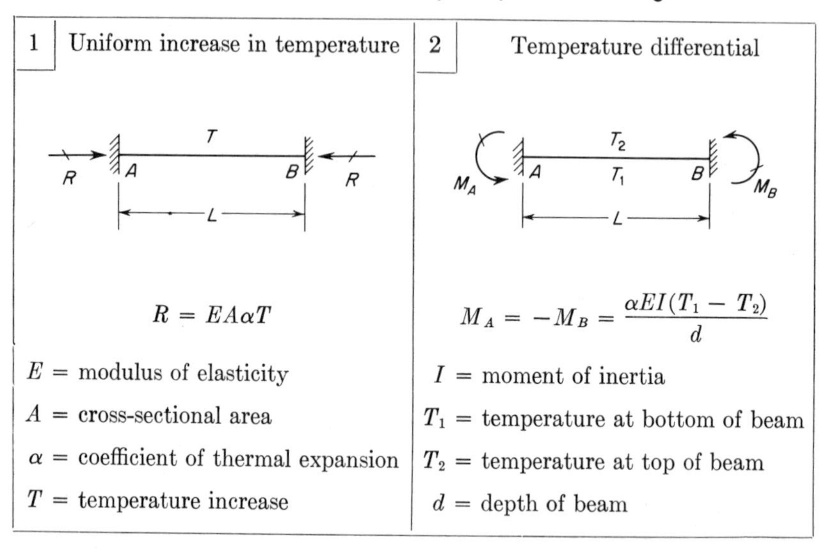

1 Uniform increase in temperature	2 Temperature differential
$R = EA\alpha T$	$M_A = -M_B = \dfrac{\alpha EI(T_1 - T_2)}{d}$
E = modulus of elasticity	I = moment of inertia
A = cross-sectional area	T_1 = temperature at bottom of beam
α = coefficient of thermal expansion	T_2 = temperature at top of beam
T = temperature increase	d = depth of beam

Table B-3 gives fixed-end actions due to prestrains in the members. A prestrain is an initial deformation of a member, causing end-actions to be developed when the ends of the member are held in the restrained positions. The simplest example of a prestrain is shown in Case 1, where member AB is assumed to have an initial length that is greater than the distance between supports by a small amount e. When the ends of the member are held in their final positions, the member will have been shortened by the distance e. The resulting fixed-end actions are the axial compressive forces shown in the table. Case 2 is a member with an initial bend in it, and the last case is a member having an initial circular curvature such that the deflection at the middle of the beam is equal to the small distance e.

Table B-4 lists formulas for fixed-end actions caused by displacements of one end of the member. Cases 1 and 2 are for axial and lateral translations of the end B of the member through the small distance Δ, while Cases 3 and 4 are for rotations. The rotation through the angle θ shown in Case 3 produces bending of the member, while the rotation through the angle ϕ in Case 4 produces torsion. Formulas for the torsion constant

TABLE B-3

Fixed-End Actions Caused by Prestrains

1 Bar with excess length

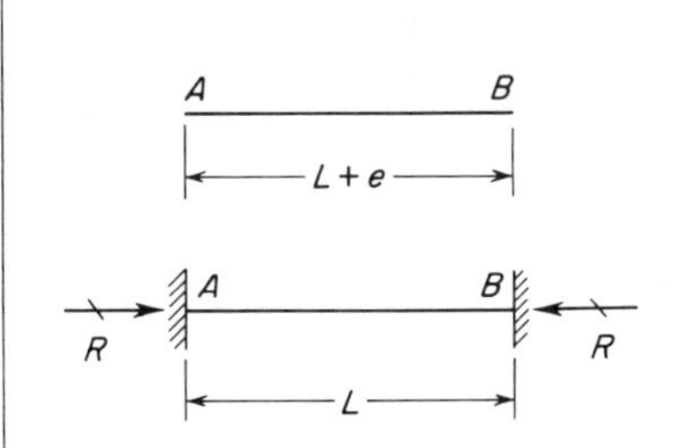

$$R = \frac{EAe}{L}$$

E = modulus of elasticity

A = cross-sectional area

e = excess length

2 Bar with a bend

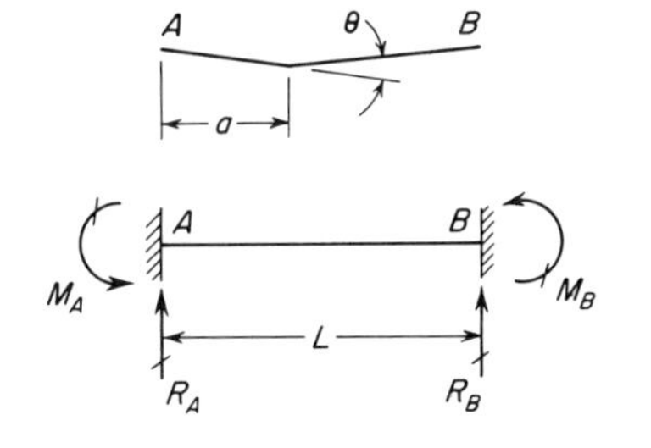

$$M_A = \frac{2EI\theta}{L^2}(2L - 3a)$$

$$M_B = \frac{2EI\theta}{L^2}(L - 3a)$$

$$R_A = -R_B = \frac{6EI\theta}{L^3}(L - 2a)$$

I = moment of inertia

θ = angle of bend

3 Initial circular curvature

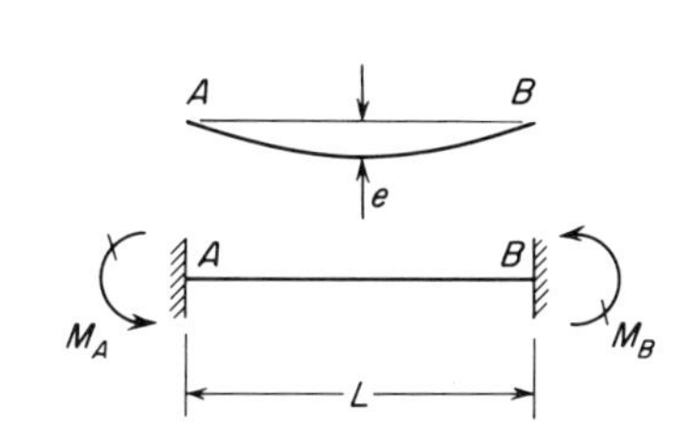

$$M_A = -M_B = \frac{8EIe}{L^2}$$

e = initial deflection at middle of bar

TABLE B-4

Fixed-End Actions Caused by End-Displacements

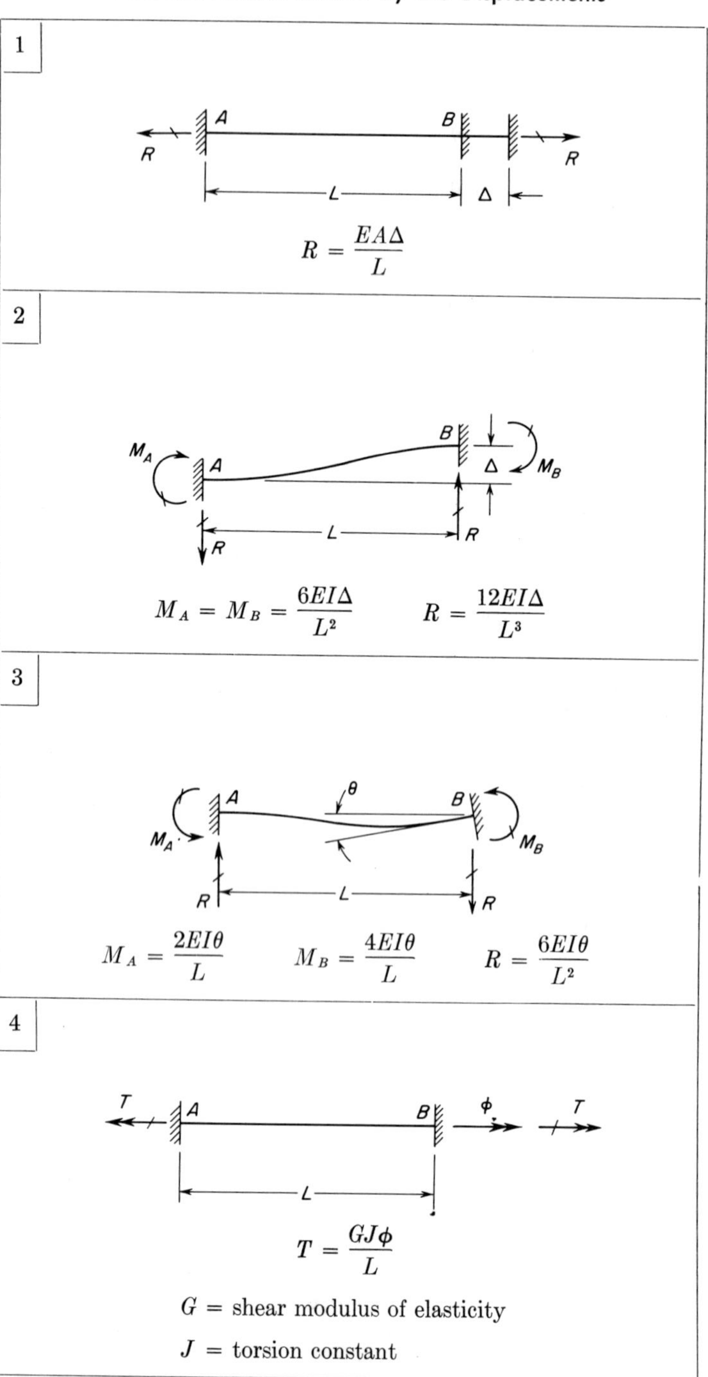

TABLE B-5

End-Actions for Truss Members

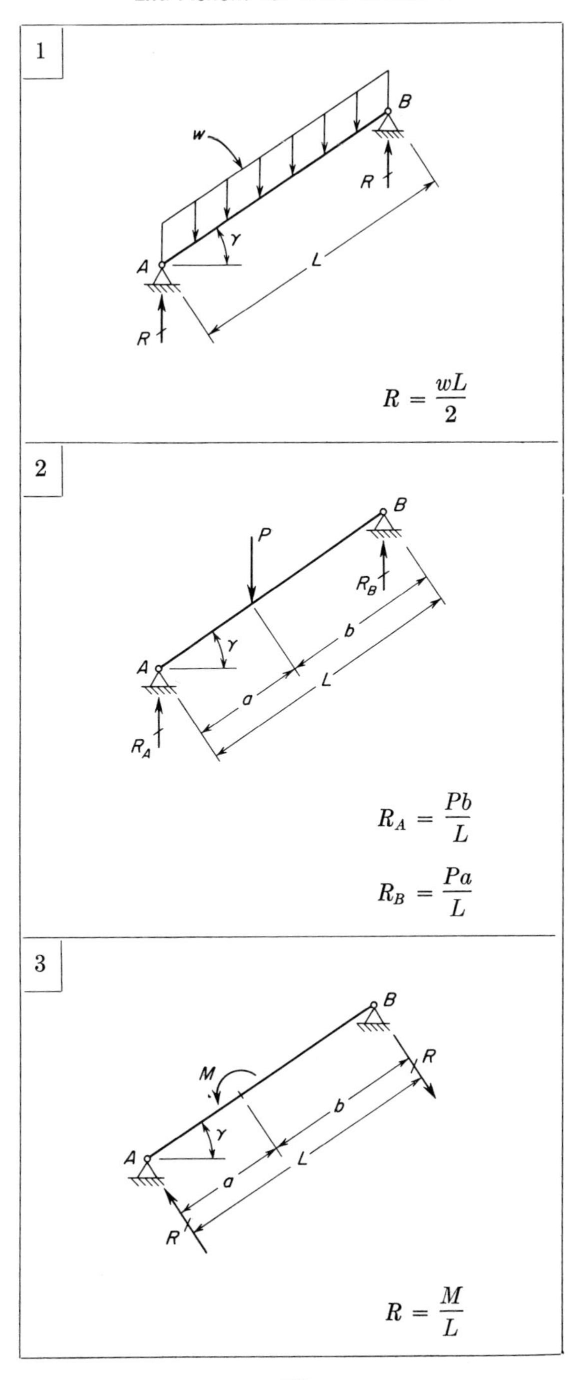

J, which appears in the formulas of Case 4, are given in Appendix C for several cross-sectional shapes.

End-actions for truss members are listed in Table B-5 for three cases of loading: a uniform load, a concentrated load, and a couple. The members shown in the figures have pinned ends that are restrained against translation but not rotation, because only joint translations are of interest in a truss analysis. The members are shown inclined at an angle γ to the horizontal, in order to have a general orientation. However, the end-actions are independent of the angle of inclination, which may have any value (including 0 and 90 degrees). For both the uniform load and the concentrated load (Cases 1 and 2) the reactions are parallel to the line of action of the load, while in Case 3 the reactions are perpendicular to the axis of the member.

If a truss member is subjected to a uniform increase in temperature, Case 1 of Table B-2 can be used; if subjected to a prestrain consisting of an increase in length, Case 1 of Table B-3 can be used; and if subjected to a displacement in the axial direction, Case 1 of Table B-4 can be used.

Appendix C

PROPERTIES OF SECTIONS

Symbols in Table

I_Z = moment of inertia of cross-section about z axis

I_Y = moment of inertia of cross-section about y axis

A = area of cross-section

J = torsion constant

f = form factor for shear

APPENDIX C

Properties of Sections

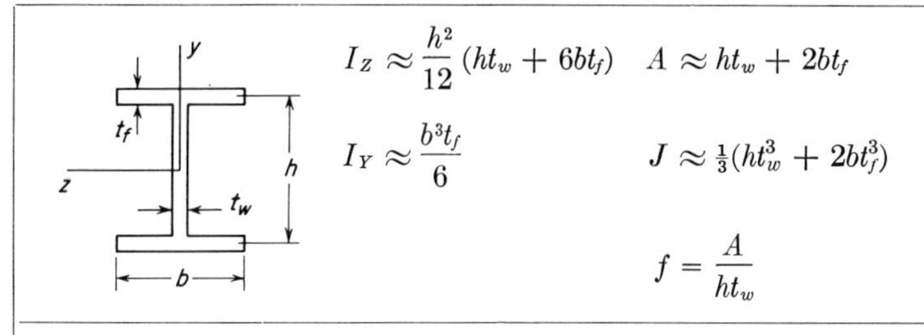

$$I_Z \approx \frac{h^2}{12}(ht_w + 6bt_f) \qquad A \approx ht_w + 2bt_f$$

$$I_Y \approx \frac{b^3 t_f}{6} \qquad J \approx \tfrac{1}{3}(ht_w^3 + 2bt_f^3)$$

$$f = \frac{A}{ht_w}$$

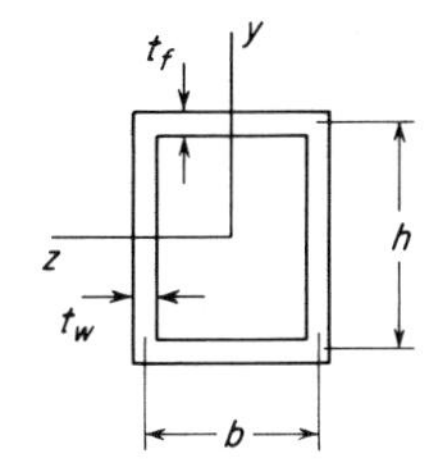

$$I_Z \approx \frac{h^2}{6}(ht_w + 3bt_f) \qquad A = 2(bt_f + ht_w)$$

$$I_Y \approx \frac{b^2}{6}(bt_f + 3ht_w) \qquad J \approx 2b^2h^2 \frac{t_f t_w}{bt_w + ht_f}$$

$$f = \frac{A}{2ht_w}$$

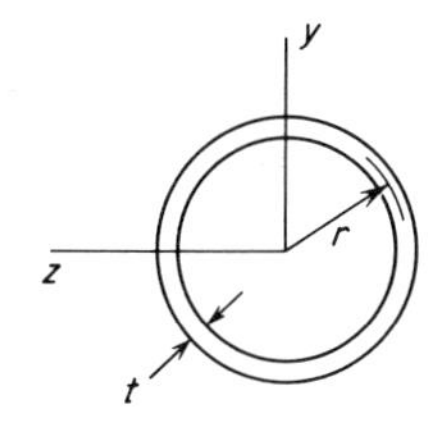

$$I_Z = I_Y \approx \pi r^3 t \qquad J \approx 2\pi r^3 t$$

$$A \approx 2\pi r t \qquad f = 2$$

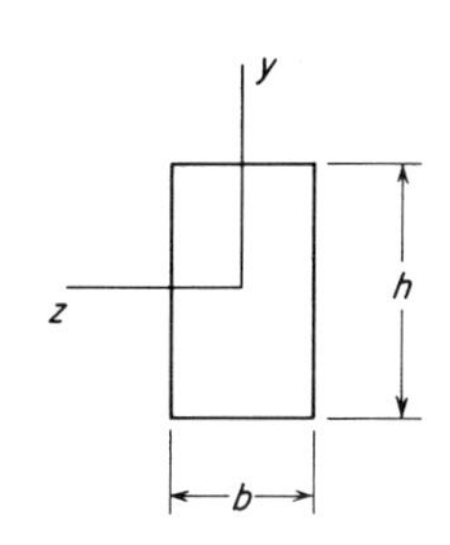

$$I_Z = \frac{bh^3}{12} \qquad J = \beta\, hb^3$$

$$I_Y = \frac{hb^3}{12} \qquad \beta \approx \frac{1}{3} - 0.21\,\frac{b}{h}\left(1 - \frac{b^4}{12h^4}\right)$$

$$A = bh \qquad f = \frac{6}{5}$$

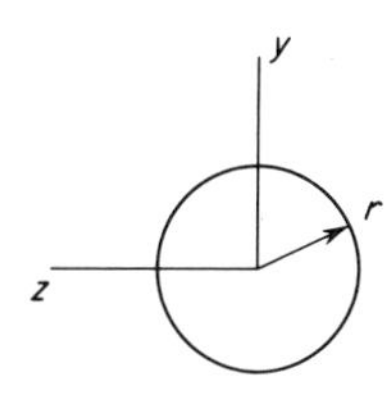

$$I_Z = I_Y = \frac{\pi r^4}{4} \qquad J = \frac{\pi r^4}{2}$$

$$A = \pi r^2 \qquad f = \frac{10}{9}$$

ANSWERS TO PROBLEMS

CHAPTER 1

1.4-1 The increase in length of the bar. **1.4-2** A horizontal force acting to the right at joint C, and a clockwise couple acting at joint C.

1.4-3 $D_{11} = \dfrac{A_1L^3}{48EI}$ $\quad D_{12} = \dfrac{A_2L^2}{16EI}$ $\quad D_1 = D_{11} + D_{12}$

1.4-4 $D_{11} = D_{31} = \dfrac{A_1L}{2EI}$ $\quad D_{21} = \dfrac{3A_1L^2}{8EI}$

1.4-5 $D_{11} = \dfrac{A_1L}{3EI}$ $\quad D_{23} = -\dfrac{A_3L^3}{32EI}$ $\quad D_{33} = \dfrac{A_3L^3}{8EI}$

1.4-6 $D_{11} = 4.83\dfrac{A_1L}{EA}$ $\quad D_{12} = D_{22} = 3.41\dfrac{A_2L}{EA}$ $\quad D_{21} = 3.41\dfrac{A_1L}{EA}$

1.7-1 **(a)** 3 **(b)** 2 **(c)** 3 **(d)** 2
1.7-2 **(a)** 2 **(b)** 7 **(c)** 4
1.7-3 **(a)** 1 **(b)** 5
1.7-4 10
1.7-5 **(a)** 3 **(b)** 9
1.7-6 **(a)** 3 **(b)** 6 **(c)** 3
1.7-7 **(a)** 1 **(b)** 8 **(c)** 5
1.7-8 **(a)** 8 **(b)** 16 **(c)** 8
1.7-9 **(a)** 0 **(b)** 9
1.7-10 **(a)** 3 **(b)** 6
1.7-11 **(a)** 21 **(b)** 12
1.7-12 **(a)** 30 **(b)** 36 **(c)** 25

CHAPTER 2

2.3-1 $Q_1 = \dfrac{PL}{8} - \dfrac{M}{4}$ $\quad Q_2 = \dfrac{PL}{8} + \dfrac{M}{4}$

2.3-2 $Q_1 = -\dfrac{31}{56}PL$ $\quad Q_2 = \dfrac{5PL}{14}$ $\quad$ **2.3-3** $Q_1 = -\dfrac{96sEI}{7L^3}$ $\quad Q_2 = \dfrac{30sEI}{7L^3}$

2.3-4 $Q_1 = \dfrac{3PL}{28}$ $\quad Q_2 = \dfrac{17P}{14}$ $\quad$ **2.3-5** $Q_1 = -\dfrac{5wL^2}{608}$ $\quad Q_2 = -\dfrac{33wL^2}{152}$

2.3-6 $Q_1 = -\dfrac{wL^2}{24}$ $\quad Q_2 = -\dfrac{wL^2}{12}$ $\quad$ **2.3-7** $Q_1 = -0.243P$ $\quad Q_2 = 0$

2.3-8 $Q_1 = -0.243P$ $\quad Q_2 = 0.172P$
2.3-9 $Q_1 = 0.0267P$ $\quad Q_2 = 0.4698P$
2.3-10 $Q_1 = 456.6$ lb $\quad Q_2 = 83.96$ lb
2.3-11 $Q_1 = -0.3$ k $\quad Q_2 = 2.1$ k $\quad Q_3 = -28.8$ in.-k

2.3-12 $F_{11} = \dfrac{L}{EA} + \dfrac{H^3}{3EI} + \dfrac{fH}{GA}$ $\quad F_{22} = \dfrac{L^3}{3EI} + \dfrac{fL}{GA} + \dfrac{H}{EA}$

Other terms in $\mathbf{F}$ are the same as in Eq. (a) of Example 4, Art. 2.3.

2.3-13 **(a)** $F_{11} = \dfrac{2L^3}{3EI}$ $\quad F_{12} = F_{23} = 0$ $\quad F_{13} = -\dfrac{L^2}{EI}$ $\quad F_{22} = \dfrac{8L^3}{3EI}$

$F_{33} = \dfrac{4L}{EI}$

(b) $F_{11} = \frac{2L^3}{3EI} + \frac{2L}{EA} + \frac{2fL}{GA}$ $\quad F_{12} = F_{23} = 0$ $\quad F_{13} = -\frac{L^2}{EI}$

$F_{22} = \frac{8L^3}{3EI} + \frac{2L}{EA} + \frac{2fL}{GA}$ $\quad F_{33} = \frac{4L}{EI}$

2.3-14 $F_{11} = \frac{8L^3}{3EI}$ $\quad F_{12} = \frac{3L^3}{EI}$ $\quad F_{13} = \frac{3L^2}{EI}$ $\quad F_{22} = \frac{20L^3}{3EI}$ $\quad F_{23} = \frac{4L^2}{EI}$

$F_{33} = \frac{4L}{EI}$ **2.3-15** $Q_1 = -\frac{P}{16}$ $\quad Q_2 = \frac{11P}{16}$

2.3-16 $Q_1 = -6.24$ k $\quad Q_2 = 5.06$ k $\quad Q_3 = 49.7$ ft-k

2.3-17 $F_{11} = \frac{2L^3}{3EI} + \frac{L^3}{GJ}$ $\quad F_{12} = \frac{L^2}{2EI} + \frac{L^2}{GJ}$ $\quad F_{13} = \frac{L^2}{2EI}$

$F_{22} = F_{33} = \frac{L}{EI} + \frac{L}{GJ}$ $\quad F_{23} = 0$ **2.3-18** $Q_1 = \frac{3P}{16}\frac{1}{2 + 3\rho}; \ \rho = \frac{EI}{GJ}$

2.3-19 $Q_1 = -5500$ in.-lb $\quad Q_2 = 529.4$ in.-lb

2.3-20 $Q_1 = -8436$ in.-lb $\quad Q_2 = 1974$ in.-lb

2.3-21 $F_{11} = F_{33} = \frac{2L^3}{3EI} + \frac{2L}{EA}$ $\quad F_{12} = 0$ $\quad F_{13} = -\frac{L}{EA}$

$F_{22} = \frac{4L^3}{3EI} + \frac{2L}{EA}$ $\quad F_{23} = \frac{2L^3}{3EI} + \frac{L}{EA}$

2.3-22 $D_{QL1} = -\frac{PL^3}{2EI}$ $\quad D_{QL2} = -\frac{PL^3}{3EI}$ $\quad D_{QL3} = 0$ $\quad F_{11} = \frac{8L^3}{3EI} + \frac{L^3}{GJ}$

$F_{12} = -F_{13} = \frac{L^3}{2EI}$ $\quad F_{22} = \frac{2L^3}{3EI} + \frac{L^3}{GJ}$ $\quad F_{23} = \frac{L^3}{2EI} + \frac{L^3}{GJ}$

$F_{33} = \frac{5L^3}{3EI} + \frac{L^3}{GJ}$ **2.4-1** $Q_1 = 0$ $\quad Q_2 = \frac{\alpha EI(T_1 - T_2)}{d}$

2.4-2 $D_{QT2} = 2D_{QT1} = \frac{\alpha L(T_1 - T_2)}{d}$ **2.4-3** $D_{QT1} = \alpha LT$ $\quad D_{QT2} = 0$

2.4-4 $Q_1 = 0$ $\quad Q_2 = -\frac{\beta EI}{L}$ **2.4-5** $D_{QP1} = 0$ $\quad D_{QP2} = -e\sqrt{2}$

2.4-6 $Q_1 = -\frac{12EIs}{L^3} + \frac{6EI\beta}{L^2}$ $\quad Q_2 = -\frac{6EIs}{L^2} + \frac{2EI\beta}{L}$

2.4-7 $D_{QR1} = -s$ $\quad D_{QR2} = 0$

2.4-8 $D_{QR1} = -\frac{2s_1}{L} + \frac{s_2}{L}$ $\quad D_{QR2} = \frac{s_1}{L} - \frac{2s_2}{L}$

2.4-9 $D_{QS1} = \alpha LT - \beta H$ $\quad D_{QS2} = -\frac{5PL^3}{48EI} - \alpha LT - s$

$D_{QS3} = -\frac{PL^2}{8EI} + \beta$

2.5-1 $\mathbf{D}_J = \frac{PL^2}{720EI}\begin{bmatrix} -16 \\ 2 \\ -7 \\ 26 \end{bmatrix}$ $\quad \mathbf{A}_R = \frac{P}{120}\begin{bmatrix} 46 \\ 129 \\ 144 \\ 41 \end{bmatrix}$

2.5-2 $\mathbf{D}_J = \frac{PL}{EA}\begin{bmatrix} -1.172 \\ -1.828 \end{bmatrix}$ $\quad \mathbf{A}_M = P\begin{bmatrix} 1.172 \\ -1.828 \\ 0 \end{bmatrix}$ $\quad \mathbf{A}_R = P\begin{bmatrix} -0.1716 \\ 1.828 \end{bmatrix}$

2.5-3 $D_{J1} = -0.000438$ in. $\quad D_{J2} = -0.00193$ in.
$D_{J3} = 0.000532$ radians $\quad A_{R1} = -0.913$ k $\quad A_{R2} = 4.03$ k
$A_{R3} = 43.6$ in.-k

2.9-1 $A_{R1} = \dfrac{5wL}{8}$ $A_{R2} = \dfrac{wL^2}{8}$ $A_{R3} = \dfrac{3wL}{8}$

2.9-2 $A_{M1} = \dfrac{11P}{16}$ $A_{M2} = \dfrac{3PL}{16}$ $A_{M3} = \dfrac{5P}{16}$

2.9-3 $A_{R1} = 2P$ $A_{R2} = \dfrac{13PL}{18}$ $A_{R3} = \dfrac{5PL}{18}$

2.9-4 $A_{M1} = \dfrac{13P}{8}$ $A_{M2} = \dfrac{PL}{2}$ $A_{M3} = \dfrac{P}{4}$ $A_{M4} = \dfrac{3PL}{8}$ $A_{R1} = -\dfrac{3P}{8}$
$A_{R2} = -\dfrac{P}{4}$

2.9-5 $A_{M1} = \dfrac{P}{3}$ $A_{M2} = \dfrac{5PL}{72}$ $A_{R1} = P$ $A_{R2} = \dfrac{7PL}{72}$

2.9-6 $A_{M1} = \dfrac{37P}{60}$ $A_{M2} = \dfrac{59PL}{360}$ $A_{M3} = \dfrac{23P}{60}$ $A_{M4} = -\dfrac{17PL}{360}$
$A_{R1} = \dfrac{49P}{120}$ $A_{R2} = \dfrac{23P}{60}$

2.9-7 $A_{M1} = \dfrac{PL}{72}$ $A_{M2} = -A_{M3} = -\dfrac{25PL}{72}$ $A_{M4} = \dfrac{5PL}{18}$ $A_{M5} = \dfrac{13PL}{18}$
$A_{M6} = \dfrac{17PL}{72}$ $A_{R1} = \dfrac{127P}{48}$ $A_{R2} = \dfrac{79P}{48}$

2.9-8 $S_{11} = S_{44} = \dfrac{4EI}{L}$ $S_{12} = S_{34} = \dfrac{2EI}{L}$ $S_{22} = S_{33} = \dfrac{6EI}{L}$ $S_{23} = \dfrac{EI}{L}$
$S_{13} = S_{14} = S_{24} = 0$

2.9-9 $S_{22} = 2S_{11} = \dfrac{8EI}{L}$ $S_{12} = S_{23} = \dfrac{2EI}{L}$ $S_{33} = 5S_{34} = \dfrac{20EI}{3L}$
$S_{44} = \dfrac{8EI}{3L}$ $S_{13} = S_{14} = S_{24} = 0$

2.9-10 $S_{11} = S_{66} = \dfrac{4EI}{L}$ $S_{22} = S_{33} = S_{44} = S_{55} = \dfrac{8EI}{L}$
$S_{12} = S_{23} = S_{34} = S_{45} = S_{56} = \dfrac{2EI}{L}$

2.9-11 $S_{11} = S_{44} = \dfrac{8EI}{L}$ $S_{24} = 2S_{12} = \dfrac{4EI}{L}$ $S_{22} = \dfrac{12EI}{L}$
$S_{23} = S_{34} = -\dfrac{24EI}{L^2}$ $S_{33} = \dfrac{96EI}{L^3}$

2.9-12 $S_{11} = S_{44} = \dfrac{12EI}{L^3}$ $S_{22} = S_{33} = \dfrac{12EI}{L}$ $S_{12} = -S_{34} = \dfrac{6EI}{L^2}$
$S_{23} = \dfrac{4EI}{L}$

2.9-13 $S_{11} = S_{55} = 2S_{12} = 2S_{45} = 166{,}667$ in.-k $S_{22} = 291{,}667$ in.-k
$S_{23} = 62{,}500$ in.-k $S_{33} = 275{,}000$ in.-k $S_{34} = 75{,}000$ in.-k
$S_{44} = 316{,}667$ in.-k

2.9-14 $A_{R1} = \dfrac{19P}{22}$ $A_{R2} = \dfrac{79PL}{264}$ $A_{R3} = \dfrac{3P}{22}$ $A_{R4} = -\dfrac{19PL}{264}$

2.9-15 $A_{M1} = \dfrac{P}{2}$ $A_{M2} = 0.433P$ $A_{M3} = \dfrac{P}{4}$

2.9-16 $A_{M1} = A_{M3} = 0.293P$ $A_{M2} = 2A_{M1}$

2.9-17 $A_{M1} = A_{M3} = 0.293P - 0.060wL$ $A_{M2} = 0.586P - 0.621wL$

2.9-18 $A_{M1} = P \quad A_{M2} = 0.586P \quad A_{M3} = -0.414P$

2.9-19 $A_{M1} = -A_{M5} = 0.236P \quad A_{M2} = 5A_{M4} = 0.417P \quad A_{M3} = 0.354P$

2.9-20 $A_{R1} = 0.446wL \quad A_{R2} = 2.153wL \quad A_{R3} = 2.968wL$
$A_{R4} = -0.446wL \quad A_{R5} = 0 \quad A_{R6} = 1.707wL$

2.9-21 $S_{11} = S_{22} = S_{33} = S_{44} = 1.354\dfrac{EA}{L} \quad S_{12} = -S_{34} = -0.354\dfrac{EA}{L}$

$S_{24} = -\dfrac{EA}{L}$

2.9-22 $S_{11} = S_{22} = S_{33} = S_{44} = 1.354\dfrac{EA}{L} \quad S_{14} = -S_{13} = -S_{34} = 0.354\dfrac{EA}{L}$

2.9-23 $S_{11} = S_{33} = S_{44} = S_{66} = S_{88} = \dfrac{253}{500}\dfrac{EA}{L} \quad S_{22} = S_{77} = \dfrac{233}{375}\dfrac{EA}{L}$

$S_{55} = S_{99} = \dfrac{179}{375}\dfrac{EA}{L} \quad S_{13} = S_{46} = S_{68} = -\dfrac{EA}{4L}$

$S_{14} = S_{18} = S_{36} = -\dfrac{32}{125}\dfrac{EA}{L}$

$S_{15} = -S_{19} = S_{24} = -S_{28} = -S_{37} = -S_{45} = S_{89} = -\dfrac{24}{125}\dfrac{EA}{L}$

$S_{25} = S_{29} = -\dfrac{18}{125}\dfrac{EA}{L} \quad S_{27} = -\dfrac{EA}{3L}$

2.9-24 $\mathbf{A}_M = \dfrac{M}{4L}\begin{bmatrix} 3 \\ 3 \\ -2L \end{bmatrix} \quad \mathbf{A}_R = \dfrac{M}{4L}\begin{bmatrix} -3 \\ -3 \\ -L \end{bmatrix}$

2.9-25 $A_{R1} = \dfrac{27P}{448} \quad A_{R2} = \dfrac{31P}{56} \quad A_{R3} = \dfrac{PL}{7} \quad A_{R4} = -\dfrac{27P}{224} \quad A_{R5} = \dfrac{3P}{2}$

$A_{R6} = \dfrac{3PL}{112}$ **2.9-26** $\mathbf{A}_M = \dfrac{wL}{28}\begin{bmatrix} -45 \\ 40 \\ 26L \end{bmatrix} \quad \mathbf{A}_R = \dfrac{3wL}{28}\begin{bmatrix} 15 \\ \\ -4 \end{bmatrix}$

2.9-27 $A_{R1} = 0.36$ k $\quad A_{R2} = 3.34$ k $\quad A_{R3} = 249.6$ in.-k $\quad A_{R4} = -0.72$ k
$A_{R5} = 8.96$ k $\quad A_{R6} = 46.3$ in.-k

2.9-28 $A_{R1} = -6.24$ k $\quad A_{R2} = 5.06$ k $\quad A_{R3} = 49.7$ ft-k

2.9-29 **(a)** $S_{11} = \dfrac{24EI}{L^3} \quad S_{12} = S_{13} = \dfrac{6EI}{L^2} \quad S_{22} = S_{33} = \dfrac{6EI}{L} \quad S_{23} = \dfrac{EI}{L}$

(b) $S_{11} = S_{44} = \dfrac{12EI}{L^3} + \dfrac{EA}{2L} \quad S_{22} = S_{55} = \dfrac{3EI}{2L^3} + \dfrac{EA}{L}$

$S_{13} = S_{46} = \dfrac{6EI}{L^2} \quad S_{14} = -\dfrac{EA}{2L}$

$S_{23} = S_{26} = -S_{35} = -S_{56} = \dfrac{3EI}{2L^2} \quad S_{25} = -\dfrac{3EI}{2L^3}$

$S_{33} = S_{66} = 6S_{36} = \dfrac{6EI}{L}$

2.9-30 $S_{11} = S_{33} = S_{44} = \dfrac{4EI}{L} \quad S_{12} = S_{23} = S_{24} = \dfrac{2EI}{L} \quad S_{22} = \dfrac{12EI}{L}$

2.9-31 $S_{22} = 2S_{11} = -2S_{12} = \dfrac{48EI_1}{H^3} \quad S_{33} = S_{44} = \dfrac{4EI_1}{H} + \dfrac{4EI_2}{L}$

$S_{55} = S_{66} = \dfrac{8EI_1}{H} + \dfrac{4EI_2}{L} \quad S_{13} = S_{14} = S_{15} = S_{16} =$

$= -S_{23} = -S_{24} = \dfrac{6EI_1}{H^2} \quad S_{34} = S_{56} = \dfrac{2EI_2}{L} \quad S_{35} = S_{46} = \dfrac{2EI_1}{H}$

2.9-32 $A_{R1} = \frac{3P}{8}\frac{9+2\eta}{4+\eta}$ $A_{R2} = -\frac{PL}{16}\frac{\eta(5+2\eta)}{(1+\eta)(4+\eta)}$

$A_{R3} = \frac{PL}{16}\frac{4\eta^2+23\eta+22}{(1+\eta)(4+\eta)}$ $\eta = \frac{GJ}{EI}$

2.9-33 $A_{M1} = \frac{P}{8}\frac{5+2\eta}{4+\eta}$ $A_{M2} = -\frac{PL}{16}\frac{(2+\eta)(5+2\eta)}{(1+\eta)(4+\eta)}$

$A_{M3} = \frac{3PL}{16}\frac{\eta}{(1+\eta)(4+\eta)}$

2.9-34 $\mathbf{D} = \frac{wL^3}{24EI(1+\eta)}\begin{bmatrix} -L(3+\eta) \\ 4 \\ -4 \end{bmatrix}$

2.9-35 $A_{R1} = \frac{P}{4}\frac{11+2\eta}{4+\eta}$ $A_{R2} = 0$ $A_{R3} = \frac{PL}{8}\frac{6+\eta}{4+\eta}$

2.9-36 $S_{11} = S_{77} = \frac{4EI}{L}$ $S_{13} = S_{46} = S_{57} = \frac{2EI}{L}$

$S_{22} = -S_{24} = -S_{35} = -S_{68} = S_{88} = \frac{GJ}{L}$

$S_{33} = S_{44} = S_{55} = S_{66} = \frac{4EI}{L} + \frac{GJ}{L}$

2.9-37 $S_{11} = -S_{14} = -S_{46} = S_{66} = \frac{GJ}{L}$ $S_{22} = S_{77} = \frac{4EI}{L}$

$S_{23} = -S_{34} = -S_{37} = -\frac{6EI}{L^2}$ $S_{25} = S_{57} = \frac{2EI}{L}$ $S_{33} = \frac{36EI}{L^3}$

$S_{44} = \frac{4EI}{L} + \frac{2GJ}{L}$ $S_{55} = \frac{8EI}{L} + \frac{GJ}{L}$

2.10-1 $\mathbf{A}_R = \frac{3\alpha EI(T_1 - T_2)}{7dL}\begin{bmatrix} 2 \\ 3L \\ -1 \\ -1 \end{bmatrix}$ **2.10-2** $\mathbf{A}_R = \frac{6EI}{7L^3}\begin{bmatrix} 11s_1 - 3s_2 \\ 6s_1L - s_2L \\ -16s_1 + 5s_2 \\ 5s_1 - 2s_2 \end{bmatrix}$

2.10-3 $A_{M1} = \frac{EI\beta}{4L^2}$ $A_{M2} = \frac{EI\beta}{12L}$ $A_{M3} = \frac{4EI\beta}{9L^2}$ $A_{M4} = -\frac{EI\beta}{6L}$

$A_{R1} = \frac{7EI\beta}{36L^2}$ $A_{R2} = -\frac{61EI\beta}{36L^2}$

2.10-4 $A_{M1} = 0$ $A_{M2} = -\frac{\alpha EI(T_1 - T_2)}{d}$ $A_{R1} = A_{R3} = 0$

$A_{R2} = -A_{R4} = -A_{M2}$ **2.10-5** $A_{R1} = 0$ $A_{R2} = -A_{R3} = -\frac{EI\beta}{L}$

2.10-6 $A_{DS1} = -\frac{PL}{8} - \frac{6EIs}{L^2}$ $A_{DS2} = \frac{PL}{12} + \frac{2EI\beta}{L}$ $A_{RS1} = \frac{P}{2} + \frac{12EIs}{L^3}$

$A_{RS2} = \frac{P}{2} + \frac{6EI\beta}{L^2}$

2.10-7 $A_{M1} = A_{M4} = -0.1745EA\alpha T$ $A_{M2} = A_{M3} = 0.1277EA\alpha T$

2.10-8 $A_{M1} = 2A_{M3} = \frac{\sqrt{3}\,EAe}{4L}$ $A_{M2} = -\frac{5EAe}{8L}$

2.10-9 $A_{DS} = -EA\left(\frac{s}{L} + \sqrt{2}\,\alpha T\right)$ $A_{MS1} = A_{MS3} = -EA\alpha T$

$A_{MS2} = -\frac{EAs}{L}$

2.10-10 $A_{DS1} = EA\left(\frac{s}{L} - 1.707\alpha T\right) \qquad A_{DS2} = EA\left(\frac{e}{L} + 1.707\alpha T\right)$

$A_{DS3} = EA\left(0.354\frac{s}{L} - 1.707\alpha T\right) \qquad A_{DS4} = A_{DS3} - \frac{EAe}{L}$

2.10-11 $A_{DS1} = A_{MS1} = -EA\alpha T \qquad A_{DS2} = \frac{EAs_1}{L} + \frac{12EIs_2}{L^3}$

$A_{DS3} = \frac{\alpha EI(T_1 - T_2)}{d} - \frac{6EIs_2}{L^2} \qquad A_{MS2} = \frac{12EIs_2}{L^3} \qquad A_{MS3} = -\frac{6EIs_2}{L^2}$

CHAPTER 3

3.3-1 $\mathbf{D}_J = \frac{PL}{16EA}\{54, -9\}$ **3.3-2** $\mathbf{D}_J = \frac{L}{2EA}\begin{bmatrix} P_1 & 3.83P_2 \\ -5.83P_1 & -P_2 \end{bmatrix}$

3.3-3 $\mathbf{D}_J = -10^{-2}\{0.351, 10.61, 1.60\}$ inches

3.4-1 $\mathbf{D}_J = \frac{1}{1152EI}\begin{bmatrix} -33wL^4 & 64PL^3 \\ 72wL^3 & -128PL^2 \end{bmatrix}$

3.4-2 $\mathbf{D}_J = -\frac{PL^2}{18EI}\{28L, 24, 58L, 33\}$

3.4-3 $\mathbf{D}_J = -\frac{PL^2}{18EI}\{4L(7 + 9\eta), 24, 2L(29 + 27\eta), 33\}$

in which $\eta = fEI/GAL^2$

3.5-1 $\mathbf{D}_J = \frac{PL^3}{EI}\{-3.18, 4.18\}$

3.5-2 $\mathbf{D}_J = -\frac{PL^3}{48EI}(11 + 288\psi)$ in which $\psi = I/AL^2$

3.6-1 $\mathbf{D}_J = \frac{L}{48EI}\{-3PL, 8M(2 + 3\rho), 3PL, -8M\}$ in which $\rho = EI/GJ$

3.9-1 $\mathbf{A}_M = P\{0.396, 0.396, 0.396, -0.604, 0.854, -0.561\}$

$\mathbf{D}_J = -\frac{PL}{EA}\{0.604, 2.311\}$

3.9-2 $\mathbf{A}_M = \{-14.14, -19.10, -28.28, 0.90, 12.86, -1.28, 20.90, 10, 10.90, 20\}$ kips **3.9-3** $\mathbf{A}_M = P\{0.500, 0.433, 0.250\} \qquad D_J = \frac{PL}{2EA}$

3.9-4 $\mathbf{A}_R = \frac{P}{264}\{228, 79L, 36, -19L\}$ **3.9-5** $D_J = -\frac{5wL^3}{144EI}$

3.9-6 $(\mathbf{A}_M)_A = \{1.935 \text{ kips}, 15.56 \text{ in.-k}\} \qquad D_J = 0.213$ in.

3.9-7 $\mathbf{A}_R = \frac{5wL}{28}\{-9, 8, 3L\} \qquad \mathbf{D}_J = \frac{wL^3}{168EI}\{45, -19\}$

3.9-8 $\mathbf{A}_R = \frac{P}{16}\{1, 6, -1, 11\} \qquad D_J = \frac{PL^2}{48EI}$

3.9-9 $\mathbf{D}_J = \frac{PL^2}{16EI(2 + 3\rho)}\{-1, 2 + 3\rho\}$ in which $\rho = EI/GJ$

CHAPTER 4

4.9-1 $\mathbf{D} = \frac{PL^2}{384EI}\{7, -53\} \qquad \mathbf{A}_R = \frac{P}{576}\{351, 93L, 1049, 427, 765, -207L\}$

$\mathbf{A}_{M1} = \frac{P}{192}\{117, 31L, 75, -10L\} \qquad \mathbf{A}_{M2} = \frac{P}{288}\{124, 15L, 308, -153L\}$

$\mathbf{A}_{M3} = \frac{P}{192}\{-63, -90L, 255, -69L\}$

4.9-2 $\mathbf{D} = -\dfrac{PL^2}{240EI}\{6, 13L\} \qquad \mathbf{A}_{\mathrm{R}} = \dfrac{P}{240}\{204, 48L, 516, 36L\}$

$\mathbf{A}_{\mathrm{M1}} = \dfrac{P}{240}\{204, 48L, 276, -84L\} \qquad \mathbf{A}_{\mathrm{M2}} = \dfrac{P}{240}\{240, 84L, 0, 36L\}$

4.9-3 $\mathbf{D} = \dfrac{PL^2}{144EI}\{-8, 23\} \qquad \mathbf{A}_{\mathrm{R}} = \dfrac{P}{144}\{24, 2L, 381, 237, -66, 34L\}$

$\mathbf{A}_{\mathrm{M1}} = \dfrac{P}{72}\{12, L, 60, -25L\} \qquad \mathbf{A}_{\mathrm{M2}} = \dfrac{P}{144}\{117, 50L, 27, 40L\}$

$\mathbf{A}_{\mathrm{M3}} = \dfrac{P}{72}\{105, 52L, -33, 17L\}$

4.9-4 $\mathbf{D} = \dfrac{PL^2}{528EI}\{-7L, 2\} \qquad \mathbf{A}_{\mathrm{R}} = \dfrac{P}{264}\{228, 79L, 36, -19L\}$

$\mathbf{A}_{\mathrm{M1}} = \dfrac{P}{264}\{228, 79L, 36, 17L\} \qquad \mathbf{A}_{\mathrm{M2}} = \dfrac{P}{264}\{-36, -17L, 36, -19L\}$

4.9-5 $\mathbf{D} = -\dfrac{wL^3}{384EI}\{16, 15L, 36\} \qquad \mathbf{A}_{\mathrm{R}} = \dfrac{wL}{4}\{1, 0, 6\}$

$\mathbf{A}_{\mathrm{M1}} = \dfrac{wL}{4}\{1, 0, 3, -L\} \qquad \mathbf{A}_{M2} = \dfrac{wL}{4}\{3, L, -1, 0\}$

4.9-6 $\mathbf{D} = \dfrac{PL^2}{416EI}\{109, -36, 9\} \qquad \mathbf{A}_{\mathrm{R}} = \dfrac{P}{416}\{646, 24, 1464, 362, -86L\}$

$\mathbf{A}_{\mathrm{M1}} = \dfrac{P}{416}\{646, 416L, -230, 22L\}$

$\mathbf{A}_{\mathrm{M2}} = \dfrac{P}{416}\{254, -22L, 578, -140L\}$

$\mathbf{A}_{\mathrm{M3}} = \dfrac{P}{416}\{470, 140L, 362, -86L\}$

4.9-7 $\mathbf{D} = \dfrac{PL^2}{2448EI}\{387L, 279, -819\}$

$\mathbf{A}_{\mathrm{R}} = \dfrac{P}{1224}\{702, -1152L, -3159, 3681, -1125L\}$

$\mathbf{A}_{\mathrm{M1}} = \dfrac{P}{612}\{351, -576L, -351, -297L\}$

$\mathbf{A}_{\mathrm{M2}} = \dfrac{P}{612}\{-261, 297L, -963, 54L\}$

$\mathbf{A}_{\mathrm{M3}} = \dfrac{P}{1224}\{-1233, -1332L, 3681, -1125L\}$

4.9-8 Nonzero elements of rearranged $\mathbf{S}_{\mathrm{J}}$ after factoring $\dfrac{EI}{2L^3}$:

$S_{J11} = 2S_{J12} = \frac{2}{3}S_{J22} = 4S_{J23} = \frac{2}{3}S_{J33} = 2S_{J34} = S_{J44} = 8L^2$

$S_{J15} = -S_{J16} = S_{J25} = -\frac{4}{3}S_{J26} = -4S_{J27} = 4S_{J36} = \frac{4}{3}S_{J37}$
$= -S_{J38} = S_{J47} = -S_{J48} = 12L$

$S_{J55} = -S_{J56} = \frac{8}{9}S_{J66} = -8S_{J67} = \frac{8}{9}S_{J77} = -S_{J78} = S_{J88} = 24$

4.9-9 Nonzero elements of rearranged $\mathbf{S}_{\mathrm{J}}$ after factoring $\dfrac{EI}{L^3}$:

$S_{J11} = 4S_{J12} = 4S_{J16} = \frac{2}{3}S_{J22} = 2S_{J24} = S_{J44} = 2S_{J66} = 8L^2$

$S_{J15} = -S_{J18} = -\frac{1}{4}S_{J23} = S_{J27} = \frac{1}{3}S_{J28} = -\frac{1}{4}S_{J34} = \frac{1}{4}S_{J48}$
$= S_{J56} = -S_{J67} = 6L$

$\frac{1}{8}S_{J33} = -\frac{1}{8}S_{J38} = S_{J55} = -S_{J57} = \frac{1}{2}S_{J77} = -S_{J78} = \frac{1}{9}S_{J88} = 12$

4.9-10 Nonzero elements of rearranged $\mathbf{S}_J$ after factoring $\dfrac{EI}{L^3}$:

$$S_{J11} = -S_{J16} = S_{J44} = -S_{J47} = \tfrac{1}{3}S_{J66} = -\tfrac{1}{2}S_{J67} = \tfrac{1}{3}S_{J77} = 12$$
$$S_{J12} = S_{J15} = S_{J26} = -\tfrac{1}{2}S_{J27} = -S_{J34} = \tfrac{1}{2}S_{J36}$$
$$= -S_{J37} = -S_{J48} = -S_{J56} = S_{J78} = 6L$$
$$S_{J22} = 3S_{J23} = 6S_{J25} = S_{J33} = 6S_{J38} = 3S_{J55} = 3S_{J88} = 12L^2$$

4.9-11 Nonzero elements of rearranged $\mathbf{S}_J$ after factoring $\dfrac{EI}{L^3}$:

$$S_{J11} = 2S_{J12} = \tfrac{1}{2}S_{J22} = 2S_{J23} = \tfrac{1}{2}S_{J33} = 2S_{J34} = \tfrac{1}{2}S_{J44} = 2S_{J45} = S_{J55} = 4L^2$$
$$S_{J16} = -S_{J17} = S_{J26} = -S_{J28} = S_{J37} = -S_{J39} = S_{J48}$$
$$= -S_{J410} = S_{J59} = -S_{J510} = 6L$$
$$S_{J66} = -S_{J67} = \tfrac{1}{2}S_{J77} = -S_{J78} = \tfrac{1}{2}S_{J88} = -S_{J89} = \tfrac{1}{2}S_{J99}$$
$$= -S_{J910} = S_{J1010} = 12$$

4.9-12 Nonzero elements of rearranged $\mathbf{S}_J$ after factoring $\dfrac{2EI}{L^3}$:

$$S_{J11} = 4S_{J13} = 4S_{J18} = \tfrac{4}{3}S_{J33} = 8S_{J34} = 2S_{J44} = 8S_{J46}$$
$$= 4S_{J66} = 2S_{J88} = 8L^2$$
$$-S_{J12} = S_{J17} = -2S_{J23} = 2S_{J24} = S_{J39} = -2S_{J310} = -2S_{J45}$$
$$= -2S_{J56} = 2S_{J610} = S_{J78} = -S_{J89} = 6L$$
$$S_{J22} = -\tfrac{3}{2}S_{J29} = -3S_{J210} = 3S_{J55} = -3S_{J510} = \tfrac{3}{2}S_{J77} = -\tfrac{3}{2}S_{J79}$$
$$= \tfrac{3}{4}S_{J99} = \tfrac{3}{2}S_{J1010} = 18$$

4.12-1 $\mathbf{D} = \dfrac{\sqrt{2}\,PL}{(1+\sqrt{2})EA_X}\{1+\sqrt{2}, 1\}$

$\mathbf{A}_R = \dfrac{P}{2(1+\sqrt{2})}$
$\{-(2+\sqrt{2}), -(2+\sqrt{2}), -2(1+\sqrt{2}), -2\sqrt{2}, -\sqrt{2}, \sqrt{2}\}$

$\mathbf{A}_{M1} = P\{-1, 0, 1, 0\}$

$\mathbf{A}_{M2} = \dfrac{P}{1+\sqrt{2}}\{-\sqrt{2}, -(1+\sqrt{2}), \sqrt{2}, -(1+\sqrt{2})\}$

$\mathbf{A}_{M3} = \dfrac{P}{1+\sqrt{2}}\{1, 0, -1, 0\}$

4.12-2 $\mathbf{D} = \dfrac{PL}{EA_X}\{1.30, 1.60\}$

$\mathbf{A}_R = P\{-3.80, 1.50, -1.00, -0.600, -0.200, 1.10\}$
$\mathbf{A}_{M1} = P\{1.30, -2.00, -1.30, 2.00\}$ $\quad\mathbf{A}_{M2} = P\{0, 1.00, 0, 1.00\}$
$\mathbf{A}_{M3} = P\{-1.60, 1.00, 1.60, 1.00\}$
$\mathbf{A}_{M4} = P\{0.500, 0.500, 0.500, 0.500\}$ $\quad\mathbf{A}_{M5} = P\{-0.500, 0, 0.500, 0\}$
$\mathbf{A}_{M6} = P\{0, 0, 0, 0\}$

4.12-3 $\mathbf{D} = \dfrac{wL^2}{EA_X}\{-1.26, 0\}$ $\quad\mathbf{A}_R = wL\{0, 1.71, -0.446, 0.446, 2.15, 2.97\}$

$\mathbf{A}_{M1} = wL\{0, 0.500, 0, 0.500\}$ $\quad\mathbf{A}_{M2} = wL\{0, 0.500, 0, 0.500\}$
$\mathbf{A}_{M3} = wL\{0.500, 0, 0.500, 0\}$ $\quad\mathbf{A}_{M4} = wL\{1.76, 0, -0.762, 0\}$
$\mathbf{A}_{M5} = wL\{-0.500, 0.500, -0.500, 0.500\}$
$\mathbf{A}_{M6} = wL\{1.131, 0.500, 0.131, 0.500\}$

4.12-4 $\mathbf{D} = \dfrac{wL^2}{EA_X}\{-0.288, -0.815, -0.731\}$

$\mathbf{A}_R = wL\{-0.0630, 0.351, 2.48, -0.288, 2.32\}$
$\mathbf{A}_{M1} = wL\{-0.288, 0.300, 0.288, 0.300\}$
$\mathbf{A}_{M2} = wL\{0, 0.300, 0, 0.300\}$ $\quad\mathbf{A}_{M3} = wL\{1.22, 0, -0.416, 0\}$

$\mathbf{A}_{M4} = wL\{1.13, 0, -0.332, 0\}$
$\mathbf{A}_{M5} = wL\{0.080, 0.300, -0.880, 0.300\}$
$\mathbf{A}_{M6} = wL\{0.985, 0.300, -0.185, 0.300\}$

4.12-5 $\mathbf{D} = \dfrac{PL}{EA_X}\{0.667, -2.00, -4.00\}$

$\mathbf{A}_R = P\{-2.40, 0.700, 0.21, 0.700, 4.79\}$
$\mathbf{A}_{M1} = P\{0.667, 1.00, -0.667, 2.00\}$ $\mathbf{A}_{M2} = P\{0, -1.00, 0, -1.00\}$
$\mathbf{A}_{M3} = P\{1.40, 0, -1.40, 0\}$ $\mathbf{A}_{M4} = P\{2.07, 0, -2.07, 0\}$
$\mathbf{A}_{M5} = P\{4.33, 0.500, -2.60, 0.500\}$

4.12-6 $\mathbf{D} = \dfrac{wL^2}{EA_X}\{-2.92, -2.92\}$

$\mathbf{A}_R = wL\{-1.40, 1.40, -1.40, 2.25, 1.40, 2.25\}$
$\mathbf{A}_{M1} = wL\{0, 0.400, 0, 0.400\}$ $\mathbf{A}_{M2} = wL\{0, 0.400, 0, 0.400\}$
$\mathbf{A}_{M3} = wL\{0.300, 0, 0.300, 0\}$ $\mathbf{A}_{M4} = wL\{0.150, 0, 0.150, 0\}$
$\mathbf{A}_{M5} = wL\{-2.05, 0.400, 1.45, 0.400\}$
$\mathbf{A}_{M6} = wL\{2.05, 0.400, -1.45, 0.400\}$

4.12-7 Nonzero elements of $\mathbf{S}_J$ after factoring $\dfrac{EA_X}{L}$:

$S_{J11} = -2S_{J15} = -2S_{J17} = -\frac{2}{3}S_{J26} = -\frac{2}{3}S_{J28} = -\frac{2}{3}S_{J36} = \frac{2}{3}S_{J38}$
$= -\frac{2}{3}S_{J45} = \frac{2}{3}S_{J47} = 0.500$
$S_{J16} = -S_{J18} = S_{J25} = -S_{J27} = S_{J46} = S_{J48} = -0.433$ $S_{J22} = 3.23$
$S_{J24} = -1.73$ $S_{J33} = -2S_{J35} = -2S_{J37} = S_{J44} = 2.60$
$S_{J55} = S_{J77} = 2.55$ $S_{J56} = S_{J66} = -S_{J78} = S_{J88} = 1.18$
$S_{J57} = -1.00$

4.12-8 Nonzero elements of $\mathbf{S}_J$ after factoring $\dfrac{EA_X}{L}$:

$S_{J11} = S_{J33} = 1.38$
$S_{J12} = -\frac{1}{2}S_{J16} = S_{J18} = -\frac{1}{2}S_{J25} = S_{J27} = -S_{J34} = \frac{1}{2}S_{J36} = -S_{J310}$
$= \frac{1}{2}S_{J45} = -S_{J49} = \frac{1}{2}S_{J57} = \frac{1}{2}S_{J59} = \frac{3}{2}S_{J68} = \frac{3}{2}S_{J610} = -0.217$
$S_{J13} = 4S_{J15} = 8S_{J17} = -\frac{8}{9}S_{J22} = \frac{4}{3}S_{J26} = \frac{8}{3}S_{J28} = 4S_{J35} = 8S_{J39}$
$= -\frac{8}{9}S_{J44} = \frac{4}{3}S_{J46} = \frac{8}{3}S_{J410} = 4S_{J58} = -4S_{J510} = 4S_{J67}$
$= -4S_{J69} = -1.00$
$S_{J55} = 1.368$ $S_{J66} = 1.79$ $S_{J77} = S_{J99} = 0.559$
$S_{J78} = -S_{J910} = 0.467$ $S_{J88} = S_{J1010} = 0.520$

4.12-9 Nonzero elements of rearranged $\mathbf{S}_J$ after factoring $\dfrac{EA_X}{L}$:

$S_{J11} = S_{J33} = S_{J77} = S_{J99} = 1.71$
$S_{J12} = S_{J27} = S_{J35} = S_{J49} = -1.00$ $S_{J22} = 2.71$
$S_{J14} = S_{J15} = -S_{J16} = -S_{J23} = S_{J29} = S_{J210} = S_{J39} = S_{J310} = S_{J46}$
$= S_{J47} = -S_{J48} = S_{J56} = -S_{J57} = S_{J58} = -S_{J66} = S_{J78}$
$= -S_{J88} = -S_{J910} = -S_{J1010} = -0.707$
$S_{J44} = S_{J55} = 2.42$

4.12-10 Nonzero elements of $\mathbf{S}_J$ after factoring $\dfrac{EA_X}{L}$:

$S_{J11} = S_{J33} = \frac{1}{3}S_{J55} = \frac{1}{3}S_{J77} = \frac{1}{2}S_{J99} = \frac{1}{2}S_{J1111} = 2.03$
$S_{J12} = \frac{4}{3}S_{J17} = -S_{J18} = -S_{J27} = \frac{3}{4}S_{J28} = -S_{J34} = \frac{4}{3}S_{J35} = S_{J36}$
$= S_{J45} = \frac{3}{4}S_{J46} = \frac{2}{3}S_{J511} = -\frac{1}{2}S_{J512} = -\frac{1}{2}S_{J611} = \frac{3}{8}S_{J612}$
$= \frac{2}{3}S_{J79} = \frac{1}{2}S_{J710} = \frac{1}{2}S_{J89} = \frac{3}{8}S_{J810} = -\frac{1}{2}S_{J910} = -\frac{3}{8}S_{J1010}$
$= \frac{1}{2}S_{J1112} = -\frac{3}{8}S_{J1212} = -0.480$
$S_{J15} = S_{J37} = \frac{1}{2}S_{J59} = \frac{1}{2}S_{J711} = -1.67$ $S_{J22} = S_{J44} = 1.89$
$S_{J24} = \frac{1}{2}S_{J68} = -1.25$ $S_{J56} = -\frac{1}{10}S_{J66} = -S_{J78} = -\frac{1}{10}S_{J88} = -0.440$

Answers to problems for Art. 4.18 are in units of kips, inches, and radians.

4.18-1 $\mathbf{D} = \{-0.000438, -0.00193, 0.000532\}$
$\mathbf{A}_{R} = \{0.913, 5.97, 227.7, -0.913, 4.03, 43.59\}$
$\mathbf{A}_{M1} = \{0.913, 5.97, 227.7, -0.913, 4.03, -87.94\}$
$\mathbf{A}_{M2} = \{4.03, 0.913, 87.94, -4.03, -0.913, 43.59\}$

4.18-2 $\mathbf{D} = \{-0.000216, -0.00358, 0.000424\}$
$\mathbf{A}_{R} = \{0.360, 3.34, 249.6, 0.360, -0.303, 28.19, -0.720, 8.96, 46.00\}$
$\mathbf{A}_{M1} = \{0.360, 3.34, 249.6, -0.360, 2.66, -151.5\}$
$\mathbf{A}_{M2} = \{-0.360, 0.303, 59.09, 0.360, -0.303, 28.19\}$
$\mathbf{A}_{M3} = \{8.96, 0.721, 92.36, -8.96, -0.721, 46.0\}$

4.18-3 $\mathbf{D} = \{0.00671, -0.00147, 0.000868\}$
$\mathbf{A}_{R} = \{-5.04, 7.79, -496.3, -4.96, 2.21, 349.5\}$
$\mathbf{A}_{M1} = \{5.04, 2.21, 357.9, -5.04, 7.79, -496.3\}$
$\mathbf{A}_{M2} = \{2.21, 4.96, 349.0, -2.21, 5.04, -117.9\}$

4.18-4 $\mathbf{D} = \{0.00851, -0.0294, -0.000744\}$
$\mathbf{A}_{R} = \{-2.84, 2.42, 74.9, -2.36, 7.80, -245.6, -4.80, 9.79, 142.9\}$
$\mathbf{A}_{M1} = \{-2.84, 2.42, 74.9, 2.84, 2.58, -85.0\}$
$\mathbf{A}_{M2} = \{2.36, 7.20, 192.0, -2.36, 7.80, -245.6\}$
$\mathbf{A}_{M3} = \{9.79, 4.80, 142.9, -9.79, 5.20, -167.0\}$

4.18-5 $\mathbf{D} = \{0, -0.000620, 0\}$
$\mathbf{A}_{R} = \{-0.323, 0, -8.55, 0, 0.434, 0, 0.323, 1.57, 8.77\}$
$\mathbf{A}_{M1} = \{-0.434, 0, 0, 0.434, 0, 0\}$
$\mathbf{A}_{M2} = \{1.52, 0.503, 8.77, 0.215, 0.497, -8.55\}$

4.18-6 $\mathbf{D} = \{0.00599, -0.0130, 0.000409\}$
$\mathbf{A}_{R} = \{-6.59, -0.890, 55.1, 0.590, 2.89, 182.1\}$
$\mathbf{A}_{M1} = \{6.59, 0.890, 91.9, -6.59, -0.890, 55.1\}$
$\mathbf{A}_{M2} = \{2.46, 1.63, 182.0, -5.29, 1.20, -91.9\}$

4.18-7 $\mathbf{D} = \{0.000507, -0.00187, -0.000159\}$
$\mathbf{A}_{R} = \{-0.677, 0.942, 17.8, -1.01, 0.342, -5.21, 1.69, 2.72, 6.47\}$
$\mathbf{A}_{M1} = \{-0.677, 0.942, 17.8, 0.677, 0.258, -4.95\}$
$\mathbf{A}_{M2} = \{1.01, 0.458, 8.70, -1.01, 0.342, -5.21\}$
$\mathbf{A}_{M3} = \{3.19, 0.276, 6.47, -2.23, 0.444, -3.75\}$

4.18-8 Nonzero elements of $\mathbf{S}_J$:
$S_{J11} = S_{J44} = 1541.7$
$S_{J13} = S_{J19} = -S_{J37} = S_{J46} = S_{J412} = -S_{J610} = -S_{J79} = -S_{J1012} = 5000.0$
$S_{J14} = \frac{1}{2}S_{J28} = \frac{1}{2}S_{J511} = -\frac{1}{2}S_{J88} = -\frac{1}{2}S_{J1111} = -1500.0$
$S_{J17} = S_{J410} = -S_{J77} = -S_{J1010} = -41.7 \qquad S_{J22} = S_{J44} = 3005.2$
$S_{J23} = S_{J26} = -S_{J35} = -S_{J56} = 1250.0 \qquad S_{J25} = -5.22$
$\frac{1}{3}S_{J33} = 2S_{J36} = S_{J39} = \frac{1}{3}S_{J66} = S_{J612} = \frac{1}{2}S_{J99} = \frac{1}{2}S_{J1212} = 4.0 \times 10^5$

4.18-9 Nonzero elements of $\mathbf{S}_J$:
$S_{J11} = S_{J22} = 3041.7$
$S_{J13} = S_{J19} = S_{J23} = S_{J26} = -S_{J35} = -S_{J37} = -S_{J79} = 5000.0$
$S_{J14} = S_{J28} = -S_{J88} = -3000.0 \qquad S_{J17} = S_{J25} = -S_{J77} = -41.7$
$\frac{1}{4}S_{J33} = S_{J36} = S_{J39} = \frac{5}{18}S_{J66} = \frac{5}{4}S_{J612} = \frac{1}{2}S_{J99} = \frac{5}{8}S_{J1212} = 4.0 \times 10^5$
$S_{J44} = 3878.65 \qquad S_{J45} = -S_{J411} = -S_{J510} = S_{J1011} = -1143.0$
$S_{J46} = S_{J412} = -S_{J610} = -S_{J1012} = 2560.0 \qquad S_{J410} = -S_{J1010} = -878.7$
$S_{J55} = 1585.0 \qquad S_{J56} = -3080.0 \qquad S_{J511} = -S_{J1111} = -1543.7$
$S_{J512} = -S_{J611} = -S_{J1112} = 1920.0$

INDEX